W9-BWN-546

# NUMERICAL METHODS FOR ENGINEERS

## With Personal Computer Applications

# NUMERICAL METHODS FOR ENGINEERS

## With Personal Computer Applications

Steven C. Chapra, Ph.D.
Professor of Civil Engineering
Texas A&M University

Raymond P. Canale, Ph.D.
Professor of Civil Engineering
The University of Michigan

McGraw-Hill Book Company

New York St. Louis San Francisco Auckland Bogotá Hamburg
Johannesburg London Madrid Mexico Montreal New Delhi
Panama Paris São Paulo Singapore Sydney Tokyo Toronto

To
Margaret and Gabriel Chapra
Helen and Chester Canale

**NUMERICAL METHODS FOR ENGINEERS WITH PERSONAL COMPUTER APPLICATIONS**



3 4 5 6 7 8 9 0 DOCDOC 8 9 8 7 6 5

ISBN 0-07-010664-9

This book was set in Souvenir Light by Interactive Composition Corporation.
The editors were Kiran Verma and David A. Damstra;
the designer was Charles A. Carson;
the production supervisor was Joe Campanella.
The drawings were done by Fine Line Illustrations, Inc.
Cover concept by Steven C. Chapra.
R. R. Donnelley & Sons Company was printer and binder.

Library of Congress Cataloging in Publication Data

Chapra, Steven C.
Numerical methods for engineers with personal computer applications.

Bibliography: p.
Includes index.
1. Engineering mathematics—Data processing. 2. Numerical calculations—Data processing. 3. Microcomputers—Programming. I. Canale, Raymond P. II. Title.
TA345.C47 1985 511′.024′62 84-12531
ISBN 0-07-010664-9

# CONTENTS

# PREFACE

For the modern engineer, "keeping pace with one's profession" inevitably involves the use of computers. There are few disciplines—or, for that matter, few routine daily activities—that do not somehow seem to interface with these powerful and rapidly evolving machines. Certainly, computers have been an ally to engineering for years, performing myriad analytical and practical tasks and expediting projects and problem solving. Logically, then, the sooner and more thoroughly an engineering student befriends his or her personal computer or terminal, the more successful the collaboration between the two can become.

But how soon and how thorough should this exposure be? Engineering educators have long recognized that early training in computer technology is important. This training has traditionally involved mainframe computers and a high-level programming language such as FORTRAN. Unfortunately, it is often difficult for students to apply their new skills to problems in other classes. This is due to a variety of factors, not the least of which are the logistics involved with utilizing most mainframe systems. As a result, many engineering students do not fully exploit the problem-solving capabilities of computers until well along in their education.

We believe that the microelectronic revolution has provided an opportunity to more effectively integrate computing in the classroom. Because of their low cost and convenience, personal computers can enhance the student engineers' capability to solve problems during their school years. However, to exploit this opportunity to its fullest, some restructuring of introductory computing courses is necessary. For instance, at Texas A&M and at the University of Michigan, a two-step approach to this restructuring has been evolving for several years. A "first computing course" is devoted to orienting the student to the available computer hardware and to developing sound programming skills. The "second computing course" is then designed to hone these skills and demonstrate how to employ the computer for engineering problem solving.

This book grew out of the second course. The subject of numerical methods was chosen as the focal point because of its many engineering applications. We strongly believe that whether engineers utilize canned or self-composed software, a solid background in numerical methods is essential for the effective application of computers for engineering problem solving. Unfortunately, numerical methods are generally introduced during the senior year or at the graduate level—years past the point when they could be useful, instructive, and creative tools for the prospective engineer.

We have, therefore, designed this book so that it can be taught at the lower as well as the upper end of the undergraduate engineering curriculum. One aspect of this design is apparent in the book's organization and scope. The book is divided into six parts. Part I deals with introductory material, including information on programming and error analysis. The remaining five parts are devoted to the areas of numerical methods that have the most direct relevance to the undergraduate engineer: roots of equations, linear algebraic equations, curve fitting (regression and interpolation), integration, and ordinary differential equations. We have excluded topics such as eigenvalues and partial differential equations, which have more direct relevance to graduate students.

Aside from the scope of the material, we have incorporated a number of additional design features to make the book more accessible to both the lower- and the upper-level undergraduate audience. These include:

1. *Boxed material.* We have endeavored to include important derivations and error analyses in order to enrich the presentation. However, such material sometimes represents a stumbling block for the beginning student. Consequently, we have sequestered the more complicated mathematical material in boxes. Many students will find that they can apply the numerical methods without completely mastering the boxed material.

2. *Introductory material and mathematical background.* Every part of the book includes an introductory section. After a brief statement of the general mathematical problem under study, motivation is provided by describing how the problem would be approached in the absence of computers and where the problem occurs in engineering practice. This treatment is followed by a review of the mathematics required to successfully master the subject at hand. For example, matrix algebra is reviewed prior to introducing linear algebraic equations, and statistics is reviewed prior to introducing regression. Finally, an outline and study objectives are listed to provide some orientation to subsequent material.

3. *Epilogues.* Just as the introduction is designed to provide motivation and orientation, we include an epilogue at the end of each part of the book to consolidate the newly acquired concepts. An important feature of this epilogue is a section devoted to the trade-offs involved in choosing the appropriate numerical methods for a particular problem. In addition, important formulas and references for advanced methods are summarized.

4. *Sequential and graphical presentations.* Each major part of the book consists of three chapters—two devoted to theory and one to case studies. Wherever possible, the theory chapters are structured sequentially; that is, the more elementary and straightforward approaches are presented first. Because many of the more advanced methods build on the simpler ones,

this development is intended to provide a sense of the evolution of the techniques. Additionally, we have developed graphical depictions to supplement the mathematical descriptions for most of the approaches in the book. We have found that this visual approach is particularly effective in providing insight to lower-level undergraduate students.

5. *Case studies.* Case studies have been included in each part of the book to demonstrate the practical utility of the numerical methods. A significant effort was made to incorporate examples from the early courses in a typical engineering curriculum. When this was not possible, the theoretical basis and motivation for the problems have been provided.

6. *Software.* A computer software package called NUMERICOMP is available that demonstrates some of the numerical methods covered in the text: bisection, Gauss elimination, Lagrange interpolation, linear regression, the trapezoidal rule, and Euler's method. These programs provide the student with computer capability for each of the major parts of the book. The software is designed for easy application. It can also be employed by students to check the results of their own programming efforts. Although the software package is optional, we feel that more rapid progress can be attained with the material when the text and the software are used together. The software is available from McGraw-Hill for both IBM-PC and APPLE II personal computers. A professional version of NUMERICOMP can be obtained directly from EnginComp Software, Inc., 15 Research Dr., Ann Arbor, MI 48103.

Finally, we have exerted a conscious effort to make this book as user-friendly as possible. Thus, we have endeavored to keep our explanations straightforward and practically oriented. Although our primary intent is to provide students with a sound introduction to numerical methods, we have the ancillary objective of making this introduction a pleasurable experience. We believe that students who enjoy numerical methods, computers, and mathematics will, in the end, make better engineers. If our book fosters an enthusiasm for these subjects, we will consider our efforts a success.

## ACKNOWLEDGMENTS

We would like to acknowledge reviews by Professors Ted Cadman (University of Maryland), Lee W. Johnson (Virginia Polytechnic Institute and State University), Richard Noble (University of Colorado), Satish Ramadhyani (Purdue University), Howard Wicke (Ohio University), and Thomas C. Young (Clarkson University). Our gratitude is extended to Texas A&M University and the University of Michigan for providing us with secretarial and graphics support and the time necessary to prepare this book. In particular, Donald McDonald and Roy Hann of Texas A&M were very supportive of this endeavor. We garnered insights and suggestions from our colleagues, Bill

Batchelor, Harry Jones, Bill Ledbetter, James Martin, and Ralph Wurbs. Jeanne Castro inspired the chapter organization graphics. Also, Vanessa Stipp, with help from Kathy Childers, Cindy Denton, and Frances Kahlich, did an excellent job of typing the manuscript.

The book was classroom-tested for four semesters with a primarily sophomore audience at Texas A&M and for two semesters with students from all undergraduate classes at the University of Michigan. During this time, many students helped us check the mathematical accuracy and enhance the readability of the book. Lisa Olson read the entire manuscript several times and prepared the FORTRAN programs. Tad Slawecki provided excellent assistance with the supplementary software. In addition, Marla Isenstein, Luis Garcia, Sijin "Tom" Lee, and Rick Thurman made noteworthy contributions.

Thanks are also due to Kiran Verma, Dave Damstra, and B. J. Clark of McGraw-Hill for their encouragement and direction. Ursula Smith did a masterful job of copy editing the manuscript. Finally, we would like to thank our families, friends, colleagues, and students who have endured, with understanding, the many hours "stolen" to complete this work.

*Steven C. Chapra*
*Raymond P. Canale*

PART ONE

# NUMERICAL METHODS AND PERSONAL COMPUTERS

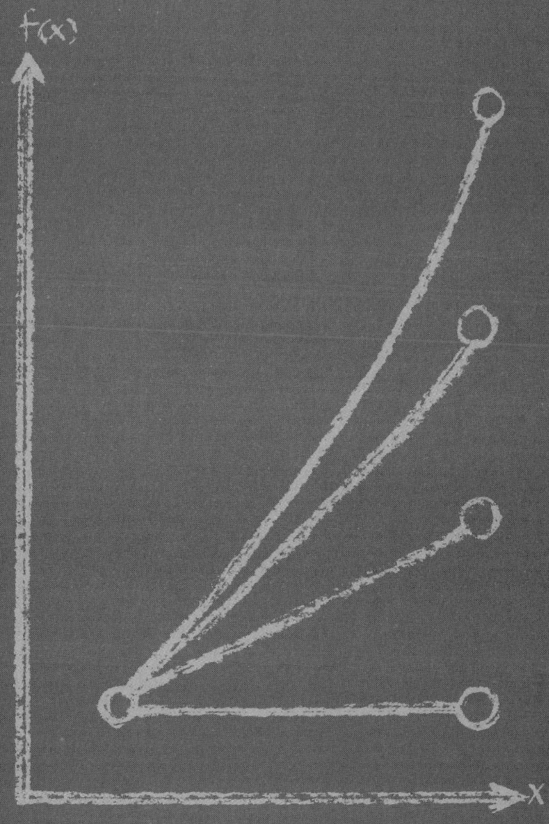


## I.1 MOTIVATION

Numerical methods are techniques by which mathematical problems are formulated so that they can be solved with arithmetic operations. Although there are many kinds of numerical methods, they have one common characteristic. Numerical methods invariably involve large numbers of tedious arithmetic calculations. It is little wonder that with the development of fast, efficient digital computers, the role of numerical methods in engineering problem solving has increased dramatically in recent years.

### I.1.1 Precomputer Methods

Beyond providing increased computational firepower, the widespread availability of computers (especially personal computers) and their partnership with numerical methods has had a significant influence on the actual engineering problem-solving process. In the precomputer era there were generally three different ways engineers approached problem solving:

1. First, solutions were derived for some problems using analytical or exact methods. These solutions were often useful and provided excellent insight into the behavior of some systems. However, analytical solutions can be derived for only a limited class of problems. These problems include those that can be approximated with linear models and those that have simple geometry and low dimensionality. Consequently, analytical solutions are of limited practical value because most real problems are nonlinear and involve complex shapes and processes.
2. Graphical solutions were used to characterize the behavior of systems. These graphical solutions usually took the form of plots or nomographs. Although graphical techniques can often be used to solve complex problems, the

results are not very precise. Furthermore, graphical solutions (without the aid of computers) are extremely tedious and awkward to implement. Finally, graphical techniques are limited to problems that can be described using three or fewer dimensions.

3. Manual calculators and slide rules were used to implement numerical methods. Although in theory such approaches should be perfectly adequate for solving complex problems, in actuality, several difficulties are encountered. Manual calculations are slow and tedious. Furthermore, consistent results are elusive because of simple blunders that arise when numerous manual tasks are performed.

During the precomputer era, significant amounts of energy were expended on the solution technique itself, rather than on problem definition and interpretation (Fig. I.1*a*). This unfortunate situation existed because so much time and drudgery were required to obtain numerical answers using precomputer techniques.

Today, computers and numerical methods provide an alternative for such complicated computations. Using computer power to obtain solu-

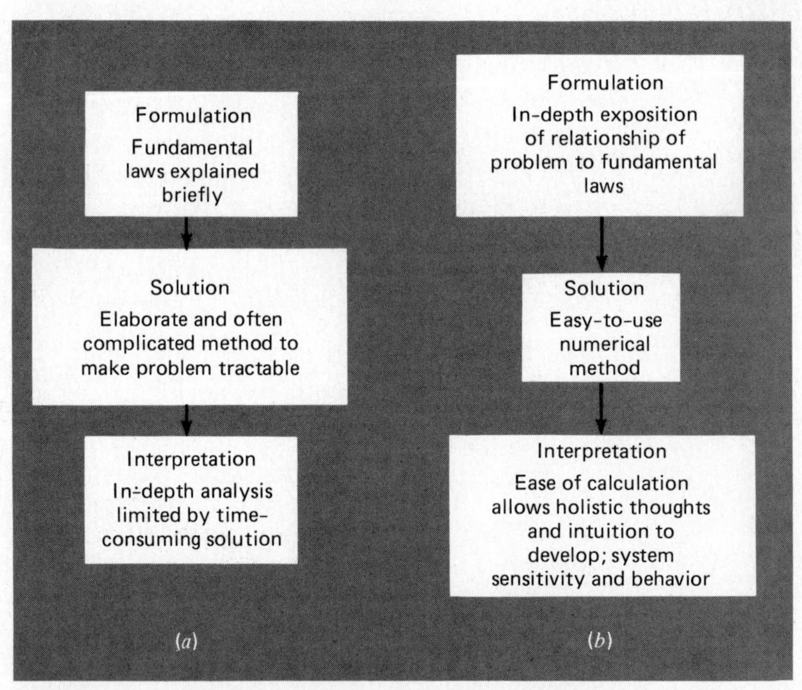


FIGURE I.1 The three phases of engineering problem solving in (*a*) the precomputer and (*b*) the computer era. The sizes of the boxes indicate the level of emphasis directed toward each phase in the classroom. Computers facilitate the implementation of solution techniques and thus allow more emphasis to be placed on the creative aspects of problem formulation and interpretation of results.

tions directly, you can approach these calculations without recourse to simplifying assumptions or inefficient techniques. Although simplifying assumptions still are extremely valuable both for problem solving and for providing insight, numerical methods represent alternatives that greatly enlarge your capabilities to confront and solve problems. As a result, more time is available for the use of your creative skills. Thus, more emphasis can be placed on problem formulation and solution interpretation and the incorporation of total system, or "holistic," awareness (Fig. I.1*b*).

### I.1.2 Numerical Methods and Engineering Practice

Since the late 1940s the widespread availability of digital computers has led to a veritable explosion in the use and development of numerical methods. At first, this growth was somewhat limited by the cost of access to large mainframe computers, and many engineers consequently continued to use simple analytical approaches in a significant portion of their consulting work. Needless to say, the recent evolution of inexpensive personal computers has given most of us ready access to powerful computational capabilities.

There are a number of additional reasons why you should study numerical methods:

1. Numerical methods are extremely powerful problem-solving tools. They are capable of handling large systems of equations, nonlinearities, and complicated geometries that are not uncommon in engineering practice and that are often impossible to solve analytically. As such, they greatly enhance your problem-solving skills.
2. During your careers, you may often have occasion to use commercially available prepackaged, or "canned," computer programs that involve numerical methods. The intelligent use of these programs often is predicated on knowledge of the basic theory underlying the methods.
3. Many problems cannot be approached using canned programs. If you are conversant with numerical methods and are adept at computer programming you will have the capability of designing your own programs to solve problems without having to buy expensive software.
4. Numerical methods are an efficient vehicle for learning to use personal computers. It is well known that an effective way to learn computer programming is to actually write computer programs. Because numerical methods are for the most part designed for implementation on computers, they are ideal for this purpose. Further,

they are especially well-suited to illustrate the power and the limitations of computers. When you successfully implement numerical methods on a personal computer and then apply them to solve otherwise intractable problems, you will be provided with a dramatic demonstration of how computers can serve your professional development. At the same time, you will also learn to acknowledge and control the errors of approximation that are part and parcel of large-scale numerical calculations.

**5.** Numerical methods provide a vehicle for you to reinforce your understanding of mathematics. Because one function of numerical methods is to reduce higher mathematics to basic arithmetic operations, they get at the "nuts and bolts" of some otherwise obscure topics. Enhanced understanding and insight can result from this alternative perspective.

## I.2 MATHEMATICAL BACKGROUND

Every part of this book requires some mathematical background. Consequently, the introductory material for each part includes a section, such as the one you are reading, on mathematical background. Because Part I itself is devoted to background material on mathematics and computers, the present section does not involve a review of a specific mathematical topic. Rather, we take this opportunity to introduce you to the types of mathematical subject areas that are covered in this book. As summarized in Fig. I.2, these are:

**1.** *Roots of equations* (*Fig. I.2a*). These problems are concerned with the value of a variable or a parameter that satisfies a single equation. These problems are especially valuable in engineering design contexts where it is often impossible to analytically solve design equations for parameters.

**2.** *Systems of linear algebraic equations* (*Fig. I.2b*). These problems are similar in spirit to roots of equations in the sense that they are concerned with values that satisfy equations. However, in contrast to satisfying a single equation, a set of values is sought that simultaneously satisfies a set of algebraic equations. Linear simultaneous equations arise in a variety of problem contexts and in all disciplines of engineering. In particular, they originate in the mathematical modeling of large systems of interconnected elements such as structures, electric circuits, and fluid networks. However, they are also encountered in other areas of numerical methods such as curve fitting.

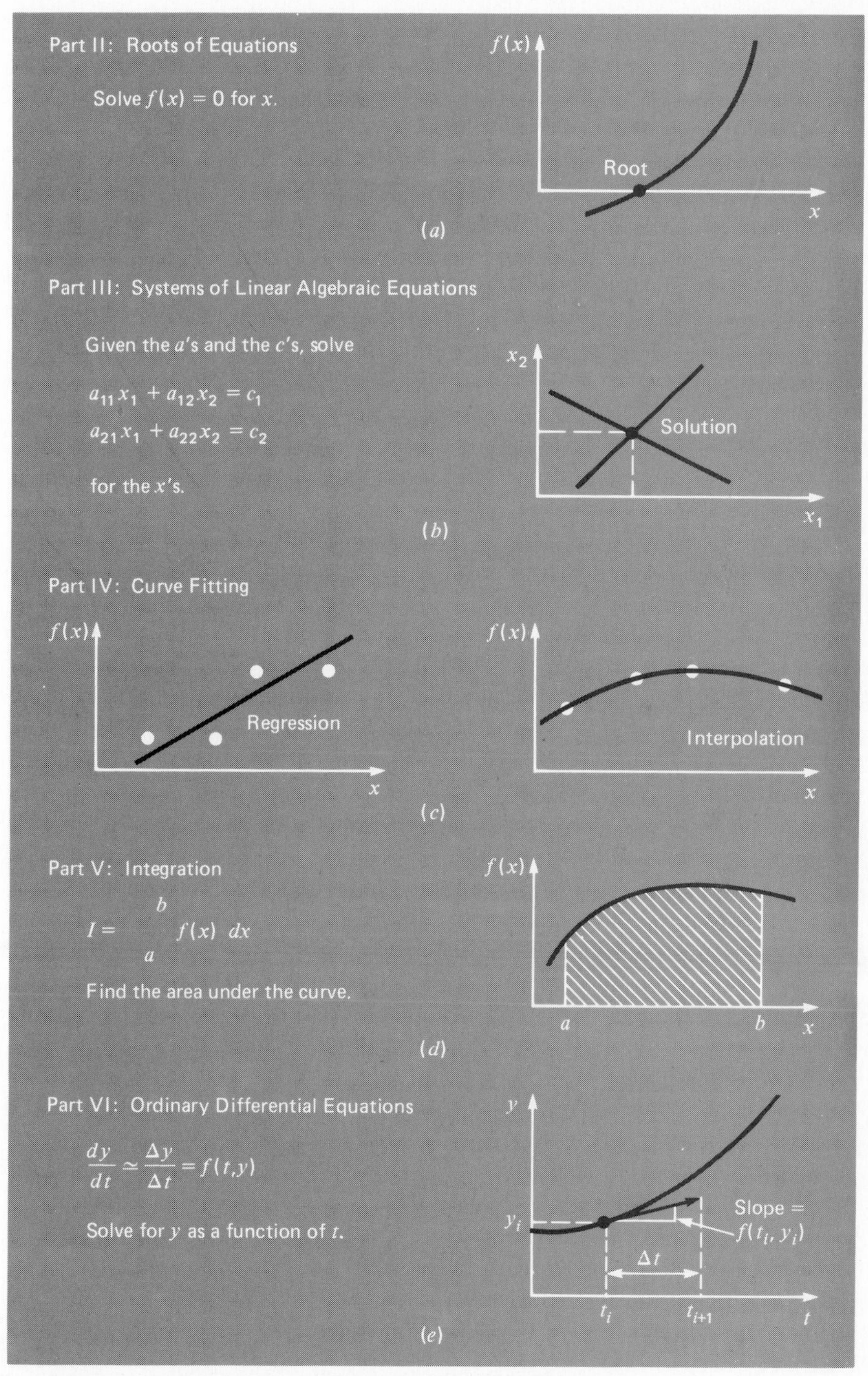


FIGURE I.2 Summary of the numerical methods covered in this book.

3. *Curve fitting* (*Fig. I.2c*). You will often have occasion to fit curves to data points. The techniques developed for this purpose can be divided into two general categories: regression and interpolation. Regression is employed where there is a significant degree of error associated with the data. Experimental results are often of this kind. For these situations, the strategy is to derive a single curve that represents the general trend of the data without necessarily matching any individual points. In contrast, interpolation is used where the objective is to determine intermediate values between relatively error-free data points. Such is usually the case for tabulated information. For these situations, the strategy is to fit a curve directly through the data points and use the curve to predict the intermediate values.

4. *Integration* (*Fig. I.2d*). As depicted, a physical interpretation of numerical integration is the determination of the area under a curve. Integration has many applications in engineering practice, ranging from the determination of the centroids of oddly shaped objects to the calculation of total quantities based on sets of discrete measurements. In addition, numerical integration formulas play an important role in the solution of rate equations.

5. *Ordinary differential* (*or rate*) *equations* (*Fig. I.2e*). Ordinary differential equations are of great significance in engineering practice. This is because many physical laws are couched in terms of the rate of change of a quantity rather than the magnitude of the quantity itself. Examples range from population forecasting models (rate of change of population) to the acceleration of a falling body (rate of change of velocity).

## I.3 ORIENTATION

Some orientation might be helpful before proceeding with our introduction to numerical methods. The following is intended as an overview of the material in Part I. In addition, some objectives have been included to focus your efforts when studying the material.

### I.3.1 Scope and Preview

Figure I.3 is a schematic representation of the material in Part I. We have designed this diagram to provide you with a global overview of this part of the book. We believe that a sense of the "big picture" is critical to developing insight into numerical methods. When reading a text, it is often possible to become lost in technical details. Whenever

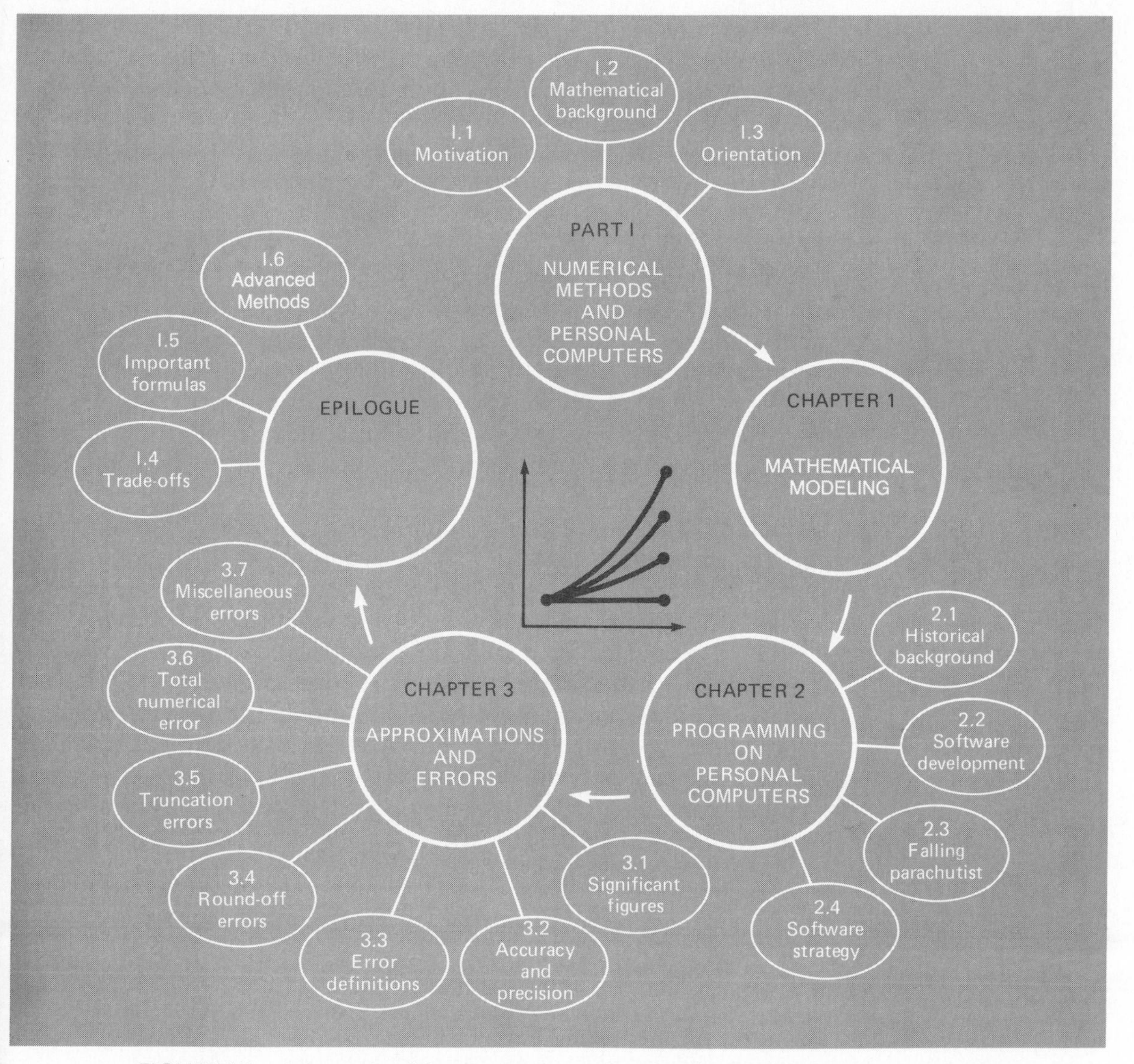


FIGURE I.3 Schematic of the organization of material in Part I: Numerical methods and personal computers.

you feel that you are losing the "big picture," refer back to Fig. I.3 to reorient yourself. Every part of this book includes a similar figure.

This figure also serves as a brief preview of the material covered in Part I. *Chapter 1* is designed to orient you to numerical methods and to provide motivation by demonstrating how these techniques can be used in the engineering modeling process. *Chapter 2* is an introduction and review of computer-related aspects of numerical methods and suggests the level of computer communication skills you should acquire to efficiently apply succeeding information. *Chapter 3* deals with the important topic of error analysis, which must be understood for the effective use of numerical methods. In addition, an *epilogue* is included that introduces the trade-offs that have such great significance for the effective implementation of numerical methods.

### I.3.2 Goals and Objectives

Study Objectives. Upon completing Part I, you should be adequately prepared to embark on your studies of numerical methods. In general, you should have gained a fundamental understanding of the importance of computers and the role of approximations and errors in the implementation and development of numerical methods. In addition to these general goals, you should have mastered each of the specific study objectives listed in Table I.1.

Computer Objectives. Upon completing Part I you should be familiar with the software (NUMERICOMP) available with this book. You should know what programs it contains and some of its computer-graphics capabilities. You should have also mastered sufficient computer skills to develop your own software for the numerical methods in this text. You should be able to develop reliable computer programs on the basis of algorithms or flowcharts. You should have the capability of saving your software on storage devices such as floppy disks or magnetic tape. Finally, you should have developed the capability to document your programs so that they may be effectively employed by users.

**TABLE I.1 Specific study objectives for Part I**

1. Understand the distinction between truncation and round-off error
2. Understand the concept of significant figures
3. Know the difference between accuracy and precision
4. Appreciate the utility of relative error
5. Know the difference between true relative error $\epsilon_t$ and approximate relative error $\epsilon_a$; realize how the latter can be employed in conjunction with a prespecified acceptable error $\epsilon_s$ in order to terminate a computation
6. Be able to relate relative error to significant figures
7. Be capable of applying the rounding rules delineated in Box 3.1
8. Understand how the Taylor series is used to approximate functions
9. Understand the approximate nature and remainder terms of the Taylor series
10. Know the relationship of finite divided differences to derivatives
11. Familiarize yourself with the trade-offs outlined in the epilogue to Part I

CHAPTER ONE

# MATHEMATICAL MODELING

Why should you master numerical methods and computer approaches to problem solving? Aside from the fact that we daily discover computers creeping into the most commonplace activities of our lives, is there some essential contribution that these machines, with their decidedly superhuman capabilities, can make to the tasks and challenges of the practicing engineer? We believe there is, and with the material in this chapter, we hope to orient and motivate you toward at least one possibility.

We begin by using the concept of mathematical modeling to help us define what we mean by numerical methods and to illustrate how they facilitate engineering problem solving. To do this, we actually develop a mathematical model of a physical process and solve it with a simple numerical method.

The physical world, in all its complexity, can appear overwhelming and unpredictable. Traditionally, the scientist's task has been to identify the reproducible patterns and laws underlying this apparent chaos. For example, on the basis of his observations, Newton formulated his second law of motion, which states that the time rate of change of momentum of a body is equal to the resultant force acting on it. Considering the exceedingly complex ways in which forces and objects interact on the earth, this law has proved to be a valid generalization.

In addition to providing insight, such laws can be applied by engineers to formulate solutions to practical problems. For instance, scientific generalizations are routinely used in the design of engineered works and products such as structures, machines, electric circuits, and synthetic chemicals. From an engineering design perspective, these generalizations are most useful when they are expressed in the form of a mathematical model.

A *mathematical model* can be broadly defined as a formulation or equation that expresses the essential features of a physical system or process in mathematical terms. Models range from simple algebraic relationships to large and complex systems of differential equations. Relying again on Newton for our example, the mathematical expression, or model, of his second law is the well-known equation

$$F = ma \tag{1.1}$$

where $F$ is the net force acting on the body (in dynes, or gram-centimeter per second squared), $m$ is the mass of the object (in grams), and $a$ is its acceleration (in centimeters per second squared).

Equation (1.1) has a number of characteristics that are typical of mathematical models of the physical world.

1. It describes a natural process or system in mathematical terms.
2. It represents an idealization and simplification of reality. That is, it ignores negligible details of the natural process and focuses on its essential manifestations. Thus, the second law does not include the effects of relativity that are of minimal importance when applied to objects and forces that interact on or about the earth's surface on scales visible to humans.
3. Finally, it yields reproducible results and, consequently, can be used for predictive purposes. For example, if the force on an object and its mass are known, Eq. (1.1) can be used to predict acceleration. Because of its simple algebraic form, Eq. (1.1) can be solved directly for

$$a = \frac{F}{m} \tag{1.2}$$

Thus, the acceleration is easily computed. However, other mathematical models of physical phenomena may be much more complex, and either cannot be solved exactly or require more sophisticated mathematical techniques than simple algebra for their solution. To illustrate a more complex model of this kind, Newton's second law can be used to determine the terminal velocity of a free-falling body near the earth's surface. Our falling body will be a parachutist as depicted in Fig. 1.1. A model for this case can

FIGURE 1.1 Schematic diagram of the forces acting on a falling parachutist. $F_D$ is the downward force due to the pull of gravity. $F_U$ is the upward force due to air resistance.

be derived by expressing the acceleration as the time rate of change of the velocity ($dv/dt$) and substituting it into Eq. (1.1) to yield

$$m\frac{dv}{dt} = F \qquad [1.3]$$

where $v$ is velocity (in centimeters per second). Thus, the mass multiplied by the rate of change of the velocity is equal to the net force acting on the body. If the net force is positive, the object will accelerate. If it is negative, the object will decelerate. If the net force is zero, the object's velocity will remain at a constant level.

For a body falling within the vicinity of the earth (Fig. 1.1), the net force is composed of two opposing forces: the downward pull of gravity $F_D$ and the upward force of air resistance $F_U$.

$$F = F_D + F_U \qquad [1.4]$$

If the downward force is assigned a positive sign, the second law can be used to formulate the force due to gravity as

$$F_D = mg \qquad [1.5]$$

where $g$ is the gravitational constant, or the acceleration due to gravity, which is approximately equal to 980 $cm/s^2$.

Air resistance can be formulated in a variety of ways. A simple approach is to assume that it is linearly proportional to velocity, as in

$$F_U = -cv \qquad [1.6]$$

where $c$ is a proportionality constant called the *drag coefficient* (in grams per second). Thus, the greater the fall velocity, the greater the upward force due to air resistance. The parameter $c$ accounts for properties of the falling object, such as shape or surface roughness, that affect air resistance. For the present case, $c$ might be a function of the type of jumpsuit or the orientation used by the parachutist during free-fall.

The net force is the difference between the downward and upward force. Therefore, Eqs. (1.3) through (1.6) can be combined to yield

$$m\frac{dv}{dt} = mg - cv \qquad [1.7]$$

or, dividing each side by $m$,

$$\frac{dv}{dt} = g - \frac{c}{m}v \qquad [1.8]$$

Equation (1.8) is a model that relates the acceleration of a falling object to the forces acting on it. It is a *differential equation* because it is written in terms of the *differential rate* of change ($dv/dt$) of the variable that we are interested in predicting. For this reason, it is sometimes referred to as a *rate equation*. However, in contrast to the solution of Newton's second law in Eq. (1.2), the exact solution of Eq. (1.8) for the velocity of the falling parachutist cannot be obtained using simple algebraic manipulations and arithmetic operations. Rather, the techniques of calculus must be applied to obtain an exact solution. For example, if the parachutist is initially at rest ($v = 0$ at $t = 0$), calculus can be used to solve Eq. (1.8) for

$$v(t) = \frac{gm}{c}[1 - e^{-(c/m)t}] \tag{1.9}$$

## EXAMPLE 1.1
### Analytical Solution to the Falling Parachutist Problem

Problem Statement: A parachutist with a mass of 68,100 g jumps out of an airplane. Use Eq. (1.9) to compute velocity prior to opening the chute. The drag coefficient $c$ is approximately equal to 12,500 g/s.

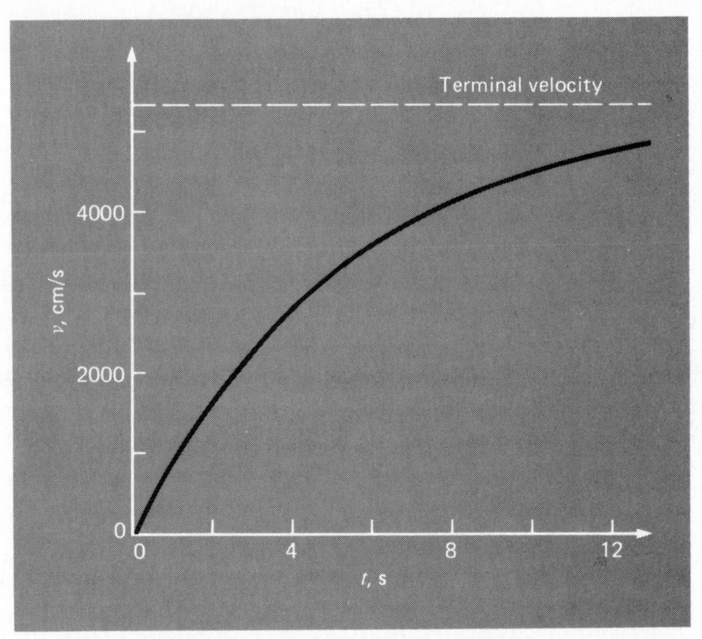


FIGURE I.2 The analytical solution to the falling parachutist problem as computed in Example 1.1. Velocity increases with time and asymptotically approaches a terminal velocity.

Solution: Inserting the parameters into Eq. (1.9) yields

$$v(t) = \frac{980(68,100)}{12,500}[1 - e^{-(12,500/68,100)t}]$$

$$= 5339.0\ (1 - e^{-0.18355t})$$

which can be used to compute

| *t*, s | *v*, cm/s |
|---|---|
| 0 | 0 |
| 2 | 1640.5 |
| 4 | 2776.9 |
| 6 | 3564.2 |
| 8 | 4109.5 |
| 10 | 4487.3 |
| 12 | 4749.0 |
| $\infty$ | 5339.0 |

According to the model, the parachutist accelerates rapidly (Fig. 1.2). A velocity of 4487.3 cm/s (100.4 mi/h) is attained after 10 s. Note also that after a sufficiently long time, a constant velocity (called the *terminal velocity*) of 5339.0 cm/s (119.4 mi/h) is reached. This velocity is constant because, after a sufficient time, the force of gravity will be in balance with the air resistance. Thus, the net force is zero and acceleration ceases.

Equation (1.9) is called an *analytical* or *exact solution* because it exactly satisfies the original differential equation. Unfortunately, there are many mathematical models that cannot be solved exactly. In many of these cases, the only alternative is to develop a *numerical solution* that approximates the exact solution. As mentioned previously, *numerical methods* are those in which the mathematical problem is reformulated so it can be solved by arithmetic operations. This can be illustrated for Newton's second law by realizing that the time rate of change of velocity can be approximated by (Fig. 1.3)

$$\frac{dv}{dt} \simeq \frac{\Delta v}{\Delta t} = \frac{v(t_{i+1}) - v(t_i)}{t_{i+1} - t_i} \tag{1.10}$$

where $\Delta v$ and $\Delta t$ are differences in velocity and time computed over finite intervals, $v(t_i)$ is velocity at an initial time $t_i$, and $v(t_{i+1})$ is velocity at some later time $t_{i+1}$. Equation (1.10) is called a *finite divided difference* at time $t_i$. It can be substituted into Eq. (1.8) to give

$$\frac{v(t_{i+1}) - v(t_i)}{t_{i+1} - t_i} = g - \frac{c}{m} v(t_i) \tag{1.11}$$

This equation can then be rearranged to yield

$$v(t_{i+1}) = v(t_i) + \left[ g - \frac{c}{m} v(t_i) \right] (t_{i+1} - t_i) \tag{1.12}$$

Thus, the differential equation [Eq. (1.8)] is transformed into an equation that can be solved algebraically for $v(t_{i+1})$. If you are given an initial value for velocity at some time $t_i$, you can easily compute $v$ at $t_{i+1}$. This new value of

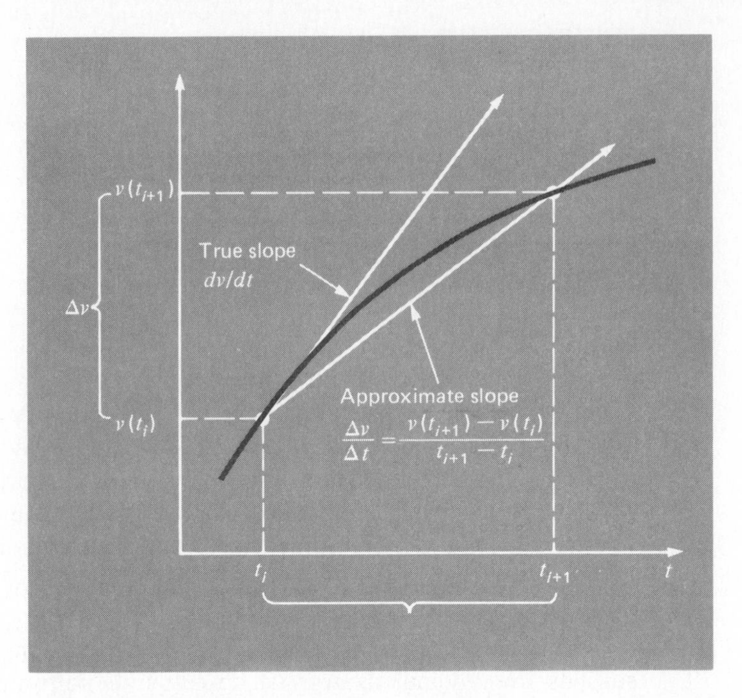


FIGURE 1.3 The use of a finite difference to approximate the first derivative of $v$ with respect to $t$.

$v$ at $t_{i+1}$ can then be used to extend the computation to $v$ at $t_{i+2}$ and so on. Thus, at any time along the way,

$$\begin{matrix}\text{New value} \\ \text{of } v\end{matrix} = \begin{matrix}\text{old value} \\ \text{of } v\end{matrix} + \begin{matrix}\text{estimated value} \\ \text{of slope}\end{matrix} \times \begin{matrix}\text{time} \\ \text{step}\end{matrix} \qquad [1.13]$$

## EXAMPLE 1.2
### Numerical Solution to the Falling Parachutist Problem

Problem Statement: Perform the same computation as in Example 1.1 but use Eq. (1.12) to compute $v(t)$ with a time increment equal to 2 s.

Solution: At the start of the calculation ($t_i = 0$), the velocity of the parachutist $v(t_i)$ is equal to zero. Using this information and the parameter values from Example 1.1, Eq. (1.12) can be used to estimate $v(t_{i+1})$ at $t_{i+1} = 2$ s.

$$v(2) = 0 + \left[980 - \frac{12{,}500}{68{,}100}(0)\right]2$$
$$= 1960 \text{ cm/s}$$

For the next interval (from $t = 2$ to 4 s), the computation is repeated with the result,

$$v(4) = 1960 + \left[980 - \frac{12{,}500}{68{,}100}(1960)\right]2$$
$$= 3200.5 \text{ cm/s}$$

The computation is continued in a similar fashion to obtain additional values as in

| *t*, s | *v*, cm/s |
|---|---|
| 0 | 0.0 |
| 2 | 1960.0 |
| 4 | 3200.5 |
| 6 | 3985.6 |
| 8 | 4482.5 |
| 10 | 4796.9 |
| 12 | 4995.9 |
| ∞ | 5339.0 |

The results are plotted in Fig. 1.4, along with the exact solution. It can be seen that the numerical method accurately captures the major features of the exact solution. However, because we have used straight-line segments to approximate a continuously curving function, there is some discrepancy between the two results. One way to minimize such discrepancies is to use a smaller computation interval. For example, applying Eq. (1.12) at 1-s

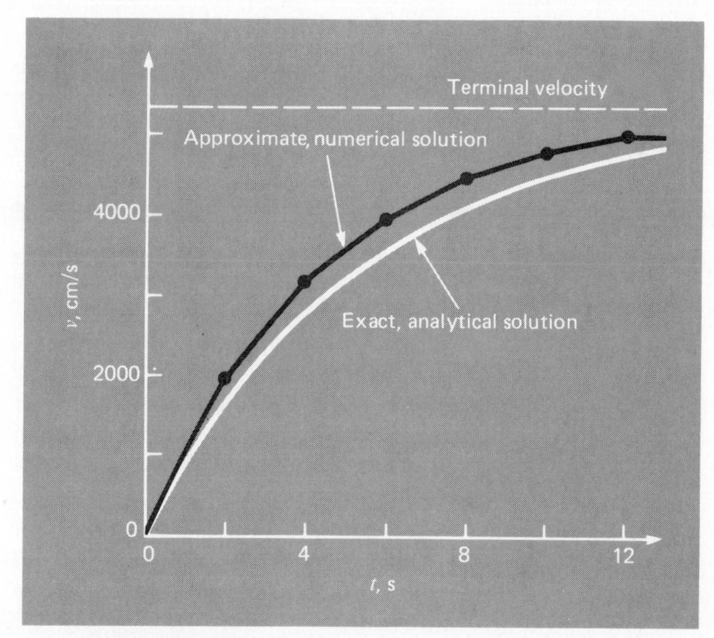


FIGURE 1.4 Comparison of the numerical and analytical solutions for the falling parachutist problem.

intervals would result in a smaller error, as the straight-line segments track closer to the true solution. Using hand calculations, the effort associated with using smaller and smaller step sizes would make such numerical solutions impractical. However, with the aid of the personal computer, large numbers of computations can be performed easily. Thus, you can accurately model the velocity of the falling parachutist without having to solve the differential equation exactly.

As in the previous example, a computational price must be paid for a more accurate numerical result. Each halving of the step size to attain more accuracy leads to a doubling of the number of computations. Thus, we see that there is a trade-off between accuracy and computational effort. Such trade-offs figure prominently in numerical methods and constitute an important theme of this book. Consequently, we have devoted the epilogue of Part I to an introduction to more of these trade-offs.

With the foregoing background, you can now begin to learn some skills that will allow you to implement numerical methods for problem solving. The next chapter is devoted to the first step in this process: learning to use your knowledge of computer languages to develop efficient and reliable software.

## PROBLEMS

**1.1** Answer true or false:

(*a*) Today more attention can be paid to problem formulation and interpretation because the computer and numerical methods facilitate the solution of engineering problems.

(*b*) The value of a variable that satisfies a single equation is called the root of the equation.

(*c*) Numerical methods are those in which a mathematical problem is reformulated so that it can be solved by arithmetic operations.

(*d*) Finite divided differences are used to represent derivatives in approximate terms.

(*e*) A physical interpretation of integration is the area under a curve.

(*f*) In the precomputer era, numerical methods were widely employed because they required little computational effort.

(*g*) Newton's second law is a good example of the fact that most physical laws are based on the rate of change of quantities rather than on their magnitudes.

(*h*) Interpolation is employed for curve-fitting problems when there is significant error associated with the data points.

(*i*) The large system of equations, nonlinearities, and complicated geometries that are common in engineering practice are easy to solve analytically.

(*j*) Math models can never be used for predictive purposes.

**1.2** Read the following problem descriptions and identify which area of numerical methods (as outlined in Fig. I.2) relates to their solution.

(*a*) You are on a survey crew and must determine the area of a field that is bounded by two roads and a meandering stream.
(*b*) You are responsible for determining the flows in a large interconnected network of pipes to distribute natural gas to a series of communities spread out over a 20-$mi^2$ area.
(*c*) For the falling parachutist problem, you must determine the value of the drag coefficient in order that a 200-lb parachutist not exceed 100 mi/h within 10 s of jumping. You must make this evaluation on the basis of the analytical solution [Eq. (1.9)]. The information will be used to design a jumpsuit.
(*d*) You are performing experiments to determine the voltage drop across a resistor as a function of current. You make measurements of voltage drop for a number of different values of current. Although there is some error associated with your data points, when you plot them they suggest a curvilinear relationship. You are to derive an equation to characterize this relationship.
(*e*) You must develop a shock-absorber system for a racing car. Newton's second law can be used to derive an equation to predict the rate of change in position of the front wheel in response to external forces. You must compute the motion of the wheel as a function of time after it hits a 6-in bump at 150 mi/h.
(*f*) You have to determine the annual revenues required over a 20-year period for an entertainment center to be built for a client. Money can presently be borrowed at an interest rate of 17.6 percent. Although the information to perform this estimate is contained in economics tables, values are only listed for interest rates of 15 and 20 percent.

**1.3** Give one example of an engineering problem where each of the following classes of numerical methods can come in handy. If possible, draw from your experiences in class and in readings or from any professional experience you have gathered to date.
(*a*) Roots of equations
(*b*) Linear algebraic equations
(*c*) Curve fitting: regression
(*d*) Curve fitting: interpolation
(*e*) Integration
(*f*) Ordinary differential equations

CHAPTER TWO

# PROGRAMMING ON PERSONAL COMPUTERS

Numerical methods combine two of the most important tools in the engineering repertoire: mathematics and computers. In fact, numerical methods can be loosely defined as computer math. Good computer programming skills will enhance your studies of numerical methods. In particular, the powers and limitations of numerical techniques are best appreciated when you use the methods in tandem with a computer to solve engineering problems.

You will have the opportunity in using this book to develop your own software. Because of the widespread availability of personal computers and magnetic storage devices, this software can be easily retained and applied throughout your career. Therefore, one of the primary objectives of this text is that you come away from it possessing useful, high-quality software.

This text has a number of features that are intended to maximize this possibility. All the numerical techniques are accompanied by material related to effective computer implementation. In addition, supplementary software is available for six of the most elementary methods discussed in this book. This software, which is compatible with widely used personal computers (IBM-PC and Apple II), can serve as a starting point for your own program library.

The present chapter provides background information that has utility if you plan to use this text as the basis for developing your own programs. It is written under the premise that you have had some prior exposure to computer programming. Because this book is not intended to support a computer programming course per se, the discussion pertains only to those aspects related to developing high-quality software for numerical methods. It is also intended to provide you with specific criteria for evaluating the quality of your efforts.

## 2.1 HISTORICAL BACKGROUND

In the broadest sense, a *computer* can be defined as any device that helps one to compute. On the basis of this definition, one of the earliest computers was the *abacus*. Developed in ancient Egypt and China, this device consists of rows of beads strung on wires in a rectangular frame (Fig. 2.1*a*). The beads

(a)

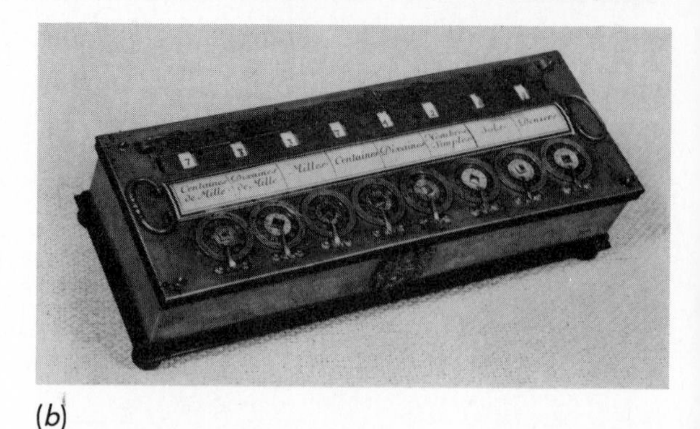

(b)

(c)

(d)

FIGURE 2.1 The evolution of computing devices: (*a*) an abacus, (*b*) Pascal's calculator, (*c*) a mainframe computer, and (*d*) a microcomputer, or personal computer (parts *b* and *c* with permission of IBM; part *d* with permission of Apple Computer, Inc.).

are used to keep track of powers of 10 (units, tens, hundreds, etc.) during the course of a computation. When employed by skilled operators, the abacus can rival a pocket calculator in speed.

Although manual devices such as the abacus certainly speed up computations, machines provide an even more powerful means for extending human calculating capability. Stimulated by the industrial revolution, seventeenth-century scientists developed the first such mechanical computers. Blaise Pascal invented an adding machine in 1642 (Fig. 2.1*b*). In the late 1600s, Gottfried Leibniz developed a mechanical calculator that could multiply and divide.

Although other computing machines were designed in the ensuing centuries, it was not until the 1940s that modern electronic computers were born. Originating primarily from military projects during World War II, they were usually one-of-a-kind devices built for research purposes. These machines, bearing names such as ENIAC and EDSAC, used vacuum tubes as their primary electronic components. Although they were expensive, slow, and often unreliable, these first-generation electronic computers demonstrated great promise for expediting large-scale data processing.

Although some first-generation machines, notably the UNIVAC, were marketed commercially, it was not until the 1960s that computers became available to significant numbers of scientists and engineers. This was due to the development of transistors and other solid-state electronic devices that supplanted vacuum tubes and resulted in computers that were, among other things, much more reliable. Although use of these "mainframe" computers burgeoned, access was somewhat limited by the fact that the machines were still much too expensive for most professionals to acquire individually. Consequently, engineers usually had to be associated with large organizations such as universities, government agencies, corporations, or consulting firms to gain access to computers.

However, in the mid-1960s and early 1970s some major technical breakthroughs dramatically altered this situation. In particular, the replacement of transistors by integrated circuits has brought tremendous computational power within the means of all practicing engineers. An integrated circuit, or IC, consists of a tiny silicon chip on which thousands of transistors are fabricated. The practical result of this technical innovation has been twofold. First, the mainframe or corporate machines are now capable of tremendous speed and storage capacity. Second, and more important in the present context, personal computers that are convenient, small, fast, and reliable are being mass-produced at reasonable prices. As expressed in an article in *Scientific American*: "Today's microcomputer at a cost of perhaps $300 has more computing capacity than the first large electronic computer, ENIAC. It is 20 times faster, has a larger memory, is thousands of times more reliable, consumes the power of a lightbulb rather than a locomotive, occupies 1/30,000 the volume and costs 1/10,000 as much. It is available by mail order or at your local hobby shop" (Noyce, 1977).

**TABLE 2.1 Comparison of typical computing systems***

| System | Significant figures | Word length, bits | Cost, $ | Computation speed, cycles/s | Storage capacity, K |
|---|---|---|---|---|---|
| Programmable calculator | 7–10 | | 25–350 | | 1–2 |
| Microcomputer | 7–10 | 7–16 | 100–5000 | $10^6$–$10^7$ | 16–256 |
| Minicomputer | 7–10 | 16–32 | 15,000–120,000 | $10^6$–$10^7$ | 128–512 |
| Mainframe | 7–14 | 32 | 100,000–10,000,000+ | $10^6$–$10^8$ | 8000–32,000 |

*Summarized from *Auerbach Computer Technology Reports,* August 1983.

Personal computers are usually lumped into one of two overlapping categories: micro- and minicomputers. *Microcomputers* are those on which the main functions of the computer are contained on a single integrated circuit chip. They typically cost on the order of a thousand dollars. *Minicomputer* is a somewhat imprecise term referring to computers that are more powerful than micros but yet are still within the means of some individuals and small firms. Both types of personal computers are in contrast to mainframe computers, or *maxicomputers*, which run in the million-dollar range and are usually owned by large-scale organizations. Table 2.1 summarizes general information related to the various types of computers.

The microelectric revolution has placed immense computational power within the reach of every professional engineer. However, no matter what type of computer you use, it only has utility if you provide it with careful instructions. These instructions are the software. The next sections contain information that will help you write your own high-quality software to implement numerical methods.

## 2.2 SOFTWARE DEVELOPMENT

The material in this chapter is organized around the five steps, delineated in Fig. 2.2, that are required to produce and maintain high-quality software. This chapter contains sections on each of these steps. This material is followed by a case study where the steps are applied to develop software for the falling parachutist problem. After assimilating this material, you should be better prepared to develop high-quality software for the methods in the remainder of the book.

### 2.2.1 Algorithm Design

We can now begin the process of developing computer programs. A *program* is merely a set of instructions to the computer. All the programs that are needed to run a particular computer are collectively called *software.*

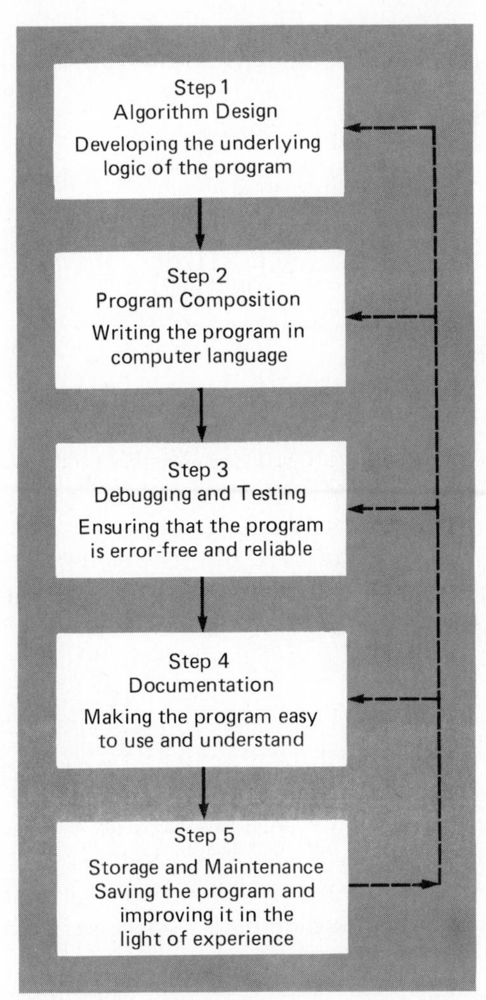


FIGURE 2.2 The five steps required to produce and maintain high-quality software. The feedback arrows indicate that the first four steps can be improved in the light of experience.

An *algorithm* is the sequence of logical steps required to perform a specific task such as solving a problem. Good algorithms have a number of characteristics. They must always end after a finite number of steps and should be general enough to deal with any type of contingency. Good algorithms must be deterministic; that is, nothing must be left to chance. The final results cannot be dependent upon who is following the algorithm. In this sense, an algorithm is analogous to a recipe. Two chefs working independently from a good recipe should end up with dishes that are identical.

Figure 2.3*a* shows an algorithm for the solution of the simple problem of adding two numbers. Two programmers working from this algorithm might develop programs exhibiting different styles. However, given the same data, the programs should yield the same results.

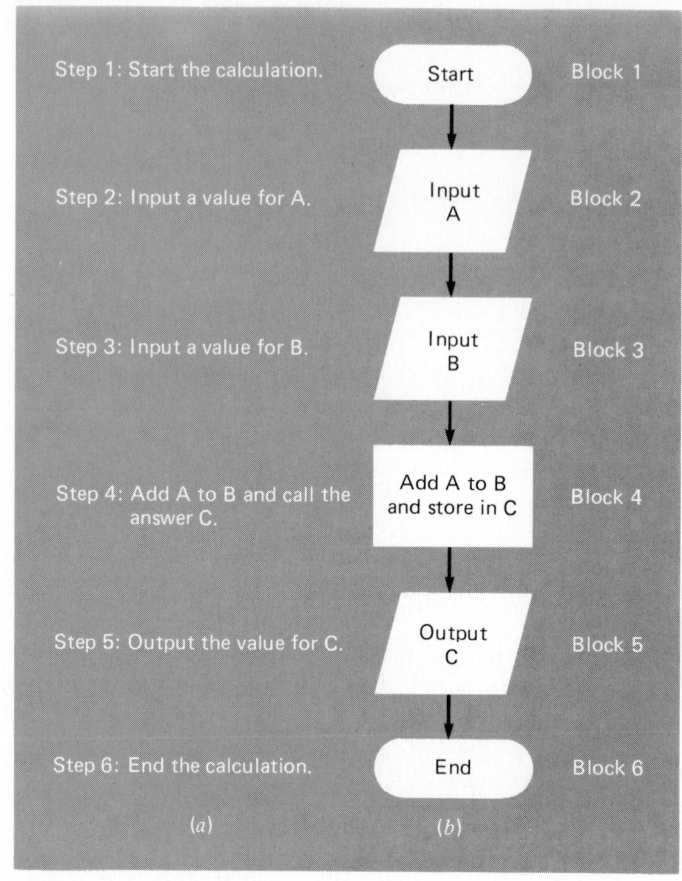


FIGURE 2.3 (*a*) Algorithm and (*b*) flow chart for the solution of a simple addition problem.

An alternative way to represent an algorithm is a *flowchart*. This is a visual or graphical representation of the algorithm that employs a series of blocks and arrows. Each block in the flowchart represents a particular operation or step in the algorithm. The arrows indicate the sequence in which the operations are implemented. Figure 2.4 illustrates eight types of blocks and arrows that conform to the most common operations required for programming on personal computers. Figure 2.3*b* shows a flowchart for the simple problem of adding two numbers. Flowcharts have particular utility for depicting complicated algorithms. For these cases, a graphical depiction can be helpful in visualizing the logical flow of the algorithm. In this text, we have included flowcharts for many of the major methods. You can use these flowcharts as the basis for developing your own programs.

### 2.2.2 Program Composition

After concocting the algorithm, the next step is to express it as a sequence of programming statements called *code*. It is important to resist the temptation

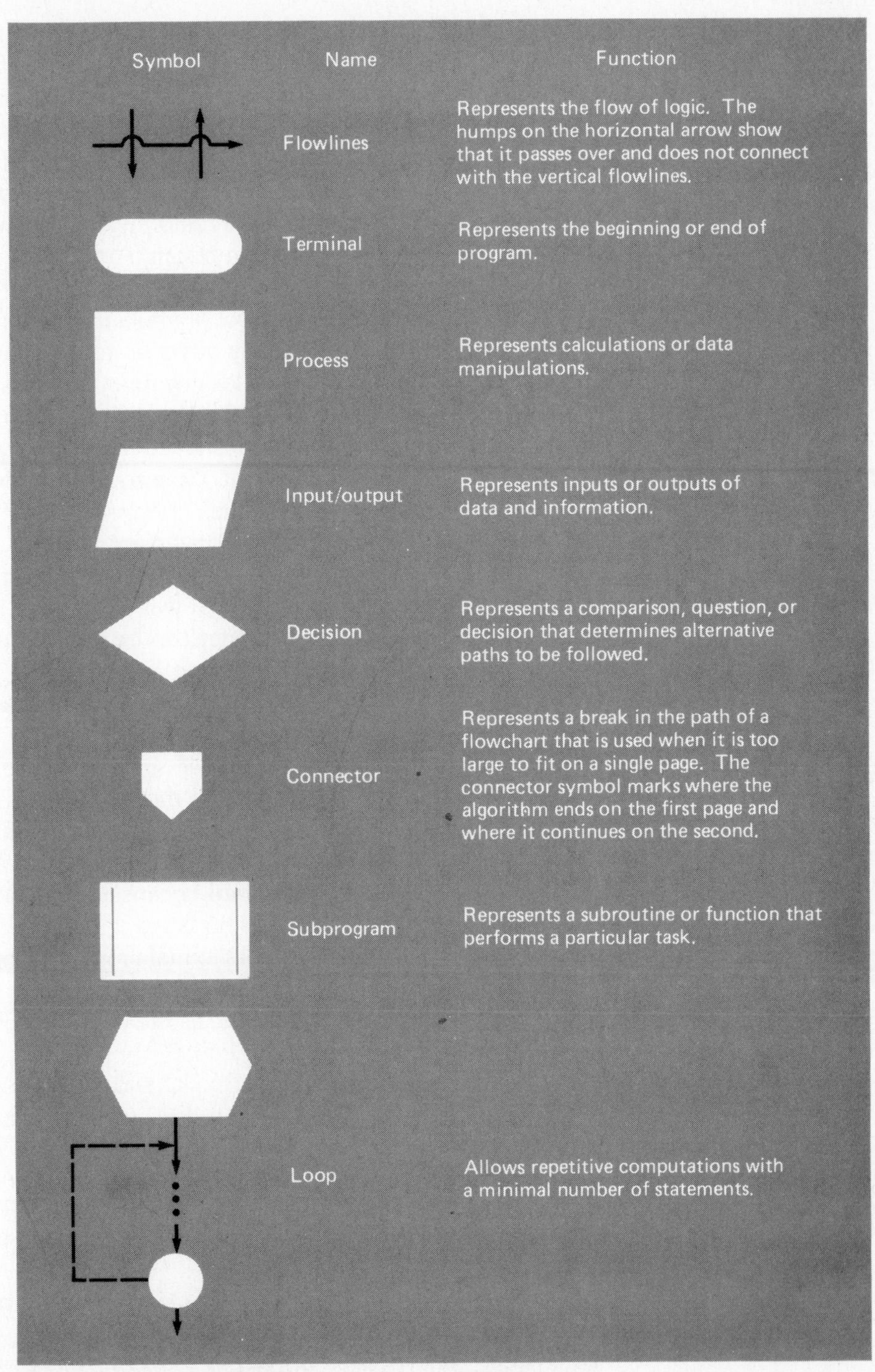


FIGURE 2.4 Symbols used in flowcharts.

to write code before the entire scope of the problem is clearly defined and the solution technique and the algorithm have been carefully designed. The most common problems encountered by inexperienced programmers usually can be traced to the premature preparation of a code that does not encompass an overall strategy or plan.

After a sound algorithm has been designed, code is written in a high-level computer language. Hundreds of high-level languages have been developed since the computer age began. Among these, three have the most relevance to personal computers—BASIC, FORTRAN, and Pascal.

*FORTRAN*, which stands for *for*mula *tran*slation, was developed in the 1950s. Because it was expressly designed for computations, it has been the most widely applied language for engineering and science.

*BASIC,* which stands for *b*eginner's *a*ll-purpose *s*ymbolic *i*nstruction *c*ode, was developed in the 1960s. It requires a small amount of memory and is relatively simple to implement. Consequently, it is the most widely used language on personal computers. However, BASIC is not as versatile as FORTRAN and is sometimes inconvenient for large, complex programs.

*Pascal,* which was named for the French scientist and philosopher Blaise Pascal, is a structured language that was developed in the 1970s. Programs written in Pascal for one computer can be easily run on another. Although Pascal is more difficult to learn than BASIC and FORTRAN, its strengths suggest that its importance will grow in the future. This is particularly true for advanced large-scale programming.

BASIC and FORTRAN are well suited for the short, simple programs that are sufficient for implementing most numerical methods in this book. Consequently, we have chosen to limit our text presentations of programs to these languages. BASIC is an obvious choice because of its widespread availability. FORTRAN is included because of its continuing significance for engineering work. Although this book emphasizes personal computers, it can also be used by those with access to mainframe machines and in conjunction with any high-level language. In this spirit, our programs and flowcharts are simple enough that they can serve as the basis for software development by those who are conversant with Pascal.

A complete description of BASIC and FORTRAN is obviously beyond the scope of this book. In addition, the number of dialects available in each language further complicates their description. For example, there are presently over 10 major dialects of BASIC. However, by limiting our discussion to the fundamentals, enough material can be covered so that you can understand and effectively implement the computer-related material in the remainder of the book.

Figure 2.5, which presents both FORTRAN and BASIC codes for adding two numbers, exhibits the major structural difference between the two languages—the labeling and spacing of code. For BASIC, each instruction is written as a line that is labeled with a number. In contrast, only those FORTRAN statements requiring identification need be labeled with a number. For

| FORTRAN | BASIC |
|---|---|

```
C       FORTRAN VERSION          100  REM    BASIC VERSION
        A=25                     110 A = 25
        B=15                     120 B = 15
        C=A+B                    130 C = A + B
        WRITE(6,1)C              140  PRINT C
    1   FORMAT(' ',F10.3)        150  END
        STOP
        END
```

column 7
column 5
column 1

FIGURE 2.5 Computer programs in FORTRAN and BASIC for the simple addition problem.

example, statement 1 of the FORTRAN version of Fig. 2.5 is called a FORMAT statement. It specifies how a particular line of input or output is to be displayed. Consequently, it must be labeled with a number in order that the computer can distinguish it from other FORMAT statements. FORTRAN statements must be numbered for a variety of other reasons, but most are usually unnumbered.

Another difference between the two languages is the spacing of each line. In BASIC, the spacing is usually unimportant. For example, line 10 could be typed as

```
10 A  =  25
10A=25
10    A    =     25
```

and the computer would interpret all versions as being equivalent. In contrast, the terms in a FORTRAN code must be aligned in specific columns. The conventions regarding the alignment stem from the fact that FORTRAN programs were originally entered into computers using punch cards. Although cards are employed less frequently today, the spacing conventions have usually been maintained.

The 80 columns of the punch card are called the card's *field*. Parts of the field are set aside for specific purposes. These are illustrated on the coding form in Fig. 2.6. A *coding form* is a piece of paper where a program can be written and checked for mistakes prior to entering it into the computer. Note that it has an 80-column field just like the punch card. Also observe that specific parts of the field are used for particular purposes.

Beyond structure, the two languages exhibit other differences as well as some strong similarities. These are both delineated in Table 2.2. This table presents a parallel comparison of six major programming elements that have direct relevance to numerical methods. These are:

1. *Constants and variables.* Conventions must be followed to express numbers and symbolic names in two languages. As should be evident from Table 2.2, this is one area where BASIC and FORTRAN differ significantly.

DATA PROCESSING CENTER
Texas A & M University

PROGRAM NUMERICAL METHODS FOR ENGINEERS | PUNCHING INSTRUCTIONS | GRAPHIC | PUNCH | PAGE OF
PROGRAMMER S.C. CHAPRA | DATE

STATEMENT — Name | Operation | Operand | Comments — Identification Square (73–80)

```
C    THIS CODING FORM IS USED TO DEMONSTRATE THE PURPOSE OF THE VARIOUS     LINE   1
C    FIELDS ON A PUNCHED CARD.                                              LINE   2
C    COMMENT CARDS, OF WHICH THE PRESENT LINE IS AN EXAMPLE, ARE            LINE   3
C    DESIGNATED BY A "C" IN COLUMN 1.                                       LINE   4
C    OTHERWISE, COLUMNS 1 THROUGH 5 ARE RESERVED FOR STATEMENT NUMBERS.     LINE   5
C    FOR EXAMPLE, THE FOLLOWING LINES ARE STATEMENTS 1, 83, AND 2ØØ6.       LINE   6

    1 FORMAT(I8)                                                            LINE   7
   83 Y=A+(B-X)**2.-5.*C/D                                                  LINE   8
 2ØØ6    Y = A + (B-X)**2. - 5.*C/D                                         LINE   9

C    COLUMNS 7 THROUGH 72 ARE RESERVED FOR THE ACTUAL STATEMENTS.           LINE  1Ø
C    THE COMPUTER IGNORES BLANK SPACES WITHIN THIS FIELD AND, THEREFORE,    LINE  11
C    SPACING CAN BE USED TO FACILITATE READING.  FOR INSTANCE. STATEMENTS   LINE  12
C    83 AND 2ØØ6 ACCOMPLISH THE SAME ALGEBRAIC MANIPULATIONS BUT            LINE  13
C    STATEMENT 2ØØ6 IS MORE READABLE DUE TO THE INCLUSION OF BLANKS.        LINE  14
C    IF A STATEMENT IS TOO LONG FOR ONE CARD, IT CAN BE CONTINUED ON        LINE  15
C    THE FOLLOWING CARD BY PLACING ANY NUMBER OTHER THAN ZERO IN COLUMN 6   LINE  16

      CONC = CONC + (THETA -XMAX) / (THETA - XMIN) * POPUL - (THETA -       LINE  17
     1   YMAX) / (THETA -YMTN) * POPUL                                      LINE  18

C    COLUMNS 73 THROUGH 8Ø CAN BE USED TO NUMBER THE LINES IN A PROGRAM     LINE  19
C    OR FOR ANY OTHER LABELING PURPOSE.                                     LINE  2Ø
```

FIGURE 2.6 Coding form.

2. *Input-output.* These are instructions whereby information is transmitted to and from the computer. Here is another area where the languages exhibit considerable differences. Although most modern dialects improve this situation, historically the input-output capabilities of BASIC have been typically quite limited. In contrast, the FORMAT statements of FORTRAN are powerful tools for labeling and spacing output. However, they are among the more difficult programming statements for the novice to master.

3. *Computations.* Mathematical operations are quite similar in the two languages. Although the nomenclature exhibits slight differences, equations written in the two languages are almost indistinguishable.

4. *Control.* These statements are used to direct the logical sequence of instructions in the program. For numerical methods, three types are usually sufficient—the GO TO statement, the logical IF, and the loop. Although there are nomenclature differences between the two languages, the statements are quite similar in operation.

5. *Subprograms.* As the name implies, these are miniprograms within the main program. They are designed to execute a statement or a set of statements that is repeated many times throughout the program. Rather than rewriting the miniprograms many times, they can be written once and then invoked with a single statement whenever they are needed. These subprograms, which include subroutines, user-defined functions, and statement functions, are another case where FORTRAN and BASIC differ significantly. The differences relate to the manner whereby information is passed between the main body of the program and the subprogram. As depicted in Table 2.2, the arguments of the FORTRAN subprograms act as windows to control the passage of information. This is an example of how FORTRAN is more complicated but, as a consequence, more powerful than BASIC.

6. *Documentation.* These statements permit the inclusion of user-oriented information in the program.

In summary, FORTRAN is somewhat more flexible and powerful but also more difficult to learn than BASIC. However, because BASIC was originally developed as a simplified version of FORTRAN, the two languages manifest many similarities. Although each has stylistic conventions that must be observed, their overall vocabulary and grammar are similar enough to permit easy translation of most programs from one language to the other. Consequently, all the computer code in this book will be presented in the parallel format of Fig. 2.5. Although this sometimes means that we will forego strengths that are peculiar to either language, it allows you to attain a working knowledge of both FORTRAN and BASIC.

**TABLE 2.2 Quick reference: Comparison of FORTRAN and BASIC. FORTRAN and BASIC, being easy to learn and practical to apply, are usually the first computer languages taught to engineering students. As with many languages, numerous dialects exist that make comprehensive descriptions difficult. The following comparison is an attempt to delineate *general* differences and similarities between FORTRAN and BASIC and should prove useful as a quick reference and refresher. Other sources can be consulted for details regarding particular dialects. This summary is focused on and limited to material that has direct relevance to the numerical methods and related software described in the text.**

| FORTRAN | BASIC |
|---|---|

**CONSTANTS AND VARIABLES**
*(means by which numbers and characters are represented throughout a program)*

**Constants**
are positive or negative numerical values, excluding commas or special symbols, that remain fixed throughout a computation.

| FORTRAN | BASIC |
|---|---|
| **Integers** are constants that do not contain decimal points: 1, −2, 100 | **Numeric constants** are integer or real numbers with optional decimal points: 1, −2.0, 0.001, 100 |
| **Real constants** contain decimal points: 1.0, −2., 0.001 | |

**Exponentials**
are constants written in scientific notation. For example, the numbers

−12000, 0.0000068, 386000000

are expressed in scientific notation as

$-12 \times 10^3, 6.8 \times 10^{-6}, 3.86 \times 10^8$

and may be written in FORTRAN and BASIC as

−12E3, 6.8E−6, 3.86E8

**Alphanumerics or character string constants**
represent letters, numbers, and symbols, which in the present text are used primarily for labeling. String constants have other applications, including use in relational expressions.

| FORTRAN | BASIC |
|---|---|
| In FORTRAN, they are enclosed, as in 'JOHN DOE', 'ENTER B' | In BASIC, they are enclosed, as in "VALUE A =", "8/5/48" |

**Numeric variables**
represent quantities that can change value. Symbolic names, which must begin with a letter and must contain no special symbols, are used for these variables.

| FORTRAN | BASIC |
|---|---|
| **Variable names** consist of one to six characters, chosen | **Variable names** consist of two characters (more in some |

**TABLE 2.2 Quick reference: Comparison of FORTRAN and BASIC (continued).**

| FORTRAN | BASIC |
|---|---|
| from A through Z and 0 through 9: | dialects) chosen from A through Z and 0 through 9: AA, X, N1 |
| **Integer variables** represent integer values and begin with letters from I through N: N, KOUNT, INDX1 | represent real or integer values. |
| **Real variables** represent real values and begin with letters A through H and O through Z: X, COUNT, VEL1 | |
| **Character or string variables** represent alphanumeric or character strings. Symbolic names are used. The treatment of string variables varies considerably among dialects. | |
| **CHARACTER statements** are of the form: CHARACTER * $n$ $var_1$,$var_2$ where $n$ is the specified length of the character strings followed by a variable list. For example, CHARACTER * 4 NAME1,NAME2 | **String variables** end with $. The length of the variable is limited by the dialect. A$, N1$ |
| **Arrays** are subscripted variables that store a set of values in one-dimensional vectors or multidimensional matrices. Sufficient storage space for a maximum number of elements is specified by | |
| **DIMENSION statement** DIMENSION A($n$), ISUM($n_1$,$n_2$) | **DIM statement** DIM A($n$), IS($n_1$,$n_2$) |
| Up to seven subscripts, which must be positive integer constants, are allowed. | The DIM statement is usually limited to a two-dimensional array; the $n$'s may be variables. |
| Nondimensioned arrays yield an error message. | Nondimensioned arrays assume a default of $n = 10$. |
| The DIMENSION statement must be placed before any executable statement. | The DIM statement must be placed before the first line where the dimensioned variable is utilized. Otherwise, it defaults to $n = 10$. Redimensioning results in an error message. |
| The DIMENSION name (e.g., A or ISUM) must conform to the element type—i.e., array A must contain real values, whereas ISUM must contain integers. | |

**TABLE 2.2 Quick reference: Comparison of FORTRAN and BASIC (continued).**

| FORTRAN | BASIC |
|---|---|
| **INPUT/OUTPUT** *(means whereby information is transmitted to and from a program)* | |
| **Format statements** specify the length and position of each piece of data that is input or output. | |
| Although dialect-specific format-free input and output statements exist, standard FORTRAN usually entails formatting. | Although dialect-specific formatting statements are evolving, early versions of BASIC usually did not employ formatting. |
| **Input** is the means whereby data is transmitted to the program. | |
| **READ statements** allow data to be entered into the program during execution:<br><br>READ *f* $var_1, var_2, \ldots, var_n$<br><br>where *f* is a formatting code that specifies the type and layout and, in some cases, the device used to input the values of $var_1, var_2, \ldots, var_n$. For example,<br><br>READ (5,2) A,B<br><br>where 2 is the number of the FORMAT statement and 5 specifies that the data will be input by a card reader. | **INPUT statements** allow data to be entered into the program during execution:<br><br>*ln* INPUT $var_1, var_2, \ldots, var_n$<br><br>where *ln* is the line number of the INPUT statement and $var_1, var_2, \ldots, var_n$ are the names of the variables for which values are to be input. For example,<br><br>10 INPUT A,B<br><br>Values for A and B must be entered on a device such as a keyboard when the statement is executed. |
| **DATA statements** are nonexecutable statements that define the initial value of a variable. They are of the general form:<br><br>DATA $var_1, \ldots, var_n/value_1, \ldots, value_n/$<br><br>where $var_i$ is the name of the variable and $value_i$ is a constant. For example,<br><br>DATA A,B,C,Z/5.,0.001,88.,1.E-6/ | **READ/DATA statements** consist of a READ statement which searches for a DATA statement that contains the values to be read, as in<br><br>10 READ A,B,C,Z<br>.<br>.<br>.<br>90 DATA 5,0.001,88,1E-6 |
| **Output** is the means whereby data is transmitted from the program. | |
| **WRITE statements** are commonly used to output data. Their general form is<br><br>WRITE *f* $var_1, \ldots, var_n$ | **PRINT statements** are commonly used to output data. Their general form is<br><br>*ln* PRINT $var_1, \ldots, var_n$ |

**TABLE 2.2 Quick reference: Comparison of FORTRAN and BASIC (continued).**

| FORTRAN | BASIC |
|---|---|
| For example, WRITE (6,2) A,B where (6,2) is a formatting code, 2 is the number of the FORMAT statement, and 6 specifies that the data will be output on a printer. | For example, 10 PRINT A,B When executed, this statement causes the values of A and B to be output on a device such as a video screen or a printer. |

**COMPUTATIONS**
*(calculations using mathematical expressions)*

**Assignment statements**
are used to assign a value to a variable:

XM=3.281

directs the computer to assign a value of 3.281 to the variable XM;

A=XM+5

directs the computer to add 5 to XM and assign the result (i.e., 8.281) to the variable A;

A=A+40

directs the computer to add 40 to A and assign the result (i.e., 48.281) as the new value of the variable A. The previous value of A is destroyed in the process. Note that, although A=A+40 is not a valid mathematical expression, it is a proper statement in a computer program. The equal sign in the assignment statement can be thought of as meaning "is replaced by," as in

A is replaced by A+40

**Arithmetic operators**
are symbols used to represent mathematical operations:

| FORTRAN | Operation | BASIC |
|---|---|---|
| + | Addition | + |
| − | Subtraction | − |
| * | Multiplication | * |
| / | Division | / |
| ** | Exponentiation | **, ↑, ∧ (as defined by dialect) |

Precedence is exponentiation, followed by multiplication and division from left to right, followed by addition and subtraction from left to right. Parenthetical or group expressions are evaluated from innermost to outermost parentheses.

$$X = \sqrt{\frac{(a + b) - r^3}{33} - \frac{y^4}{45}}$$

| FORTRAN | BASIC |
|---|---|
| X=(((A+B)−R**3)/33−Y**4/45)**.5 | X=(((A+B)−R∧3)/33−Y∧4/45)∧.5 |

**TABLE 2.2 Quick reference: Comparison of FORTRAN and BASIC (continued).**

| **FORTRAN** | | **BASIC** |
|---|---|---|

**CONTROL**
*(directs flow of program through transfer or reassignment)*

---

**GO TO statements**
specify an unconditional transfer that skips to a specified line number:

```
GO TO 200
```

---

**Logical operators**
are used to compare values of two expressions:

| FORTRAN | | BASIC |
|---|---|---|
| .EQ. | Equal to | = |
| .NE. | Not equal to | < > |
| .LT. | Less than | < |
| .LE. | Less than or equal to | < = |
| .GT. | Greater than | > |
| .GE. | Greater than or equal to | > = |
| .AND. | Logical | AND |
| .OR. | | OR |

---

**Logical IF statements**
specify a conditional transfer or reassignment predicated on the validity of a logical expression:

FORTRAN:

```
IF(N.GT.1.OR.N.LT.3)N=2
IF(N.GE.1) GO TO 10
```

BASIC:

```
IF(N>1)OR(N<3)THEN N=2
IF N>=1 THEN 10
```

In the above examples, if the logical expression is true, the transfer or assignment is executed. For the first example, if N is greater than 1 or less than 3, then N is set equal to 2 and control passes to the next line. For the second, if N is greater than or equal to 1, the program transfers to line 10. In either case, if the expression is false, the transfer or reassignment is not executed and control passes to the next line.

---

**Loops**
allow repetitive computations and manipulations with a minimal number of statements.

**Logical IF loops**
perform repetitive computations that are controlled on the basis of a LOGICAL IF statement:

FORTRAN:

```
10 X=Y(I)*Z(I-1)
   IF(X.LT.0)GO TO 50
   I=I+1
   GO TO 10
50 X=-X
```

BASIC:

```
10 X=Y(I)*Z(I-1)
20 IF X<0 THEN 50
30 I=I+1
40 GO TO 10
50 X=-X
```

**TABLE 2.2 Quick reference: Comparison of FORTRAN and BASIC (continued).**

| FORTRAN | BASIC |
|---|---|

**Indexed loops**

| DO loops | FOR/NEXT loops |
|---|---|
| DO *ln* I=*j*,*n*,*k* | FOR I=*j* TO *n* STEP *k* |
| ⋮ | ⋮ |
| *ln* CONTINUE | *ln* NEXT I |

where *ln* is the line number of the last statement in the loop, *j* is the initial value of the counter, *n* is the final or terminal value, and *k* is the increment by which I is increased to progress from *j* to *n*. After completion of loop, *I* has a value of $n + k$, when *I* is a multiple of *n*.

**SUBPROGRAMS: FUNCTIONS AND SUBROUTINES**
*(execute a statement or set of statements that are repeated many times throughout a program)*

**Intrinsic functions**
are built-in or library functions that perform commonly employed mathematical or trigonometric operations:

| FORTRAN | | BASIC |
|---|---|---|
| SIN | Sine | SIN |
| COS | Cosine | COS |
| TAN | Tangent | TAN |
| ALOG or LOG | Base *e* or natural logarithm | LOG |
| ALOG10 or LOG10 | Base 10 or common logarithm | |
| EXP | Exponential | EXP |
| SQRT | Square root | SQR |
| ABS | Absolute value | ABS |
| INT | Greatest integer value that is less than or equal to *x* | INT |

where *x* is the function argument. Note that this list is not comprehensive. Exceptions and additional intrinsic functions exist in particular dialects.

**User-defined functions**
are functions that are defined by the programmer.

**Statement functions**
are of the general form:

*name*$(x_1, \ldots, x_n) = f$

where *name* is the name of the function (a variable name); $x_1, \ldots, x_n$ are dummy variables that must be nonsubscripted; and *f* is an arithmetic expression which is a function of $x_1, \ldots, x_n$.

Statement functions are located prior to the first executable statement.

**DEF statements**
are of the general form:

*ln* DEF FN$a(x) = f$

where *ln* is the line number, *a* is any letter of the alphabet, *x* is a numeric variable (nonsubscripted), and *f* is an arithmetic expression that is a function of *x*.

DEF statements are located prior to the first place of execution.

**TABLE 2.2 Quick reference: Comparison of FORTRAN and BASIC (continued).**

| FORTRAN | BASIC |
|---|---|
| Several arguments may be passed into a statement function. Other variables in the function have the same value as they had in the main program at the point the statement function was invoked.<br>`TRIG(X,Y)=SIN(X)−LOG(Y)`<br>`A=5`<br>`B=10`<br>`S=TRIG(A,B)` | Only one argument may be passed into a DEF statement. Other variables in the statement have the same value as they had in the main program at the point the user-defined function was invoked.<br>`10 DEF FNT(X)=SIN(X)−LOG(B)`<br>`70 A=5`<br>`80 B=10`<br>`90 S=FNT(A)` |
| **Function subprograms** are similar in execution to statement functions but, as the name implies, differ in that they are programs—that is, they may consist of several lines. The function subprogram is of the general form:<br>FUNCTION *name*($x_1$, . . . $x_2$)<br>⋮<br>*name* = *f*<br>RETURN<br>where all values are the same as for the statement function defined above.<br>`A=5`<br>`B=10`<br>`S=TRIG (A,B)`<br>⋮<br>`END`<br>`FUNCTION TRIG(X,Y)`<br>`TRIG=SIN(X)−LOG(Y)`<br>`RETURN`<br>Note that constants and variables that are not passed as arguments must be either defined within the function subprogram or passed via a common statement. | |

**Subroutines**

are subprograms consisting of a group of statements that perform a particular task. They contain a RETURN statement to transfer results back to the statement where the subroutine was called.

**TABLE 2.2 Quick reference: Comparison of FORTRAN and BASIC (continued).**

| FORTRAN | BASIC |
|---|---|
| Subroutines are invoked by a CALL statement of the form:<br>CALL *name* ($arg_1$,$arg_2$, . . . ,$arg_n$)<br>where *name* is the subroutine name, and $arg_1$,$arg_2$, . . . ,$arg_n$ are *n* arguments (either variables or constants) that are passed to the subroutine. | Subroutines are invoked by a GOSUB statement of the form:<br>$ln_1$ GOSUB $ln_2$<br>where $ln_1$ is the line number of the GOSUB statement, and $ln_2$ is the line number where the subroutine begins. |
| The subroutine itself is located after the main program and begins with a SUBROUTINE statement of the form:<br>SUBROUTINE *name* ($arg_1$,$arg_2$, . . . ,$arg_n$)<br>where *name* is the same as the *name* of the CALL statement. | The first line of the subroutine can be located anywhere in the program. |

Once within the subroutine, the statements are executed in sequence until a RETURN statement is encountered, whereupon control is transferred back to the statement following the line where the subroutine was invoked.

| FORTRAN | BASIC |
|---|---|
| Only values that appear as the subroutine's arguments are passed to and from the subroutine: | All values are passed to and from the subroutine: |

```
      :
CALL SUM (X,Y,Z)
      :
END
SUBROUTINE SUM (A,B,C)
C=A+B
RETURN
```

```
  :
200 GOSUB 800
  :
500 END
800 Z=X+Y
850 RETURN
```

Note that constants and variables that are not passed as arguments must be either defined within the subroutine or passed via a COMMON statement.

**DOCUMENTATION**
*(allows user-oriented information to be included in a program)*

Documentation statements are nonexecutable instructions.

| FORTRAN | BASIC |
|---|---|
| **Comment statements** consist of the character C or the symbol * in column 1 followed by a message:<br>C any message may be typed here | **REM statements** consist of REM followed by a message:<br>10 REM any message may be typed here |

### 2.2.3 Debugging and Testing

After writing the program code, you then must test it for errors, which are called *bugs*. The process of locating the bugs and correcting them is called *debugging*. Several types of errors can occur when programming in any language. *Syntax errors* violate the rules of the language such as spelling, number formation, line numbers, and other conventions specific to each language. These errors are often the result of typing mistakes. For example, the BASIC statement

```
30 A = 5/(0.2 + 4 * SIN (2 * Y)
```

would result in an immediate syntax error because the parentheses are not correctly paired.

Errors more difficult to detect are associated with program construction and logic and may occur without syntax interruptions. Thus, great care should be exercised to ensure that the program actually does what is intended. For example, suppose you wish to sum the integers between 1 and 10 and then divide by 10 (that is, calculate the average). Your codes in FORTRAN and in BASIC would be

**FORTRAN**

```
   S = 0
   DO 40 I = 1, 10
   S = S + I
40 CONTINUE
   A = S/I
   WRITE (6, 1)A
```

**BASIC**

```
10 S = 0
20 FOR I = 1 TO 10
30 S = S + I
40 NEXT I
50 A = S/I
60 PRINT A
```

and results in A = 5, whereas the expected result is A = 5.5. The syntax is perfect, but an error in logic is present that the computer can never detect because it does not have information regarding your intent. One way to eliminate this type of error is to print values during program development for variables not required in the final form of the program. For example, if you had written

```
WRITE  (6,1)  A,I                    60 PRINT A,I
```

with the results that A = 5 and I = 11, you probably would have noticed the error of intent and realized that the value of I is incremented upon exit from the loop.

Errors of this type are often difficult to detect in large or complex programs. Thus, it is good practice to manually check the results of the program output if possible and to test for special cases. This is done with pencil, paper, and calculator. Errors associated with program logic or intent, rather than with grammar, are called *semantic errors*. These usually occur during the execution of the program and are also referred to as *run-time errors*. The technique of printing the value of intermediate variables to check the logic of a program is absolutely necessary to avoid semantic errors in large programs.

The debugging and testing of programs is facilitated by employing good coding style. This may involve designing your program so that it consists of several small parts. This type of programming style is sometimes referred to as *modular programming*. Specific, readily identified tasks are performed by each part. Subroutines are appropriate vehicles for such modularization. The main (or calling) program may then be simply a master control that orchestrates each of the parts in a logical fashion. In this way, if your program does not perform properly, you will be able to isolate the location of the problem more rapidly. For example, subroutines could be assigned the following tasks:

1. Read data.
2. Display data.
3. Input or prompt user for information.
4. Perform numerical algorithm.
5. Display results as a table.
6. Display results as a graph.

Each of these subroutines performs a limited and isolated task that can be programmed and debugged separately. This makes the overall job much easier compared to approaching all tasks simultaneously.

After the module tests, the total program should be subjected to a complete system test. For a numerical methods program, a series of computations would be performed and compared with cases where the exact answer is known a priori. Sometimes analytical solutions are available for this purpose. Such was the case for the falling parachutist (recall Examples 1.1 and 1.2). In other instances, the programmer must perform manual calculations with a pocket calculator to verify that the program yields reliable results. In any event, the object is to subject the program to a wide enough variety of tests to ensure that it will perform reliably under all possible operating conditions. Only at this point is the program ready to be used to solve engineering problems.

### 2.2.4 Documentation

After the program has been debugged and tested, it must be documented. Documentation is the addition of comments to allow the user to implement the program more easily. Remember that along with other people who might employ your software, you also are a "user." Although a program may seem simple and clear to you when it is fresh in your mind, after the passing of time the same code may seem inscrutable. Therefore, sufficient documentation must be included to allow you or other users to immediately understand and implement your program.

This task has both internal and external aspects. Internal documentation consists of discussion or explanations inserted throughout the program code to describe how various sections of the program work. It is critical in those cases where the program is to be modified. Such documentation should be added as soon as any part of the program is finished, rather than after the entire program is completed, to avoid losing the original design concept that went into the program development. Internal documentation is significantly improved by the use of good mnemonic names for variables. These names may be more difficult to code than shorter names, but their advantage of being more informative usually makes the additional effort worthwhile. Good mnemonic names imply the use of conventional or standard names or common abbreviations for variables.

External documentation relates to instructions in the form of messages and supplementary printed matter designed to assist the user in implementing your software. The printed messages are intended to make the output attractive and "user-friendly." This involves the effective use of spaces, blank lines, or special characters to illuminate the logical sequence and structure of the program output. Attractive output simplifies the detection of errors and enhances the communication of program results.

The supplementary printed matter can range from a single sheet to a comprehensive user's manual. Figure 2.7 is an example of a simple documentation form that we recommend you prepare for every program you develop. These forms can be maintained in a notebook to provide a quick reference for your program library. The user's manual for your computer is an example of comprehensive documentation. This manual tells you how to run your computer system and disk-operating programs.

### 2.2.5 Storage and Maintenance

The final steps in program development are storage and maintenance. *Maintenance* involves improvements or changes made in the program which evolve through its application to real problems. In the long run, such changes may make the program easier to use or more applicable to a larger class of problems. Maintenance is facilitated by good documentation.

*Storage* relates to the manner in which your software is retained for later use. Before the advent of personal computers, there were no simple ways to save working copies of software. Code listings could, of course, be saved but had to be retyped if they were used at a later date. Decks of punch cards could also be retained, but for a program of any magnitude they were unwieldy and susceptible to deterioration.

As mentioned at the beginning of this chapter, magnetic storage devices have greatly enhanced the ability to retain software. One common storage device, the *floppy disk*, is depicted in Fig. 2.8. Floppy disks are an inexpensive medium for storing software and data. Although they have great utility, floppy disks also have disadvantages. For one thing, they have slow access time. For another, they must be handled and stored with care. Because they can be inadvertently erased, you should always maintain a backup copy of

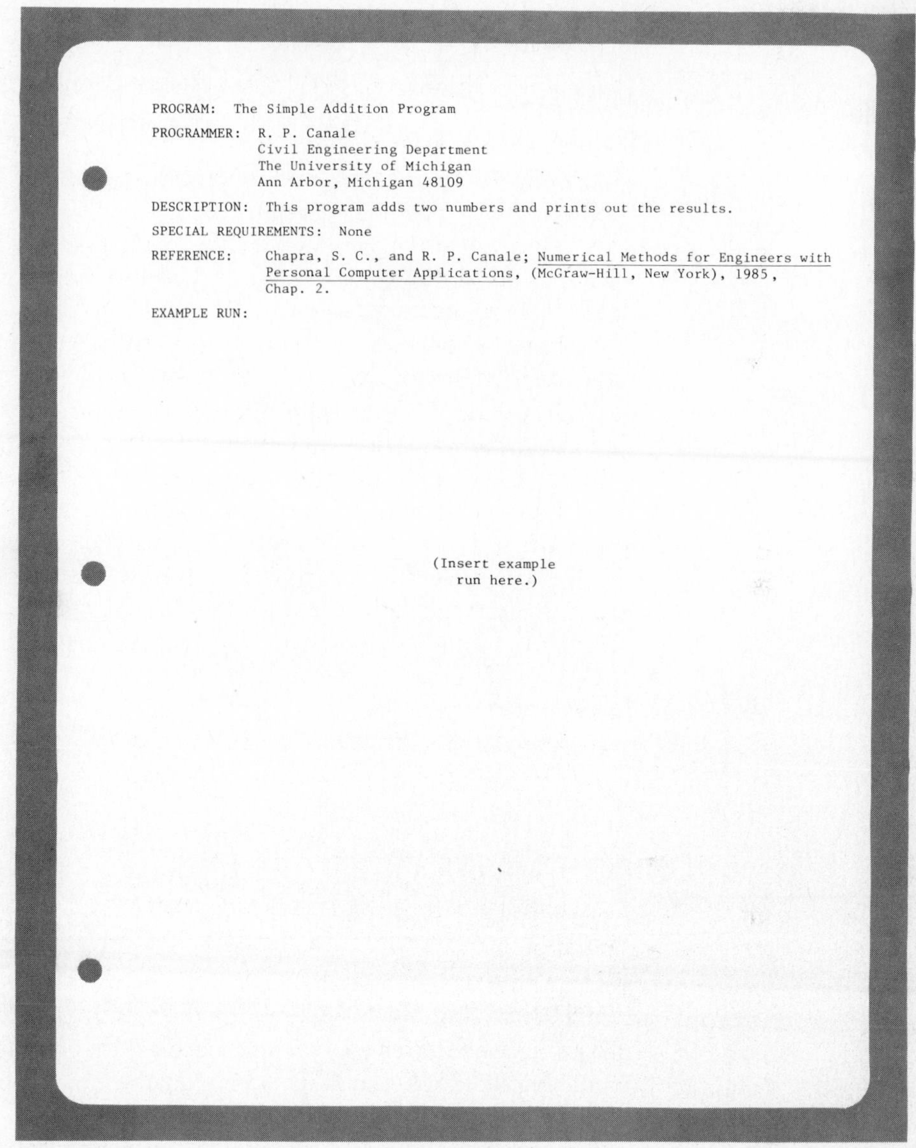

PROGRAM: The Simple Addition Program

PROGRAMMER: R. P. Canale
Civil Engineering Department
The University of Michigan
Ann Arbor, Michigan 48109

DESCRIPTION: This program adds two numbers and prints out the results.

SPECIAL REQUIREMENTS: None

REFERENCE: Chapra, S. C., and R. P. Canale; Numerical Methods for Engineers with Personal Computer Applications, (McGraw-Hill, New York), 1985, Chap. 2.

EXAMPLE RUN:

(Insert example run here.)

FIGURE 2.7 A simple one-page format for program documentation. This page could be stored in a binder along with a printout of the program code.

your disk. In addition, when you complete a computer program, you should immediately make a printout of the code and store it with your documentation. Such printouts could prove valuable in the unlikely, but possible, event that both disk copies of your software were destroyed.

FIGURE 2.8 A floppy disk.

## 2.3 SOFTWARE DEVELOPMENT FOR THE FALLING PARACHUTIST PROBLEM

We will now use the material from the previous sections to write BASIC and FORTRAN computer programs for the falling parachutist problem. These programs make an ideal example because they contain all the elements—input-output, loops, decisions, computations, and subprograms—that constitute the software in the remainder of the text.

Recall that the falling parachutist problem amounted to the solution of Eq. (1.12):

$$v(t_{i+1}) = v(t_i) + \left[g - \frac{c}{m}v(t_i)\right]\Delta t \qquad [2.1]$$

where $v(t_{i+1})$ is the velocity at a later time $t_{i+1}$, $v(t_i)$ is the velocity at the present time $t_i$, $g$ is the acceleration of gravity (equal to 980 cm/s$^2$), $c$ is the drag coefficient, $m$ is the mass of the parachutist, and $\Delta t = t_{i+1} - t_i$. The term in the brackets is actually the value for the rate of change of velocity with respect to time [Eq. (1.8)]. If the initial velocity of the parachutist $v(t_i)$ is known, Eq. (2.1) can be solved repetitively for values of $v(t_{i+1})$, as was done in Example 1.2.

With this information as background, we can now devise an algorithm for the problem. At this point, we could develop a very detailed algorithm. However, in actual practice, this is rarely done. Rather, we start with a simple general version, with details added thoughtfully and sequentially to expand the definition. Then when the final scope is clearly outlined, we can proceed with writing the program. This method of starting with the general and progressing toward the specific is called the *top-down approach* to programming. Among other things, it is efficient because it is usually much easier to eliminate errors if algorithms and programs are written in simple steps and checked as we proceed.

A very simple algorithm to perform the same calculation as Example 1.2 can be written in words as: input data, compute velocity, print out the answer, and repeat until you have calculated as many values as you require. This algorithm can then be expressed more formally as a flowchart. Figure 2.9 provides a detailed procedure for the implementation of the computation. The flowchart consists of three sets of statements:

1. Input of variables and constants
2. Initialization of variables
3. A loop for calculating and outputting answers

On the basis of this flowchart, the program code can now be written. The FORTRAN and BASIC versions are depicted in Fig. 2.10. Notice that for the BASIC version, increments of 10 are used to label the statement numbers. This is done because we anticipate that additional statements will be inserted in our subsequent refinement of the program.

Although the foregoing exercise is certainly a valid program for the falling parachutist problem, it by no means exploits the total possibilities of programming in either FORTRAN or BASIC. To demonstrate how additional statements can be employed to develop a user-friendly version, we will now refine the program.

Many of the following modifications and additions represent more efficient and communicative programming technique. However, some of the material is intended for didactic purposes to demonstrate the use of certain statements. The following discussion relates directly to the BASIC version. Because the programs in Fig. 2.11 are written in parallel, it is a simple matter to extend the discussion to the FORTRAN version.

The program in Fig. 2.11 has a number of new features. The major modifications are:

1. *The program now computes velocity for three different values of the drag coefficient and mass.* The ability to perform repetitive calculations is one

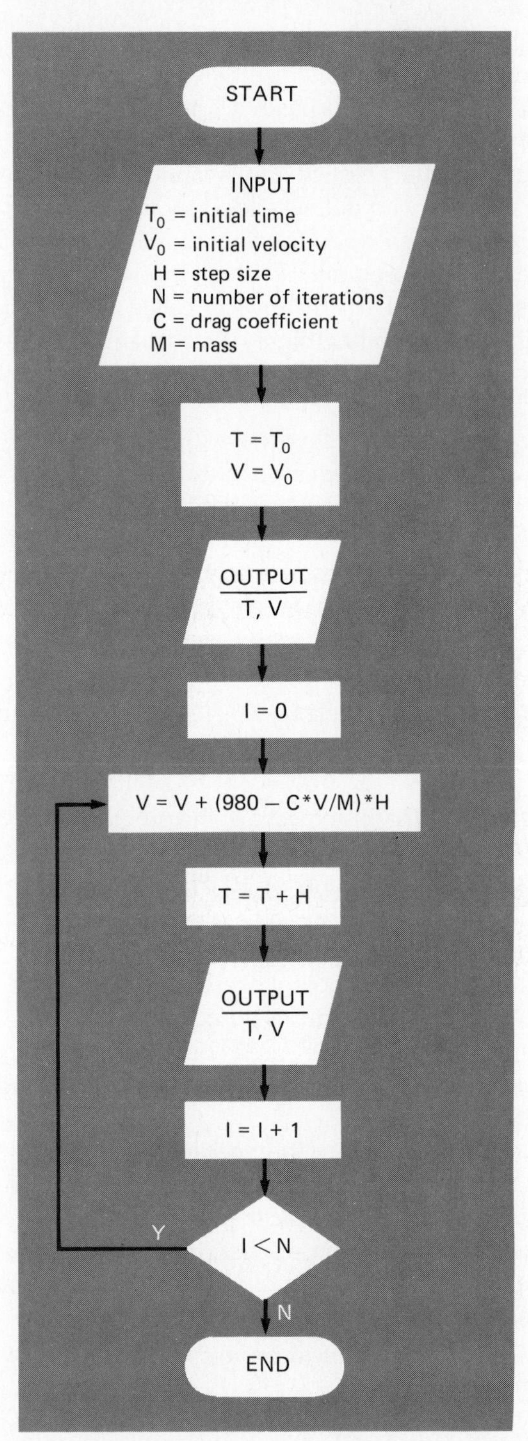


FIGURE 2.9 Flowchart for a simple program for the falling parachutist problem.

FORTRAN

```
      TO=0
      VO=0
      H=2
      N=10
      C=12500
      M=68100
      T=TO
      V=VO
      WRITE(6,1)T,V
    1 FORMAT(2(' ',F10.3))
      I=0
  200 V=V+(980-C*V/M)*H
      T=T+H
      WRITE(6,1)T,V
      I=I+1
      IF(I.LT.N)GO TO 200
      STOP
      END
```

BASIC

```
100 TO = 0                        TO = initial time
110 VO = 0                        VO = initial velocity
120 H = 2                         H = step size
130 N = 10                        N = number of steps
140 C = 12500                     c = drag coefficient
150 M = 68100                     M = mass
160 T = TO
170 V = VO
180  PRINT T,V                    [Eq. (1.2)]
190 I = 0
200 V = V + (980 - C * V / M) * H
210 T = T + H
220  PRINT T,V
230 I = I + 1
240  IF I < N THEN 200
250  END
```

FIGURE 2.10 FORTRAN and BASIC programs for the falling parachutist problem. These programs duplicate the computation that was performed manually in Example 1.2.

of the strengths of computers. In engineering design, it is often useful to perform a calculation several times with different coefficient values in order to assess the sensitivity of the model to these changes. This is done for the present case by performing the computation of Example 1.2 with the drag coefficient varied $\pm$ 10 percent. Thus, the three cases used in the program are for the original drag coefficient (12,500 g/s), the drag coefficient +10 percent (13,750 g/s), and the drag coefficient −10 percent (11,250 g/s). The repetitive computation is accomplished by merely adding a loop (lines 3080 to 3390). Each time the program passes through the loop a different drag coefficient is used to compute velocity. Notice also that the drag coefficient and the mass are treated as subscripted variables C(K) and M(K). Consequently, they are dimensioned in line 3040.

2. *The program now has a more sophisticated iterative scheme.* Aside from the addition of the large loop for the three cases of $c$ and $m$ (lines 3080 to 3390), we have also used two loops for the actual computation of $v$. This was done because we might not want to print out an answer after every time step. This would be particularly true if we were using a very small step size, say 0.01 s, to obtain a more accurate result. For a computation from $t = 0$ to 20 s, this would amount to 20/0.01, or 2000 numbers. Because we only require a value for every 2 s to obtain a reasonable characterization of the parachutist's fall, we have used two nested loops so that the program prints out results at intermediate times. A *nested loop* is one that is included inside another loop. For our example, the internal loop (lines 3320 to 3350) performs the calculation using the desired step size (line 2100). After NC iterations of this loop (where NC is calculated internally at line 3050), an answer is printed. This procedure is repeated NP times (where NP is calculated internally at line 3060) by

## FORTRAN

```
C     USER-FRIENDLY PROGRAM IN FORTRAN
C      FOR THE FALLING PARACHUTIST
C
C     SC CHAPRA
C     CIVIL ENGINEERING
C     TEXAS A&M UNIVERSITY
C     COLLEGE STATION, TEXAS  77843
C
C ***********************************
C       FUNCTION TO EVALUATE DV/DT
C ***********************************
        FUNCTION DVDT(C,V,M)
        REAL M
        DVDT=980-C*V/M
        RETURN
        END
C ***********************************
C           MAIN PROGRAM
C ***********************************

       CALL LABEL
       CALL INPUT(T0,T1,V0,H,P)
       CALL CALC(T0,T1,V0,H,P)
       STOP
       END
C ***********************************
C      SUBROUTINE TO LABEL OUTPUT
C ************************************

        SUBROUTINE LABEL
        WRITE(6,1)
      1 FORMAT('-','SOLUTION FOR VELOCITY OF FALLING PARACHUTIST')
        RETURN
        END
C ************************************
C      SUBROUTINE TO INPUT VALUES
C ************************************

        SUBROUTINE INPUT(T0,T1,V0,H,P)
C          INITIAL TIME (SEC)
        READ(5,2)T0
C          FINAL TIME (SEC)
        READ(5,2)T1
C          INITIAL VELOCITY (CM/SEC)
        READ(5,2)V0
C          STEP SIZE (SEC)
        READ(5,2)H
C          PRINT INTERVAL (SEC)
        READ(5,2)P
      2 FORMAT(F6.2)
C          CHECK STEP SIZE AND PRINT INTERVAL
        IF(P.GE.H.AND.P.NE.0)GO TO 2220
        WRITE(6,3)
      3 FORMAT('-','PRINT INTERVAL MUST BE GREATER THAN OR EQUAL TO
     C              STEP SIZE AND NOT EQUAL ZERO')
 2220   RETURN
        END
C ************************************
C  SUBROUTINE TO PERFORM COMPUTATIONS
C ************************************

        SUBROUTINE CALC(T0,T1,V0,H,P)
        REAL M
        DIMENSION C(20),M(20)
        NC=INT(P/H)
        NP=INT((T1-T0)/P)
C          LOOP TO COMPUTE V FOR DIFFERENT C AND M
        DO 3370 K=1,20
C          INPUT DRAG COEFFICIENT
        READ(5,4)C(K)
        IF(C(K).EQ.0.)GO TO 3390
C          INPUT MASS
        READ(5,4)M(K)
      4 FORMAT(F10.0)
C          CHECK FOR ZERO MASS
        IF(M(K).GT.0.)GO TO 3220
        WRITE(6,5)
      5 FORMAT('-','MASS MUST BE GREATER THAN ZERO')
        GO TO 3390
C          INITIALIZE TIME AND VELOCITY
 3220   T=T0
        V=V0
        WRITE(6,6)
      6 FORMAT(' ',4X,'T (SEC)',10X,'V (CM/SEC)')
        WRITE(6,7)T,V
C          PRINT LOOP
          DO 3360 I=1,NP
C            CALCULATION LOOP
            DO 3340 J=1,NC
            V=V+DVDT(C(K),V,M(K))*H
            T=T+H
 3340       CONTINUE
          WRITE(6,7)T,V
    7 FORMAT(2X,2(F10.3,10X))
 3360     CONTINUE
 3370   CONTINUE
 3390   RETURN
        END
```

## BASIC

```
100  REM      USER-FRIENDLY PROGRAM IN BASIC
110  REM      FOR THE FALLING PARACHUTIST
120  REM
130  REM      S.C. CHAPRA
140  REM      CIVIL ENGINEERING
150  REM      TEXAS A&M UNIVERSITY
160  REM      COLLEGE STATION,TX  77843
170  REM
180  REM **********************************
190  REM       FUNCTION TO COMPUTE DV/DT
200  REM **********************************
210  DEF  FN D(T) = 980 - C(K) * V / M(K)
220  REM **********************************
230  REM       MAIN PROGRAM
240  REM **********************************
250  GOSUB 1000
260  GOSUB 2000
270  GOSUB 3000
280  END
1000  REM **********************************
1010  REM       SUBROUTINE TO LABEL OUTPUT
1020  REM **********************************
1030  PRINT
1040  PRINT
1050  PRINT "     SOLUTION FOR VELOCITY"
1060  PRINT "    OF FALLING PARACHUTIST"
1070  PRINT
1080  PRINT
1090  RETURN
2000  REM **********************************
2010  REM       SUBROUTINE TO
2020  REM       INPUT VALUES
2030  REM **********************************
2040  INPUT "INITIAL TIME(SEC)=";T0
2050  PRINT
2060  INPUT "FINAL TIME(SEC)=";T1
2070  PRINT
2080  INPUT "INITIAL VELOCITY(CM/SEC)=";V0
2090  PRINT
2100  INPUT "STEP-SIZE(SEC)=";H
2110  PRINT
2120  INPUT "PRINT INTERVAL(SEC)=";P
2130  PRINT
2140  REM CHECK STEP-SIZE AND PRINT INTERVAL
2150  IF P >  = H AND P <  > 0 THEN 2220
2160  PRINT
2170  PRINT "PRINT INTERVAL MUST BE GREATER"
2180  PRINT "THAN OR EQUAL TO STEP-SIZE"
2190  PRINT "AND CANNOT BE ZERO"
2200  PRINT
2210  GOTO 2100
2220  RETURN
3000  REM **********************************
3010  REM       SUBROUTINE TO
3020  REM       PERFORM COMPUTATIONS
3030  REM **********************************
3040  DIM C(20),M(20)
3050 NC = P / H
3060 NP = (T1 - T0) / P
3070  REM LOOP TO COMPUTE V FOR DIFFERENT C AND M
3080  FOR K = 1 TO 20
3090  PRINT "DRAG COEFFICIENT(G/SEC)"
3100  PRINT "(TO TERMINATE COMPUTATION"
3110  INPUT "ENTER ZERO)=";C(K)
3120  IF C(K) = 0 THEN 3400
3130  PRINT
3140  INPUT "MASS(G)=";M(K)
3150  PRINT
3160  REM CHECK FOR ZERO MASS
3170  IF M(K) > 0 THEN 3230
3180  PRINT "MASS CANNOT BE LESS THAN"
3190  PRINT "OR EQUAL TO ZERO"
3200  PRINT
3210  GOTO 3140
3220  REM INITIALIZE TIME AND VELOCITY
3230 T = T0
3240 V = V0
3250  PRINT
3260  PRINT "T(SEC)           V(CM/SEC)"
3270  PRINT
3280  PRINT T,V
3290  REM PRINT LOOP
3300  FOR I = 1 TO NP
3310  REM CALCULATION LOOP
3320  FOR J = 1 TO NC
3330 V = V +  FN D(T) * H
3340 T = T + H
3350  NEXT J
3360  PRINT T,V
3370  NEXT I
3380  PRINT
3390  NEXT K
3400  RETURN
```

FIGURE 2.11 User-friendly versions in FORTRAN and BASIC of the falling parachutist program.

the outer loop (lines 3300 to 3370). Also note that rather than specifying the total number of steps (N as specified by line 130 in the simple versions of Fig. 2.10), we now merely input the initial and final times (lines 2040 and 2060) and use lines 3050 and 3060 to determine the appropriate number of steps internally.

3. *The program now includes a more descriptive labeling scheme.* These include documentation statements at the beginning of the program, output messages for example, lines 1030 through 1080, and more descriptive inputs for example, lines 2040 through 2130.

4. *The program is modular.* Note that the program consists of a series of subroutines that perform well-defined tasks. The main program serves as a master control to orchestrate each of the parts in a logical fashion.

5. *Diagnostics are included to signal the user that an error has been made. Diagnostics* are statements in the program that print out a descriptive message to the user if an error has occurred. Lines 3160 to 3210 represent a diagnostic to check whether mass is zero. If this were true, the equation in line 210 would include division by zero. If the mass is equal to or less than zero, line 3170 transfers to line 3180 where the message:

   MASS CANNOT BE LESS THAN OR EQUAL TO ZERO

   is printed. Similarly, line 2150 tests to ensure that the print interval is greater than the step size. If not, the messages on lines 2170 through 2190 are printed and the program transfers back to line 2100 and prompts for a new step size.

The foregoing are but five of several modifications that have been made to increase the program's capabilities. You should go through the code line by line to understand how each of the statements contributes to the total program. Figure 2.12 presents an actual run. In Fig. 2.12 we intentionally input an erroneous print interval to demonstrate the diagnostic capabilities of the program. Study of these runs with Fig. 2.11 should provide some insight into ways to set up your output in a clear and descriptive fashion.

## 2.4 SOFTWARE STRATEGY

This book will provide you with a diverse array of computer media for converting the theory of numerical methods into a practical tool to solve engineering problems. These media include (1) disks that contain software, (2) computer code, (3) algorithms, and (4) flowcharts. The purpose of this section is to discuss the way each of these media complement each other in the book. The overall strategy is illustrated in Fig. 2.13.

You may have purchased this book with a supplementary computer

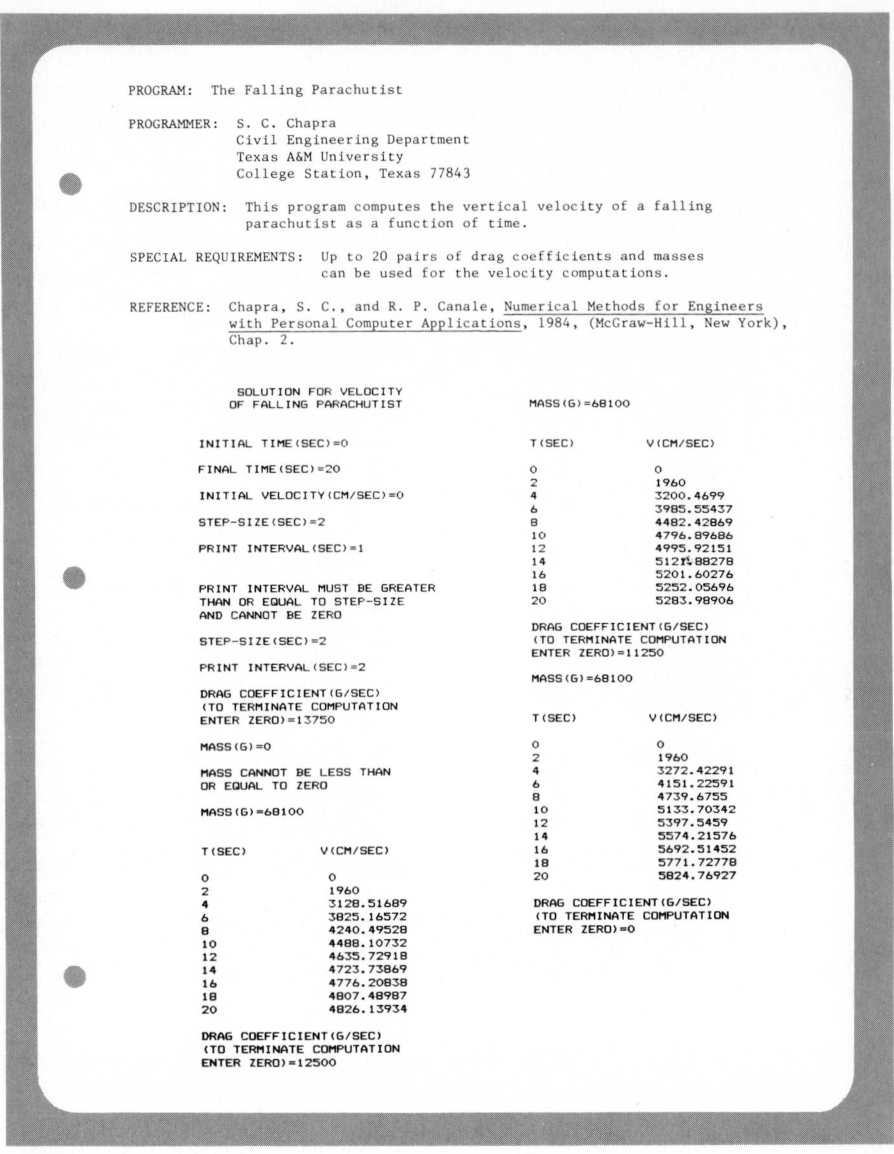

```
PROGRAM:  The Falling Parachutist

PROGRAMMER:  S. C. Chapra
             Civil Engineering Department
             Texas A&M University
             College Station, Texas 77843

DESCRIPTION:  This program computes the vertical velocity of a falling
              parachutist as a function of time.

SPECIAL REQUIREMENTS:  Up to 20 pairs of drag coefficients and masses
                       can be used for the velocity computations.

REFERENCE:  Chapra, S. C., and R. P. Canale, Numerical Methods for Engineers
            with Personal Computer Applications, 1984, (McGraw-Hill, New York),
            Chap. 2.
```

```
   SOLUTION FOR VELOCITY
  OF FALLING PARACHUTIST

INITIAL TIME(SEC)=0

FINAL TIME(SEC)=20

INITIAL VELOCITY(CM/SEC)=0

STEP-SIZE(SEC)=2

PRINT INTERVAL(SEC)=1


PRINT INTERVAL MUST BE GREATER
THAN OR EQUAL TO STEP-SIZE
AND CANNOT BE ZERO

STEP-SIZE(SEC)=2

PRINT INTERVAL(SEC)=2

DRAG COEFFICIENT(G/SEC)
(TO TERMINATE COMPUTATION
ENTER ZERO)=13750

MASS(G)=0

MASS CANNOT BE LESS THAN
OR EQUAL TO ZERO

MASS(G)=68100


T(SEC)          V(CM/SEC)

0               0
2               1960
4               3128.51689
6               3825.16572
8               4240.49528
10              4488.10732
12              4635.72918
14              4723.73869
16              4776.20838
18              4807.48987
20              4826.13934

DRAG COEFFICIENT(G/SEC)
(TO TERMINATE COMPUTATION
ENTER ZERO)=12500

MASS(G)=68100


T(SEC)          V(CM/SEC)

0               0
2               1960
4               3200.4699
6               3985.55437
8               4482.42869
10              4796.89686
12              4995.92151
14              5121.88278
16              5201.60276
18              5252.05696
20              5283.98906

DRAG COEFFICIENT(G/SEC)
(TO TERMINATE COMPUTATION
ENTER ZERO)=11250

MASS(G)=68100


T(SEC)          V(CM/SEC)

0               0
2               1960
4               3272.42291
6               4151.22591
8               4739.6755
10              5133.70342
12              5397.5459
14              5574.21576
16              5692.51452
18              5771.72778
20              5824.76927

DRAG COEFFICIENT(G/SEC)
(TO TERMINATE COMPUTATION
ENTER ZERO)=0
```

FIGURE 2.12 Documentation for the user-friendly version of the falling parachutist program, including an example run.

disk. This disk, which is called NUMERICOMP, will run on either an IBM-PC (or compatible) or an APPLE II computer. The disk contains six computer programs written in BASIC. The six programs cover bisection, Gauss elimination, linear regression, Lagrange interpolation, trapezoidal rule, and Euler's method. The programs represent a collection of simple but useful numerical methods for each part of the book. With very little preparation you can use NUMERICOMP to solve problems. This is because the programs are written in a very user-friendly style and prompt you for all the information

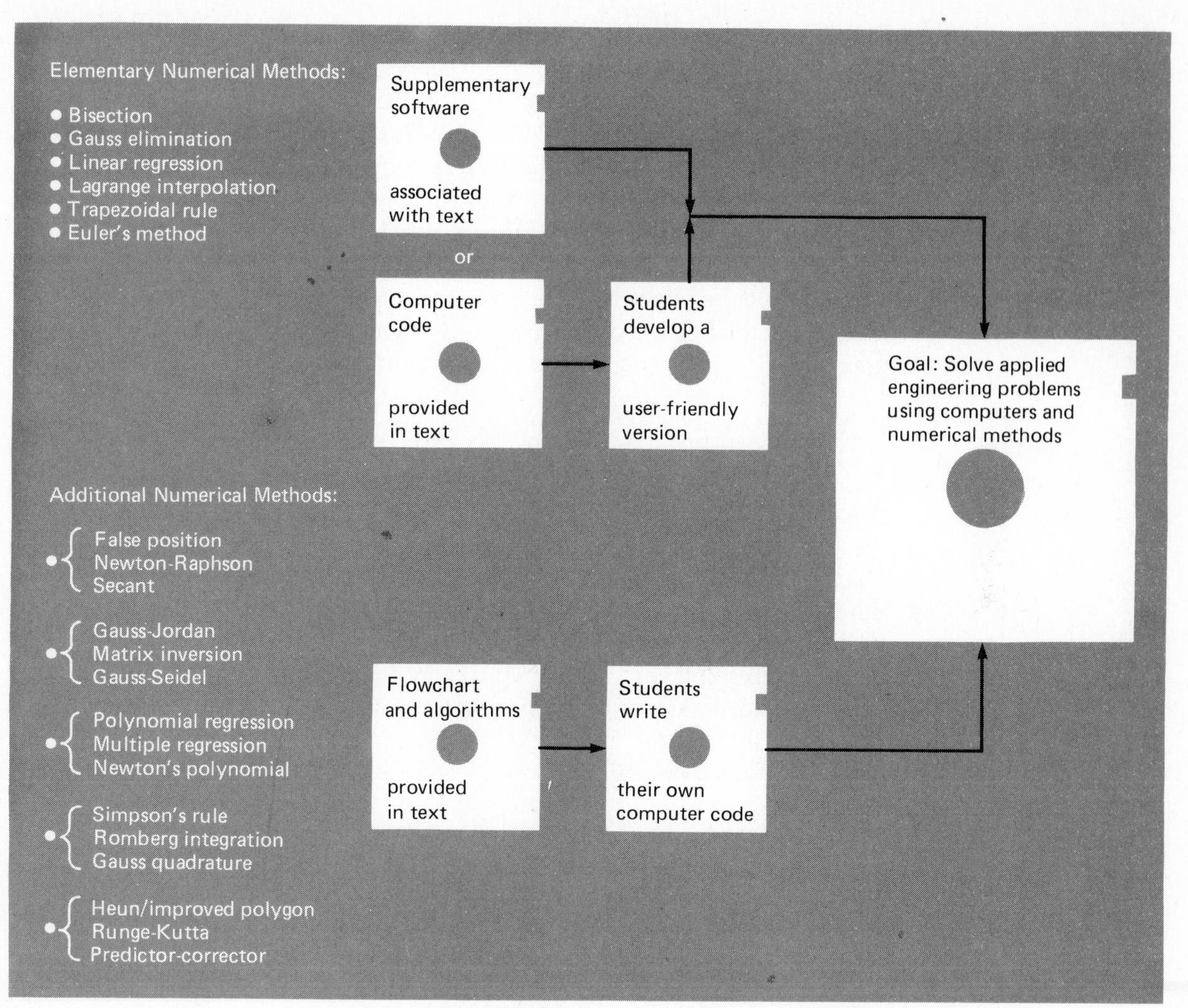


FIGURE 2.13 Strategy employed in text to integrate personal computers and numerical methods for solving applied engineering problems.

required for operation. In addition to having immediate utility, the disk provides you with a concrete example of well-written software that you can use as a model for your own work. Finally, the software can be used to check the accuracy of the results of your programming efforts.

Each of the programs is illustrated fully in the relevant chapter in the book. The illustrations will show you the design of the input and output screens, the data required, the results of the computation, and a plot of the results. These illustrations are generated by using NUMERICOMP to solve an actual problem. Some homework problems are included in each chapter to reinforce your capabilities to use the disk on your own microcomputer.

You are also provided in the text with FORTRAN and BASIC programming codes for these same methods. These programs contain the fundamental algorithms with simple input and output schemes and little documentation. Thus they are not user-friendly. One of your tasks will be to modify these programs so that they are more friendly, using your own resources and individual style. Once this is completed, you will have a tool that is roughly equivalent to the supplementary software.

The six programs on the NUMERICOMP disk are for the most fundamental methods in each part of the book. They are not necessarily the most computationally efficient or most general. Therefore, you have been provided with either flowcharts or algorithms for most of the other numerical methods in the book. You can use these flowcharts and algorithms along with your own programming skills to write programs for any of the other methods.

## EXAMPLE 2.1
### Computer Graphics

Problem Statement: The purpose of this example is to familiarize you with the programs in the optional software (NUMERICOMP) available with the text and to use NUMERICOMP's graphics capabilities to plot functions. If you purchased this book without software, then you should explore ways to perform similar tasks on your own computer. This may be accomplished with your system's software, or it may require that you develop your own. The ability to plot functions is important because the problem-solving utility of numerical methods is tremendously enhanced when used in coordination with computer graphics.

Solution: Insert the NUMERICOMP software disk into the disk drive and run the program according to the directions in the *User's Manual.*

The screen should produce a pattern similar to that in Fig. 2.14*a*. This is simply a title page. Enter RETURN to continue. The screen should change to a program selection main menu as shown in Fig. 2.14*b*. The menu contains a list of six programs along with an option to terminate the session. We will use each of these programs at appropriate places in the text after you have been introduced to the theory of each method. For now we will use the computer-graphics option within the BISECTION program to plot the velocity of the falling parachutist as a function of time. To do this, simply enter the BISECTION program by selecting option 1. The screen should automatically produce a pattern similar to that of Fig. 2.14*c* after some movement of the disk. You will only need to use options 1, 3, and 4 to plot functions. Select option 1 to enter the function using Eq. (1.9) with $m = 68{,}100$ g, $c = 12{,}500$ g/s, and $g = 980$ cm/s$^2$ (Fig. 2.14*d*). Return to the main menu and select option 3 to plot the function. Before the plot can be constructed, it is necessary to supply minimum and maximum values for both $x$ and $f(x)$ that correspond to time and velocity for this case. The values for $x$ and $f(x)$ in the first column are default values (in this case, zero). Try various values for the $x$ and $f(x)$ axes (including negative values) to gain

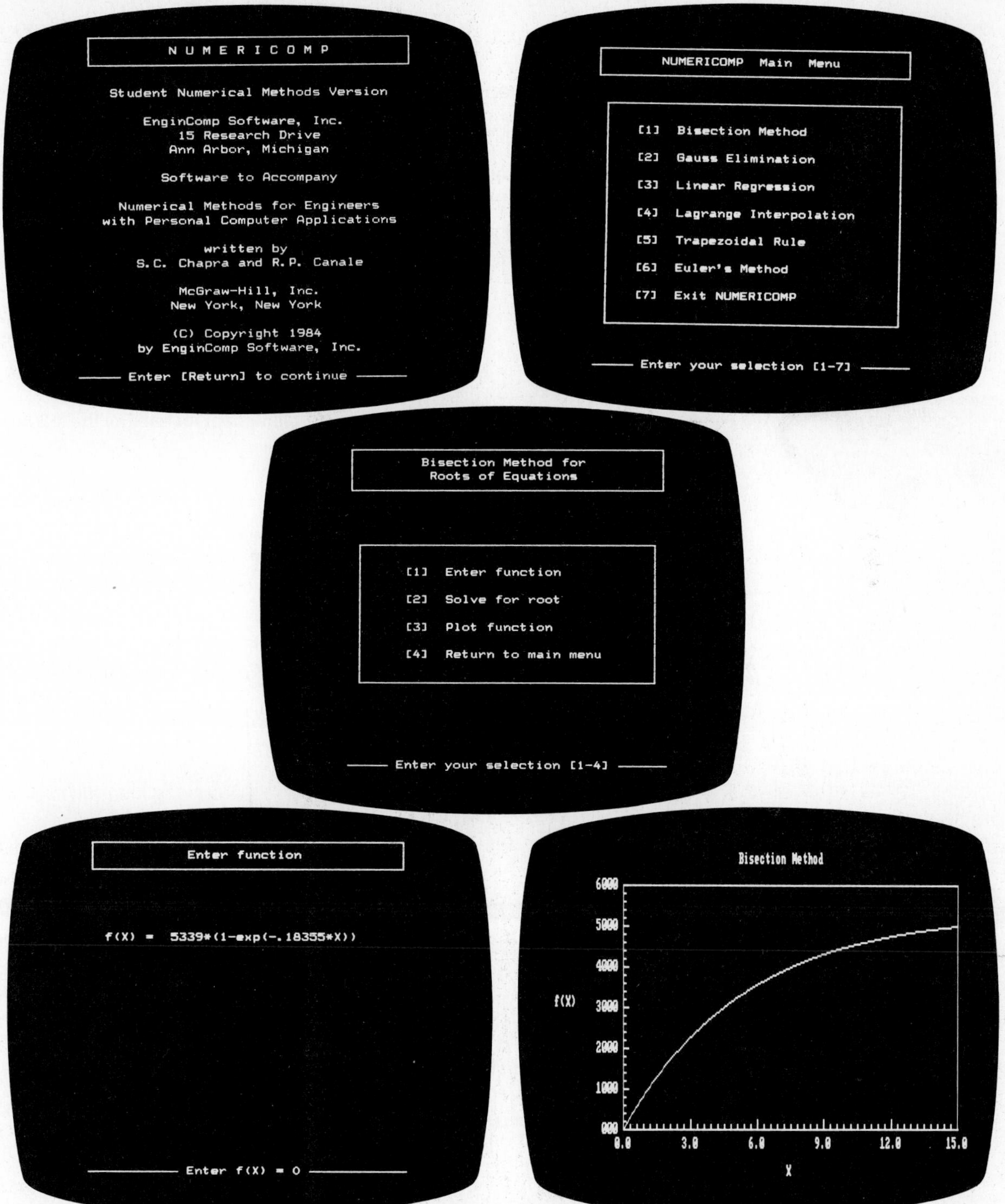

FIGURE 2.14 (*a*) Title screen for the NUMERICOMP software accompanying the text. (*b*) The main menu of NUMERICOMP. (*c*) The menu for BISECTION. (*d*) Screen showing how a function is entered using BISECTION; here the function is Eq. (1.9), which computes the velocity of the falling parachutist. (*e*) Screen showing a plot of velocity versus time for the parachutist, as computed with NUMERICOMP.

familiarity with the plot option design and operation. A well-proportioned plot that displays the behavior of the velocity as a function of time is shown in Fig. 2.14e.

The plot option of your software will have many other uses as you pursue your goal of understanding and applying numerical methods and computers to solve engineering problems. We will explore these uses in appropriate sections of the text.

## PROBLEMS

**2.1** Write out the equivalent BASIC or FORTRAN statement for each of the following:

(*a*) $W = \dfrac{(x_1 - x_2)^2}{3x_1}$

(*b*) $y = \dfrac{x|\sin x|}{x - 1}$

(*c*) $x = \dfrac{-b - \sqrt{b^2 - 4c}}{2a}$

(*d*) If A and Z are of the same sign then replace Z with Q.

**2.2** Write the BASIC or FORTRAN statements to perform the following operation

$$s = \Sigma\, x_i^{\,2}$$

for $i = 3, 6, 9, \ldots, 21$.

**2.3** Given the following program:

```
10 A = 10.1
20 B = 3.1416
30 Z = 1.1
40 PRINT X1
```

what will the output be, if the following expressions are inserted between lines 30 and 40?

(*a*) 35 X1 = A**Z/B
(*b*) 35 X1 = A**(Z/B)
(*c*) 35 X1 = A*B − B**B/Z + 2*Z
(*d*) 35 X1 = ((A*Z) − B/Z)**Z)/(B − Z)
(*e*) 32 J = INT(A**Z/B − Z)
36 X1 = J*A

**2.4** Given the program in Prob. 2.3, write the code for line 35 that will evaluate the following algebraic expressions:

$$x_1 = \frac{a^2 - 4\sqrt{b}}{z}$$

$$x_1 = a - \sqrt{z/5 + 6(a + z)^{2/3}} - \frac{7}{b}$$

**2.5** The figure for this problem shows a page from an automobile logbook. Each row conforms to a stop at a service station at which time the car's tank is filled with gasoline. The page has columns for the date, the car's mileage as read from the odometer, the amount of gasoline, and the cost.

Write a computer program that is designed to input data of this sort and then compute the miles per gallon and the dollars per mile conforming to each interval between fill-ups. Have the program print out a table with three columns conforming to the date, the miles per gallon, and the dollars per gallon.

| Date | Mileage | Gasoline purchased, gal | Gasoline cost, $ | mi/gal | $/mi |
|---|---|---|---|---|---|
| 4/12/83 | 43,364.2 | 12.6 | 15.12 | | |
| 4/19/83 | 43,632.8 | 10.7 | 13.38 | | |
| 4/29/83 | 43,907.5 | 11.4 | 13.57 | | |
| 5/7/83 | 44,150.4 | 9.2 | 12.42 | | |
| 5/13/83 | 44,461.4 | 12.8 | 16.00 | | |
| 5/15/83 | 44,828.2 | 10.6 | 13.57 | | |
| 5/16/83 | 45,222.8 | 10.9 | 13.08 | | |
| 5/20/83 | 45,472.3 | 8.4 | 10.84 | | |

FIGURE PROB. 2.5

**2.6** An amount of money $P$ is invested in an account where interest is compounded at the end of the period. The future worth $F$ yielded at an interest rate $i$ after $n$ periods may be determined from the following formulation:

$$F = P(1 + i)^n$$

Write a program which will calculate the future worth of an investment. The input to the program should include the initial investment $P$, the interest rate $i$ (as a decimal), and the number of years $n$ for which the future worth is to be calculated. The output should include these values also. The output should also include, in a labeled table format, the future worth for each year up to and including the $n$th year. Run the program for $P = \$1000.00$, $i = 0.1$, and $n = 20$ years.

**2.7** Write a program to calculate the real roots of a quadratic equation

$$ax^2 + bx + c = 0$$

where $a$, $b$, and $c$ are real coefficients. The formula used to calculate the roots is the quadratic formula

$$x = \frac{-b \pm \sqrt{b^2 - 4ac}}{2a}$$

Note that if the quantity within the square root sign is negative, the root is complex. Also, division by zero occurs if $a = 0$. Design your program so that it deals with these contingencies by printing out an error message. Also, include a liberal sprinkling of documentation and labeled output statements to make the program "friendly." Allow the computation to be for different values of $a$, $b$, and $c$ repeated as many times as the user requires. Perform test runs for the cases:

(a) $a = 1$ $b = 4$ $c = 2$
(b) $a = 0$ $b = -4$ $c = 2.3$
(c) $a = 1$ $b = 2$ $c = 2.3$

**2.8** The exponential function $e^x$ can be evaluated by the following infinite series:

$$e^x = 1 + x + \frac{x^2}{2} + \frac{x^3}{3!} + \frac{x^4}{4!} + \cdots$$

Write a program to implement this formula so that it computes the values of $e^x$ as each term in the series is added. In other words, compute and print in sequence

$$e^x \simeq 1$$
$$e^x \simeq 1 + x$$
$$e^x \simeq 1 + x + \frac{x^2}{2}$$
$$\vdots$$

up to the order term of your choosing. For each of the above, compute the percent relative error as

$$\% \text{ error} = \frac{\text{true} - \text{series approximation}}{\text{true}} 100\%$$

Use the library function for $e^x$ in your computer to determine the "true" value. Have the program print out the series approximation and the error at each step. Employ a user-defined function to compute the error. Use loops to simplify the computation as much as possible. As a test case, employ the program to compute exp(0.5) for up to and including the term $x^{20}/20!$. Interpret your results.

**2.9** Economic formulas are available to compute annual payments for loans. Suppose that you borrow an amount of money $P$ and agree to repay it in $n$ annual payments at an interest rate of $i$. The formula to compute the annual payment $A_1$ is

$$A_1 = P\frac{i(1 + i)^n}{(1 + i)^n - 1}$$

Write a computer program to compute $A_1$. Test it with $P = \$10{,}000$ and an interest rate of 20 percent ($i = 0.20$). Set up the program so that you can evaluate as many values of $n$ as you like. Compute results for $n = 1, 2, 3, 4$, and 5.

**2.10** Aside from computing annual payments for a loan, as was done in Prob. 2.9 economic formulas can also be employed to determine annual payments corresponding to other types of cash flow. For example, suppose that you had an expense that increased at a constant rate $G$ as time increased. Such a payment is called an *arithmetic gradient series.* The economic formula to compute an equivalent annual payment for this type of cash flow is

$$A_2 = G\left[\frac{1}{i} - \frac{n}{(1 + i)^n - 1}\right]$$

Now, suppose that you take out a loan of $P = \$10{,}000$ at an interest rate of $i = 0.20$ and buy a new personal computer system. The maintenance cost for this computer increases according to the arithmetic gradient series at a rate of $G = \$50$/yr/yr. Aside from these two costs (that is, negative cash flows for the loan repayment and maintenance), you also gain benefits or positive cash flow from your ownership of the system. Your annual consulting profit and enjoyment from the computer can be valued at an annual worth of $A_3 = \$4000$. Therefore, the net worth $A_N$ of owning the machine on an annual basis can be calculated as benefits minus costs, or

$$A_N = A_3 - A_1 - A_2$$

Thus, if $A_N$ is positive, the computer is making you money on an annual basis. If $A_N$ is negative, you are losing money.

Develop, debug, test, and document a computer program to calculate $A_N$. Design the program so that the user can input different values for $P$, $i$, $G$, $A_3$, and $n$. Use the program to estimate $A_N$ for your new computer system for values of $n = 1, 2, 3, 4$, and 5. That is, evaluate its worth if it's owned for 1 through 5 years. Plot $A_N$ versus $n$ (use your computer to make the plot if possible). Determine how long you must own it in order to begin to make money. (*Note:* Additional background information for this problem can be obtained from the first case study in Chap. 6.)

**2.11** Implement the program in Fig. 2.11 on your computer. Make the necessary modifications so that it is compatible with the dialect used on your machine. Once the program is on your computer, test it by duplicating the computation in Fig. 2.12. Repeat the computation but use step sizes of 1 and 0.5. Compare your results with the analytical solution obtained previously in Example 1.1. Does a smaller step size make the results better or worse? Explain your results.

**2.12** The following algorithm is designed to determine a grade for a course that consists of quizzes, homework, and a final exam:

Step 1: Input course number and name.
Step 2: Input weighting factors for quizzes (WQ), homework (WH), and the final exam (WF).
Step 3: Input quiz grades and determine an average quiz grade (AQ).
Step 4: Input homework grades and determine an average homework grade (AH).
Step 5: If this is a final grade, go to step 8. If not, continue.
Step 6: Determine average grade AG according to

$$AG = \frac{WQ * AQ + WH * AH}{WQ + WH}$$

Step 7: Go to step 10.
Step 8: Input final exam grade (FE).
Step 9: Determine average grade AG according to

$$AG = \frac{WQ * AQ + WH * AH + WF * FE}{WQ + WH + WF}$$

Step 10: Print out course number, name, and average grade.
Step 11: Stop the computation.

(*a*) Write a program based on this algorithm.
(*b*) Debug and test it using the data: WQ = 35; WH = 25; WF = 40; quizzes = 100, 98, 83, 76, 100; homework = 96, 94, 83, 100, 77; and final exam = 88.
(*c*) Prepare a short documentation for the program.

**2.13** The figure for this problem shows the back of a checking account statement. The bank has developed this sheet to help you balance your checkbook. If you look at it closely, you will realize that it is an algorithm. Develop, debug, and document a computer program to balance your checkbook based on the scheme outlined in the figure. Use the numbers contained on the figure to test your program.

**2.14** Write, debug, and document a computer program to determine statistics for your favorite sport. Pick anything from football to jogging to bowling. If you play intramural sports make one up for your team. Design the program so that it is user-friendly and provides valuable and interesting information to anyone (for example, a coach or a player) who might use it to evaluate athletic performance.

**2.15** Use the plot option on the BISECTION program (on the NUMERICOMP software disk) to plot several functions of your choice. Try polynomial and transcendental functions whose behavior may be difficult to visualize in advance of plotting. Utilize several choices for both the $x$ and $y$ axes to facilitate your exploration. Make permanent copies of the plots if you have a printer.

**2.16** Attain the capability to plot functions in a manner similar to that of the BISECTION program. Use your own computer and its system software if appropriate. If your computer does not have system programs to help you, then write your

*THE AREA BELOW IS PROVIDED TO HELP YOU BALANCE YOUR CHECKBOOK*

CHECKS OUTSTANDING — NOT CHARGED TO ACCOUNT

| NO. | $ | |
|---|---|---|
| 453 | 25 | 00 |
| 458 | 5 | 68 |
| 460 | 14 | 33 |
| 461 | 150 | 00 |
| 463 | 16 | 74 |
| 464 | 9 | 32 |
| 465 | 44 | 15 |
| 466 | 50 | 00 |
| | | |
| | | |
| | | |
| | | |
| | | |
| | | |
| | | |
| | | |
| | | |
| | | |
| | | |
| | | |
| TOTAL | $ | |

MONTH April 19 84

NEW BALANCE AS SHOWN ON THIS STATEMENT $ 643.59

ADD DEPOSITS NOT CREDITED IN THIS STATEMENT 250.00

22.15

TOTAL $

SUBTRACT TOTAL CHECKS OUTSTANDING

YOUR CHECKBOOK BALANCE $ 600.52

AFTER SUBTRACTING CURRENT MONTH'S SERVICE CHARGE AND ADDING INTEREST EARNED (EARNER ACCOUNTS ONLY) TO YOUR BALANCE

FIGURE PROB. 2.13

own using the capabilities of your computer. Save this program on a magnetic storage disk. Document this program after careful debugging and testing. Be prepared to update this program as you apply it to various problems covered in other parts of the book.

**2.17** Learn how to obtain permanent copies of your plots from Prob. 2.16 if you have a printer.

CHAPTER THREE

# APPROXIMATIONS AND ERRORS

Because so many of the methods in this book are straightforward in description and application, it would be very tempting at this point for us to proceed directly to the main body of the text and teach you how to use these techniques. However, errors are so intrinsic to the understanding and effective use of numerical methods that we have chosen to devote the present chapter to this topic.

The importance of error was introduced in our discussion of the falling parachutist in Chap. 1. Recall that we determined the velocity of a falling parachutist by both analytical and numerical methods. Although the numerical technique yielded estimates that were close to the exact, analytical solution, there was a discrepancy, or *error*, due to the fact that the numerical method involved an approximation.

Such errors are characteristic of most of the techniques described in this book. This statement at first might seem contrary to what one normally conceives of as sound engineering. Students and practicing engineers constantly strive to limit errors in their work. When taking examinations or doing homework problems, you are penalized, not rewarded, for your errors. In professional practice, errors can be costly and sometimes catastrophic. If a structure or device fails, lives can be lost.

Although perfection is a laudable goal, it is rarely, if ever, attained. For example, despite the fact that the model developed from Newton's second law is an excellent approximation, it would never in practice exactly predict the parachutist's fall. A variety of factors such as winds and slight variations in air resistance would result in deviations from the prediction. If these deviations are systematically high or low, then we might need to develop a new model. However, if they are randomly distributed and tightly grouped around the prediction, then the deviations might be considered negligible and the model deemed adequate. Numerical approximations can introduce similar discrepancies into the analysis. Again, the question is: How much error is tolerable?

The present chapter covers basic topics related to the identification, quantification, and minimization of these errors. General information concerned with the quantification of error is reviewed in the first sections. This is followed by sections on the two major forms of numerical error: round-off and truncation error. *Round-off error* is due to the fact that computers can only represent quantities with a finite number of digits. *Truncation error* is the discrepancy introduced by the fact that the numerical method employs an approximation to represent exact mathematical operations and quantities. Finally, we briefly discuss errors not directly connected with the numerical methods themselves. These include blunders, formulation or model errors, and data uncertainty.

## 3.1 SIGNIFICANT FIGURES

This book deals almost exclusively with approximations connected with the manipulation of numbers. Consequently, before discussing the errors associated with numerical methods, it is useful to review basic concepts related to approximate representation of the numbers themselves.

Whenever we employ a number in a computation, we must have assurance that it can be used with confidence. For example, Fig. 3.1 depicts a speedometer and odometer from an automobile. Visual inspection of the speedometer indicates that the car is traveling between 48 and 49 km/h. Because the indicator is higher than the midpoint between the markers on the gauge, we can say with assurance that the car is traveling at approximately 49 km/h. We have confidence in this result because two or more reasonable individuals reading this gauge would arrive at the same conclusion. However, let us say that we insist that the speed be estimated to one decimal place. For this case, one person might say 48.7, whereas another might say 48.8 km/h. Therefore, because of the limits of this instrument, only the first two digits can be used with confidence. Estimates of the third digit (or higher) must be

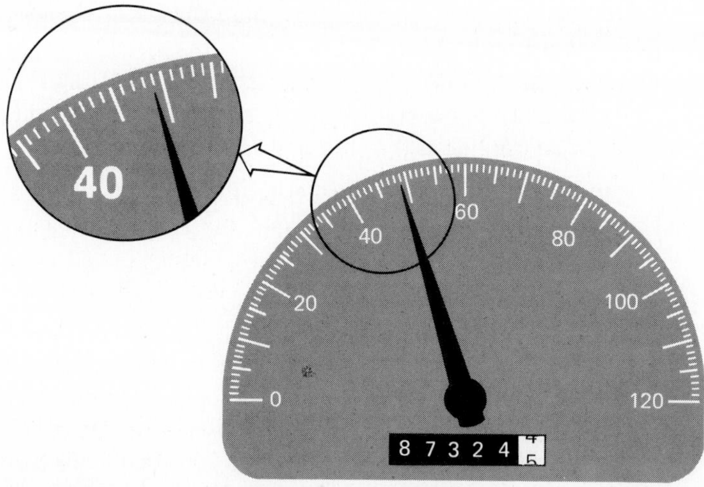


FIGURE 3.1 An automobile speedometer and odometer illustrating the concept of a significant figure.

viewed as suspect. It would be ludicrous to claim, on the basis of this speedometer, that the automobile is traveling at 48.7642138 km/h. In contrast, the odometer provides up to six certain digits. From Fig. 3.1, we can conclude that the car has traveled slightly less than 87,324.5 km during its lifetime. In this case, the seventh digit (and higher) is uncertain.

The concept of a *significant figure*, or digit, has been developed to formally designate the reliability of a numerical value. The number of significant figures is the number of certain digits, plus one estimated digit that may be used with confidence. For example, the speedometer and odometer in Fig. 3.1 yield estimates of up to three and seven significant figures, respectively. Zeros are not always significant figures because they may be necessary just to locate a decimal point. The numbers

0.00001845

0.0001845

0.001845

all have four significant figures. When trailing zeros are used in large numbers, it is not clear how many, if any, of the zeros are significant. For example, at face value, the number 45,300 may have three, four, or five significant digits, depending on whether the zeros are known with confidence. Such uncertainty can be resolved by using scientific notation where $4.53 \times 10^4$, $4.530 \times 10^4$, and $4.5300 \times 10^4$ designate that the number is known to three, four, and five significant figures.

The concept of significant figures has two important implications for our study of numerical methods:

1. As introduced in the falling parachutist problem, numerical methods yield approximate results. We must therefore develop criteria to specify how confident we are in our approximate result. One way to do this is in terms of significant figures. For example, we might decide that our approximation is acceptable if it is correct to four significant figures—that is, we are confident that the first four digits are true.

2. Although quantities such as $\pi$, $e$, or $\sqrt{7}$ represent specific quantities, they cannot be expressed exactly by a limited number of digits. For example, the quantity $\pi$ is equal to

   3.14159265358979323846264 3 . . .

   ad infinitum. Because personal computers only retain on the order of 10 significant figures (machines typically range from 7 to 14, as shown in Table 2.1), such numbers can never be represented exactly. The omission of the remaining significant figures is called *round-off error.*

Both round-off error and the use of significant figures to express our confidence in a numerical result will be explored in detail in subsequent sections. In addition, the concept of significant figures will have relevance to our definition of accuracy and precision in the next section.

## 3.2 ACCURACY AND PRECISION

The errors associated with both calculations and measurements can be characterized with regard to their precision and accuracy. *Precision* refers to (1) the number of significant figures representing a quantity or (2) the spread in repeated readings of an instrument measuring a particular physical property. *Accuracy* refers to the proximity of an approximate number or measurement to the true value it is supposed to represent.

These concepts can be illustrated graphically using an analogy from marksmanship. The bullet holes on each target in Fig. 3.2 can be thought of as the predictions of a numerical technique, whereas the target's bull's-eye represents the truth. *Inaccuracy* (also called *bias*) is defined as systematic deviation from the truth. Thus, although the shots in Fig. 3.2*c* are more tightly grouped than in Fig. 3.2*a*, the two cases are equally biased because they are both centered on the upper left quadrant of the target. *Precision*, on the other

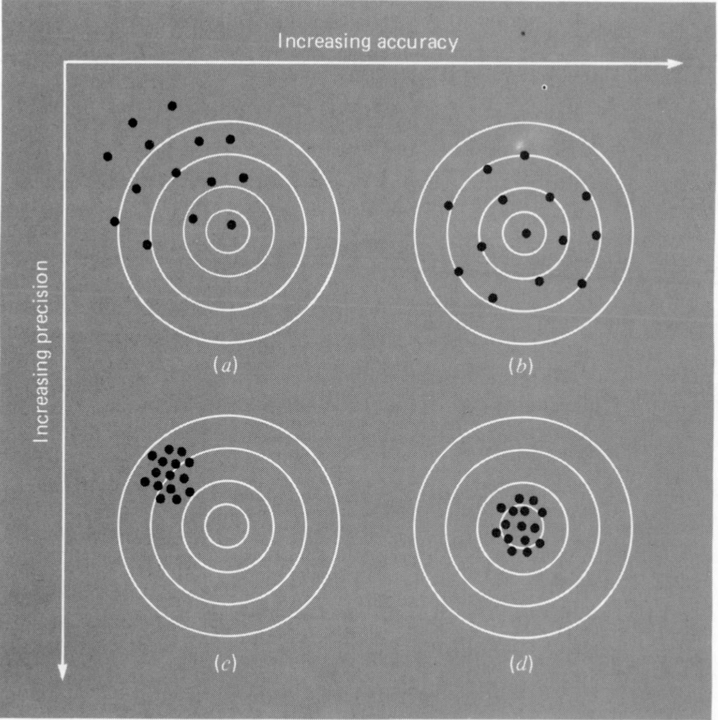


FIGURE 3.2 An example from markmanship illustrating the concepts of accuracy and precision. (*a*) Inaccurate and imprecise; (*b*) accurate and imprecise; (*c*) inaccurate and precise; (*d*) accurate and precise.

hand, refers to the magnitude of the scatter. Therefore, although Fig. 3.2*b* and 3.2*d* are equally accurate (that is, centered on the bull's-eye), the latter is more precise because the shots are tightly grouped.

Numerical methods should be sufficiently accurate or unbiased to meet the requirements of a particular engineering problem. They also should be precise enough for adequate engineering design. In this book, we will use the collective term *error* to represent both the inaccuracy and imprecision of our predictions. With these concepts as background, we can now discuss the factors that contribute to the error of numerical computations.

## 3.3 ERROR DEFINITIONS

*Numerical errors* arise from the use of approximations to represent exact mathematical operations and quantities. These include *truncation errors,* which result when approximations are used to represent an exact mathematical procedure, and *round-off errors,* which result when approximate numbers are used to represent exact numbers. For both types, the relationship between the exact, or true, result and the approximation can be formulated as

$$\text{True value} = \text{approximation} + \text{error} \tag{3.1}$$

By rearranging Eq. (3.1), we find that the numerical error is equal to the discrepancy between the truth and the approximation, as in

$$E_t = \text{true value} - \text{approximation} \tag{3.2}$$

where $E_t$ is used to designate the exact value of the error. The subscript $t$ is included to designate that this is the "true" error. This is in contrast to other cases, as described shortly, where an "approximate" estimate of the error must be employed.

A shortcoming of this definition is that it takes no account of the order of magnitude of the value under examination. For example, an error of a centimeter is much more significant if we are measuring a rivet than a bridge. One way to account for the magnitudes of the quantities being evaluated is to normalize the error to the true value, as in

$$\text{Fractional relative error} = \frac{\text{error}}{\text{true value}}$$

where, as specified by Eq. (3.2), error = true value − approximation. The relative error can also be multiplied by 100 percent in order to express it as

$$\epsilon_t = \frac{\text{true error}}{\text{true value}} 100\% \tag{3.3}$$

where $\epsilon_t$ designates the *true percent relative error.*

## EXAMPLE 3.1
### Calculation of Errors

Problem Statement: Suppose that you have the task of measuring the lengths of a bridge and a rivet and come up with 9999 and 9 cm, respectively. If the true values are 10,000 and 10 cm, respectively, compute (*a*) the error and (*b*) the percent relative error for each case.

Solution: (*a*) The error for measuring the bridge is [Eq. (3.2)]

$$E_t = 10{,}000 - 9999 = 1 \text{ cm}$$

and for the rivet it is

$$E_t = 10 - 9 = 1 \text{ cm}$$

(*b*) The percent relative error for the bridge is [Eq. (3.3)]

$$\epsilon_t = \frac{1}{10{,}000} 100\% = 0.01\%$$

and for the rivet it is

$$\epsilon_t = \frac{1}{10} 100\% = 10\%$$

Thus, although both measurements have an error of 1 cm, the relative error for the rivet is much greater. We would conclude that we have done an adequate job of measuring the bridge, whereas our estimate for the rivet leaves something to be desired.

Notice that for Eqs. (3.2) and (3.3), $E$ and $\epsilon$ are subscripted with a $t$ to signify that the error is normalized to the true value. In Example 3.1, we were provided with this value. However, in actual situations such information is rarely available. For numerical methods, the true value will only be known when we deal with functions that can be solved analytically. Such will typically be the case when we investigate the theoretical behavior of a particular technique. However, in real-world applications, we will obviously not know the true answer a priori. For these situations, an alternative is to normalize the error using the best available estimate of the true value, that is, to the approximation itself, as in

$$\epsilon_a = \frac{\text{approximate error}}{\text{approximation}} 100\% \qquad [3.4]$$

where the subscript $a$ signifies that the error is normalized to an approximate value. Note also that for real-world applications, Eq.(3.2) cannot be used to calculate the error term for Eq. (3.4). One of the challenges of numerical methods is to determine error estimates in the absence of knowledge regarding the true value. For example, certain numerical methods use an *iterative approach* to compute answers. In such an approach, a present approximation is made on the basis of a previous approximation. This process is performed repeatedly, or iteratively, in order to successively compute better and better approximations. For such cases, the error is often estimated as the difference between previous and present approximations. Thus, percent relative error is determined according to

$$\epsilon_a = \frac{\text{present approximation} - \text{previous approximation}}{\text{present approximation}} 100\% \quad [3.5]$$

This and other approaches for expressing errors will be elaborated on in subsequent chapters.

The sign of Eqs. (3.2) through (3.5) may be either positive or negative. If the approximation is greater than the true value (or the previous approximation is greater than the present approximation), the error is negative; if the approximation is less than the true value, the error is positive. Also, for Eqs. (3.3) to (3.5), the denominator may be less than zero, which can also lead to a negative error. Often, when performing computations, we may not be concerned with the sign of the error but are interested in whether the absolute value is lower than a prespecified tolerance $\epsilon_s$. Therefore, it is often useful to employ the absolute value of Eqs. (3.2) through (3.5). For such cases, the computation is repeated until

$$|\epsilon_a| < \epsilon_s \quad [3.6]$$

If this relationship holds, our result is assumed to be within the prespecified acceptable level $\epsilon_s$.

It is also convenient to relate these errors to the number of significant figures in the approximation. It can be shown (Scarborough, 1966) that if the following criterion is met, we can be assured that the result is correct to *at least* $n$ significant figures.

$$\epsilon_s = (0.5 \times 10^{2-n})\,\% \quad [3.7]$$

## EXAMPLE 3.2
### Error Estimates for Iterative Methods

Problem Statement: In mathematics, functions can often be represented by infinite series. For example, the exponential function can be computed using

$$e^x = 1 + x + \frac{x^2}{2!} + \frac{x^3}{3!} + \frac{x^4}{4!} + \cdots \quad [E3.2.1]$$

Thus, as more terms are added in sequence, the approximation becomes a better and better estimate of the true value of $e^x$. Equation (E3.2.1) is called a *Maclaurin series expansion.*

Starting with the simplest version, $e^x = 1$, add terms one at a time in order to estimate $e^{0.5}$. After each new term is added, compute the true and approximate percent relative errors with Eqs. (3.3) and (3.5), respectively. Note, that the true value is $e^{0.5} = 1.648721271$. Add terms until the absolute value of the approximate error estimate $\epsilon_a$ falls below a prespecified error criterion $\epsilon_s$ conforming to three significant figures.

Solution: First, Eq. (3.7) can be employed to determine the error criterion that ensures a result that is correct to *at least* three significant figures:

$$\epsilon_s = (0.5 \times 10^{2-3})\% = 0.05\%$$

Thus, we will add terms to the series until $\epsilon_a$ falls below this level.

The first estimate is simply equal to Eq. (E3.2.1) with a single term. Thus, the first estimate is equal to 1. The second estimate is then generated by adding the second term, as in

$$e^x \simeq 1 + x$$

or for $x = 0.5$

$$e^{0.5} \simeq 1 + 0.5 = 1.5$$

This represents a true percent relative error of [Eq. (3.3)]

$$\epsilon_t = \frac{1.648721271 - 1.5}{1.648721271} 100\% = 9.02\%$$

Equation (3.5) can be used to determine an approximate estimate of the error, as in

$$\epsilon_a = \frac{1.5 - 1}{1.5} 100\% = 33.3\%$$

Because $\epsilon_a$ is not less than the required value of $\epsilon_s$, we would continue the computation by adding another term, $x^2/2!$, and repeating the error calculations. The process is continued until $\epsilon_a < \epsilon_s$. The entire computation can be summarized as

| Terms | Result | $\epsilon_t$% | $\epsilon_a$% |
|---|---|---|---|
| 1 | 1 | 39.3 | |
| 2 | 1.5 | 9.02 | 33.3 |
| 3 | 1.625 | 1.44 | 7.69 |
| 4 | 1.645833333 | 0.175 | 1.27 |
| 5 | 1.648437500 | 0.0172 | 0.158 |
| 6 | 1.648697917 | 0.00142 | 0.0158 |

Thus, after six terms are included, the approximate error falls below $\epsilon_s = 0.05\%$, and the computation is terminated. However, notice that rather than three significant figures, the result is accurate to five! This is due to the fact that, for this case, both Eqs. (3.5) and (3.7) are conservative. That is, they ensure that the result is *at least* as good as they specify. Although as discussed in Chap. 5, this is not always the case for Eq. (3.5), it is true most of the time.

With the preceding definitions as background, we can now proceed to the two types of error connected directly with numerical methods. These are round-off and truncation errors.

## 3.4 ROUND-OFF ERRORS

As mentioned previously, round-off errors originate from the fact that computers retain only a fixed number of significant figures during a calculation. Computers do this in different ways. For example, if it only retains seven significant figures, the computer might store and use $\pi$ as $\pi = 3.141592$, with omitted terms resulting in a round-off error of [Eq. (3.2)]

$$E_t = 0.00000065\ldots$$

The above manipulation is but one way in which computers round off numbers. This technique of retaining only the first seven terms was originally dubbed "truncation" in computer jargon. We prefer to call it *chopping* to distinguish it from the truncation errors discussed in the next section. A shortcoming of chopping is that it ignores the remaining terms in the complete decimal representation. For example, the eighth significant figure in the present case is 6. Consequently, $\pi$ is actually more accurately represented by 3.141593 than by the 3.141592 that was obtained by chopping, because the former is closer to the true value. This can be seen by realizing that, if $\pi$ is approximated by $\pi = 3.141593$, the round-off error is reduced to

$$E_t = 0.00000035\ldots$$

Computers can be developed to round numbers according to rounding rules of the sort used above. However, this adds to computation costs, and consequently some machines use simple chopping. This approach is justified under the supposition that the number of significant figures on most computers is large enough that round-off error based on chopping is usually negligible. This assumption is supported by the following example.

### EXAMPLE 3.3
Effect of Round-Off Error on the Falling Parachutist Computation

Problem Statement: Repeat the computation in Example 1.2, using three, four, five, and six significant figures as determined by chopping.

**TABLE 3.1 Comparison of the falling parachutist problem using different numbers of significant figures and a step size of 2 s. The computation uses chopping to limit the number of significant figures.**

| | VELOCITY, cm/s (significant figures) | | | |
|---|---|---|---|---|
| **Time, s** | **3** | **4** | **5** | **6** |
| 0 | 0 | 0 | 0.0 | 0.0 |
| 2 | 1960 | 1960 | 1960.0 | 1960.00 |
| 4 | 3200 | 3200 | 3200.4 | 3200.46 |
| 6 | 3980 | 3985 | 3985.5 | 3985.54 |
| 8 | 4470 | 4482 | 4482.3 | 4482.41 |
| 10 | 4780 | 4796 | 4796.8 | 4796.88 |
| 12 | 4980 | 4995 | 4995.8 | 4995.91 |

Solution: Using three significant figures, $v(2)$ would be computed as in Example 1.2:

$$v(2) = 1960$$

For three significant figures, the value of $v(4) = 3200.5$ would have to be chopped to 3200. The computation would be continued as

$$v(6) = 3980$$

$$v(8) = 4470$$

$$v(10) = 4780$$

$$v(12) = 4980$$

The other computations are performed and are contained in Table 3.1.

The numerical value for $t = 12$ s using 10 significant figures is 4995.921508. Therefore, relative round-off errors of 0.32, 0.018, 0.0024, and 0.00023 percent are created by using three, four, five, and six significant figures.

The round-off error associated with Example 3.3 is negligible for all cases when compared to the truncation error at $t = 12$ s which is (see Examples 1.1 and 1.2)

$$\epsilon_t = \frac{4749.0 - 4995.9}{4749.0} 100 = -5.20\%$$

Because most personal computers carry from 7 to 14 significant figures, round-off errors would seem to be unimportant. However, there are two reasons why they can be critical in some numerical methods:

**1.** Certain methods require extremely large numbers of arithmetic manipulations to arrive at answers. In addition, these computations are often interdependent. That is, the later calculations are dependent on the earlier ones. Consequently, even though an individual round-off error could be very small, the cumulative effect over the course of a large computation can be significant.

**2.** The effect of round off may be magnified when performing algebraic manipulations simultaneously employing very large and very small numbers. Because this is the case in certain numerical techniques, round-off error can sometimes be important.

EXAMPLE 3.4
The Importance of Significant Figures in Algebraic Manipulations

Problem Statement: Determine the difference of two large numbers, 32,981,108.1234 and 32,981,107.9989. Then repeat the computation but increase the minuend by 0.001 percent.

Solution:
The difference of the original numbers is

$$\begin{array}{r} 32{,}981{,}108.1234 \\ -32{,}981{,}107.9989 \\ \hline 0.1245 \end{array}$$

Now, increasing the minuend by 0.001 percent gives 32,981,437.9345, and the difference becomes

$$\begin{array}{r} 32{,}981{,}437.9345 \\ -32{,}981{,}107.9989 \\ \hline 329.9356 \end{array}$$

which is considerably different from the original result. Thus, a seemingly negligible modification of the minuend leads to a major discrepancy in the difference.

The types of errors delineated above can lead to difficulties for certain numerical methods. These will be discussed in subsequent sections of the book.

### 3.4.1 Rules for Rounding

Rules for rounding numbers in manual computations are reviewed in Box 3.1 and illustrated in Example 3.5. These rules are not normally applied when performing extensive computer calculations. However, because we will be performing numerous manual calculations throughout the text, we have included these rules as a point of reference for subsequent calculations.

BOX 3.1 Rules for Rounding

The following rules provide general guidelines for rounding numbers when performing manual calculations.

1. When rounding, the significant digits are retained, whereas the nonsignificant are discarded (Fig. B3.1). The last retained digit is rounded up if the first discarded digit is larger than 5. Otherwise the last retained digit should be unchanged. If the first discarded digit is 5, or is 5 followed by zeros, the last retained digit should be increased by 1 only if it is odd.
2. For addition and subtraction, round so that the last retained digit in the answer corresponds to the most significant last retained digit in the numbers that are being added and subtracted. Note that a digit in the hundredths column is more significant than one in the thousandths column.
3. For multiplication and division, round so that the number of significant figures in the result equals the smallest number of significant figures contained by a quantity in the operation.
4. For combinations of arithmetic operations, there are two general cases. We can add or subtract the results of multiplications or divisions:

$$\begin{pmatrix}\text{Multiplication}\\\text{or}\\\text{division}\end{pmatrix} \begin{matrix}+\\-\end{matrix} \begin{pmatrix}\text{multiplication}\\\text{or}\\\text{division}\end{pmatrix}$$

Alternatively, we can multiply or divide the results of additions or subtractions:

$$\begin{pmatrix}\text{Addition}\\\text{or}\\\text{subtraction}\end{pmatrix} \begin{matrix}\times\\\div\end{matrix} \begin{pmatrix}\text{addition}\\\text{or}\\\text{subtraction}\end{pmatrix}$$

In both cases, the parenthetical operations are performed and the results are rounded before proceeding with the other operation, rather than only rounding the final result.

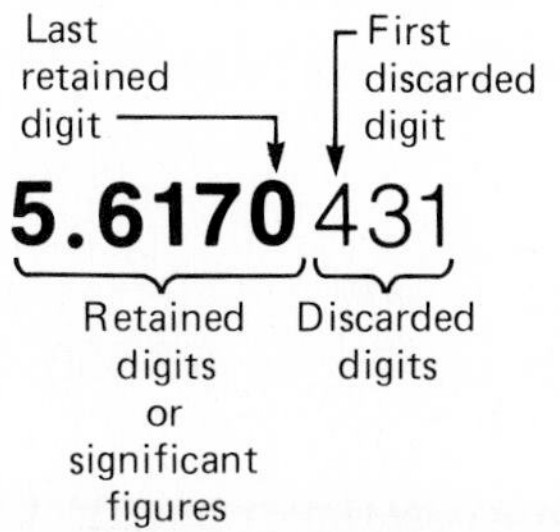


FIGURE B3.1 An illustration of the retained and discarded digits for a number with five significant figures.

EXAMPLE 3.5
Illustrations of Rounding Rules

The following are intended to illustrate the rounding rules delineated in Box 3.1.

**1.** *Rounding off:*

5.6723 $\longrightarrow$ 5.67 3 significant figures

10.406 $\longrightarrow$ 10.41 4 significant figures

7.3500 $\longrightarrow$ 7.4 2 significant figures

$$88.21650 \longrightarrow 88.216 \qquad \text{5 significant figures}$$

$$1.25001 \longrightarrow 1.3 \qquad \text{2 significant figures}$$

2. *Addition and subtraction* (*Note*: The most significant last retained digits are in boldface):
(*a*) Evaluate 2.2 − 1.768

$$2.\mathbf{2} - 1.768 = 0.432 \longrightarrow 0.4$$

(*b*) Evaluate $4.68 \times 10^{-7} + 8.3 \times 10^{-4} - 228 \times 10^{-6}$
The evaluation of this calculation is facilitated by expressing it so that all numbers are raised to the same exponent:

$$0.00468 \times 10^{-4} + 8.\mathbf{3} \times 10^{-4} - 2.28 \times 10^{-4}$$

In this way, it can be seen clearly that the 3 is the most significant last retained digit, and the answer is therefore rounded, as in

$$6.02468 \times 10^{-4} \longrightarrow 6.0 \times 10^{-4}$$

3. *Multiplication and division:*
(*a*) Evaluate $0.0642 \times 4.8$

$$0.0642 \times 4.8 = 0.30816 \longrightarrow 0.31$$

(*b*) Evaluate $945 \div 0.3185$

$$\frac{945}{0.3185} = 2967.032967\ldots \longrightarrow 2970$$

4. *Combinations:*
(*a*) Evaluate $[15.2 \times (2.8 \times 10^{-4})] + [(8.456 \times 10^{-4}) \div 0.177]$
First, perform the multiplication and division within the brackets:

$$[4.256 \times 10^{-3}] + [4.777401\ldots \times 10^{-3}]$$

Now, before adding, round off the bracketed quantities:

$$[4.\mathbf{3} \times 10^{-3}] + [4.78 \times 10^{-3}]$$

and then add and round the result:

$$9.08 \times 10^{-3} \longrightarrow 9.1 \times 10^{-3}$$

(*b*) Evaluate $\dfrac{6.740 \times 10^{-5} - 8.7 \times 10^{-7}}{2.672 \times 10^{3} + 5.8}$

Before performing addition and subtraction, express the numbers in the numerator and the denominator so that they are raised to the same exponent.

$$\frac{67\mathbf{4} \times 10^{-7} - 8.7 \times 10^{-7}}{2.67\mathbf{2} \times 10^{3} + 0.0058 \times 10^{3}}$$

Now perform the addition and subtraction:

$$\frac{665.3 \times 10^{-7}}{2.6778 \times 10^{3}}$$

and round off:

$$\frac{665 \times 10^{-7}}{2.678 \times 10^{3}}$$

Finally, divide and round:

$$2.483196\ldots \times 10^{-8} \longrightarrow 2.48 \times 10^{-8}$$

## 3.5 TRUNCATION ERRORS

*Truncation errors* are those that result from using an approximation in place of an exact mathematical procedure. For example, in Chap. 1 we approximated the derivative of velocity of a falling parachutist by a finite-divided-difference equation of the form [Eq. (1.10)]

$$\frac{dv}{dt} \simeq \frac{\Delta v}{\Delta t} = \frac{v(t_{i+1}) - v(t_i)}{t_{i+1} - t_i} \tag{3.8}$$

A truncation error was introduced into the numerical solution because the difference equation only approximates the true value of the derivative (Fig. 1.3). In order to gain insight into the properties of such errors, we now turn to a mathematical formulation that is used widely in numerical methods to express functions in an approximate fashion—the Taylor series.

### 3.5.1 The Taylor Series

For Example 3.2, we employed an infinite series to evaluate a function at a specific value of the independent variable $x$. In a similar spirit, the Taylor series provides a formulation for predicting a function value at $x_{i+1}$ in terms of the function value and its derivatives at a nearby point $x_i$.

Rather than presenting the Taylor series as a whole, we can gain more insight into its behavior by building it term by term. For example, the first term in the series is

$$f(x_{i+1}) \simeq f(x_i) \tag{3.9}$$

This relationship, which is called the *zero-order approximation*, indicates that the value of $f$ at the new point is the same as the value at the old point. This result makes intuitive sense because if $x_i$ and $x_{i+1}$ are close to each other it is likely that the new value is probably similar to the old value.

Equation (3.9) provides a perfect estimate if the function being approximated is, in fact, a constant. However, if the function changes at all over the interval, additional terms of the Taylor series are required to provide a better estimate. For example, the *first-order approximation* is developed by adding another term to yield

$$f(x_{i+1}) \simeq f(x_i) + f'(x_i)(x_{i+1} - x_i) \qquad [3.10]$$

The additional first-order term consists of a slope $f'(x_i)$ multiplied by the distance between $x_i$ and $x_{i+1}$. Thus, the expression is now in the form of a straight line and is capable of predicting an increase or decrease of the function between $x_i$ and $x_{i+1}$.

Although Eq. (3.10) can predict a change, it is only exact for a straight-line, or *linear*, trend. Therefore, a *second-order* term is added to the series in order to capture some of the curvature that the function might exhibit:

$$f(x_{i+1}) \simeq f(x_i) + f'(x_i)(x_{i+1} - x_i) + \frac{f''(x_i)}{2!}(x_{i+1} - x_i)^2 \qquad [3.11]$$

In a similar manner, additional terms can be included to develop the complete Taylor series expansion.

$$f(x_{i+1}) = f(x_i) + f'(x_i)(x_{i+1} - x_i) + \frac{f''(x_i)}{2!}(x_{i+1} - x_i)^2 + \frac{f'''(x_i)}{3!}(x_{i+1} - x_i)^3 + \cdots + \frac{f^{(n)}(x_i)}{n!}(x_{i+1} - x_i)^n + R_n \qquad [3.12]$$

Note that because Eq. (3.12) is an infinite series, an equal sign replaces the approximate sign that was used in Eqs. (3.9) to (3.11). A remainder term is included to account for all terms from $n + 1$ to infinity:

$$R_n = \frac{f^{(n+1)}(\xi)}{(n+1)!}(x_{i+1} - x_i)^{n+1} \qquad [3.13]$$

where the subscript $n$ connotes that this is the remainder for the $n$th-order approximation and $\xi$ is a value of $x$ that lies somewhere between $x_i$ and $x_{i+1}$. The introduction of the $\xi$ is so important that we will devote an entire section (Sec. 3.5.2) to its derivation. For the time being, it is sufficient to recognize that there is such a value that provides an exact estimate of the error.

It is often convenient to simplify the Taylor series by defining a step size $h = x_{i+1} - x_i$ and expressing Eq. (3.12) as

$$f(x_{i+1}) = f(x_i) + f'(x_i)h + \frac{f''(x_i)}{2!}h^2 + \frac{f'''(x_i)}{3!}h^3 + \cdots + \frac{f^{(n)}(x_i)}{n!}h^n + R_n \qquad [3.14]$$

where the remainder term is now

$$R_n = \frac{f^{(n+1)}(\xi)}{(n+1)!}h^{n+1} \qquad [3.15]$$

## EXAMPLE 3.6
### Taylor Series Approximation of a Polynomial

Problem Statement: Use zero- through fourth-order Taylor series expansions to approximate the function

$$f(x) = -0.1x^4 - 0.15x^3 - 0.5x^2 - 0.25x + 1.2$$

from $x_i = 0$ with $h = 1$. That is, predict the function's value at $x_{i+1} = 1$.

Solution: Because we are dealing with a known function, we can compute values for $f(x)$ between 0 and 1. The results (Fig. 3.3) indicate that the function starts at $f(0) = 1.2$ and then curves downward to $f(1) = 0.2$. Thus, the true value that we are trying to predict is 0.2.

The Taylor series approximation with $n = 0$ is [Eq. (3.9)]

$$f(x_{i+1}) \simeq 1.2$$

Thus, as in Fig. 3.3, the zero-order approximation is a constant. Using this formulation results in a truncation error [recall Eq. (3.2)] of

$$E_t = 0.2 - 1.2 = -1.0$$

at $x = 1$.

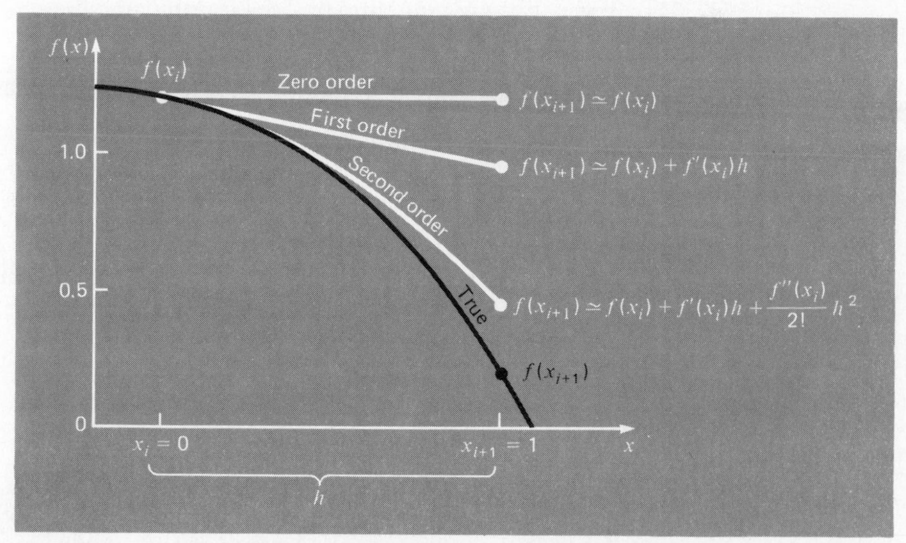


FIGURE 3.3 The approximation of $f(x) = -0.1x^4 - 0.15x^3 - 0.5x^2 - 0.25x + 1.2$ at $x = 1$ by zero-order, first-order, and second-order Taylor series expansions.

For $n = 1$, the first derivative must be determined and evaluated at $x_i = 0$:

$$f'(0) = -0.4(0.0)^3 - 0.45(0.0)^2 - 1.0(0.0) - 0.25 = -0.25$$

Therefore, the first-order approximation is [Eq. (3.10)]

$$f(x_{i+1}) \simeq 1.2 - 0.25h$$

which can be used to compute $f(1) = 0.95$. Consequently, the approximation begins to capture the downward trajectory of the function in the form of a sloping straight line (Fig. 3.3). This results in a reduction of the truncation error to

$$E_t = 0.2 - 0.95 = -0.75$$

at $x = 1$. For $n = 2$, the second derivative is evaluated at $x_i = 0$:

$$f''(0) = -1.2(0.0)^2 - 0.9(0.0) - 1.0 = -1.0$$

Therefore, according to Eq. (3.11)

$$f(x_{i+1}) \simeq 1.2 - 0.25h - 0.5h^2$$

and, substituting $h = 1$

$$f(1) \simeq 0.45$$

The inclusion of the second derivative now adds some downward curvature resulting in an improved estimate, as in Fig. 3.3. The truncation error is reduced further to $0.2 - 0.45 = -0.25$.

Additional terms would improve the approximation even more. In fact, the inclusion of the third and the fourth derivatives results in exactly the same equation as we started with:

$$f(x_{i+1}) \simeq 1.2 - 0.25h - 0.5h^2 - 0.15h^3 - 0.10h^4$$

where the remainder term is

$$R_4 = \frac{f^{(5)}(\xi)}{5!}h^5$$

Thus, because the fifth derivative of a fourth-order polynomial is zero, $R_4 = 0$. Consequently, the Taylor series expansion to the fourth derivative yields an exact estimate at $x_{i+1} = 1$:

$$f(1) \simeq 1.2 - 0.25(1) - 0.5(1)^2 - 0.15(1)^3 - 0.10(1)^4 = 0.2$$

In general, the $n$th-order Taylor series expansion will be exact for an $n$th-order polynomial. For other differentiable and continuous functions, such as exponentials and sinusoids, a finite number of terms will probably not yield an exact estimate. Each additional term will contribute some improvement, however slight, to the approximation. This behavior will be demonstrated in Example 3.7. Only if an infinite number of terms are added, will the series yield an exact result.

Although the above is true, the practical value of Taylor series expansions is that, in most cases, the inclusion of only a few of the terms will result in an approximation that is close enough to the true value for practical purposes. The assessment of how many terms are required to get "close enough" is based on the remainder term of the expansion. Recall that the remainder term is of the general form of Eq. (3.15). This relationship has two major drawbacks. First, $\xi$ is not known exactly but merely lies somewhere between $x_i$ and $x_{i+1}$. Second, in order to evaluate Eq. (3.15), we need to determine the $(n + 1)$th derivative of $f(x)$. To do this, we need to know $f(x)$. However, if we knew $f(x)$, there would be no reason to perform the Taylor series expansion in the first place!

Despite this dilemma, Eq. (3.15) is still useful for gaining insight into truncation errors. This is because we *do* have control over the term $h^{n+1}$ in the equation. In other words, we can choose how far away from $x_i$ we want to evaluate $f(x)$, and we can control the number of terms we include in the expansion. Consequently, Eq. (3.15) is usually expressed as

$$R_n = 0(h^{n+1})$$

where the nomenclature $0(h^{n+1})$ means that the truncation error is of the order of $h^{n+1}$. That is, the error is proportional to the step size $h$ to the $n + 1$ power. Although this approximation implies nothing regarding the magnitude of the derivatives that multiply $h^{n+1}$, it is extremely useful in judging the relative error of numerical methods based on Taylor series expansions. For example, if the error is $0(h)$, halving the step size will halve the error. On the other hand, if the error is $0(h^2)$, halving the step size will quarter the error.

In general, we can usually assume that the truncation error is decreased by the addition of terms to the Taylor series. In addition, if $h$ is sufficiently small, the first and other lower-order terms usually account for a disproportionately high percent of the error. Thus, only a few terms are required to obtain an adequate estimate. This property is illustrated by the following example.

## EXAMPLE 3.7
### Use of Taylor Series Expansion to Approximate a Function with an Infinite Number of Derivatives

Problem Statement: Use Taylor series expansions with $n = 0$ to 6 to approximate

$$f(x) = \cos x$$

at $x_{i+1} = \pi/3$ (60°) on the basis of the value of $f(x)$ and its derivatives at $x_i = \pi/4$ (45°). Note that this means that $h = \pi/3 - \pi/4 = \pi/12$.

Solution: As with Example 3.6, our knowledge of the true function means that we can determine the correct value of $f(\pi/3) = 0.5$.

The zero-order approximation is [Eq. (3.9)]

$$f(\pi/3) \simeq \cos(\pi/4) = 0.707106781$$

which represents a percent relative error of

$$\epsilon_t = \frac{0.5 - 0.707106781}{0.5} 100\% = -41.4\%$$

For the first-order approximation, we add the first derivative term where $f'(x) = -\sin x$:

$$f\left(\frac{\pi}{3}\right) \simeq \cos\left(\frac{\pi}{4}\right) - \sin\left(\frac{\pi}{4}\right)\left(\frac{\pi}{12}\right) = 0.521986659$$

which has $\epsilon_t = -4.40$ percent.

For the second-order approximation, we add the second derivative term where $f''(x) = -\cos x$:

$$f\left(\frac{\pi}{3}\right) \simeq \cos\left(\frac{\pi}{4}\right) - \sin\left(\frac{\pi}{4}\right)\left(\frac{\pi}{12}\right) - \frac{\cos(\pi/4)}{2}\left(\frac{\pi}{12}\right)^2$$
$$= 0.497754491$$

with $\epsilon_t = 0.449$ percent. Thus, the inclusion of additional terms results in an improved estimate.

The process can be continued and the results compiled, as in Table 3.2. Notice that the derivatives never go to zero as was the case with the polynomial in Example 3.6. Therefore, each additional term results in some improvement in the estimate. However, also notice how most of the improvement comes with the initial terms. For this case, by the time we have added the third-order term, the error is reduced to $2.62 \times 10^{-2}$ percent, which means that we have attained 99.9738 percent of the true value. Consequently, although the addition of more terms will reduce the error further, the improvement becomes negligible.

**TABLE 3.2 Taylor series approximation of $f(x) = \cos x$ at $x_{i+1} = \pi/3$ using a base point of $x_i = \pi/4$. Values are shown for various orders ($n$) of approximation.**

| Order $n$ | $f^{n}(x)$ | $f(\pi/3)$ | $\epsilon_t$ |
|---|---|---|---|
| 0 | $\cos x$ | 0.707106781 | −41.4 |
| 1 | $-\sin x$ | 0.521986659 | −4.4 |
| 2 | $-\cos x$ | 0.497754491 | 0.449 |
| 3 | $\sin x$ | 0.499869147 | $2.62 \times 10^{-2}$ |
| 4 | $\cos x$ | 0.500007551 | $-1.51 \times 10^{-3}$ |
| 5 | $-\sin x$ | 0.500000304 | $-6.08 \times 10^{-5}$ |
| 6 | $-\cos x$ | 0.499999988 | $2.40 \times 10^{-6}$ |

### 3.5.2 The Remainder for the Taylor Series Expansion

Before demonstrating how the Taylor series is actually used to estimate numerical errors, we must explain why we included the argument $\xi$ in Eq. (3.15). Rather than present a general, mathematical derivation, we will develop a simpler exposition based on a somewhat more visual interpretation. Then we can extend this specific case to the more general formulation.

Suppose that we truncated the Taylor series expansion [Eq. (3.14)] after the zero-order term to yield

$$f(x_{i+1}) \simeq f(x_i)$$

A visual depiction of this zero-order prediction is shown in Fig. 3.4. The remainder, or error, of this prediction, which is also shown in the illustration, consists of the infinite series of terms that were truncated:

$$R_0 = f'(x_i)h + \frac{f''(x_i)}{2!}h^2 + \frac{f'''(x_i)}{3!}h^3 + \cdots$$

It is obviously inconvenient to deal with the remainder in this infinite series format. One simplification might be to truncate the remainder itself, as in

$$R_0 \simeq f'(x_i)h \qquad [3.16]$$

Although, as stated in the previous section, lower-order derivatives usually account for a greater share of the remainder than the higher-order terms, this result is still inexact because of the neglected second- and higher-order terms.

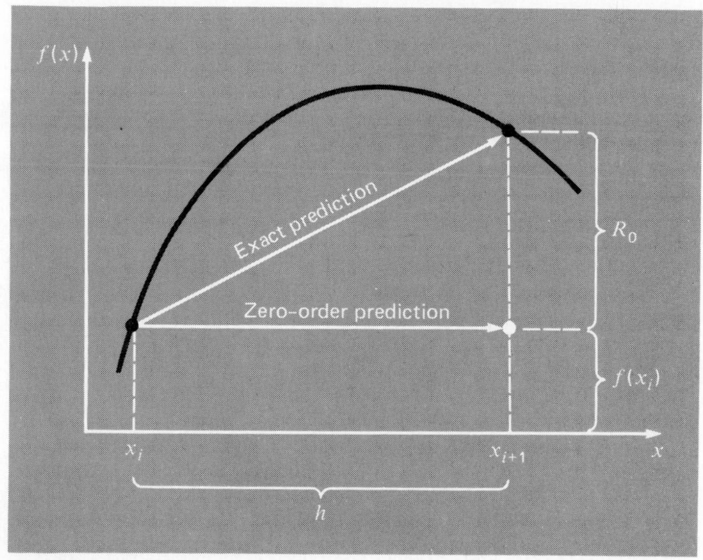


FIGURE 3.4 Graphical depiction of a zero-order Taylor series prediction and remainder.

This "inexactness" is implied by the approximate equality symbol ($\cong$) employed in Eq. (3.16).

An alternative simplification that transforms the approximation to an equivalence is based on a graphical insight. Notice that in Fig. 3.4 the error $R_0$ could be determined if we knew the location of the exact value. Obviously, this value is not known because otherwise there would be no need for a Taylor series expansion. However, the mean-value theorem of calculus provides a way to recast the problem to partially circumvent this dilemma.

The *mean-value theorem* states that if a function $f(x)$ and its first derivative are continuous over an interval from $x_i$ to $x_{i+1}$, then there exists at least one point on the function that has a slope, designated by $f'(\xi)$, that is parallel to the line joining $f(x_i)$ and $f(x_{i+1})$. The parameter $\xi$ marks the $x$ value where this slope occurs (Fig. 3.5). A physical illustration of this theorem is seen in the fact that if you travel between two points with an average velocity, there will be at least one moment during the course of the trip when you will be moving at that average velocity.

By invoking this theorem it is simple to realize that, as illustrated in Fig. 3.5, the slope $f'(\xi)$ is equal to the rise $R_0$ divided by the run $h$, or

$$f'(\xi) = \frac{R_0}{h}$$

which can be rearranged to give

$$R_0 = f'(\xi)h \tag{3.17}$$

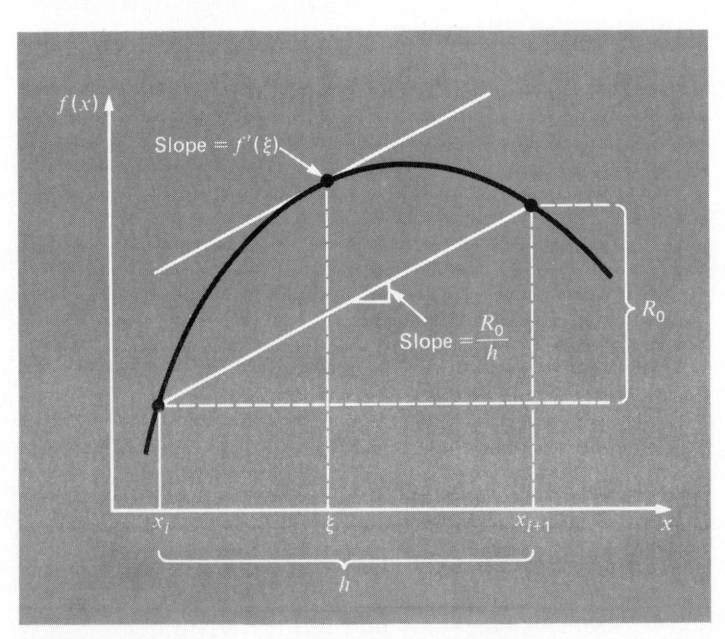


FIGURE 3.5 Graphical depiction of the mean-value theorem.

Thus, we have derived the zero-order version of Eq. (3.15). The higher-order versions are merely a logical extension of the reasoning used to derive Eq. (3.17), based on the general form of the extended mean-value theorem (Thomas and Finney, 1979). Thus, the first-order version is

$$R_1 = \frac{f''(\xi)}{2!}h^2 \tag{3.18}$$

For this case, the value of $\xi$ conforms to the $x$ value corresponding to the second derivative that makes Eq. (3.18) exact. Similar higher-order versions can be developed from Eq. (3.15).

### 3.5.3 Using the Taylor Series to Estimate Truncation Errors

Although the Taylor series will be extremely useful in estimating truncation errors throughout this book, it may not be clear to you how the expansion can actually be applied to numerical methods. In fact, we have already done so in our example of the falling parachutist. Recall that the objective of both Examples 1.1 and 1.2 was to predict velocity as a function of time. That is, we were interested in determining $v(t)$. As specified by Eq. (3.12), $v(t)$ can be expanded in a Taylor series:

$$v(t_{i+1}) = v(t_i) + v'(t_i)(t_{i+1} - t_i) + \frac{v''(t_i)}{2!}(t_{i+1} - t_i)^2 + \cdots + R_n \tag{3.19}$$

Now let us truncate the series after the first derivative term:

$$v(t_{i+1}) = v(t_i) + v'(t_i)(t_{i+1} - t_i) + R_1 \tag{3.20}$$

Equation (3.20) can be solved for

$$v'(t_i) = \underbrace{\frac{v(t_{i+1}) - v(t_i)}{t_{i+1} - t_i}}_{\text{First-order approximation}} - \underbrace{\frac{R_1}{t_{i+1} - t_i}}_{\text{Truncation error}} \tag{3.21}$$

The first part of Eq. (3.21) is exactly the same relationship that was used to approximate the derivative in Example 1.2 [Eq. (1.10)]. However, because of the Taylor series approach, we have also obtained an estimate of the truncation error associated with this approximation of the derivative. Using Eqs. (3.13) and (3.21) yields

$$\frac{R_1}{t_{i+1} - t_i} = \frac{v''(\xi)}{2!}(t_{i+1} - t_i) \tag{3.22}$$

or

$$\frac{R_1}{t_{i+1} - t_i} = 0(t_{i+1} - t_i) \tag{3.23}$$

Thus, the estimate of the derivative [Eq. (1.10) or the first part of Eq. (3.21)] has a truncation error of order $t_{i+1} - t_i$. In other words, the error of our derivative approximation should be proportional to the step size. Consequently, if we halve the step size, we would expect to halve the error of the derivative.

### 3.5.4 Numerical Differentiation

Equation (3.21) is given a formal label in numerical methods—it is called a *finite divided difference*. It can be represented generally as

$$f'(x_i) = \frac{f(x_{i+1}) - f(x_i)}{x_{i+1} - x_i} + 0(x_{i+1} - x_i) \qquad [3.24]$$

or

$$f'(x_i) = \frac{\Delta f_i}{h} + 0(h) \qquad [3.25]$$

where $\Delta f_i$ is referred to as the *first forward difference* and $h$ is called the *step size*, that is, the length of the interval over which the approximation is made. It is termed a "forward" difference because it utilizes data at $i$ and $i+1$ to estimate the derivative (Fig. 3.6*a*). The entire term $\Delta f_i/h$ is referred to as a *first finite divided difference.*

This forward divided difference is but one of many than can be developed from the Taylor series to approximate derivatives numerically. For example, *backward* and *centered difference* approximations of the first derivative can be developed in a fashion similar to the derivation of Eq. (3.24). The former utilizes data at $x_{i-1}$ (Fig. 3.6*b*), whereas the latter uses information that is equally spaced around the point at which the derivative is estimated (Fig. 3.6*c*). More accurate approximations of the first derivative can be developed by including higher-order terms of the Taylor series. Finally, all the above versions can also be developed for second, third, and higher derivatives. The following sections provide brief summaries illustrating how each of these is derived.

**Backward Difference Approximation of the First Derivative.** The Taylor series can be expanded backwards to calculate a previous value on the basis of a present value, as in

$$f(x_{i-1}) = f(x_i) - f'(x_i)h + \frac{f''(x_i)}{2}h^2 - \cdots \qquad [3.26]$$

Truncating this equation after the first derivative and rearranging yields

$$f'(x_i) \simeq \frac{f(x_i) - f(x_{i-1})}{h} = \frac{\nabla f_i}{h} \qquad [3.27]$$

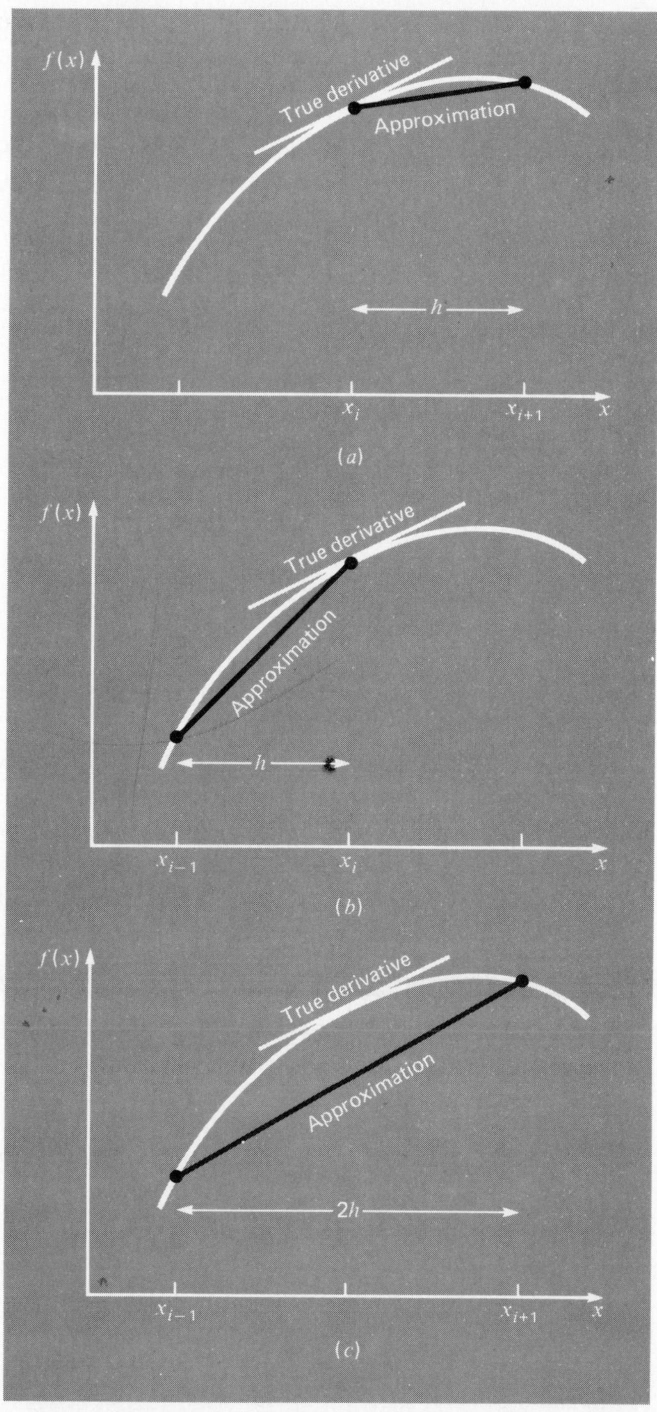


FIGURE 3.6 Graphical depiction of (*a*) forward, (*b*) backward, and (*c*) centered finite-divided-difference approximations of the first derivative.

where the error is $0(h)$ and $\nabla f_i$ is referred to as the *first backward difference.* See Fig. 3.6*b* for a graphical representation.

Centered Difference Approximation of the First Derivative. A third way to approximate the first derivative is to subtract Eq. (3.26) from the forward Taylor series expansion:

$$f(x_{i+1}) = f(x_i) + f'(x_i)h + \frac{f''(x_i)}{2}h^2 + \cdots \qquad [3.28]$$

to yield

$$f(x_{i+1}) - f(x_{i-1}) = 2f'(x_i)h + \frac{f'''(x_i)}{3}h^3 + \cdots$$

which can be solved for

$$f'(x_i) = \frac{f(x_{i+1}) - f(x_{i-1})}{2h} - \frac{f'''(x_i)}{6}h^2 + \cdots$$

or

$$f'(x_i) = \frac{f(x_{i+1}) - f(x_{i-1})}{2h} + 0(h^2) \qquad [3.29]$$

Equation (3.29) is a *centered* (or *central*) *difference* representation of the first derivative. Notice that the truncation error is of the order of $h^2$ in contrast to the forward and backward approximations that were of the order of $h$. Consequently, the Taylor series analysis yields the practical information that the centered difference is a more accurate representation of the derivative (Fig. 3.6*c*). For example, if we halve the step size using a forward or backward difference, we would approximately halve the truncation error, whereas for the central difference, the error would be quartered.

Finite Difference Approximations of Higher Derivatives. Besides first derivatives, the Taylor series expansion can be used to derive numerical estimates of higher derivatives. To do this, we write a forward Taylor series expansion for $f(x_{i+2})$ in terms of $f(x_i)$:

$$f(x_{i+2}) = f(x_i) + f'(x_i)(2h) + \frac{f''(x_i)}{2}(2h)^2 + \cdots \qquad [3.30]$$

Equation (3.28) can be multiplied by 2 and subtracted from Eq. (3.30) to give

$$f(x_{i+2}) - 2f(x_{i+1}) = -f(x_i) + f''(x_i)h^2 + \cdots$$

which can be solved for

$$f''(x_i) = \frac{f(x_{i+2}) - 2f(x_{i+1}) + f(x_i)}{h^2} + 0(h) \qquad [3.31]$$

This relationship is called the *second forward finite divided difference*. Similar manipulations can be used to derive backward and centered versions. Backward, forward, and centered difference approximations of third- and higher-order derivatives can also be developed (Figs. 3.7 through 3.9). In all cases, the centered difference yields a more accurate estimate.

Higher-Accuracy Difference Formulas. All the above estimates truncated the Taylor series estimates after only a few terms. Higher-accuracy formulas can be developed by including additional terms. For example, the forward expansion [Eq. (3.28)] can be solved for

$$f'(x_i) = \frac{f(x_{i+1}) - f(x_i)}{h} - \frac{f''(x_i)}{2}h + 0(h^2) \qquad [3.32]$$

In contrast to Eq. (3.24), we can retain the second derivative term by substituting Eq. (3.31) into Eq. (3.32) to yield

First Derivative — Error

$$f'(x_i) = \frac{f(x_i) - f(x_{i-1})}{h} \qquad 0(h)$$

$$f'(x_i) = \frac{3f(x_i) - 4f(x_{i-1}) + f(x_{i-2})}{2h} \qquad 0(h^2)$$

Second Derivative

$$f''(x_i) = \frac{f(x_i) - 2f(x_{i-1}) + f(x_{i-2})}{h^2} \qquad 0(h)$$

$$f''(x_i) = \frac{2f(x_i) - 5f(x_{i-1}) + 4f(x_{i-2}) - f(x_{i-3})}{h^2} \qquad 0(h^2)$$

Third Derivative

$$f'''(x_i) = \frac{f(x_i) - 3f(x_{i-1}) + 3f(x_{i-2}) - f(x_{i-3})}{h^3} \qquad 0(h)$$

$$f'''(x_i) = \frac{5f(x_i) - 18f(x_{i-1}) + 24f(x_{i-2}) - 14f(x_{i-3}) + 3f(x_{i-4})}{2h^3} \qquad 0(h^2)$$

Fourth Derivative

$$f''''(x_i) = \frac{f(x_i) - 4f(x_{i-1}) + 6f(x_{i-2}) - 4f(x_{i-3}) + f(x_{i-4})}{h^4} \qquad 0(h)$$

$$f''''(x_i) = \frac{3f(x_i) - 14f(x_{i-1}) + 26f(x_{i-2}) - 24f(x_{i-3}) + 11f(x_{i-4}) - 2f(x_{i-5})}{h^4} \qquad 0(h^2)$$

FIGURE 3.7 Backward finite-divided-difference formulas: Two versions are presented for each derivative. The latter incorporates more terms of the Taylor series expansion and is, consequently, more accurate.

First Derivative | Error

$$f'(x_i) = \frac{f(x_{i+1}) - f(x_i)}{h} \qquad 0(h)$$

$$f'(x_i) = \frac{-f(x_{i+2}) + 4f(x_{i+1}) - 3f(x_i)}{2h} \qquad 0(h^2)$$

Second Derivative

$$f''(x_i) = \frac{f(x_{i+2}) - 2f(x_{i+1}) + f(x_i)}{h^2} \qquad 0(h)$$

$$f''(x_i) = \frac{-f(x_{i+3}) + 4f(x_{i+2}) - 5f(x_{i+1}) + 2f(x_i)}{h^2} \qquad 0(h^2)$$

Third Derivative

$$f'''(x_i) = \frac{f(x_{i+3}) - 3f(x_{i+2}) + 3f(x_{i+1}) - f(x_i)}{h^3} \qquad 0(h)$$

$$f'''(x_i) = \frac{-3f(x_{i+4}) + 14f(x_{i+3}) - 24f(x_{i+2}) + 18f(x_{i+1}) - 5f(x_i)}{2h^3} \qquad 0(h^2)$$

Fourth Derivative

$$f''''(x_i) = \frac{f(x_{i+4}) - 4f(x_{i+3}) + 6f(x_{i+2}) - 4f(x_{i+1}) + f(x_i)}{h^4} \qquad 0(h)$$

$$f''''(x_i) = \frac{-2f(x_{i+5}) + 11f(x_{i+4}) - 24f(x_{i+3}) + 26f(x_{i+2}) - 14f(x_{i+1}) + 3f(x_i)}{h^4} \qquad 0(h^2)$$

FIGURE 3.8 Forward finite-divided difference formulas: Two versions are presented for each derivative. The latter incorporates more terms of the Taylor series expansion and is, consequently, more accurate.

$$f'(x_i) = \frac{f(x_{i+1}) - f(x_i)}{h} - \frac{f(x_{i+2}) - 2f(x_{i+1}) + f(x_i)}{2h^2}h + 0(h^2)$$

or, by collecting terms,

$$f'(x_i) = \frac{-f(x_{i+2}) + 4f(x_{i+1}) - 3f(x_i)}{2h} + 0(h^2)$$

Notice that inclusion of the second derivative term has improved the accuracy to $0(h^2)$. Similar improved versions can be developed for the backward and centered formulas as well as for the approximations of the higher derivatives. The formulas are summarized in Figs. 3.7 through 3.9. The following example illustrates their utility for estimating derivatives.

First Derivative

$$f'(x_i) = \frac{f(x_{i+1}) - f(x_{i-1})}{2h} \qquad O(h^2)$$

$$f'(x_i) = \frac{-f(x_{i+2}) + 8f(x_{i+1}) - 8f(x_{i-1}) + f(x_{i-2})}{12h} \qquad O(h^4)$$

Second Derivative

$$f''(x_i) = \frac{f(x_{i+1}) - 2f(x_i) + f(x_{i-1})}{h^2} \qquad O(h^2)$$

$$f''(x_i) = \frac{-f(x_{i+2}) + 16f(x_{i+1}) - 30f(x_i) + 16f(x_{i-1}) - f(x_{i-2})}{12h^2} \qquad O(h^4)$$

Third Derivative

$$f'''(x_i) = \frac{f(x_{i+2}) - 2f(x_{i+1}) + 2f(x_{i-1}) - f(x_{i-2})}{2h^3} \qquad O(h^2)$$

$$f'''(x_i) = \frac{-f(x_{i+3}) + 8f(x_{i+2}) - 13f(x_{i+1}) + 13f(x_{i-1}) - 8f(x_{i-2}) + f(x_{i-3})}{8h^3} \qquad O(h^4)$$

Fourth Derivative

$$f''''(x_i) = \frac{f(x_{i+2}) - 4f(x_{i+1}) + 6f(x_i) - 4f(x_{i-1}) + f(x_{i-2})}{h^4} \qquad O(h^2)$$

$$f''''(x_i) = \frac{-f(x_{i+3}) + 12f(x_{i+2}) - 39f(x_{i+1}) + 56f(x_i) - 39f(x_{i-1}) + 12f(x_{i-2}) - f(x_{i-3})}{6h^4} \qquad O(h^4)$$

FIGURE 3.9 Centered finite-divided-difference formulas: Two versions are presented for each derivative. The latter incorporates more terms of the Taylor series expansion and is, consequently, more accurate.

## EXAMPLE 3.8
## Finite-Divided-Difference Approximations of Derivatives

Problem Statement: Use forward and backward difference approximations of $O(h)$ and a centered difference approximation of $O(h^2)$ to estimate the first derivative of

$$f(x) = -0.1x^4 - 0.15x^3 - 0.5x^2 - 0.25x + 1.2$$

at $x = 0.5$ using a step size $h = 0.5$. Repeat the computation using $h = 0.25$. Note that the derivative can be calculated directly as

$$f'(x) = -0.4x^3 - 0.45x^2 - 1.0x - 0.25$$

and can be used to compute the true value as $f'(0.5) = -0.9125$.

Solution: For $h = 0.5$, the function can be used to determine

$$x_{i-1} = 0 \qquad f(x_{i-1}) = 1.2$$
$$x_i = 0.5 \qquad f(x_i) = 0.925$$
$$x_{i+1} = 1.0 \qquad f(x_{i+1}) = 0.2$$

This data can be used to compute the forward divided difference [Eq. (3.24)]:

$$f'(0.5) \simeq \frac{0.2 - 0.925}{0.5} = -1.45 \qquad \epsilon_t = -58.9\%$$

the backward divided difference [Eq. (3.27)]:

$$f'(0.5) \simeq \frac{0.925 - 1.2}{0.5} = -0.55 \qquad \epsilon_t = 39.7\%$$

and the centered divided difference [Eq. (3.29)]:

$$f'(0.5) \simeq \frac{0.2 - 1.2}{1.0} = -1.0 \qquad \epsilon_t = -9.6\%$$

For $h = 0.25$, the data is

$$x_{i-1} = 0.25 \qquad f(x_{i-1}) = 1.10351563$$
$$x_i = 0.50 \qquad f(x_i) = 0.925$$
$$x_{i+1} = 0.75 \qquad f(x_{i+1}) = 0.63632813$$

which can be used to compute the forward divided difference:

$$f'(0.5) \simeq \frac{0.63632813 - 0.925}{0.25} = -1.155 \qquad \epsilon_t = -26.5\%$$

the backward divided difference:

$$f'(0.5) \simeq \frac{0.925 - 1.10351563}{0.25} = -0.714 \qquad \epsilon_t = 21.7\%$$

and the centered divided difference:

$$f'(0.5) \simeq \frac{0.63632813 - 1.10351563}{0.5} = -0.934 \qquad \epsilon_t = -2.4\%$$

For both step sizes, the centered difference approximation is more accurate than forward or backward differences. Also, as predicted by the Taylor series analysis, halving the step size halves the error of the backward and forward differences and quarters the error of the centered difference.

This section has covered only some of the ways in which Taylor series prove useful in numerical analysis. However, the material should provide you with an initial indication of their value in estimating and controlling truncation error. Many of the numerical methods in this book are based on representing complicated mathematical expressions by simpler, lower-order approximations. Because it provides a framework for separating lower- and higher-order components, the Taylor series expansion will prove valuable throughout this text as a vehicle for gaining insight into numerical methods.

## 3.6 TOTAL NUMERICAL ERROR

The total numerical error is the summation of the truncation and round-off errors. From the falling parachutist problem (Example 3.3), we discovered that the only way to minimize round-off errors is to increase the numbers of significant figures of the computer. Further, we have noted that round-off error will *increase* as the number of computations in an analysis increases. In contrast, Example 3.8 demonstrated that the derivative estimate can be improved by decreasing the step size. Because a decrease in step size leads to an increase in computations, the truncation errors are *decreased* as the number of computations increases. Therefore, we are faced by the following dilemma: the strategy for decreasing one component of the total error leads to an increase of the other component. In a computation, we could conceivably decrease the step size to minimize truncation errors only to discover that in doing so, the round-off error begins to dominate the solution and the total error grows! Thus, our remedy becomes our problem (Fig. 3.10). One challenge that we face is to determine an appropriate step size for a particular computation. We would like to choose a large step size in order to decrease the amount of calculations and round-off errors without incurring the penalty of a large truncation error. If the total error is as shown in Fig. 3.10, the

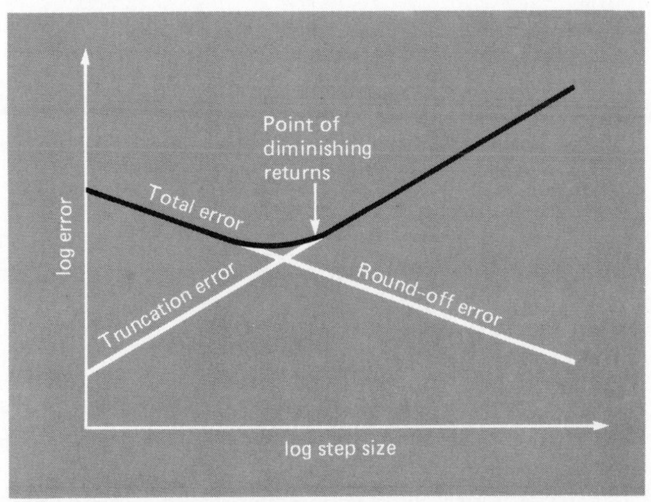


FIGURE 3.10 A graphical depiction of the trade-off between round-off and truncation error that sometimes comes into play in the course of a numerical method. The point of diminishing returns, where round-off error begins to negate the benefits of step-size reduction, is shown.

challenge is to identify the point of diminishing returns where round-off error begins to negate the benefits of step-size reduction.

In actual cases, however, such situations are relatively uncommon because most computers carry enough significant figures that round-off errors do not predominate. Nevertheless, they sometimes do occur and suggest a sort of "numerical uncertainty principle" that places an absolute limit on the accuracy that may be obtained using certain computerized numerical methods.

Because of these shortcomings, we have limitations on our ability to estimate errors. Consequently, error estimation in numerical methods is, to a certain extent, an art that depends in part on trial-and-error solution and the intuition and experience of the analyst.

Although the present chapter has concentrated on one type of numerical problem—the solution of an ordinary differential equation—the foregoing conclusions have general relevance to many of the other techniques in the book. However, it must be stressed that although the subject is, to a certain point, an art, there are a variety of methods that the analyst can use to quantify and control the errors in a computation. The elaboration of these techniques will play a prominent role in the following pages.

## 3.7 BLUNDERS, FORMULATION ERRORS, AND DATA UNCERTAINTY

Although the following sources of error are not directly connected with most of the numerical methods in this book, they can sometimes have great impact on the success of a modeling effort. Thus, they must always be kept in mind when applying numerical techniques in the context of real-world problems.

### 3.7.1 Blunders

We are all familiar with gross errors, or blunders. In the early years of computers, erroneous numerical results could sometimes be attributed to malfunctions of the computer itself. Today, this source of error is highly unlikely, and most blunders must be attributed to human imperfection.

Blunders can occur at any stage of the mathematical modeling process and can contribute to all the other components of error. They can only be avoided by sound knowledge of fundamental principles and by the care with which you approach and design your solution to a problem.

Blunders are usually disregarded in discussions of numerical methods. This is no doubt due to the fact that, try as we may, gross errors are to a certain extent unavoidable. However, we believe that there are a number of ways in which their occurrence can be minimized. In particular, the good programming habits that were outlined in Chap. 2 are extremely useful for mitigating programming blunders. In addition, there are usually simple ways to check whether a particular numerical method is working properly. Throughout this book, we discuss ways to check the results of numerical calculations.

### 3.7.2 Formulation Errors

Formulation, or model, errors relate to bias that can be ascribed to incomplete mathematical models. An example of a negligible formulation error is the fact that Newton's second law does not account for relativistic effects. This does not detract from the adequacy of the solution in Example 1.1 because these errors are minimal on the time and space scales of a falling parachutist.

However, suppose that air resistance is not linearly proportional to fall velocity, as in Eq. (1.6), but is a function of the square of velocity. If this were the case, both the analytical and numerical solutions obtained in the first chapter would be erroneous because of formulation error. Further consideration of formulation error is included in some of the case studies in the remainder of the book. You should be cognizant of these problems and realize that, if you are working with a poorly conceived model, no numerical method will provide adequate results.

### 3.7.3 Data Uncertainty

Errors sometimes enter into an analysis because of uncertainty in the physical data upon which a model is based. For instance, suppose we wanted to test the falling parachutist model by having an individual make repeated jumps and then measuring his or her velocity after a specified time interval. Uncertainty would undoubtedly be associated with these measurements, as the parachutist would fall faster during some jumps than during others. These errors can exhibit both inaccuracy and imprecision. If our instruments consistently underestimate or overestimate the velocity, we are dealing with an inaccurate, or biased, device. On the other hand, if the measurements are randomly high and low, we are dealing with a question of precision.

Measurement errors can be quantified by summarizing the data with one or more well-chosen *statistics* that convey as much information as possible regarding specific characteristics of the data. These descriptive statistics are most often selected to represent (1) the location of the center of the distribution of the data and (2) the degree of spread of the data. As such, they provide a measure of the bias and imprecision, respectively. We will return to the topic of characterizing data uncertainty in Chap. 10.

Although you must be cognizant of blunders, formulation errors, and uncertain data, the numerical methods used for building models can be studied, for the most part, independently of these errors. Therefore, for most of this book, we will assume that we have not made gross errors, we have a sound model, and we are dealing with errorless measurements. Under these conditions, we can study numerical errors without complicating factors.

# PROBLEMS

**3.1** How many significant figures are in each of the following numbers?

(a) $0.84 \times 10^2$
(b) 84.0
(c) 70
(d) 70.0
(e) 7
(f) 0.04600
(g) 0.00460
(h) $8.00 \times 10^3$
(i) $8.0 \times 10^3$
(j) 8000

**3.2** Round off the following numbers to three significant figures.

(a) 8.755
(b) $0.368124 \times 10^2$
(c) 4225.0002
(d) $5.445 \times 10^3$
(e) 0.999500

**3.3** Carry out the following additions and subtractions and write the results to the correct number of significant figures.

(a) $0.00423 + (25.1 \times 10^{-3}) + (10.322 \times 10^{-2})$
(b) $5068 - 2.4$
(c) $(4.68 \times 10^6) - (8.2 \times 10^2)$
(d) $(9.8 \times 10^{-6}) - (8.696 \times 10^{-5})$
(e) $(7.7 \times 10^{-5}) - (5.409 \times 10^{-6}) + (7.0 \times 10^{-4})$

**3.4** Carry out the following multiplications and divisions and write the results to the correct number of significant figures.

(a) $(8.38 \times 10^5) \times (6.9 \times 10^{-5})$
(b) $(8.38 \times 10^4) \times (6.90 \times 10^{-4})$
(c) $87{,}619/(0.00871 \times 99{,}999)$
(d) $(2.06 \times 111)/888$
(e) $\dfrac{(0.4000 \times 0.02000)}{(0.01000 \times 0.800)}$

**3.5** Carry out the following combined operations and write the results to the correct number of significant figures.

(a) $6.80\,(4.0 \times 10^{-6}) - 22\,(8.06 \times 10^{-9})$
(b) $(14 \times 10^{-3} + 555 - 80.8) \times (2.0001 - 0.004)$
(c) $\dfrac{486 \times 10^{-6} - 4.45 \times 10^{-5}}{(7.777 \times 10^3) + 9.6}$
(d) $\dfrac{4.81 \times 10^{-3}}{(6.9134 \times 10^3) + 32.26} - 6.7845 \times 10^{-6}$
(e) $\dfrac{58.6\,(12 \times 10^{-6}) - (208 \times 10^{-6})\,(1801)}{468.94 \times 10^{-6}}$

**3.6** In Example 3.2, we used the infinite series

$$f(x) = 1 + x + \frac{x^2}{2!} + \frac{x^3}{3!} + \cdots$$

to approximate $e^x$.

(a) Prove that this Maclaurin series expansion is a special case of the Taylor series expansion [Eq. (3.14)] with $x_i = 0$ and $h = x$.
(b) Use the Taylor series to estimate $f(x) = e^{-x}$ at $x_{i+1} = 2$ for three separate

cases: $x_i = 0.5$, 1.0, and 1.5. Employ the zero-, first-, second-, and third-order versions and compute the $|\epsilon_t|$ for each case.

**3.7** The Maclaurin series expansion for cos $x$ is

$$\cos x = 1 - \frac{x^2}{2!} + \frac{x^4}{4!} - \frac{x^6}{6!} + \frac{x^8}{8!} - \cdots$$

Starting with the simplest version, $\cos x = 1$, add terms one at a time in order to estimate $\cos(\pi/3)$. After each new term is added, compute the true and approximate percent relative errors. Use your pocket calculator to determine the true value. Add terms until the absolute value of the approximate error estimate falls below an error criterion conforming to two significant figures.

**3.8** Perform the same computation as in Prob. 3.7, but use the Maclaurin series expansion for sin $x$

$$\sin x = x - \frac{x^3}{3!} + \frac{x^5}{5!} - \frac{x^7}{7!} + \cdots$$

to estimate $\sin(\pi/2)$.

**3.9** Use zero- through third-order Taylor series expansions to predict $f(3)$ for

$$f(x) = 25x^3 - 6x^2 + 7x - 88$$

using a base point at $x = 2$. Compute the true percent relative error $\epsilon_t$ for each approximation.

**3.10** Use zero- through fourth-order Taylor series expansions to predict $f(4)$ for $f(x) = \ln x$ using a base point at $x = 2$. Compute the true percent relative error $\epsilon_t$ for each approximation.

**3.11** Use zero- through fourth-order Taylor series expansions to predict $f(2)$ for $f(x) = e^{-x}$ using a base point at $x = 1$. Compute the percent relative error $\epsilon_t$ for each approximation.

**3.12** Use forward and backward difference approximations of $O(h)$ and a centered difference approximation of $O(h^2)$ to estimate the first derivative of the function examined in Prob. 3.9. Evaluate the derivative at $x = 2.5$ using a step size of $h = 0.25$. Compare your results with the true value of the derivative at $x = 2.5$. Interpret your results on the basis of the remainder term of the Taylor series expansion.

**3.13** Use forward, backward, and centered difference approximations of $O(h^2)$ to estimate the second derivative of the function examined in Prob. 3.9. Perform the evaluation at $x = 2.6$ using a step size of $h = 0.2$. Compare your estimates with the true value of the second derivative at $x = 2.6$. Interpret your results on the basis of the remainder term of the Taylor series expansion.

# EPILOGUE: PART I

## I.4 TRADE-OFFS

Numerical methods are scientific in the sense that they represent systematic techniques for solving mathematical problems. However, there is a certain degree of art, subjective judgment, and compromise associated with their effective use in engineering practice. For each problem, you may be confronted with several alternative numerical methods and many different types of computers. Thus, the elegance and efficiency of different approaches to problems is highly individualistic and correlated with your ability to choose wisely between options. Unfortunately, as with any intuitive process, the factors influencing this choice are difficult to communicate. Only by experience can these skills be fully comprehended and honed. However, because these skills play such a prominent role in the effective implementation of the methods, we have included this section as an introduction to some of the trade-offs that you must consider when selecting a numerical method and the tools for implementing the method. Although it is not expected that, on first reading, you will appreciate all the following issues, it is hoped that the discussion will influence your orientation when approaching subsequent material. Also, it is hoped that you will refer back to this material when you are confronted with choices and trade-offs in the remainder of the book.

Figure I.4 illustrates seven different factors, or trade-offs, that should be considered when selecting a numerical method for a particular problem.

1. Type of Mathematical Problem. As delineated previously in Fig. I.2, several types of mathematical problems are discussed in this book:

**a.** Roots of equations

**b.** Systems of simultaneous linear algebraic equations

**c.** Curve fitting

**d.** Numerical integration

**e.** Ordinary differential equations

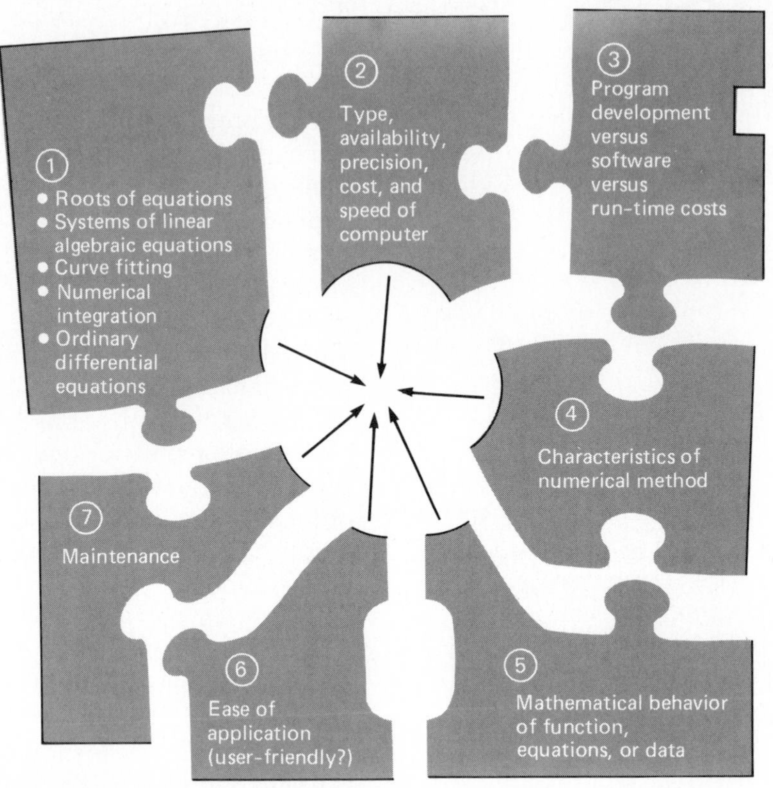


FIGURE I.4 Seven considerations when choosing a numerical method for the solution of engineering problems.

You will probably be introduced to the applied aspects of numerical methods by confronting a problem in one of the above areas. Numerical methods will be required because the problem cannot be solved efficiently using analytical techniques. You should be cognizant of the fact that your professional activities will eventually involve problems in all the above areas. Thus, the study of numerical methods and selection of automatic computation equipment should, at the minimum, consider these basic types of problems. More advanced problems may require capabilities of handling solutions of simultaneous nonlinear algebraic equations, multiple variable curve fitting, parameter optimization, linear programming, eigenvalue problems, and partial differential equations. These areas typically demand greater computation power and advanced methods not covered in this text. Other references such as Carnahan, Luther, and Wilkes (1969); Hamming (1973); and Ralston and Rabinowitz (1978) should be consulted for problems beyond the scope of this book. In addition, at the end of each part of this text, we include a brief summary and references for advanced methods to provide you with avenues for pursuing further studies of numerical methods.

2. Type, Availability, Precision, Cost, and Speed of Computer. You may have the option of working with four different computation tools (recall Table 2.1). These range from pocket calculators to large mainframe computers. Of course, any of the tools can be used to implement any numerical method (including simple paper and pencil, which are not included in the table). It is usually not a question of ultimate capability but rather of cost, convenience, speed, dependability, repeatability, and precision. Although each of the tools listed in Table 2.1 will continue to have utility, the recent rapid advances in the performance of personal computers have already had a major impact on the engineering profession. We expect this revolution will spread as technological improvements continue, because personal computers offer an excellent compromise in convenience, cost, precision, speed, and storage capacity. Furthermore, they can be readily applied to most practical engineering problems. The techniques in this book were, therefore, expressly chosen to be compatible with this class of computers.

3. Program Development Cost versus Software Cost versus Run-Time Cost. Once the types of mathematical problems to be solved have been identified and the computer system has been selected, it is appropriate to consider software and run-time costs. Software development may represent a substantial effort in many engineering projects and may therefore be a significant cost. In this regard, it is particularly important that you be very well acquainted with the theoretical and practical aspects of the relevant numerical methods. Professionally developed software may be available for a limited number of engineering problems at a considerable cost. However, these programs should be used with great care because you will not ordinarily be intimately familiar with the program logic. Alternatively, general utility software of low cost (such as that associated with this book) is available to implement numerical methods that may be readily adapted to a wide variety of problems. Program development costs and software costs may be recovered during execution if the programs are efficiently written and well tested.

4. Characteristics of the Numerical Method. When computer hardware and software costs are high, or if computer availability is limited (for example, on some timeshare systems), it pays to choose carefully the numerical method to suit the situation. On the other hand, if the problem is still at the exploratory stage and computer access and cost are no problem, it may be appropriate for you to select a numerical method that always works but may not be the most computationally efficient. The numerical methods available to solve any particular type of problem involve the types of trade-offs just discussed and others:

a. *Number of initial guesses or starting points.* Some of the numerical methods for finding roots of equations or solving differential equations require the user to specify initial guesses or starting points. Simple methods usually require one value, whereas complicated methods may require more than one value. You must consider the trade-offs; the advantages of complicated methods that are computationally efficient may be offset by the requirement for multiple starting points. You must use your experience and judgment for each particular problem.

b. *Rate of convergence.* Certain numerical methods converge more rapidly than others. However, this rapid convergence may require more initial guesses and more complex programming than a method with slower convergence. Again, you must use your judgment in selecting a method. Faster is not always better!

c. *Stability.* Some numerical methods for finding roots of equations or solutions for systems of linear equations may diverge rather than converge on the correct answer for certain problems. Why would you tolerate this possibility when confronted with design or planning problems? The answer is that these methods may be highly efficient when they work. Thus, trade-offs again emerge. You must decide if your problem requirements justify the effort needed to apply a method that may not always converge.

d. *Accuracy and precision.* Some numerical methods are simply more accurate or precise than others. Good examples are the various equations available for numerical integration. Usually, the performance of low-accuracy methods can be improved by decreasing the step size or increasing the number of applications over a given interval. Is it better to use a low-accuracy method with small step sizes or a high-accuracy method with large step sizes? This question must be addressed on a case-by case basis considering the additional factors such as cost and ease of programming. In addition, you must also be concerned with round-off errors when you are using multiple applications of low-accuracy methods and the number of computations becomes large. Here the significant figures handled by the computer may be the deciding factor.

e. *Breadth of application.* Some numerical methods can only be applied to a limited class of problems or to problems that satisfy certain mathematical restrictions. Other methods are not affected by such restrictions. You must evaluate whether it is worth your effort to develop programs that employ techniques that are appropriate for only a limited number of problems. The fact that such techniques may be widely used suggests that they have advantages that will often outweigh their disadvantages. Obviously, trade-offs are occurring.

*f. Special requirements.* Some numerical techniques attempt to increase accuracy and rate of convergence using additional or special information. An example would be to use estimated or theoretical values of errors to improve accuracy. However, these improvements are generally not achieved without some inconvenience in terms of added computer costs or increased program complexity.

*g. Programming effort required.* Efforts to improve rates of convergence, stability, and accuracy can be creative and ingenious. When improvements can be made without increasing the programming complexity, they may be considered elegant and will probably find immediate use in the engineering profession. However, if they require more complex programs, you are again faced with a trade-off situation that may or may not favor the new method.

It is clear that the above discussion concerning a choice of numerical methods reduces to one of cost and accuracy. The costs are those involved with computer time and program development. Appropriate accuracy is a question of professional ethics and judgment.

5. Mathematical Behavior of the Function, Equation, or Data. In selecting a particular numerical method, type of computer, and type of software, you must consider the complexity of your functions, equations, or data. Simple equations and smooth data may be appropriately handled by simple numerical algorithms and inexpensive computers. The opposite is true for complicated equations and data exhibiting discontinuities.

6. Ease of Application (User-Friendly?). Some numerical methods are easy to apply, others are difficult. This may be a consideration when choosing one method over another. This same idea applies to decisions regarding program development costs versus professionally developed software. It may take considerable effort to convert a difficult program to one that is user-friendly. Ways to do this were introduced in Chap. 2 and are elaborated throughout the book. In addition, the NUMERICOMP software accompanying this text is an example of user-friendly programming.

7. Maintenance. Programs for solving engineering problems require maintenance because during application, difficulties invariably occur. Maintenance may require changing the program code or expanding the documentation. Simple programs and numerical algorithms are simpler to maintain.

The chapters that follow involve the development of various types of numerical methods for various types of mathematical problems.

Several alternative methods will be given in each chapter. These various methods (rather than a single method chosen by the authors) are presented because there is no single "best" method. There are no "best" methods because there are many trade-offs that must be considered when applying the methods to practical problems. A table that highlights the trade-offs involved in each method will be found at the end of each part of the book. This table should assist you in selecting the appropriate numerical procedure for your particular problem context.

**TABLE I.2 Summary of important information presented in Part I.**

**Error Definitions**

True error $E_t = \text{true value} - \text{approximation}$

True percent relative error $\epsilon_t = \dfrac{\text{true value} - \text{approximation}}{\text{true value}}\,100\%$

Approximate percent relative error $\epsilon_a = \dfrac{\text{present approximation} - \text{previous approximation}}{\text{present approximation}}\,100\%$

Stopping criterion Terminate computation when

$$\epsilon_a < \epsilon_s$$

where $\epsilon_s$ is the desired percent relative error specified directly or calculated in terms of the desired number of significant figures $n$

$$\epsilon_s = (0.5 \times 10^{2-n})\%$$

**Taylor series**

Taylor series expansion

$$f(x_{i+1}) = f(x_i) + f'(x_i)h + \frac{f''(x_i)}{2!}h^2 + \frac{f'''(x_i)}{3!}h^3 + \cdots + \frac{f^{(n)}(x_i)}{n!}h^n + R_n$$

where

Remainder

$$R_n = \frac{f^{(n+1)}(\xi)}{(n+1)!}h^{n+1}$$

or

$$R_n = 0(h^{n+1})$$

**Numerical Differentiation**

First forward finite divided difference

$$f'(x_i) = \frac{f(x_{i+1}) - f(x_i)}{h} + 0(h)$$

(Other divided differences are summarized in Figs. 3.7 through 3.9.)

## I.5 IMPORTANT RELATIONSHIPS AND FORMULAS

Table I.2 summarizes important information that was presented in Part I. The table can be consulted to quickly access important relationships and formulas. The epilogue of each part of the book will contain such a summary.

## I.6 ADVANCED METHODS AND ADDITIONAL REFERENCES

The epilogue of each part of the book will also include a section designed to facilitate and encourage your further studies of numerical methods. This section will reference other books on the subject as well as material related to more advanced methods.*

To extend the background provided in Part I, numerous manuals on computer programming are available. It would be difficult to reference all the excellent books and manuals pertaining to specific languages and computers. In addition, you probably already have material from your previous exposure to programming. However, if this is your first experience with computers, Bent and Sethares (1982) provide a good general introduction to BASIC. McCracken (1965), Merchant (1979), and Merchant and Sturgul (1977) are all useful books on FORTRAN. Your instructor and fellow students should also be able to advise you regarding good reference books for the machines and languages available at your school.

As for error analysis, any good introductory calculus book will include supplementary material related to subjects such as the Taylor series expansion. Texts by Swokowski (1979) and Thomas and Finney (1979) provide very readable discussions of these subjects.

Finally, although we hope that our book serves you well, it is always good to consult other sources when trying to master a new subject. Ralston and Rabinowitz (1978) and Carnahan, Luther, and Wilkes (1969) provide comprehensive discussions of most numerical methods, including many advanced methods that are beyond our scope. Other enjoyable books on the subject are Gerald and Wheatley (1984), James, Smith, and Wolford (1977), Stark (1970), Rice (1983), Hornbeck (1975) and Cheney and Kincaid (1980).

*Books are referenced only by author here; a complete bibliography will be found at the back of this text.

PART TWO

# ROOTS OF EQUATIONS

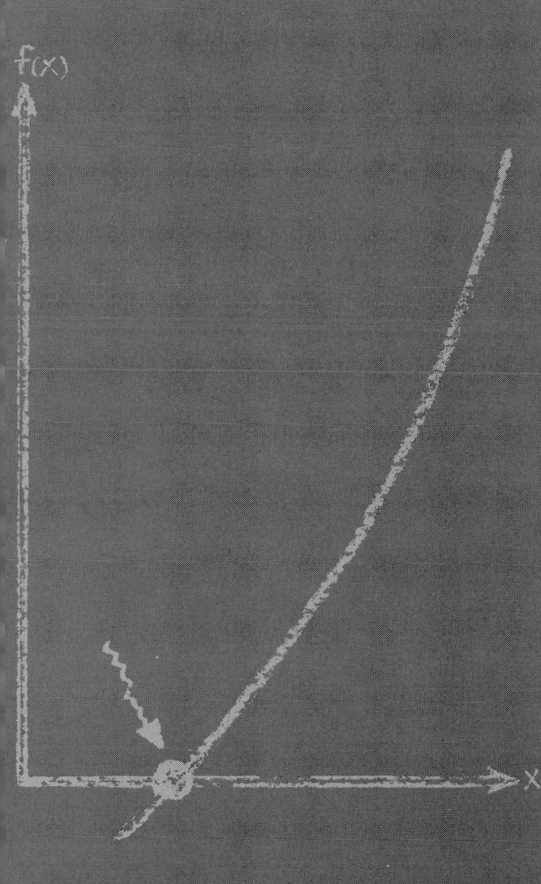


## II.1 MOTIVATION

Years ago, you learned to use the *quadratic formula*

$$x = \frac{-b \pm \sqrt{b^2 - 4ac}}{2a} \qquad \text{[II.1]}$$

to solve

$$f(x) = ax^2 + bx + c = 0 \qquad \text{[II.2]}$$

The values calculated with Eq. (II.1) are called the "roots" of Eq. (II.2). They represent the values of $x$ that make Eq. (II.2) equal to zero. Thus, we can define the *root* of an equation as the value of $x$ that makes $f(x) = 0$. For this reason, roots are sometimes called the *zeros* of the equation.

Although the quadratic formula is handy for solving Eq. (II.2), there are many other functions for which the root cannot be determined so easily. For these cases, the numerical methods described in Chaps. 4 and 5 provide efficient means to obtain the answer.

### II.1.1 Precomputer Methods for Determining Roots

Before the advent of digital computers, there were a number of ways to solve for roots of algebraic and transcendental equations. For some cases, the roots could be obtained by direct methods, as was done with Eq. (II.1). Although there were equations like this that could be solved directly, there were many more that could not. For example, even an apparently simple function such as $f(x) = e^{-x} - x$ cannot be solved analytically. In such instances, the only alternative is an approximate solution technique.

One method to obtain an approximate solution is to plot the function and determine where it crosses the $x$ axis. This point, which represents the $x$ value for which $f(x) = 0$, is the root. Graphical techniques are discussed at the beginning of Chaps. 4 and 5.

Although graphical methods are useful for obtaining rough estimates of roots, they are limited because of their lack of precision. An alternative approach is to use trial and error. This "technique" consists of guessing a value of $x$ and evaluating whether $f(x)$ is zero. If not (as is almost always the case), another guess is made, and $f(x)$ is again evaluated to determine whether the new value provides a better estimate of the root. The process is repeated until a guess is obtained that results in an $f(x)$ that is close to zero.

Such haphazard methods are obviously inefficient and inadequate for the requirements of engineering practice. The techniques described in Part II represent alternatives that are also approximate but employ systematic strategies to home in on the true root. In addition, they are ideally suited for implementation on personal computers. As elaborated in the following pages, the combination of these systematic methods and computers makes the solution of most applied roots of equations problems a simple and efficient task.

### II.1.2 Roots of Equations and Engineering Practice

Although they arise in other problem contexts, roots of equations frequently occur in the area of engineering design. Table II.1 lists a number of fundamental principles that are routinely used in design work. Mathematical equations or models derived from these principles are employed to predict dependent variables as a function of independent variables and parameters. Note that in each case, the dependent variables reflect the state or performance of the system, whereas the parameters represent its properties or composition.

An example of such a model is the equation, derived from Newton's second law, used in Chap. 1 for the parachutist's velocity:

$$v = \frac{gm}{c}[1 - e^{-(c/m)t}] \qquad [II.3]$$

where velocity $v$ is the dependent variable; time $t$ is the independent variable; and the gravitational constant $g$, the drag coefficient $c$, and the mass $m$ are parameters. If the parameters are known, Eq. (II.3) can be used to predict the parachutist's velocity as a function of time. Such computations can be performed directly because $v$ is expressed *explicitly* as a function of time. That is, it is isolated on one side of the equal sign.

However, suppose that we had to determine the drag coefficient for a parachutist of a given mass to attain a prescribed velocity in a set time

**TABLE II.1 Fundamental principles used in engineering design problems**

| Fundamental principle | Dependent variable | Independent variable | Parameters |
|---|---|---|---|
| Heat balance | Temperature | Time and position | Thermal properties of material and geometry of system |
| Material balance | Concentration or quantity of mass | Time and position | Chemical behavior of material, mass transfer coefficients, and geometry of system |
| Force balance | Magnitude and direction of forces to establish equilibrium | Time and position | Strength of material, structural properties, and system configuration |
| Energy balance | Changes in the kinetic- and potential-energy states of the system | Time and position | Thermal properties, mass of material, and system geometry |
| Newton's laws of motion | Acceleration, velocity, or location | Time and position | Mass of material, system geometry, and dissipative parameters such as friction or drag |
| Kirchhoff's laws | Currents and voltages in electric circuit | Time | Electrical properties of systems such as resistance, capacitance, and inductance |

period. Although Eq. (II.3) provides a mathematical representation of the interrelationship among the model variables and parameters, it cannot be solved explicitly for the drag coefficient. Try it. There is no way to rearrange the equation so that $c$ is isolated on one side of the equal sign. In such cases, $c$ is said to be *implicit.*

This represents a real dilemma, because many engineering design problems involve specifying the properties or composition of a system (as represented by its parameters) in order to ensure that it performs in a desired manner (as represented by its variables). Thus, these problems often require the determination of implicit parameters.

The solution to the dilemma is provided by numerical methods for roots of equations. To solve the problem using numerical methods, it is conventional to reexpress Eq. (II.3). This is done by subtracting the dependent variable $v$ from both sides of the equation to give

$$f(c) = \frac{gm}{c}[1 - e^{-(c/m)t}] - v \qquad [II.4]$$

The value of $c$ that makes $f(c) = 0$ is, therefore, the root of the equation. This value also represents the drag coefficient that solves the design problem.

Part II of this book deals with a variety of numerical and graphical methods for determining roots of relationships such as Eq. (II.4). These techniques can be applied to engineering design problems that are based on the fundamental principles outlined in Table II.1 as well as to many other problems confronted routinely in engineering practice.

## II.2 MATHEMATICAL BACKGROUND

For most of the subject areas in this book, there is usually some prerequisite mathematical background needed to successfully master the topic. For example, the concepts of error estimation and the Taylor series expansion discussed in Chap. 3 have direct relevance to our discussion of roots of equations. Additionally, prior to this point we have mentioned the terms "algebraic" and "transcendental" equations. It might be helpful to formally define these terms and discuss how they relate to the scope of this part of the book.

By definition, a function given by $y = f(x)$ is *algebraic* if it can be expressed in the form

$$f_n y^n + f_{n-1} y^{n-1} + \cdots + f_1 y + f_0 = 0 \qquad [II.5]$$

where the $f$'s are polynomials in $x$. *Polynomials* are a simple class of algebraic functions that are represented generally by

$$f(x) = a_0 + a_1 x + \cdots + a_n x^n \qquad [II.6]$$

where the $a$'s are constants. Some specific examples are

$$f(x) = 1 - 2.37x + 7.5x^2 \qquad [II.7]$$

and

$$f(x) = 5x^2 - x^3 + 7x^6 \qquad [II.8]$$

A *transcendental* function is one that is nonalgebraic. These include trigonometric, exponential, logarithmic, and other, less familiar, functions. Examples are

$$f(x) = e^{-x} - x \tag{II.9}$$

$$f(x) = \sin x \tag{II.10}$$

$$f(x) = \ln x^2 - 1 \tag{II.11}$$

Roots of equations may be either real or complex. A simple example of complex roots is the case where the term $b^2 - 4ac$ in Eq. (II.1) is negative. For instance, given the second-order polynomial,

$$f(x) = 4x^2 - 16x + 17$$

Eq. (II.1) can be used to determine that the roots are

$$x = \frac{16 \pm \sqrt{(-16)^2 - 4(4)\,(17)}}{2(4)} = \frac{16 \pm \sqrt{-16}}{8}$$

Therefore, one root is

$$x = 2 + \tfrac{1}{2}i$$

and the other is

$$x = 2 - \tfrac{1}{2}i$$

where $i = \sqrt{-1}$.

Although there are cases where complex roots of nonpolynomials are of interest, such situations are less common than for polynomials. As a consequence, the standard methods for locating roots typically fall into two somewhat related but primarily distinct problem areas:

1. *The determination of the real roots of algebraic and transcendental equations.* These techniques are designed to determine the value of a single root on the basis of foreknowledge of its approximate location.

2. *The determination of all real and complex roots of polynomials.* These methods are specifically designed for polynomials. They systematically determine all the roots of the polynomial rather than a single real root given an approximate location.

In the present book, we focus on the first problem area. Because they are somewhat beyond the scope of our book, we will not discuss the methods expressly designed for polynomials. However, in the epilogue at the end of Part II, we provide references for these techniques.

# II.3 ORIENTATION

Some orientation is helpful before proceeding to the numerical methods for determining roots of equations. The following is intended to give you an overview of the material in Part II. In addition, some objectives have been included to help you focus your efforts when studying the material.

## II.3.1 Scope and Preview

Figure II.1 is a schematic representation of the organization of Part II. Examine this figure carefully, starting at the top and working clockwise.

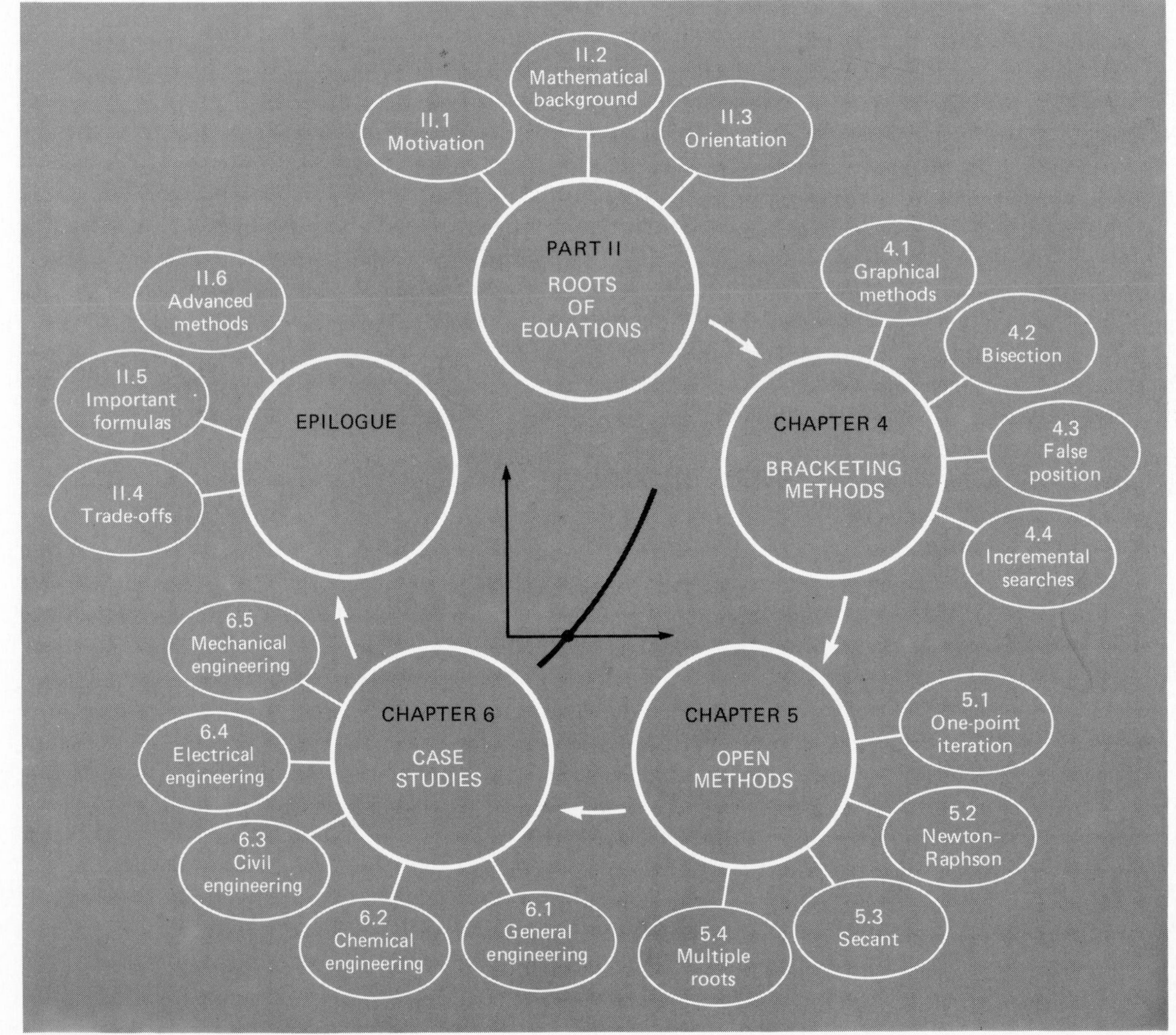


FIGURE II.1 Schematic of the organization of material in Part II: Roots of Equations.

After the present introduction, *Chapter 4* is devoted to *bracketing methods* for finding roots. These methods start with guesses that bracket, or contain, the root and then systematically reduce the width of the bracket. Two specific methods are covered: *bisection* and *false position.* Graphical methods are used to provide visual insight into the techniques. Special formulations are developed to help you determine how much computational effort is required to estimate the root to a prespecified level of precision.

*Chapter 5* covers *open methods.* These methods also involve systematic trial-and-error iterations but do not require that the initial guesses bracket the root. We will discover that these methods are usually more computationally efficient than bracketing methods, but that they do not always work. *One-point iteration, Newton-Raphson,* and *secant* methods are described. Graphical methods are used to provide geometric insight into cases where the open methods do not work. Formulas are developed that provide an idea of how fast open methods home in on the root.

*Chapter 6* extends the above concepts to actual engineering problems. Case studies are used to illustrate the strengths and weaknesses of each method and provide insight into the application of the techniques in professional practice. The case studies in Chap. 6 also highlight the trade-offs (as discussed in Part I) associated with the various methods.

An epilogue is included at the end of Part II. It contains a detailed comparison of the methods discussed in Chaps. 4 and 5. This comparison includes a description of trade-offs related to the proper use of each technique. This section also provides a summary of important formulas, along with references for some numerical methods that are beyond the scope of this text.

Automatic computation capability is integrated into Part II in a number of ways. First, user-friendly NUMERICOMP software for the bisection method is available for the Apple-II and the IBM-PC. But computer codes for bisection, using both FORTRAN and BASIC, are also given directly in the text. This provides you with the opportunity to copy and enhance the code for implementation on your own personal or mainframe computer. Flowcharts or algorithms are included for most of the other methods described in the text. This material can form the basis for a comprehensive software package that you can develop and apply to a number of engineering problems.

### II.3.2 Goals and Objectives

*Study objectives.* After completing Part II, you should have sufficient information to successfully approach a wide variety of engineering problems dealing with roots of equations. In general, you should have mastered the techniques, have learned to assess their reliability, and be capable of choosing the best method (or methods) for any particular problem. In addition to these general goals, the specific concepts in Table II.2 should be assimilated for a comprehensive understanding of the material in Part II.

*Computer objectives.* The book provides you with software, simple computer programs, algorithms, and flowcharts to implement the techniques discussed in Part II. All have utility as learning tools.

The optional software is user-friendly. It contains the bisection method to determine the real roots of algebraic and transcendental equations. The graphics associated with NUMERICOMP will enable you to easily visualize the behavior of the function being analyzed. The software can be used to conveniently determine roots of equations to any desired degree of precision. NUMERICOMP is easy to apply in solving many practical problems and can be used to check the results of any computer programs you may develop yourself.

FORTRAN and BASIC programs for the bisection method and for simple one-point iteration are also supplied directly in the text. In addition, general algorithms or flowcharts are provided for most of the other methods in Part II. This information will allow you to expand your software library to include programs that are more efficient than the bisection method. For example, you may also want to have your own software for the false-position, Newton-Raphson, and secant techniques, which are usually more efficient than bisection.

**TABLE II.2 Specific study objectives for Part II**

1. Understand the graphical interpretation of a root
2. Know the graphical interpretation of the false-position method and why it is usually superior to the bisection method
3. Understand the difference between bracketing and open methods for root location
4. Understand the concepts of convergence and divergence. Use the two-curve graphical method to provide a visual manifestation of the concepts
5. Know why bracketing methods always converge, whereas open methods may sometimes diverge
6. Realize that convergence of open methods is more likely if the initial guess is close to the true root
7. Understand the concepts of linear and quadratic convergence and their implications for the efficiencies of the one-point iteration and Newton-Raphson methods
8. Know the fundamental difference between the false-position and secant methods and how it relates to convergence
9. Understand the problems posed by multiple roots and the modifications available to mitigate them

CHAPTER FOUR

# BRACKETING METHODS

This chapter on roots of equations deals with methods that exploit the fact that a function typically changes sign in the vicinity of a root. These techniques are called *bracketing methods* because two initial guesses for the root are required. As the name implies, these guesses must "bracket," or be on either side of, the root. The particular methods described herein employ different strategies to systematically reduce the width of the bracket and, hence, home in on the correct answer.

As a prelude to these techniques, we will briefly discuss graphical methods for depicting functions and their roots. Beyond their utility for providing rough guesses, graphical techniques are also useful for visualizing the properties of the functions and the behavior of the various numerical methods.

## 4.1 GRAPHICAL METHODS

A simple method for obtaining an estimate of the root of the equation $f(x) = 0$ is to make a plot of the function and observe where it crosses the $x$ axis. This point, which represents the $x$ value for which $f(x) = 0$, provides a rough approximation of the root.

### EXAMPLE 4.1
The Graphical Approach

Problem Statement: Use the graphical approach to obtain an approximate root of the function $f(x) = e^{-x} - x$.

Solution: The following values are computed:

| $x$ | $f(x)$ |
|---|---|
| 0.0 | 1.000 |
| 0.2 | 0.619 |
| 0.4 | 0.270 |
| 0.6 | −0.051 |
| 0.8 | −0.351 |
| 1.0 | −0.632 |

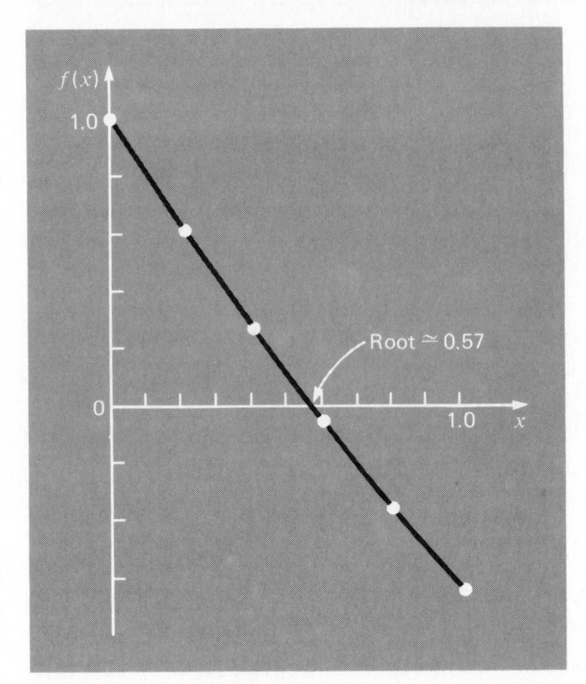


FIGURE 4.1 Illustration of the graphical approach for solving algebraic and transcendental equations. Plot of $f(x) = e^{-x} - x$ versus $x$. The root corresponds to the $x$ value where $f(x) = 0$, that is, the point where the function crosses the $x$ axis. Visual inspection of the plot provides a rough estimate of 0.57.

These points are plotted in Fig. 4.1. The resulting curve crosses the $x$ axis between 0.5 and 0.6. Visual inspection of the plot provides a rough estimate of the root of 0.57. This is close to the true root of 0.56714329. . . , which must be determined by numerical methods. The validity of the visual estimate can be checked by substituting it into the original equation to yield

$$f(0.57) = e^{-0.57} - 0.57 = -0.0045$$

which is close to zero.

Graphical techniques are of limited practical value because they are not precise. However, graphical methods can be utilized to obtain rough estimates of roots. These estimates can be employed as starting guesses for numerical methods discussed in this and the next chapter. For example, the NUMERICOMP computer software that accompanies this text allows you to plot the function over a specified range. This plot can be used to select guesses that bracket the root prior to implementing the numerical method. The plotting option greatly enhances the utility of the software.

Aside from providing rough estimates of the root, graphical interpretations are important tools for understanding the properties of the functions and anticipating the pitfalls of the numerical methods. For example, Fig.

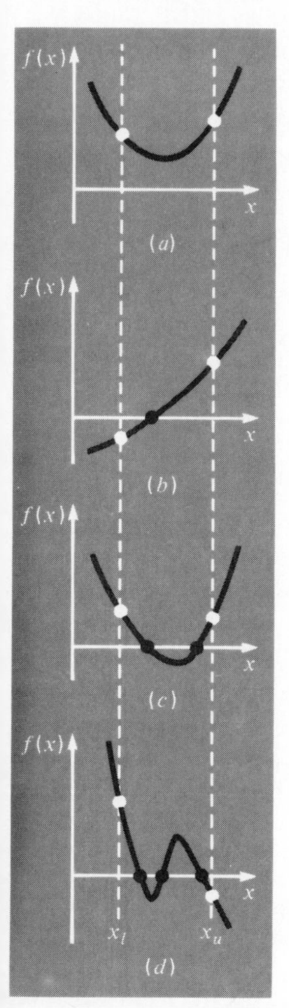


FIGURE 4.2
Illustration of a number of general ways that a root may occur in an interval prescribed by a lower bound $x_l$ and an upper bound $x_u$. Parts (*a*) and (*c*) indicate that if both $f(x_l)$ and $f(x_u)$ have the same sign, either there will be no roots or there will be an even number of roots within the interval. Parts (*b*) and (*d*) indicate that if the function has different signs at the end points, there will be an odd number of roots in the interval.

4.2 shows a number of ways in which roots can occur in an interval prescribed by a lower bound $x_l$ and an upper bound $x_u$. Figure 4.2*b* depicts the case where a single root is bracketed by negative and positive values of $f(x)$. However, Fig. 4.2*d*, where $f(x_l)$ and $f(x_u)$ are also on opposite sides of the $x$ axis, shows three roots occurring within the interval. In general, if $f(x_l)$ and $f(x_u)$ have opposite signs, there are an odd number of roots in the interval. As indicated by Fig. 4.2*a* and *c*, if $f(x_l)$ and $f(x_u)$ have the same sign, there are no roots or an even number of roots between the values.

Although these generalizations are usually true, there are cases where they do not hold. For example, *multiple roots*, that is, functions that are tangential to the $x$ axis (Fig. 4.3*a*), and discontinuous functions (Fig. 4.3*b*) can violate these principles. An example of a function that has a multiple root is the cubic equation $f(x) = (x - 2)(x - 2)(x - 4)$. Notice that $x = 2$ makes two terms in this polynomial equal to zero. Hence, $x = 2$ is called a multiple root. At the end of Chap. 5, we will present techniques that are expressly designed to locate multiple roots.

The existence of cases of the type depicted in Fig. 4.3 makes it difficult to develop general computer algorithms guaranteed to locate all the roots in an interval. However, when used in conjunction with graphical approaches, the methods described in the following sections are extremely useful for solving many roots of equations problems confronted routinely by engineers and applied mathematicians.

## EXAMPLE 4.2
### Use of Computer Graphics to Locate Roots

Problem Statement: Computer graphics can expedite and inform your efforts to locate roots of equations. The present example was developed using the NUMERICOMP software available with the text. However, the insights and issues raised are relevant to computer graphics in general.

The function

$$f(x) = \sin 10x + \cos 3x$$

has several roots over the range from $x = -5$ to $x = 5$. Use computer graphics to gain insight into the behavior of this function.

Solution: As previously illustrated in Example 2.1, NUMERICOMP can be used to generate plots. Figure 4.4*a* is a plot of $f(x)$ from $x = -5$ to $x = 5$. This plot suggests the presence of several roots, including a possible double root at about $x = 4.2$ where $f(x)$ appears to be tangent to the $x$ axis. A more detailed picture of the behavior of $f(x)$ is obtained by changing the plotting range from $x = 3$ to $x = 5$, as shown in Fig. 4.4*b*. Finally, in Fig. 4.4*c*, the vertical scale is narrowed further to $f(x) = -0.15$ to $f(x) = 0.15$ and the horizontal scale narrowed to $x = 4.2$ to $x = 4.3$. This plot shows clearly that a double root does not exist in this region and that in fact there are two distinct roots at about $x = 4.229$ and $x = 4.264$.

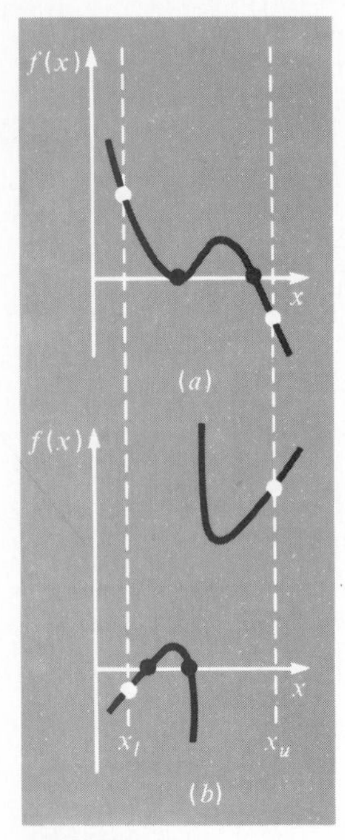


FIGURE 4.3
Illustration of some exceptions to the general cases depicted in Fig. 4.2. (*a*) Multiple root that occurs when the function is tangential to the *x* axis. For this case, although the end points are of opposite signs, there are an even number of roots for the interval. (*b*) Discontinuous function where end points of opposite sign also bracket an even number of roots. Special strategies are required for determining the roots for these cases.

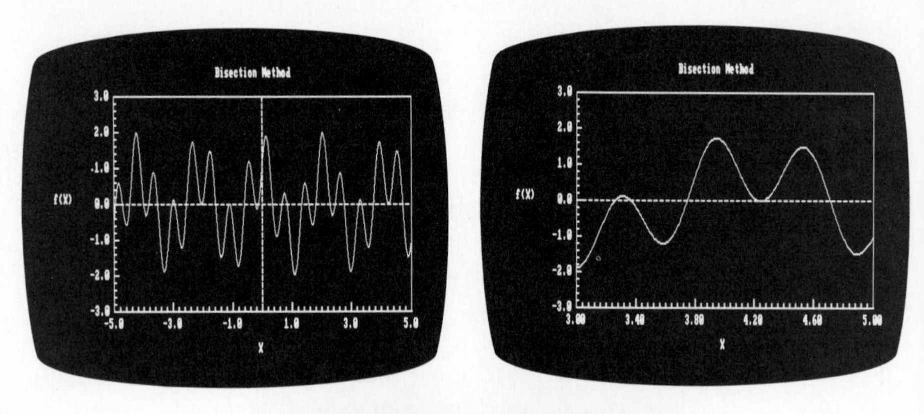


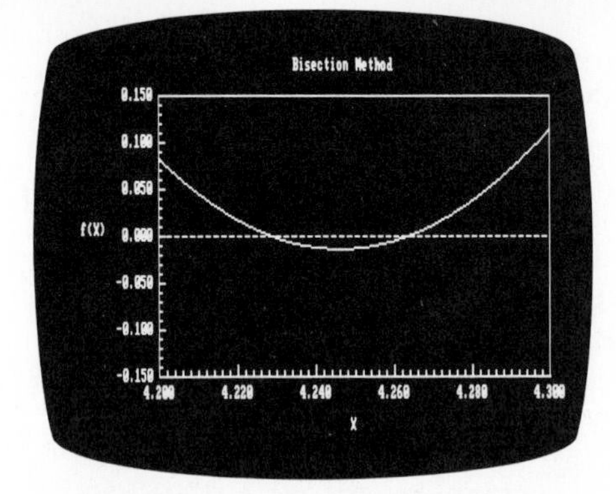


FIGURE 4.4 The progressive enlargement of $f(x) = \sin 10x + \cos 3x$ by the computer. Such interactive graphics permits the analyst to determine that two roots exist between $x = 4.2$ and $x = 4.3$.

Computer graphics will have great utility in your studies of numerical methods. This capability will also find many other applications in your other classes and professional activities as well.

## 4.2 THE BISECTION METHOD

When applying the graphical technique in Example 4.1, you have observed (Fig. 4.1) that $f(x)$ changed sign on opposite sides of the root. In general, if $f(x)$ is real and continuous in the interval from $x_l$ to $x_u$ and $f(x_l)$ and $f(x_u)$ have opposite signs, that is,

$$f(x_l)\,f(x_u) < 0 \tag{4.1}$$

then there is at least one real root between $x_l$ and $x_u$.

*Incremental search methods* capitalize on this observation by locating an interval where the function changes sign. Then the location of the sign change (and consequently, the root) is identified more precisely by dividing the interval into a number of subintervals. Each of these subintervals is searched to locate the sign change. The process is repeated and the root estimate refined by dividing the subintervals into finer increments. We will return to the general topic of incremental searches in Sec. 4.4.

The *bisection method*, which is alternatively called binary chopping, interval halving, or Bolzano's method, is one type of incremental search method in which the interval is always divided in half. If a function changes sign over an interval, the function value at the midpoint is evaluated. The location of the root is then determined as lying at the midpoint of the subinterval within which the sign change occurs. The process is repeated to obtain refined estimates. An algorithm for bisection is listed in Fig. 4.5, and a graphical depiction of the method is provided in Fig. 4.6.

Step 1: Choose lower $x_l$ and upper $x_u$ estimates for the root, so that the function changes sign over the interval. This can be checked by ensuring that $f(x_l)\,f(x_u) < 0$.

Step 2: A first estimate of the root, $x_r$, is determined by

$$x_r = \frac{x_l + x_u}{2}$$

Step 3: Make the following evaluations to determine in which subinterval the root lies:

*a.* If $f(x_l)f(x_r) < 0$, the root lies in the first subinterval. Therefore, set $x_u = x_r$ and continue to step 4.

*b.* If $f(x_l)f(x_r) > 0$, the root lies in the second subinterval. Therefore, set $x_l = x_r$ and continue to step 4.

c. If $f(x_l)f(x_r) = 0$, the root equals $x_r$ and terminate the computation.

Step 4: Calculate a new estimate of the root by

$$x_r = \frac{x_l + x_u}{2}$$

Step 5: Decide if your new estimate is accurate enough to meet your requirements. If "yes," then terminate the computation. If "no," then return to step 3.

FIGURE 4.5 Algorithm for bisection.

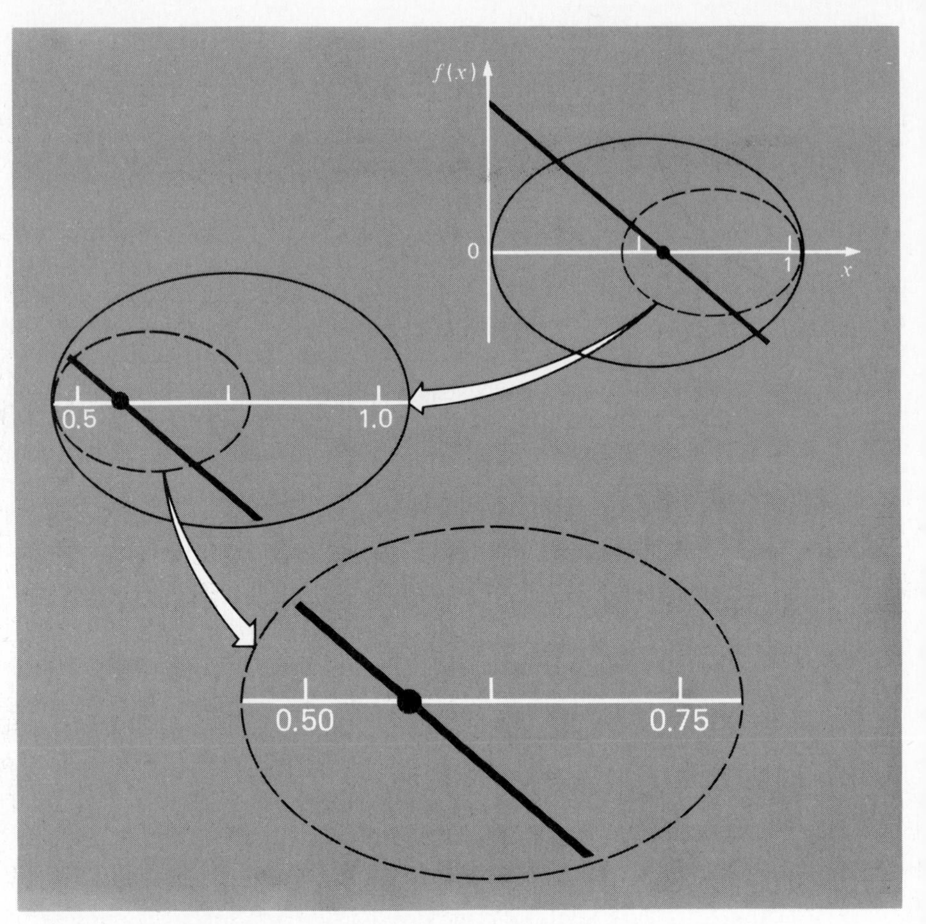


FIGURE 4.6 A graphical depiction of the bisection method. This plot conforms to the first three iterations from Example 4.3.

## EXAMPLE 4.3
### Bisection

Problem Statement: Use bisection to determine the root of $f(x) = e^{-x} - x$.

Solution: From the graph of the function (Fig. 4.1), recall that the root lies between 0 and 1. Therefore, the initial interval can be chosen from $x_l = 0$ to $x_u = 1$. Consequently, the initial estimate of the root lies at the midpoint of this interval:

$$x_r = \frac{0 + 1}{2} = 0.5$$

This estimate represents an error of (the true value is 0.56714329. . .)

$$E_t = 0.56714329 - 0.5 = 0.06714329$$

or, in relative terms,

$$|\epsilon_t| = \left|\frac{0.06714329}{0.56714329}\right| 100\% = 11.8\%$$

where the subscript $t$ designates that the error is with reference to the true value. Now we compute

$$f(0)\,f(0.5) = (1)(0.10653) = 0.10653$$

which is greater than zero, and consequently no sign change occurs between $x_l$ and $x_r$. Therefore, the root is located in the interval between $x = 0.5$ and 1.0. The lower bound is redefined as $x_l = 0.5$, and the root estimate for the second iteration is calculated as

$$x_r = \frac{0.5 + 1.0}{2} = 0.75 \qquad |\epsilon_t| = 32.2\%$$

The process can be repeated to obtain refined estimates. For example, the third iteration is

$$f(0.5)\,f(0.75) = -0.030 < 0$$

Therefore, the root is between 0.5 and 0.75:

$$x_u = 0.75$$

$$x_r = \frac{0.5 + 0.75}{2} = 0.625 \qquad |\epsilon_t| = 10.2\%$$

And the fourth iteration is

$$f(0.5)\,f(0.625) = -0.010 < 0$$

Therefore, the root is between 0.5 and 0.625:

$$x_u = 0.625$$

$$x_r = \frac{0.5 + 0.625}{2} = 0.5625 \qquad |\epsilon_t| = 0.819\%$$

The method can be repeated to achieve more refined estimates. A visual depiction of the first three iterations is found in Fig. 4.6.

In the previous example, you may have noticed that the true error does not decrease with each iteration. However, the interval within which the root is located is halved with each step in the process. As discussed in the next section, the interval width provides an exact estimate of the upper bound of the error for the bisection method.

### 4.2.1 Termination Criteria and Error Estimates

We ended Example 4.3 with the statement that the method could be continued in order to obtain a refined estimate of the root. We must now develop an objective criterion for deciding when to terminate the method.

An initial suggestion might be to end the calculation when the error falls below some prespecified level. For instance, in Example 4.3, the relative error dropped from 11.8 to 4.69 percent during the course of the computation. We might decide that we should terminate when the error drops below, say, 0.1 percent. This strategy is flawed because the error estimates in the example were based on knowledge of the true root of the function. Such would not be the case in an actual situation because there would be no point in using the method if we already knew the root.

Therefore, we require an error estimate that is not contingent on foreknowledge of the root. As developed previously in Sec. 3.3, an approximate relative error $\epsilon_a$ can be calculated as in [recall Eq. (3.5)]

$$|\epsilon_a| = \left| \frac{x_r^{\text{new}} - x_r^{\text{old}}}{x_r^{\text{new}}} \right| 100\% \qquad [4.2]$$

where $x_r^{\text{new}}$ is the root for the present iteration and $x_r^{\text{old}}$ is the root from the previous iteration. The absolute value is used because we are usually concerned with the magnitude of $\epsilon_a$ rather than with its sign. When $|\epsilon_a|$ becomes less than a prespecified stopping criterion $\epsilon_s$, the computation is terminated.

#### EXAMPLE 4.4
Error Estimates for Bisection

Problem Statement: Use Eq. (4.2) to estimate the error for the iterations of Example 4.3.

Solution: The first two estimates of the root for Example 4.3 were 0.5 and 0.75. Substituting these values into Eq. (4.2) yields

$$|\epsilon_a| = \left| \frac{0.75 - 0.5}{0.75} \right| 100\% = 33.3\%$$

Recall that the true error for the root estimate of 0.75 is 32.2 percent. Thus, $\epsilon_a$ is greater than $\epsilon_t$. This behavior is manifested for the other iterations

| Iteration | $x_r$ | $\lvert\epsilon_t\rvert$ % | $\lvert\epsilon_a\rvert$ % |
|---|---|---|---|
| 1 | 0.5 | 11.8 | |
| 2 | 0.75 | 32.2 | 33.3 |
| 3 | 0.625 | 10.2 | 20.0 |
| 4 | 0.5625 | 0.819 | 11.1 |
| 5 | 0.59375 | 4.69 | 5.3 |

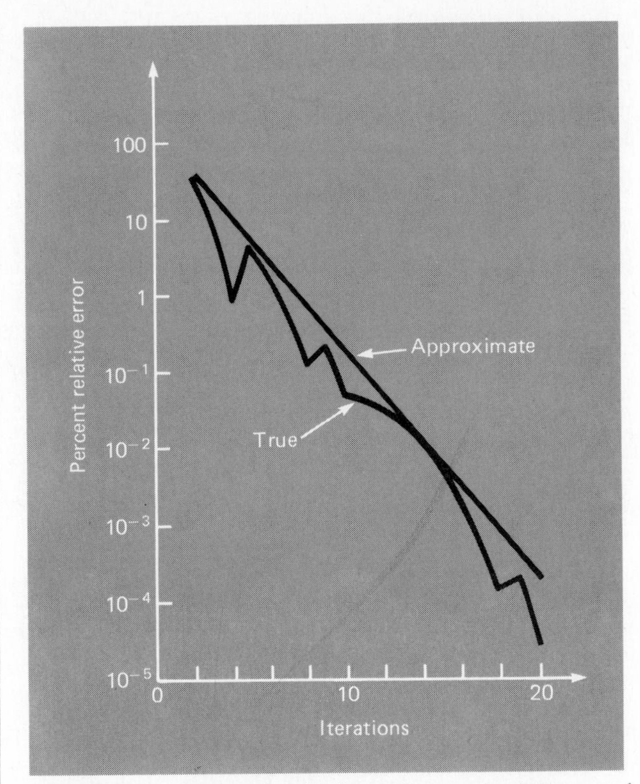


FIGURE 4.7 Errors for the bisection method. True and estimated errors are plotted versus the number of iterations.

These results, along with those for subsequent iterations, are summarized in Fig. 4.7. The "ragged" nature of the true error is due to the fact that, for bisection, the true root can lie anywhere within the bracketing interval. The true and approximate errors are close when the interval happens to be centered on the true root. They are far apart when the true root falls at either end of the interval.

Although the approximate error does not provide an exact estimate of the true error, Fig. 4.7 suggests that $\epsilon_a$ captures the general downward trend of $\epsilon_t$. In addition, the plot exhibits the extremely attractive characteristic that $\epsilon_a$ is always greater than $\epsilon_t$. Thus, when $\epsilon_a$ falls below $\epsilon_s$, the computation could be terminated with confidence that the root is known to be *at least* as accurate as the prespecified acceptable level.

Although it is always dangerous to draw general conclusions from a single example, it can be demonstrated that $\epsilon_a$ will always be greater than $\epsilon_t$ for the bisection method. This is due to the fact that each time an approximate root is located using bisection as $x_r = (x_l + x_u)/2$, we know that the true root lies somewhere within an interval of $(x_u - x_l)/2 = \Delta x/2$. There-

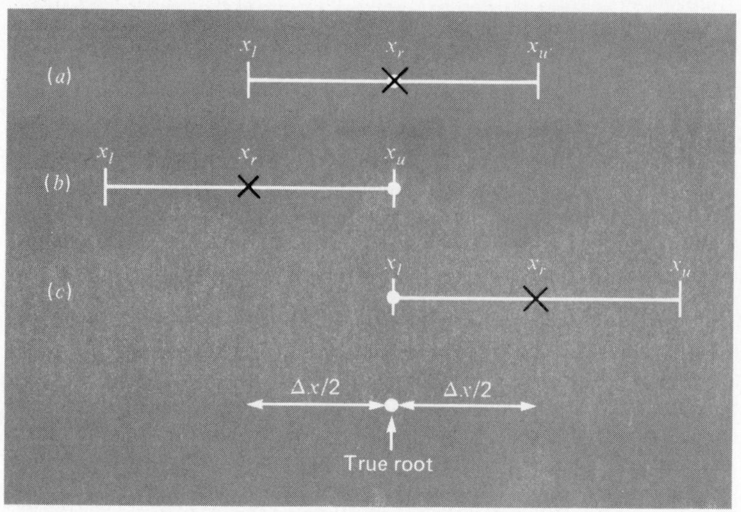


FIGURE 4.8 Three ways in which the interval may bracket the root. In (*a*) the true value lies at the center of the interval, whereas in (*b*) and (*c*) the true value lies near the extreme. Notice that the discrepancy between the true value and the midpoint of the interval never exceeds half the interval length, or $\Delta x/2$.

fore, the root must lie within $\pm\Delta x/2$ of our estimate (Fig. 4.8). For instance, when Example 4.3 was terminated, we could make the definitive statement that

$$x_r = 0.5625 \pm 0.0625$$

Because $\Delta x/2 = x_r^{\text{new}} - x_r^{\text{old}}$ (Fig. 4.9), Eq. (4.2) provides an exact upper bound on the true error. For this bound to be exceeded, the true root

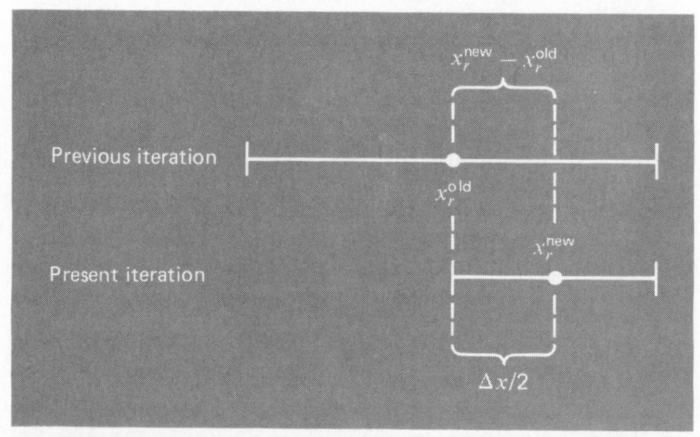


FIGURE 4.9 Graphical depiction of why the error estimate for bisection ($\Delta x/2$) is equivalent to the root estimate for the present iteration ($x_r^{\text{new}}$) minus the root estimate for the previous iteration ($x_r^{\text{old}}$).

would have to fall outside the bracketing interval, which, by definition, could never occur for the bisection method. As illustrated in a subsequent example (Example 4.7), other root-locating techniques do not always behave as nicely. Although bisection is generally slower than other methods, the neatness of its error analysis is certainly a positive aspect that could make it attractive for certain engineering applications.

### 4.2.2 Computer Program for the Bisection Method

The algorithm in Fig. 4.5 can now be expanded into the computer program in Fig. 4.10. The program uses a function (line 100) to make root location and function modification more efficient. In addition, line 200 is included to check whether division by zero occurs during the error evaluation. Such would be the case when the bracketing interval is centered on zero. For this situation, Eq. (4.2) becomes infinite. If this occurs, the program skips over the error evaluation for that iteration.

The program in Fig. 4.10 is not user-friendly; it is strictly designed to come up with the answer. In Prob. 4.13 at the end of this chapter, you will have the task of making this skeletal computer code easier to use and understand. An example of a user-friendly program for finding roots of equations is included in the NUMERICOMP software associated with this text. The following example demonstrates the use of this software for root location. It also provides a good reference for assessing and testing your own software.

**FORTRAN**

```
      F(X)=EXP(-X)-X
      READ(5,1)XL,XU,ES,IM
    1 FORMAT(3F10.0,I5)
      AA=F(XL)*F(XU)
      IF(AA.GE.0)GO TO 310
      XR=(XL+XU)/2
      DO 240 NI=2,IM
      AA=F(XL)*F(XR)
      IF(AA.EQ.0)GO TO 300
      IF(AA.LT.0)XU=XR
      IF(AA.GT.0)XL=XR
      XN=(XL+XU)/2
      IF(XN.EQ.0)GO TO 230
      EA=ABS((XN-XR)/XN)*100
      IF(EA.LT.ES)GO TO 280
  230 XR=XN
  240 CONTINUE
      WRITE(6,2)
    2 FORMAT(' ','ROOT NOT REACHED')
      WRITE(6,3)XR,EA
    3 FORMAT(' ',2F10.3)
      GO TO 310
  280 WRITE(6,4)XN,EA,NI
    4 FORMAT(' ',2F10.3,I5)
      GO TO 310
  300 WRITE(6,5)XR
    5 FORMAT(' ','EXACT ROOT = ',F10.3)
  310 STOP
      END
```

**BASIC**

```
100  DEF  FN F(X) = EXP ( - X) -
     X
110  INPUT XL,XU,ES,IM
120  IF  FN F(XL) * FN F(XU) > =
     0 THEN 310
130 XR = (XL + XU) / 2
140  FOR NI = 2 TO IM
150 AA = FN F(XL) * FN F(XR)
160  IF AA = 0 THEN 300
170  IF AA < 0 THEN XU = XR
180  IF AA > 0 THEN XL = XR
190 XN = (XL + XU) / 2
200  IF XN = 0 THEN 230
210 EA = ABS ((XN - XR) / XN) *
     100
220  IF EA < ES THEN 280
230 XR = XN
240  NEXT NI
250  PRINT "ROOT NOT REACHED"
260  PRINT XR,EA
270  GOTO 310
280  PRINT XN,EA,NI
290  GOTO 310
300  PRINT "EXACT ROOT =";XR
310  END
```

(Function for which root is to be found)

XL, XU = lower and upper guesses
ES = acceptable error %
IM = maximum iterations

(Check to determine whether XL and XU bracket a root)

XR = initial estimate of the root

(Evaluation to determine which subinterval contains the root)

XN = new estimate of the root

EA = error estimate %

(Error check)

FIGURE 4.10 Computer program for the bisection method.

## EXAMPLE 4.5
### Root Location Using the Computer

Problem Statement: A user-friendly computer program to implement the bisection method is contained on the NUMERICOMP software associated with the text.

We can use this software to solve a design problem associated with the falling parachutist example discussed in Chap. 1. As you recall, the velocity of the parachutist is given as the following function of time:

$$v(t) = \frac{gm}{c}[1 - e^{-(c/m)t}] \qquad \text{[E4.5.1]}$$

where $v$ is the velocity in centimeters per second, $g$ is the gravitational constant of 980 cm/s$^2$, $m$ is the mass of the parachutist equal to 68,100 g, and $c$ is the drag coefficient. In Example 1.1 you calculated the velocity of the parachutist as a function of time for given values of $m$, $c$, and $g$. However, suppose you wished to control the motion of the parachutist so that a prespecified velocity is achieved after a given time of free-fall. In this case, you must select an appropriate value of $c$ to satisfy these design requirements because $m$, $g$, $t$, and $v$ are fixed. An inspection of Eq. (E4.5.1) reveals that $c$ cannot be explicitly calculated in terms of the known values. Therefore, it is necessary to employ a numerical method such as bisection. Assume that you would like the parachutist to attain a velocity of 4000 cm/s after a period of 7 s. Thus, you must determine a value of $c$ such that

$$0 = f(c) = \frac{gm}{c}[1 - e^{-(c/m)t}] - v \qquad \text{[E4.5.2]}$$

with $t = 7$ s and $v = 4000$ cm/s.

Solution: In order to implement the BISECTION method, it is required that we obtain an initial interval that brackets the value of $c$ that satisfies Eq. (E4.5.2). It is convenient to select this interval in association with the BISECTION plot option on the disk (option 3). The program prompts you for minimum and maximum values for both $x$ and $f(x)$ and produces the plot shown in Fig. 4.11*a* after you have entered the plot dimensions. It is seen that a root exists between 10,000 g/s and 15,000 g/s.

The BISECTION program prompts you for a limit on the maximum number of iterations, a convergence error $\epsilon_s$, and upper and lower bounds for the root. These inputs, along with a calculated root of 11643.14 g/s, are shown in Fig. 4.11*b*. Note that an estimated value of the root with an error of less than $\epsilon_s$ was obtained in 16 iterations. Furthermore, the computer displays an error check of

$$f(11643.14) = 1.025391 \times 10^{-2}$$

to confirm the results. If the required accuracy was not achieved within the

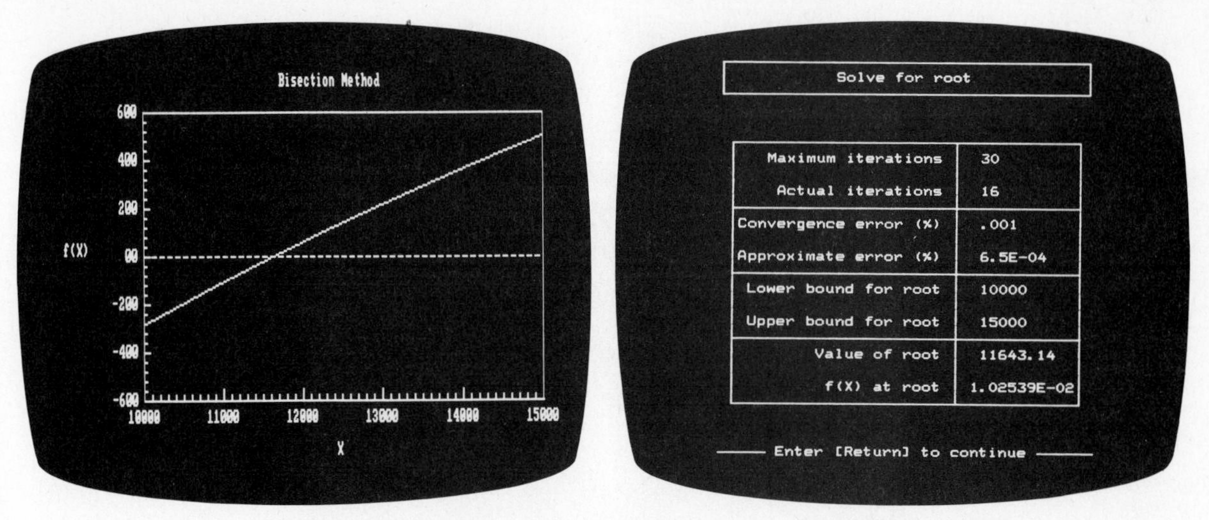


FIGURE 4.11 (*a*) Plot of Eq. (E4.5.2) and (*b*) results using BISECTION to determine the drag coefficient for the falling parachutist.

specified number of iterations, the solution algorithm would have terminated after 30 iterations.

These results are based on a simple algorithm for the BISECTION method with user-friendly input and output routines. The algorithm employed is similar to that shown in Fig. 4.10. You should be able to write your own program for the bisection method. If you have obtained our software, you can use it as a model and use it to check the adequacy of your own program.

## 4.3 THE FALSE-POSITION METHOD

Although bisection is a perfectly valid technique for determining roots, its "brute-force" approach is relatively inefficient. False position is an improved alternative based on a graphical insight.

A shortcoming of the bisection method is that in dividing the interval from $x_l$ to $x_u$ into equal halves, no account is taken of the magnitudes of $f(x_l)$ and $f(x_u)$. For example, if $f(x_l)$ is much closer to zero than $f(x_u)$, it is likely that the root is closer to $x_l$ than to $x_u$ (Fig. 4.12). An alternative method that exploits this graphical insight is to join the points by a straight line. The intersection of this line with the $x$ axis represents an improved estimate of the root. The fact that the replacement of the curve by a straight line gives a "false position" of the root is the origin of the name, *method of false position*, or, in Latin, *regula falsi*. It is also called the linear interpolation method.

With the use of similar triangles (Fig. 4.12), the intersection of the straight line and the $x$ axis can be estimated as

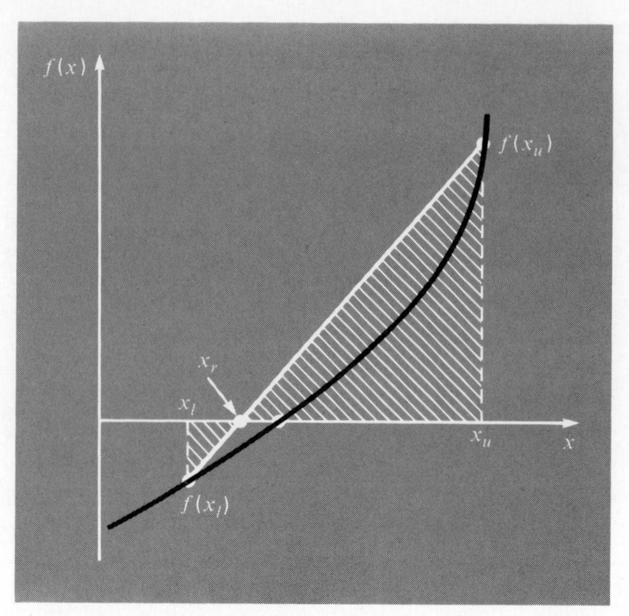


FIGURE 4.12 A graphical depiction of the method of false position. Similar triangles used to derive the formula for the method are shaded.

$$\frac{f(x_l)}{x_r - x_l} = \frac{f(x_u)}{x_r - x_u} \tag{4.3}$$

which can be solved for (see Box 4.1 for details).

BOX 4.1 Derivation of the Method of False Position

Cross-multiply Eq. (4.3) to yield

$$f(x_l)(x_r - x_u) = f(x_u)(x_r - x_l)$$

Collect terms and rearrange:

$$x_r[f(x_l) - f(x_u)] = x_u f(x_l) - x_l f(x_u)$$

Divide by $f(x_l) - f(x_u)$:

$$x_r = \frac{x_u f(x_l) - x_l f(x_u)}{f(x_l) - f(x_u)}$$

This is one form of the method of false position. Note that it allows the computation of the root $x_r$ as a function of the lower and upper guesses $x_l$ and $x_u$. It can be put in an alternative form by expanding it:

$$x_r = \frac{f(x_l)x_u}{f(x_l) - f(x_u)} - \frac{f(x_u)x_l}{f(x_l) - f(x_u)}$$

then adding and subtracting $x_u$ on the right-hand side:

$$x_r = x_u + \frac{f(x_l)x_u}{f(x_l) - f(x_u)} - x_u - \frac{f(x_u)x_l}{f(x_l) - f(x_u)}$$

Collecting terms yields

$$x_r = x_u + \frac{f(x_u)x_u}{f(x_l) - f(x_u)} - \frac{f(x_u)x_l}{f(x_l) - f(x_u)}$$

or

$$x_r = x_u - \frac{f(x_u)(x_l - x_u)}{f(x_l) - f(x_u)}$$

which is the same as Eq. (4.4). We use this form because it is directly comparable with the secant method discussed in Chap. 5.

$$x_r = x_u - \frac{f(x_u)(x_l - x_u)}{f(x_l) - f(x_u)} \qquad [4.4]$$

This is the *false-position formula.* The value of $x_r$ computed with Eq. (4.4) then replaces whichever of the two initial guesses, $x_l$ or $x_u$, yields a function value with the same sign as $f(x_r)$. In this way the values of $x_l$ and $x_u$ always bracket the true root. The process is repeated until the root is estimated adequately. The algorithm is identical to the one for bisection (Fig. 4.6) with the exception that Eq. (4.4) is used for steps 2 and 4. In addition, the same stopping criterion [Eq. (4.2)] is used to terminate the computation.

## EXAMPLE 4.6
### False Position

Problem Statement: Use the false-position method to determine the root of $f(x) = e^{-x} - x$. The correct answer is 0.56714329.

Solution: As in Example 4.3, initiate the computation with guesses of $x_l = 0$ and $x_u = 1$.
First iteration:

$$x_l = 0 \qquad f(x_l) = 1$$

$$x_u = 1 \qquad f(x_u) = -0.63212$$

$$x_r = 1 - \frac{-0.63212(0 - 1)}{1 - (-0.63212)} = 0.6127$$

The true relative error can be estimated by

$$|\epsilon_t| = \left| \frac{0.56714329 - 0.6127}{0.56714329} \right| 100\% = 8.0\%$$

Second iteration:

$$f(x_l)f(x_r) = -0.0708$$

Therefore, the root lies in the first subinterval, and $x_r$ becomes the upper limit for the next iteration, $x_u = 0.6127$.

$$x_l = 0 \qquad f(x_l) = 1$$

$$x_u = 0.6127 \qquad f(x_u) = -0.0708$$

$$x_r = 0.6127 - \frac{-0.0708(0 - 0.6127)}{1 - (-0.0708)} = 0.57219 \qquad |\epsilon_t| = 0.89\%$$

An approximate error estimate can be computed as

$$|\epsilon_a| = \left|\frac{0.57219 - 0.6127}{0.57219}\right| 100\% = 7.08\%$$

Additional iterations can be performed to refine the estimate of the root.

A feeling for the relative efficiency of the bisection and false-position methods can be appreciated by referring to Fig. 4.13, where we have plotted the true percent relative errors for Examples 4.3 and 4.6. Note how the error for false position decreases much faster than for bisection because of the more efficient scheme for root location in the false-position method.

Recall in the bisection method that the interval between $x_l$ and $x_u$ grew smaller during the course of a computation. Therefore, the interval, as defined by $\Delta x/2 = |x_u - x_l|/2$ provided a measure of the error for these approaches. This is not the case for the method of false position because one of the initial guesses may stay fixed throughout the computation as the other guess converges on the root. For instance, in Example 4.4 the lower guess

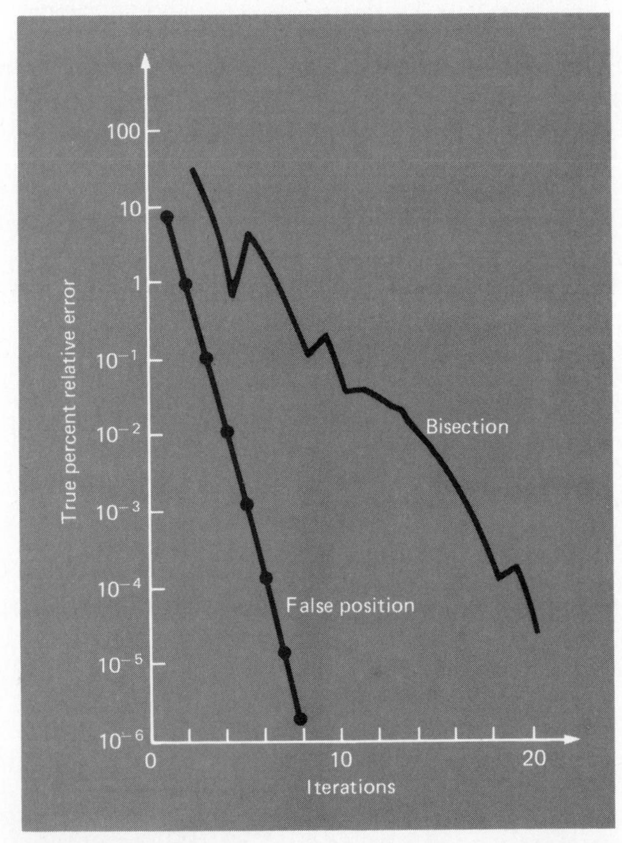


FIGURE 4.13 Comparison of relative errors in the bisection and false-position methods for $f(x) = e^x - x$.

$x_l$ remained at zero, while $x_u$ converged on the root. For such cases, the interval does not shrink but rather approaches a constant value.

Example 4.6 suggests that Eq. (4.2) represents a very conservative error criterion. In fact, Eq. (4.2) actually constitutes an approximation of the discrepancy of the *previous* iteration. This is due to the fact that for a case such as Example 4.6, where the method is converging quickly (for example, the error is being reduced nearly an order of magnitude per iteration), the root for the present iteration $x_r^{new}$ is a much better estimate of the true value than the result of the previous iteration $x_r^{old}$. Thus, the quantity in the numerator of Eq. (4.2) actually represents the discrepancy of the previous iteration. Consequently, we are assured that satisfaction of Eq. (4.2) ensures that the root will be known with greater accuracy than the prescribed tolerance. However, as described in the next section, there are cases where false position converges slowly. For these cases, Eq. (4.2) becomes unreliable, and an alternative stopping criterion must be developed.

### 4.3.1 Pitfalls of the False-Position Method

Although the false-position method would seem to always be the bracketing method of preference, there are cases where it performs poorly. In fact, as in the following example, there are certain cases where bisection yields superior results.

#### EXAMPLE 4.7
A Case Where Bisection Is Preferable to False Position

Problem Statement: Use bisection and false position to locate the root of

$$f(x) = x^{10} - 1$$

between $x = 0$ and 1.3.

Solution: Using bisection, the results can be summarized as

| Iteration | $x_l$ | $x_u$ | $x_r$ | $\lvert\epsilon_t\rvert$ % | $\lvert\epsilon_a\rvert$ % |
|---|---|---|---|---|---|
| 1 | 0 | 1.3 | 0.65 | 35 | |
| 2 | 0.65 | 1.3 | 0.975 | 2.5 | 33.3 |
| 3 | 0.975 | 1.3 | 1.1375 | 13.8 | 14.3 |
| 4 | 0.975 | 1.1375 | 1.05625 | 5.6 | 7.7 |
| 5 | 0.975 | 1.05625 | 1.015625 | 1.6 | 4.0 |

Thus, after five iterations, the true error is reduced to less than 2 percent. For false position, a very different outcome is obtained

| Iteration | $x_l$ | $x_u$ | $x_r$ | $\|\epsilon_t\|$ % | $\|\epsilon_a\|$ % |
|---|---|---|---|---|---|
| 1 | 0 | 1.3 | 0.09430 | 90.6 | |
| 2 | 0.09430 | 1.3 | 0.18176 | 81.8 | 48.1 |
| 3 | 0.18176 | 1.3 | 0.26287 | 73.7 | 30.9 |
| 4 | 0.26287 | 1.3 | 0.33811 | 66.2 | 22.3 |
| 5 | 0.33811 | 1.3 | 0.40788 | 59.2 | 17.1 |

After five iterations, the true error has only been reduced to about 59 percent. In addition, note that $|\epsilon_a| < |\epsilon_t|$. Thus, the approximate error is misleading. Insight into these results can be gained by examining a plot of the function. As in Fig. 4.14, the curve violates the premise upon which false position was based—that is, if $f(x_l)$ is much closer to zero than $f(x_u)$, then the root is closer to $x_l$ than to $x_u$ (recall Fig. 4.12). Because of the shape of the present function, the opposite is true.

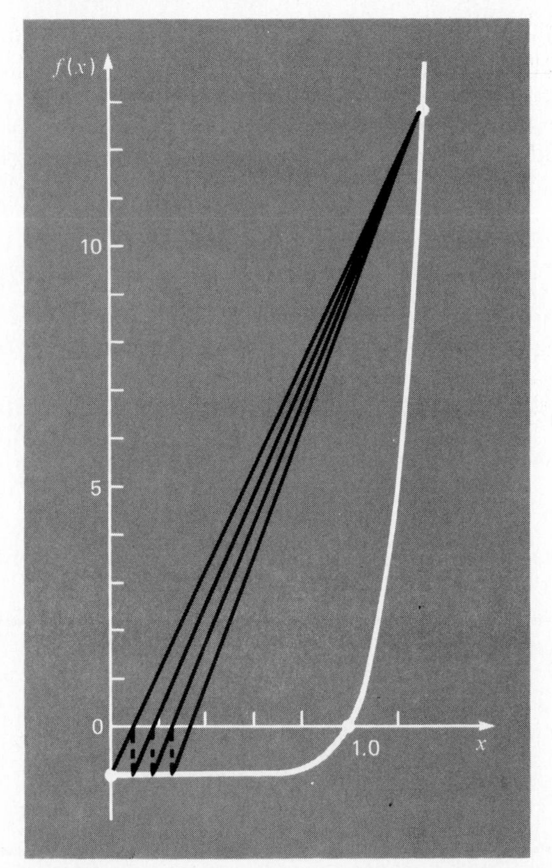


FIGURE 4.14 Plot of $f(x) = x^{10} - 1$, illustrating the slow convergence of the false-position method.

The foregoing example illustrates that blanket generalizations regarding root-location methods are usually not possible. Although a method such as false position is usually superior to bisection, there are invariably special cases that violate the general conclusion. Therefore, in addition to using Eq. (4.2), the results can be checked by substituting the root estimate into the original equation and determining whether the result is close to zero. Such a check should be incorporated into all computer programs for root location.

### 4.3.2 Computer Program for the False-Position Method

A computer program for the false-position method can be developed directly from the bisection code in Fig. 4.10. The only modification is to substitute Eq. (4.4) for lines 130 and 190. In addition, the zero check suggested in the last section should also be incorporated into the code.

## 4.4 INCREMENTAL SEARCHES AND DETERMINING INITIAL GUESSES

Besides checking an individual answer, you must determine whether all possible roots have been located. As mentioned previously, a plot of the function is usually very useful in guiding you in this task. Another option is to incorporate an *incremental search* at the beginning of the computer program. This consists of starting at one end of the region of interest and then making function evaluations at small increments across the region. When the function changes sign, it is assumed that a root falls within the increment. The $x$ values at the beginning and the end of the increment can then serve as the initial guesses for one of the bracketing techniques described in this chapter.

A potential problem with an incremental search is the choice of the increment length. If the length is too small, the search can be very time-consuming. On the other hand, if the length is too great, there is a possibility that closely spaced roots might be missed (Fig. 4.15). The problem is compounded by the possible existence of multiple roots. A partial remedy for such cases is to compute the first derivative of the function $f'(x)$ at the beginning and the end of each interval. If the derivative changes sign, it suggests that a minimum or maximum may have occurred and that the interval should be examined more closely for the existence of a possible root.

Although such modifications or the employment of a very fine increment can alleviate the problem, it should be clear that brute-force methods such as incremental search are not foolproof. You would be wise to supplement such automatic techniques with any other information that provides insight into the location of the roots. Such information can be found in plotting and in understanding the physical problem from which the equation originated.

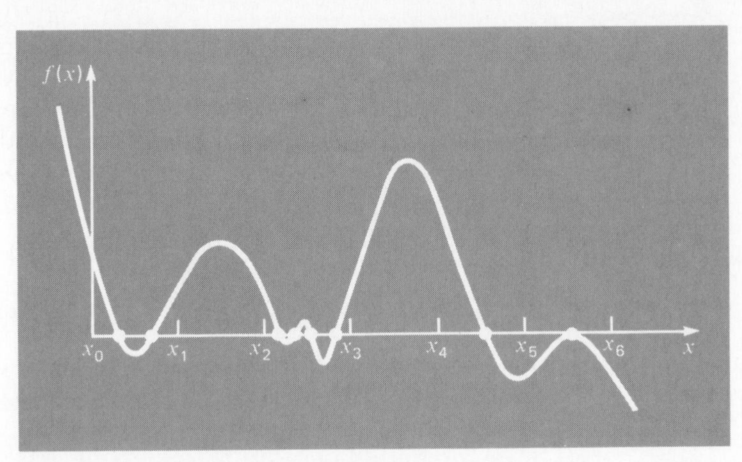


FIGURE 4.15 Cases where roots could be missed because the increment length of the search procedure is too large. Note that the last root is multiple and would be missed regardless of increment length.

# PROBLEMS

## Hand Calculations

**4.1** Determine the real roots of

$$f(x) = -0.874x^2 + 1.75x + 2.627$$

(*a*) Graphically.
(*b*) Using the quadratic formula.
(*c*) Using three iterations of the bisection method to determine the highest root. Employ initial guess of $x_l = 2.9$ and $x_u = 3.1$. Compute the estimated error $\epsilon_a$ and the true error $\epsilon_t$ after each iteration.

**4.2** Determine the real roots of

$$f(x) = -2.1 + 6.21x - 3.9x^2 + 0.667x^3$$

(*a*) Graphically.
(*b*) Using bisection to locate the lowest root. Employ initial guesses of $x_l = 0.4$ and $x_u = 0.6$ and iterate until the estimated error $\epsilon_a$ falls below a level of $\epsilon_s = 4$ percent.

**4.3** Determine the real roots of

$$f(x) = -23.33 + 79.35x - 88.09x^2 + 41.6x^3 - 8.68x^4 + 0.658x^5$$

(*a*) Graphically.
(*b*) Using bisection to determine the highest root to $\epsilon_s = 1\%$. Employ initial guesses of $x_l = 4.5$ and $x_u = 5$.
(*c*) Perform the same computation as in (*b*) but use the false-position method.

**4.4** Determine the real roots of

$$f(x) = 9.36 - 21.963x + 16.2965x^2 - 3.70377x^3$$

(*a*) Graphically.
(*b*) Using the false-position method with a value of $\epsilon_s$ corresponding to three significant figures to determine the lowest root.

**4.5** Locate the first nontrivial root of $\tan x = 1.1x$ where $x$ is in radians. Use a graphical technique and bisection with the initial interval from 0.1 to 0.6. Perform the computation until $\epsilon_a$ is less than $\epsilon_s = 10$ percent. Also perform an error check by substituting your final answer into the original equation.

**4.6** Determine the real root of $\ln x = 0.5$.
(*a*) Graphically.
(*b*) Using three iterations of the bisection method, with initial guesses of $x_l = 1$ and $x_u = 2$.
(*c*) Using three iterations of the false-position method, with the same initial guesses as in (*b*).

**4.7** Determine the real root of

$$f(x) = \frac{1 - 0.6x}{x}$$

(*a*) Analytically.
(*b*) Graphically.
(*c*) Using three iterations of the false-position method and initial guesses of 1.5 and 2.0. Compute the approximate error $\epsilon_a$ and the true error $\epsilon_t$ after each iteration.

**4.8** Find the positive square root of 10 using the false-position method to within $\epsilon_s = 0.5$ percent. Employ initial guesses of $x_l = 3$ and $x_u = 3.2$.

**4.9** Find the smallest positive root of the function ($x$ is in radians)

$$x^2 |\sin x| = 4$$

using the false-position method. To locate the region in which the root lies, first plot this function for values of $x$ between 0 and 4. Perform the computation until $\epsilon_a$ falls below $\epsilon_s = 1$ percent. Check your final answer by substituting it into the initial function.

**4.10** Find the positive real root of

$$f(x) = x^4 - 8.6x^3 - 35.51x^2 + 464x - 998.46$$

using the false-position method. Use a plot to make your initial guesses, and perform the computation to within $\epsilon_s = 0.1$ percent.

**4.11** Determine the real root of

$$f(x) = x^3 - 100$$

(*a*) Analytically.
(*b*) With the false-position method to within $\epsilon_s = 0.1$ percent.

**4.12** The velocity of a falling parachutist is given by

$$v = \frac{gm}{c}[1 - e^{-(c/m)t}]$$

where $g = 980$. For a parachutist of mass $m = 75{,}000$ g, compute the drag coefficient $c$ so that velocity is $v = 3600$ cm/s at $t = 6$ s. Use the false-position method to determine $c$ to a level of $\epsilon_s = 0.1$ percent.

## Computer-Related Problems

**4.13** Reprogram Fig. 4.10 so that it is user-friendly. Among other things:
(*a*) Place documentation statements throughout the program to identify what each section is intended to accomplish.
(*b*) Label the input and output.
(*c*) Add a test to ensure that the root guesses $x_l$ and $x_u$ bracket the root.
(*d*) Add an answer check that substitutes the root estimate into the original function to check whether the final result is close to zero.

**4.14** Test the program you developed in Prob. 4.13 by duplicating the computations from Example 4.3.

**4.15** Use the program you developed in Prob. 4.13 to repeat Probs. 4.1 through 4.6.

**4.16** Repeat Probs. 4.14 and 4.15, except use the NUMERICOMP software available with the text. Use the plotting capabilities of this program to verify your results.

**4.17** Use the NUMERICOMP software to find the real roots of two polynomial functions of your choice. Plot the function over a range you specify to obtain upper and lower bounds on the roots.

**4.18** Repeat Prob. 4.17 except use two transcendental functions of your choice.

**4.19** This problem uses only the graphics capability of the NUMERICOMP software available with the text. The software plots the function over smaller and smaller intervals to increase the number of significant figures to which a root can be estimated. Start with $f(x) = e^{-x} \sin(10x)$. Plot the function with a full-scale range of $x = 0$ to $x = 2.5$. Estimate the root. Plot the function again with $x = 0.5$ to $x = 1.0$. Estimate the root. Finally, plot the function over a range of 0.6 to 0.7. This permits you to estimate the root to two significant figures.

**4.20** Develop a user-friendly program for the false-position method based on Sec. 4.3.2. Test the program by duplicating Example 4.6.

**4.21** Use the program you developed in Prob. 4.20 to duplicate the computation from Example 4.7. Perform a number of runs of 5, 10, 15, and more iterations until the true percent relative error falls below 0.1 percent. Plot the true and approximate percent relative errors versus number of iterations on semilog paper. Interpret your results.

# CHAPTER FIVE

# OPEN METHODS

For the bracketing methods in the previous chapter, the root is located within an interval prescribed by a lower and an upper bound. Repeated application of these methods always results in closer estimates of the true value of the root. Such methods are said to be *convergent* because they move closer to the truth as the computation progresses (Fig. 5.1*a*).

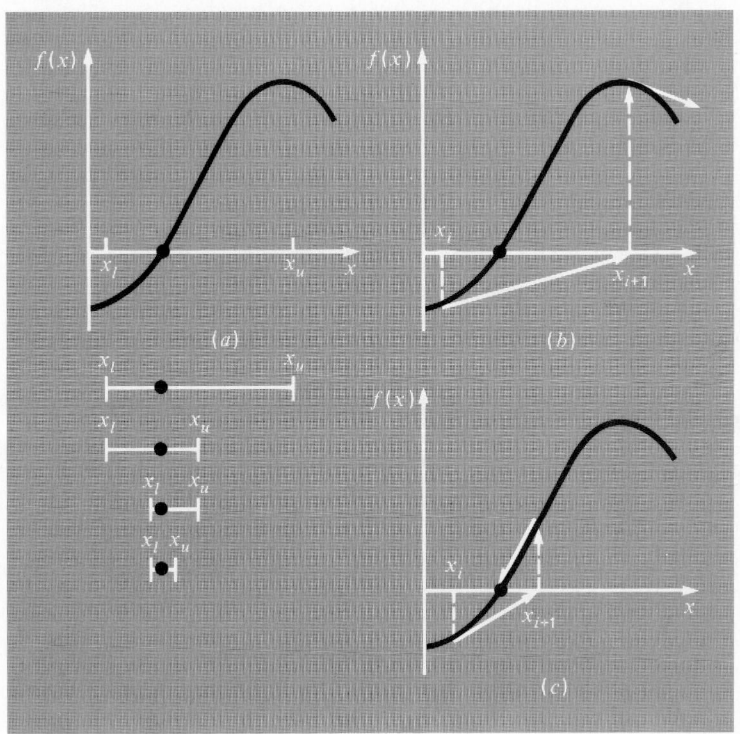


FIGURE 5.1 Graphical depiction of the fundamental difference between the (*a*) bracketing and (*b*) and (*c*) open methods for root location. In (*a*), which is the bisection method, the root is constrained within the interval prescribed by $x_l$ and $x_u$. In contrast, for the open method depicted in (*b*) and (*c*), a formula is used to project from $x_i$ to $x_{i+1}$ in an iterative fashion. Thus, the method can either (*b*) diverge or (*c*) converge rapidly, depending on the value of the initial guess.

In contrast, the *open methods* described in the present chapter are based on formulas that require a single value of $x$ or two values that do not necessarily bracket the root. As such, they sometimes *diverge* or move away from the true root as the computation progresses (Fig. 5.1*b*). However, when the open methods converge (Fig. 5.1*c*) they usually do so much more quickly than the bracketing methods. We will begin our discussion of open techniques with a simple version that is useful for illustrating their general form and also for demonstrating the concept of convergence.

## 5.1 SIMPLE ONE-POINT ITERATION

As mentioned above, open methods employ a formula to predict an estimate of the root. Such a formula can be developed for simple one-point iteration by rearranging the function $f(x) = 0$ so that $x$ is on the left-hand side of the equation:

$$x = g(x) \tag{5.1}$$

This transformation can be accomplished either by algebraic manipulation or by simply adding $x$ to both sides of the original equation. For example,

$$x^2 - 2x + 3 = 0$$

can be simply manipulated to yield

$$x = \frac{x^2 + 3}{2}$$

whereas $\sin x = 0$ would be put into the form of Eq. (5.1) by adding $x$ to both sides to yield

$$x = \sin x + x$$

The utility of Eq. (5.1) is that it provides a formula to predict a value of $x$ as a function of $x$. Thus, given an initial guess at the root $x_i$, Eq. (5.1) can be used to compute a new estimate $x_{i+1}$, as expressed by the iterative formula

$$x_{i+1} = g(x_i) \tag{5.2}$$

As with other iterative formulas in this book, the approximate error for this equation can be determined using the error estimator [Eq. (3.5)]:

$$|\epsilon_a| = \left| \frac{x_{i+1} - x_i}{x_{i+1}} \right| 100\%$$

EXAMPLE 5.1
Simple One-Point Iteration

Problem Statement: Use simple one-point iteration to locate the root of $f(x) = e^{-x} - x$.

Solution: The function can be separated directly and expressed in the form of Eq. (5.2) as $x_{i+1} = e^{-x_i}$. Starting with an initial guess of $x_0 = 0$, this iterative equation can be applied to compute:

| Iteration, $i$ | $x_i$ | $\lvert\epsilon_t\rvert$% | $\lvert\epsilon_a\rvert$% |
|---|---|---|---|
| 0 | 0 | 100 | |
| 1 | 1.000000 | 76.3 | 100.0 |
| 2 | 0.367879 | 35.1 | 171.8 |
| 3 | 0.692201 | 22.1 | 46.9 |
| 4 | 0.500473 | 11.8 | 38.3 |
| 5 | 0.606244 | 6.89 | 17.4 |
| 6 | 0.545396 | 3.83 | 11.2 |
| 7 | 0.579612 | 2.20 | 5.90 |
| 8 | 0.560115 | 1.24 | 3.48 |
| 9 | 0.571143 | 0.705 | 1.93 |
| 10 | 0.564879 | 0.399 | 1.11 |

Thus, each iteration brings the estimate closer to the true value of the root, or 0.56714329.

### 5.1.1 Convergence

Notice that the exact relative error for each iteration of Example 5.1 is roughly proportional (by a factor of about 0.5 to 0.6) to the error from the previous iteration. This property, called *linear convergence*, is characteristic of one-point iteration. Box 5.1 presents a theoretical basis for this observation.

#### BOX 5.1 Convergence of One-Point Iteration

From studying Fig. 5.3, it should be clear that one-point iteration converges if, in the region of interest $|g'(x)| < 1$. In other words, convergence occurs if the magnitude of the slope of $g(x)$ is less than the slope of the line $f(x) = x$. This observation can be demonstrated theoretically. Recall that the approximate equation is

$$x_{i+1} = g(x_i)$$

Suppose that the true solution is

$$x_r = g(x_r)$$

Subtracting these equations yields

$$x_r - x_{i+1} = g(x_r) - g(x_i) \qquad \text{[B5.1.1]}$$

In calculus, there is a principle called the *mean-value theorem* (Sec. 3.5.2). It states that if a function $g(x)$ and its first derivative are continuous over an interval $a \le x \le b$, then there exists at least one value of $x = \xi$ within the interval such that

$$g'(\xi) = \frac{g(b) - g(a)}{b - a} \qquad \text{[B5.1.2]}$$

The right-hand side of this equation is the slope of the line joining $g(a)$ and $g(b)$. Thus, the mean-value theorem states that there is at least one point between $a$ and $b$ that has a slope, designated by $g'(\xi)$, which is parallel to the line joining $g(a)$ and $g(b)$ (Fig. 3.5).

Now, if we let $a = x_i$ and $b = x_r$, the right-hand side of Eq. (B5.1.1) can be expressed as

$$g(x_r) - g(x_i) = (x_r - x_i)\, g'(\xi)$$

where $\xi$ is somewhere between $x_i$ and $x_r$. This result can then be substituted into Eq. (B5.1.1) to yield

$$x_r - x_{i+1} = (x_r - x_i)\, g'(\xi) \qquad \text{[B5.1.3]}$$

If the true error for iteration $i$ is defined as

$$E_{t,i} = x_r - x_i$$

then Eq. (B5.1.3) becomes

$$E_{t,i+1} = g'(\xi)\, E_{t,i}$$

Consequently, if $|g'(\xi)| < 1$, the errors decrease with each iteration. For $|g'(\xi)| > 1$ the errors grow. Notice also that if the derivative is positive, the errors will be positive, and hence, the iterative solution will be monotonic (Fig. 5.3*a* and *c*). If the derivative is negative, the errors will oscillate (Fig. 5.3*b* and *d*).

An offshoot of the analysis is that it also demonstrates that when the method converges, the error is roughly proportional to and less than the error of the previous step. For this reason, simple one-point iteration is said to be *linearly convergent.*

Aside from the "rate" of convergence, we must comment at this point about the "possibility" of convergence. The concepts of convergence and divergence can be depicted graphically. Recall that in Sec. 4.1, we graphed

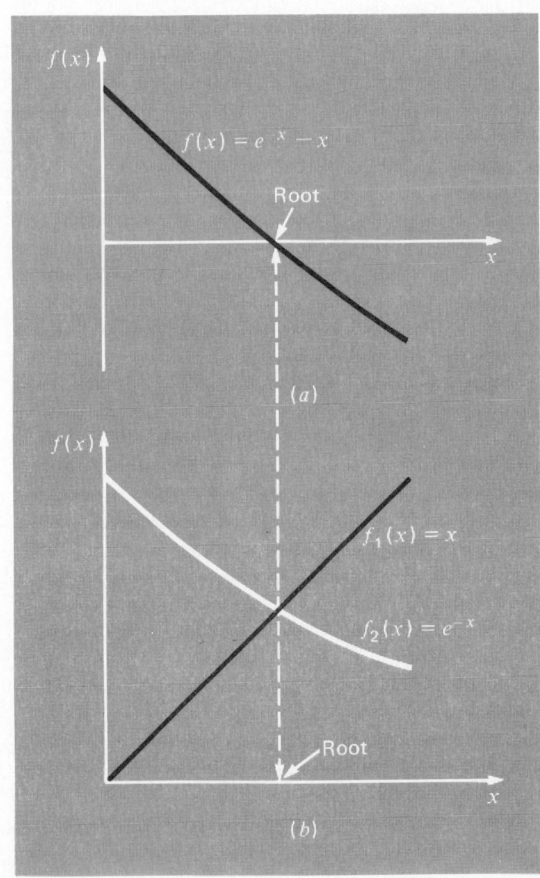


FIGURE 5.2 Two alternative graphical methods for determining the root of $f(x) = e^{-x} - x$. (*a*) Root at the point where it crosses the $x$ axis; (*b*) root at the intersection of the component functions.

a function in order to visualize its structure and behavior (Example 4.1). This function is replotted in Fig. 5.2*a*. An alternative graphical approach is to separate the equation $f(x) = 0$ into two component parts, as in

$$f_1(x) = f_2(x)$$

Then the two equations

$$y_1 = f_1(x) \qquad [5.3]$$

and

$$y_2 = f_2(x) \qquad [5.4]$$

can be plotted separately (Fig. 5.2*b*). The $x$ values corresponding to the intersections of these functions represent the roots of $f(x) = 0$.

## EXAMPLE 5.2
### The Two-Curve Graphical Method

Problem Statement: Separate the equation $e^{-x} - x = 0$ into two parts and determine its root graphically.

Solution: Reformulate the equation as $y_1 = x$ and $y_2 = e^{-x}$. The following values can be computed:

| $x$ | $y_1$ | $y_2$ |
|---|---|---|
| 0.0 | 0.0 | 1.000 |
| 0.2 | 0.2 | 0.819 |
| 0.4 | 0.4 | 0.670 |
| 0.6 | 0.6 | 0.549 |
| 0.8 | 0.8 | 0.449 |
| 1.0 | 1.0 | 0.368 |

These points are plotted on Fig. 5.2*b*. The intersection of the two curves indicates a root estimate of $x = 0.57$, which corresponds to the point where the single curve in Fig. 5.2*a* crosses the $x$ axis.

The two-curve method can now be used to illustrate the convergence and divergence of one-point iteration. First, Eq. (5.1) can be reexpressed as a pair of equations: $y_1 = x$ and $y_2 = g(x)$. These two equations can then be plotted separately. As was the case with Eqs. (5.3) and (5.4), the roots of $f(x) = 0$ correspond to the abscissa value at the intersection of the two curves. The function $y_1 = x$ and four different shapes for $y_2 = g(x)$ are plotted in Fig. 5.3.

For the first case (Fig. 5.3*a*), the initial guess of $x_0$ is used to determine the corresponding point on the $y_2$ curve $[x_0, g(x_0)]$. The point $[x_1, x_1]$ is

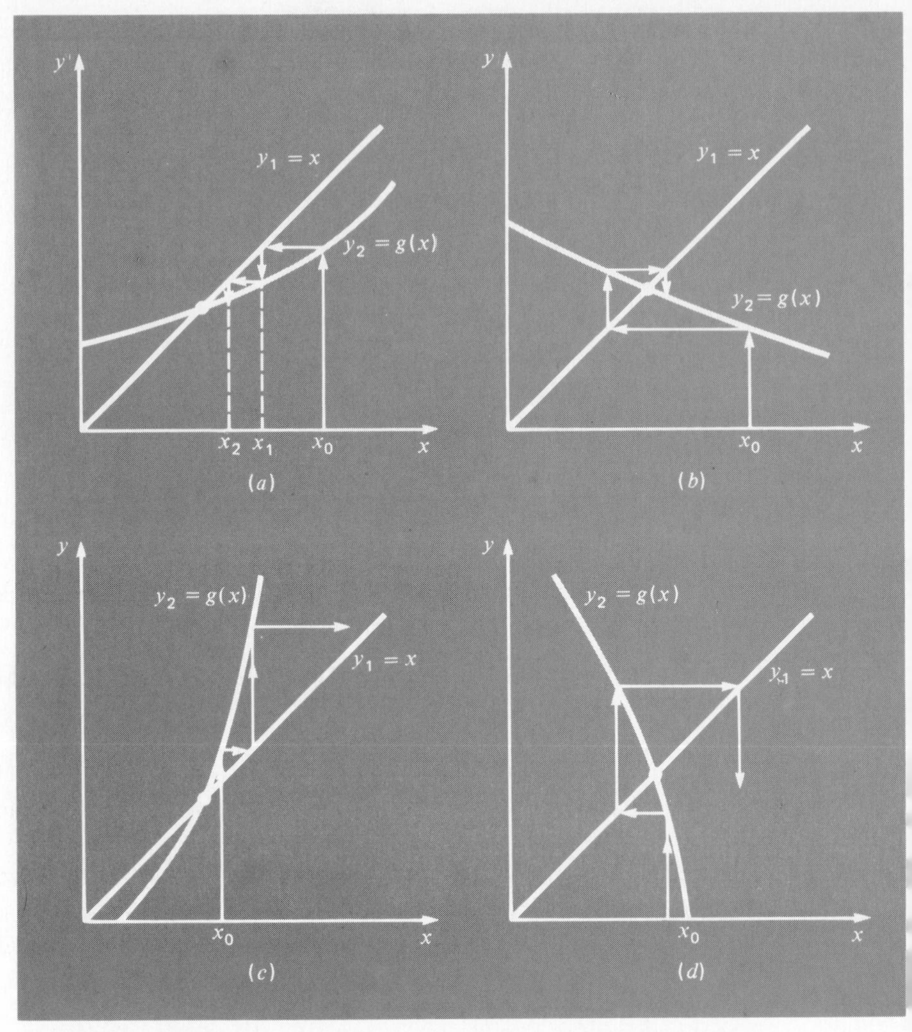


FIGURE 5.3 Graphical depiction of (*a*) and (*b*) convergence and (*c*) and (*d*) divergence of simple one-point iteration. Graphs (*a*) and (*c*) are called monotone patterns, whereas (*b*) and (*d*) are called oscillating or spiral patterns. Note that convergence occurs when $|g'(x)| < 1$.

located by moving left horizontally to the $y_1$ curve. These movements are equivalent to the first iteration in the one-step method:

$$x_1 = g(x_0)$$

Thus, in both the equation and in the plot, a starting value of $x_0$ is used to obtain an estimate of $x_1$. The next iteration consists of moving to $[x_1, g(x_1)]$ and then to $[x_2, x_2]$. This iteration is equivalent to the equation

$$x_2 = g(x_1)$$

The solution in Fig. 5.3*a* is *convergent* because the estimates of *x* move closer

FORTRAN

```
      F(X)=EXP(-X)
      READ(5,1)XR,ES,IM
    1 FORMAT(2F10.0,I5)
      DO 180 NI=1,IM
      XN=F(XR)
      IF(XN.EQ.0)GO TO 170
      EA=ABS((XN-XR)/XN)*100
      IF(EA.LE.ES)GO TO 210
  170 XR=XN
  180 CONTINUE
      WRITE(6,2)
    2 FORMAT(' ','ROOT NOT REACHED')
      NI=NI-1
  210 WRITE(6,3)XN,EA,NI
    3 FORMAT(' ',2F10.3,I5)
      STOP
      END
```

BASIC

```
100  DEF  FN F(X) =  EXP ( - X)          (Function to compute root estimate)
110  INPUT XR,ES,IM                      XR = initial guess
120  FOR NI = 1 TO IM                    ES = acceptable error %
130 XN =  FN F(XR)                       IM = maximum iterations
140  IF XN = 0 THEN 170                  XN = root estimate
150 EA =  ABS ((XN - XR) / XN) *
      100                                EA = error estimate %
160  IF EA <  = ES THEN 210              (Error check)
170 XR = XN
180  NEXT NI
190  PRINT "ROOT NOT REACHED"
200 NI = NI - 1
210  PRINT XN,EA,NI
220  END
```

FIGURE 5.4 Computer program for one-point iteration. Note that this general algorithm is similar to those for other open methods.

to the root with each iteration. The same is true for Fig. 5.3*b*. However, this is not the case for Figs. 5.3*c* and *d*, where the iterations diverge from the root. Notice that convergence occurs only when the absolute value of the slope of $y_2 = g(x)$ is less than the slope of $y_1 = x$, that is, when $|g'(x)| < 1$. Box 5.1 provides a theoretical derivation of this result.

### 5.1.2 Computer Program for One-Point Iteration

The computer algorithm for one-point iteration is extremely simple. It consists of a loop to iteratively compute new estimates along with a logical statement to check whether the termination criterion has been met. Figure 5.4 presents the FORTRAN and BASIC codes for the algorithm. Other open methods can be programmed in a similar way, the major modification being to change the iterative formula (statement 130).

## 5.2 THE NEWTON-RAPHSON METHOD

Perhaps the most widely used of all root-locating formulas is the Newton-Raphson equation (Fig. 5.5). If the initial guess at the root is $x_i$, a tangent can be extended from the point $[x_i, f(x_i)]$. The point where this tangent crosses the $x$ axis usually represents an improved estimate of the root.

The Newton-Raphson method can be derived on the basis of this geometrical interpretation (an alternative method based on the Taylor series is described in Box 5.2). As in Fig. 5.5, the first derivative at $x_i$ is equivalent to the slope:

$$f'(x_i) = \frac{f(x_i) - 0}{x_i - x_{i+1}} \qquad [5.5]$$

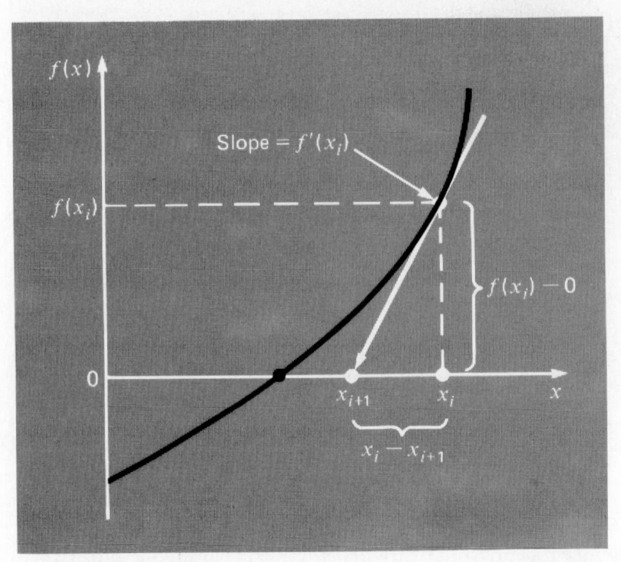


FIGURE 5.5 Graphical depiction of the Newton-Raphson method. A tangent to the function at $x_i$ [that is, $f'(x_i)$] is extrapolated down to the $x$ axis to provide an estimate of the root at $x_{i+1}$.

which can be rearranged to yield

$$x_{i+1} = x_i - \frac{f(x_i)}{f'(x_i)} \tag{5.6}$$

which is called the *Newton-Raphson formula.*

## EXAMPLE 5.3
### Newton-Raphson Method

Problem Statement: Use the Newton-Raphson method to estimate the root of $e^{-x} - x$ employing an initial guess of $x_0 = 0$.

Solution: The first derivative of the function can be evaluated as

$$f'(x) = -e^{-x} - 1$$

which can be substituted along with the original function into Eq. (5.6) to give

$$x_{i+1} = x_i - \frac{e^{-x_i} - x_i}{-e^{-x_i} - 1}$$

Starting with an initial guess of $x_0 = 0$, this iterative equation can be applied to compute:

| Iteration, $i$ | $x_i$ | $\lvert\epsilon_t\rvert$% |
|---|---|---|
| 0 | 0 | 100 |
| 1 | 0.500000000 | 11.8 |
| 2 | 0.566311003 | 0.147 |
| 3 | 0.567143165 | 0.0000220 |
| 4 | 0.567143290 | $<10^{-8}$ |

Thus, the approach rapidly converges on the true root. Notice that the relative error at each iteration decreases much faster than it does in simple one-point iteration (compare with Example 5.1).

### 5.2.1 Termination Criteria and Error Estimates

As with other root-location methods, Eq. (3.5) can be used as a termination criterion. In addition, however, the Taylor series derivation of the method (Box 5.2) provides theoretical insight regarding the rate of convergence as expressed by $E_{i+1} = 0(E_i^2)$. Thus the error should be roughly proportional to the square of the previous error. In other words, the number of significant figures of accuracy approximately doubles with each iteration. This behavior is examined in the following example.

#### BOX 5.2 Derivation and Error Analysis of the Newton-Raphson Method from a Taylor Series Expansion

Aside from the geometric derivation [Eqs. (5.5) and (5.6)], the Newton-Raphson method may also be developed from the Taylor series expansion. This alternative derivation is useful in that it provides insight into the rate of convergence of the method.

Recall from Chap. 3 that the Taylor series expansion can be represented as

$$f(x_{i+1}) = f(x_i) + f'(x_i)(x_{i+1} - x_i) + \frac{f''(\xi)}{2}(x_{i+1} - x_i)^2 \quad \text{[B5.2.1]}$$

where $\xi$ lies somewhere in the interval from $x_i$ to $x_{i+1}$. An approximate version is obtainable by truncating the series after the first derivative term:

$$f(x_{i+1}) \simeq f(x_i) + f'(x_i)(x_{i+1} - x_i)$$

At the intersection with the $x$ axis, $f(x_{i+1})$ would be equal to zero, or

$$0 \simeq f(x_i) + f'(x_i)(x_{i+1} - x_i) \quad \text{[B5.2.2]}$$

which can be solved for

$$x_{i+1} = x_i - \frac{f(x_i)}{f'(x_i)}$$

which is identical to Eq. (5.6). Thus, we have derived the Newton-Raphson formula using a Taylor series.

Aside from the derivation, the Taylor series can also be used to estimate the error of the formula. This can be done by realizing that if the complete Taylor series were employed, an exact result would be obtained. For this situation $x_{i+1} = x_r$, where $x_r$ is the true value of the root. Substituting this value along with $f(x_r) = 0$ into Eq. (B5.2.1) yields

$$0 = f(x_i) + f'(x_i)(x_r - x_i) + \frac{f''(\xi)}{2}(x_r - x_i)^2 \quad \text{[B5.2.3]}$$

Equation (B5.2.2) can be subtracted from Eq. (B5.2.3) to give

$$0 = f'(x_i)(x_r - x_{i+1}) + \frac{f''(\xi)}{2}(x_r - x_i)^2 \quad [B5.2.4]$$

Now, realize that the error is equal to the discrepancy between $x_{i+1}$ and the true value $x_r$, as in

$$E_{t,i+1} = x_r - x_{i+1}$$

and Eq. (B5.2.4) can be expressed as

$$0 = f'(x_i) E_{t,i+1} + \frac{f''(\xi)}{2} E_{t,i}^2 \quad [B5.2.5]$$

If we assume convergence, both $x_i$ and $\xi$ should eventually be approximated by the root $x_r$, and Eq. (B5.2.5) can be rearranged to yield

$$E_{t,i+1} \simeq \frac{-f''(x_r)}{2f'(x_r)} E_{t,i}^2 \quad [B5.2.6]$$

According to Eq. (B5.2.6), the error is roughly proportional to the square of the previous error. This means that the number of correct decimal places approximately doubles with each iteration. Such behavior is referred to as *quadratic convergence*. Example 5.4 manifests this property.

## EXAMPLE 5.4
### Error Analysis of Newton-Raphson Method

Problem Statement: As deduced in Box 5.2, the Newton-Raphson method is quadratically convergent. That is, the error is roughly proportional to the square of the previous error, as in

$$E_{t,i+1} \simeq -\frac{f''(x_r)}{2f'(x_r)} E_{t,i}^2 \quad [E5.4.1]$$

Examine this formula and see if it applies to the results of Example 5.3.

Solution: The first derivative of $f(x) = e^{-x} - x$ is

$$f'(x) = -e^{-x} - 1$$

which can be evaluated at $x_r = 0.56714329$ as

$$f'(0.56714329) = -1.56714329$$

The second derivative is

$$f''(x) = e^{-x}$$

which can be evaluated as

$$f''(0.56714329) = 0.56714329$$

These results can be substituted into Eq. (E5.4.1) to yield

$$E_{t,i+1} \simeq -\frac{0.56714329}{2(-1.56714329)} E_{t,i}^2$$

or

$$E_{t,i+1} \simeq 0.18095\, E_{t,i}^2$$

From Example 5.3, the initial error was $E_{t,0} = 0.56714329$, which can be substituted into the error equation to predict

$$E_{t,1} \simeq 0.18095(0.56714329)^2 = 0.0582$$

which is close to the true error of = 0.06714329. For the next iteration,

$$E_{t,2} \simeq 0.18095(0.06714329)^2 = 0.0008158$$

which also compares favorably with the true error of 0.0008323. For the third iteration,

$$E_{t,3} \simeq 0.18095(0.0008323)^2 = 0.000000125$$

which is exactly the error obtained in Example 5.3. The error estimate improves in this manner because as we come closer to the root, $x_i$ and $\xi$ are better approximated by $x_r$ [recall our assumption in going from Eq. (B5.2.5) to Eq. (B5.2.6) in Box 5.2]. Finally,

$$E_{t,4} \simeq 0.18095(0.000000125)^2 = 2.83 \times 10^{-15}$$

Thus, this example illustrates that the error of the Newton-Raphson method for this case is, in fact, roughly proportional (by a factor of 0.18095) to the square of the error of the previous iteration.

### 5.2.2 Pitfalls of the Newton-Raphson Method

Although the Newton-Raphson method is usually very efficient, there are situations where it performs poorly. A special case—multiple roots—will be addressed at the end of this chapter. However, even when dealing with simple roots, difficulties sometimes arise, as in the following example.

EXAMPLE 5.5
Example of a Slowly Converging Function Using the Newton-Raphson Method

Problem Statement: Determine the positive root of $f(x) = x^{10} - 1$ using the Newton-Raphson method and an initial guess of $x = 0.5$.

Solution: The Newton-Raphson formula for this case is

$$x_{i+1} = x_i - \frac{x_i^{10} - 1}{10x_i^9}$$

which can be used to compute

| **Iteration** | $x_i$ |
|---|---|
| 0 | 0.5 |
| 1 | 51.65 |
| 2 | 46.485 |
| 3 | 41.8365 |
| 4 | 37.65285 |
| 5 | 33.887565 |

Thus, after the first poor prediction, the technique is converging on the true root of 1, but at a very slow rate.

Aside from slow convergence due to the nature of the function, other difficulties can arise, as illustrated in Fig. 5.6. For example, Fig. 5.6*a* depicts

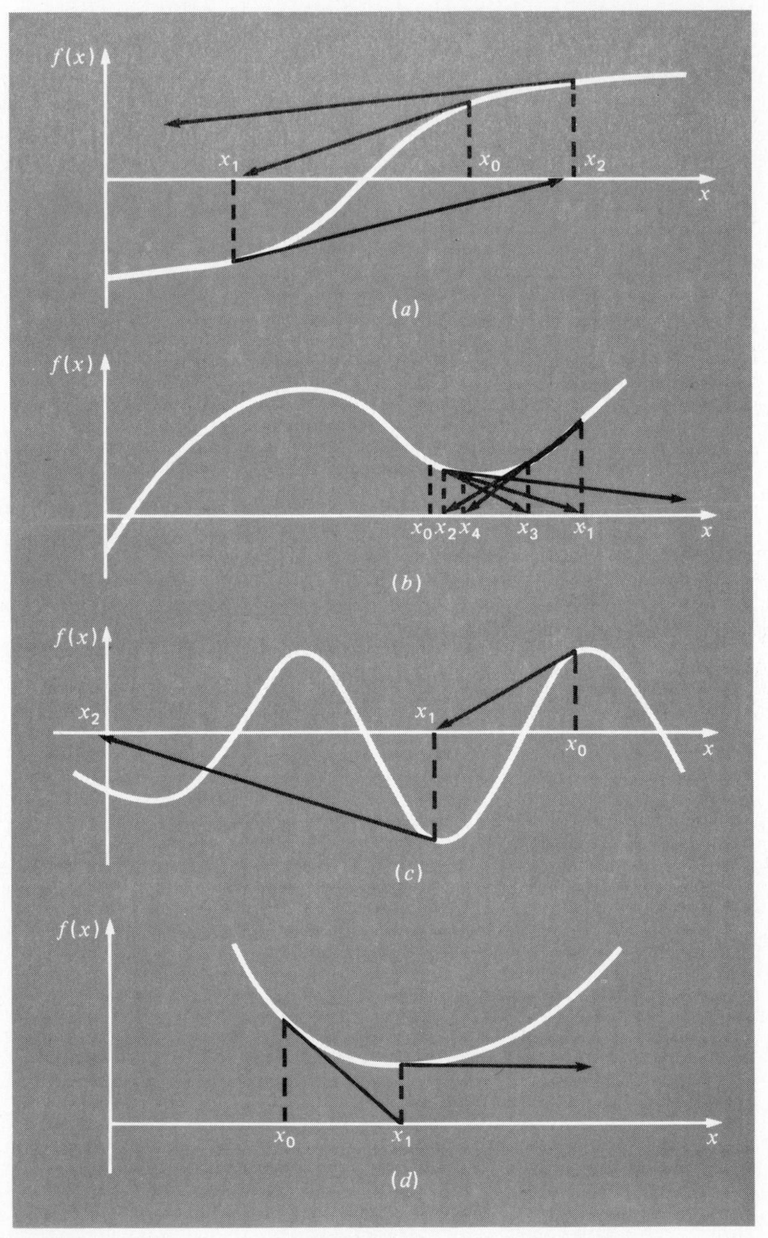


FIGURE 5.6 Four cases where the Newton-Raphson method exhibits poor convergence.

the case where an inflection point—that is, $f''(x) = 0$—occurs in the vicinity of a root. Notice that iterations beginning at $x_0$ progressively diverge from the root. Figure 5.6*b* illustrates the tendency of the Newton-Raphson technique to oscillate around a local maximum or minimum. Such oscillations may persist, or, as in Fig. 5.6*b*, a near-zero slope is reached whereupon the solution is sent far from the area of interest. Figure 5.6*c* shows how an initial guess that is close to one root can jump to a location several roots away. This tendency to move away from the area of interest is due to the fact that near-zero slopes are encountered. Obviously, a zero slope $[f'(x) = 0]$ is a real disaster as it causes division by zero in the Newton-Raphson formula [Eq. (5.6)]. Graphically (Fig. 5.6*d*), it means that the solution shoots off horizontally and *never* hits the $x$ axis.

The only remedy for these situations is to have an initial guess that is close to the root. This knowledge is, of course, predicated on knowledge of the physical problem setting or devices such as graphs that provide insight into the behavior of the solution. It also suggests that good computer software should be designed to recognize slow convergence or divergence. The next section addresses some of these issues.

### 5.2.3 Computer Program for the Newton-Raphson Method

A computer program for the Newton-Raphson method is readily obtained by substituting Eq. (5.6) for line 130 in Fig. 5.4. Note, however, that the program must also be modified to compute the first derivative. This can be simply accomplished by the inclusion of a user-defined function.

Additionally, in light of the foregoing discussion of potential problems of the Newton-Raphson method, the program would be improved by incorporating a number of additional features:

1. If possible, a plotting routine should be included in the program.
2. At the end of the computation, the final root estimate should always be substituted into the original function to compute whether the result is close to zero. This check partially guards against those cases where slow or oscillating convergence may lead to a small value of $\epsilon_a$, while the solution is still far from a root.
3. The program should always include an upper limit on the number of iterations to guard against oscillating, slowly convergent, or divergent solutions that could persist interminably.

## 5.3 THE SECANT METHOD

A potential problem in implementing the Newton-Raphson method is the evaluation of the derivative. Although this is not inconvenient for polynomials

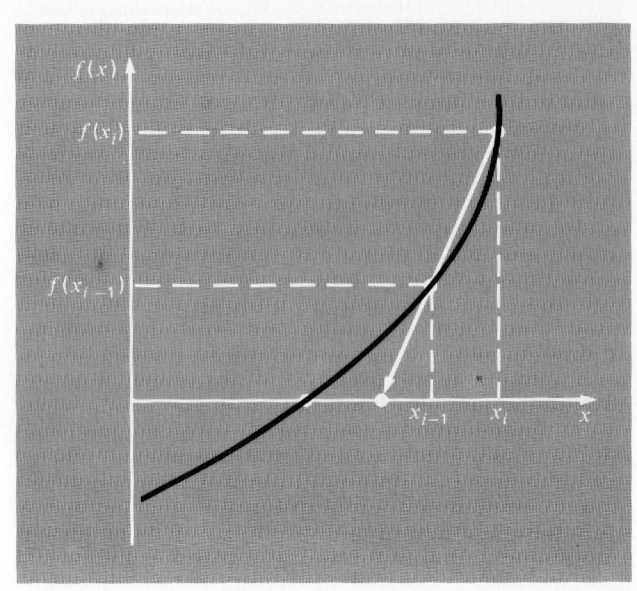


FIGURE 5.7 Graphical depiction of the secant method. This technique is similar to the Newton-Raphson technique (Fig. 5.5) in the sense that an estimate of the root is predicted by extrapolating a tangent of the function to the $x$ axis. However, the secant method uses a difference rather than a derivative to estimate the slope.

and many other functions, there are certain functions whose derivatives may be extremely difficult to evaluate. For these cases, the derivative can be approximated by a finite divided difference, as in (Fig. 5.7)

$$f'(x_i) \simeq \frac{f(x_{i-1}) - f(x_i)}{x_{i-1} - x_i}$$

This approximation can be substituted into Eq. (5.6) to yield the following iterative equation:

$$x_{i+1} = x_i - \frac{f(x_i)(x_{i-1} - x_i)}{f(x_{i-1}) - f(x_i)} \qquad [5.7]$$

Equation (5.7) is the formula for the *secant method.* Notice that the approach requires two initial estimates of $x$. However, because $f(x)$ is not required to change signs between the estimates, it is not classified as a bracketing method.

## EXAMPLE 5.6
The Secant Method

Problem Statement: Use the secant method to estimate the root of $f(x) = e^{-x} - x$. Start with initial estimates of $x_{-1} = 0$ and $x_0 = 1.0$.

Solution: Recall that true root is 0.56714329 . . .
First iteration:

$$x_{-1} = 0 \qquad f(x_{-1}) = 1.00000$$
$$x_0 = 1 \qquad f(x_0) = -0.63212$$
$$x_1 = 1 - \frac{-0.63212(0-1)}{1-(-0.63212)} = 0.61270 \qquad |\epsilon_t| = 8.0\%$$

Second iteration:

$$x_0 = 1 \qquad f(x_0) = -0.63212$$
$$x_1 = 0.61270 \qquad f(x_1) = -0.07081$$

(Note that both estimates are now on the same side of the root.)

$$x_2 = 0.61270 - \frac{-0.07081\,(1-0.61270)}{-0.63212-(-0.07081)} = 0.56384$$
$$|\epsilon_t| = 0.58\%$$

Third iteration:

$$x_1 = 0.61270 \qquad f(x_1) = -0.07081$$
$$x_2 = 0.56384 \qquad f(x_2) = 0.00518$$

$$x_3 = 0.56384 - \frac{0.00518\,(0.61270-0.56384)}{-0.07081-(0.00518)} = 0.56717$$
$$|\epsilon_t| = 0.0048\%$$

### 5.3.1 The Difference Between the Secant and False-Position Methods

Note the similarity between the secant method and the false-position method. For example, Eqs. (5.7) and (4.4) are identical on a term-by-term basis. Both use two initial estimates to compute an approximation of the slope of the function that is used to project to the $x$ axis for a new estimate of the root. However, a critical difference between the methods relates to how one of the initial values is replaced by the new estimate. Recall that, in the false-position method, the latest estimate of the root replaces whichever of the original values yielded a function value with the same sign as $f(x_r)$. Consequently, the two estimates always bracket the root. Therefore, for all practical purposes, the method always converges because the root is kept within the bracket. In contrast, the secant method replaces the values in strict sequence, with the new value $x_{i+1}$ replacing $x_i$ and $x_i$ replacing $x_{i-1}$. As a result, the two values can sometimes lie on the same side of the root. For certain cases, this can lead to divergence.

## EXAMPLE 5.7
### Comparison of Convergence of the Secant and False-Position Techniques

Problem Statement: Use the false-position and the secant methods to estimate the root of $f(x) = \ln x$. Start the computation with values of $x_l = x_{i-1} = 0.5$ and $x_u = x_i = 5.0$.

Solution: For the false-position method the use of Eq. (4.4) and the bracketing criterion for replacing estimates results in the following iterations:

| Iteration | $x_l$ | $x_u$ | $x_r$ |
|---|---|---|---|
| 1 | 0.5 | 5.0 | 1.8546 |
| 2 | 0.5 | 1.8546 | 1.2163 |
| 3 | 0.5 | 1.2163 | 1.0585 |

As can be seen (Figs. 5.8*a* and *c*), the estimates are converging on the true root = 1.

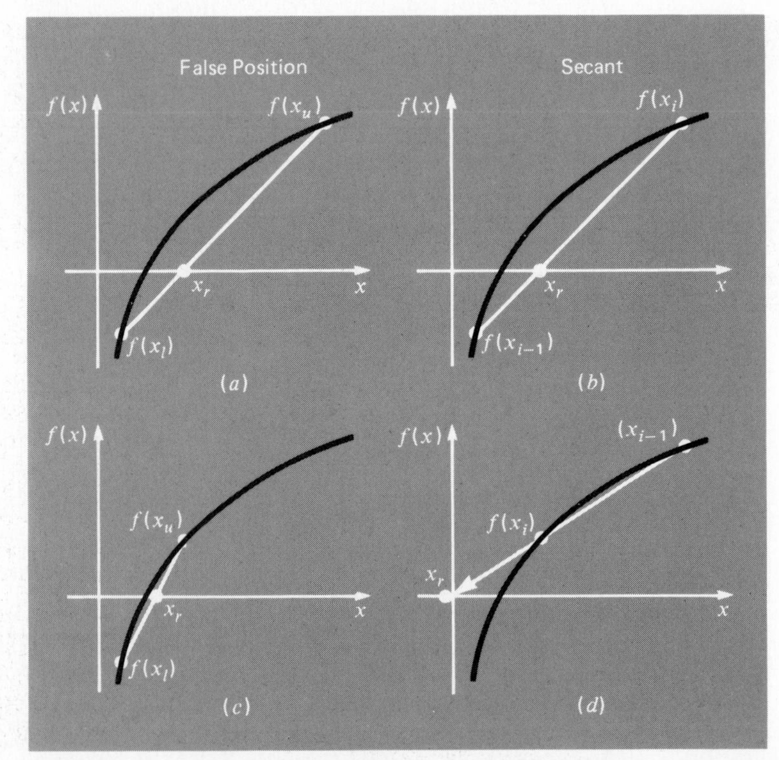


FIGURE 5.8 Comparison of the false-position and the secant methods. The first iterations (*a*) and (*b*) for both techniques are identical. However, for the second iterations (*c*) and (*d*), the points used differ. As a consequence, the secant method can diverge, as indicated in (*d*).

For the secant method, using Eq. (5.7) and the sequential criterion for replacing estimates results in:

| Iteration | $x_{i-1}$ | $x_i$ | $x_{i+1}$ |
|---|---|---|---|
| 1 | 0.5 | 5.0 | 1.8546 |
| 2 | 5.0 | 1.8546 | −0.10438 |

As in Fig. 5.8*d*, the approach is divergent.

Although the secant method may be divergent, when it converges it usually does so at a quicker rate than the false-position method does. For instance, Fig. 5.9, which is based on Examples 4.3, 4.6, 5.3, and 5.6, demonstrates the superiority of the secant method in this regard. The inferiority of the false-position method is due to the fact that one end stays fixed in order to maintain the bracketing of the root. This property, which is an advantage in that it prevents divergence, is a shortcoming with regard to the

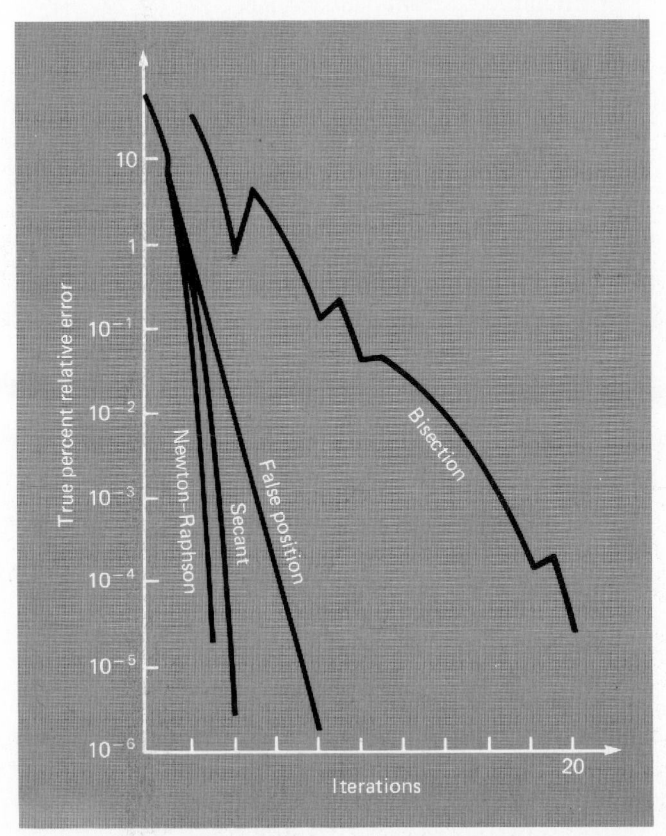


FIGURE 5.9 Comparison of the true percent relative errors $\epsilon_t$ for the methods to determine the roots of $f(x) = e^{-x} - x$.

rate of convergence; it makes the finite-difference estimate a less accurate approximation of the derivative.

### 5.3.2 Computer Program for the Secant Method

As with the other open methods, a computer program for the secant method is simply obtained by modifying line 110 so that two initial guesses are input and by substituting Eq. (5.7) for line 130 in Fig. 5.4. In addition, the options suggested in Sec. 5.2.3 for the Newton-Raphson method can also be applied to good advantage for the secant program.

## 5.4 MULTIPLE ROOTS

A *multiple root* corresponds to a point where a function is tangential to the $x$ axis. For example, a *double root* results from

$$f(x) = (x - 3)(x - 1)(x - 1) \qquad [5.8]$$

or, multiplying terms,

$$f(x) = x^3 - 5x^2 + 7x - 3 \qquad [5.9]$$

The equation has a double root because one value of $x$ makes two terms in Eq. (5.8) equal to zero. Graphically, this corresponds to the curve touching the $x$ axis tangentially at the double root. Examine Fig. 5.10*a* at $x = 1$. Notice that the function touches the axis but does not cross it at the root.

A *triple root* corresponds to the case where one $x$ value makes three terms in an equation equal to zero, as in

$$f(x) = (x - 3)(x - 1)(x - 1)(x - 1)$$

or, multiplying terms,

$$f(x) = x^4 - 6x^3 + 12x^2 - 10x + 3$$

Notice that the graphical depiction (Fig. 5.10*b*) again indicates that the function is tangential to the axis at the root but that for this case the axis is crossed. In general, odd multiple roots cross the axis, whereas even ones do not. For example, the quadruple root in Fig. 5.10*c* does not cross the axis.

Multiple roots pose a number of difficulties for many of the numerical methods described in Part II:

1. The fact that the function does not change sign at an even multiple root precludes the use of the reliable bracketing methods that were discussed in Chap. 4. Thus, of the methods covered in this book, you are limited to the open methods that may diverge.
2. Another possible problem is related to the fact that not only $f(x)$ but also $f'(x)$ goes to zero at the root. This poses problems for both the Newton-

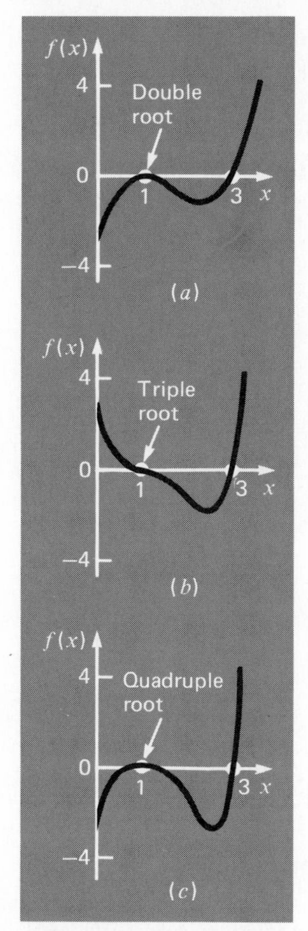


FIGURE 5.10
Examples of multiple roots that are tangential to the $x$ axis. Notice that the function does not cross the axis on either side of even multiple roots ($a$) and ($c$), whereas it crosses the axis for odd cases ($b$).

Raphson and secant methods, which both contain the derivative (or its estimate) in the denominator of their respective formulas. This could result in division by zero when the solution converges very close to the root. A simple way to circumvent these problems is based on the fact that it can be demonstrated theoretically (Ralston and Rabinowitz, 1978) that $f(x)$ will always reach zero before $f'(x)$. Therefore, if a zero check for $f(x)$ is incorporated into the computer program, the computation can be terminated before $f'(x)$ reaches zero.

**3.** It can be demonstrated that the Newton-Raphson and secant methods are linearly, rather than quadratically, convergent for multiple roots (Ralston and Rabinowitz, 1978). Modifications have been proposed to alleviate this problem. Ralston and Rabinowitz (1978) have indicated that a slight change in the formulation returns it to quadratic convergence, as in

$$x_{i+1} = x_i - m\frac{f(x_i)}{f'(x_i)}$$

where $m$ is the *multiplicity* of the root (that is, $m = 2$ for a double root, $m = 3$ for a triple root, etc.). Of course, this may be an unsatisfactory alternative because it hinges on foreknowledge of the multiplicity of the root.

Another alternative, also suggested by Ralston and Rabinowitz (1978), is to define a new function $u(x)$, that is, the ratio of the function to its derivative, as in

$$u(x) = \frac{f(x)}{f'(x)} \qquad [5.10]$$

It can be shown that this function has roots at all the same locations as the original function. Therefore, Eq. (5.10) can be substituted into Eq. (5.6) in order to develop an alternative form of the Newton-Raphson method:

$$x_{i+1} = x_i - \frac{u(x_i)}{u'(x_i)} \qquad [5.11]$$

Equation (5.10) can be differentiated to give

$$u'(x) = \frac{f'(x)\,f'(x) - f(x)\,f''(x)}{[f'(x)]^2} \qquad [5.12]$$

Equations (5.10) and (5.12) can be substituted into Eq. (5.11) and the result simplified to yield

$$x_{i+1} = x_i - \frac{f(x_i)\,f'(x_i)}{[f'(x_i)]^2 - f(x_i)\,f''(x_i)} \qquad [5.13]$$

## EXAMPLE 5.8

### Modified Newton-Raphson Method for Multiple Roots

Problem Statement: Use both the standard and modified Newton-Raphson methods to evaluate the multiple root of Eq. (5.9), with an initial guess of $x_0 = 0$.

Solution: The first derivative of Eq. (5.9) is $f'(x) = 3x^2 - 10x + 7$, and, therefore, the standard Newton-Raphson method for this problem is [Eq. (5.6)]

$$x_{i+1} = x_i - \frac{x_i^3 - 5x_i^2 + 7x_i - 3}{3x_i^2 - 10x_i + 7}$$

which can be solved iteratively for

| $i$ | $x_i$ | $\|\epsilon_t\|$% |
|---|---|---|
| 0 | 0 | 100 |
| 1 | 0.428571429 | 57 |
| 2 | 0.685714286 | 31 |
| 3 | 0.832865400 | 17 |
| 4 | 0.913328983 | 8.7 |
| 5 | 0.955783293 | 4.4 |
| 6 | 0.977655101 | 2.2 |

As anticipated, the method is linearly convergent toward the true value of 1.0.

For the modified method, the second derivative is $f''(x) = 6x - 10$, and the iterative relationship is [Eq. (5.13)]

$$x_{i+1} = x_i - \frac{(x_i^3 - 5x_i^2 + 7x_i - 3)\,(3x_i^2 - 10x_i + 7)}{(3x_i^2 - 10x_i + 7)^2 - (x_i^3 - 5x_i^2 + 7x_i - 3)\,(6x_i - 10)}$$

which can be solved for

| $i$ | $x_i$ | $\|\epsilon_t\|$% |
|---|---|---|
| 0 | 0 | 100 |
| 1 | 1.105263158 | 11 |
| 2 | 1.003081664 | 0.31 |
| 3 | 1.000002382 | 0.00024 |

Thus, the modified formula is quadratically convergent. We can also use both methods to search for the single root at $x = 3$. Using an initial guess of $x_0 = 4$ gives the following results:

| $i$ | Standard, $\lvert\epsilon_t\rvert$ | Modified, $\lvert\epsilon_t\rvert$ |
|---|---|---|
| 0 | 4 (33%) | 4 (33%) |
| 1 | 3.4 (13%) | 2.636363637 (12%) |
| 2 | 3.1 (3.3%) | 2.820224720 (6.0%) |
| 3 | 3.008695652 (0.29%) | 2.961728211 (1.3%) |
| 4 | 3.000074641 ($2.5 \times 10^{-3}$%) | 2.998478719 (0.051%) |
| 5 | 3.000000006 ($2 \times 10^{-7}$%) | 2.999997682 ($7.7 \times 10^{-5}$%) |

Thus, both methods converge quickly, with the standard method being somewhat more efficient.

The above example illustrates the trade-offs involved in opting for the modified Newton-Raphson method. Although it is much preferable for multiple roots, it is somewhat less efficient and requires much more computational effort than the standard method for simple roots. It should be noted that a modified version of the secant method suited for multiple roots can also be developed by substituting Eq. (5.10) into Eq. (5.7). The resulting formula is (Ralston and Rabinowitz, 1978)

$$x_{i+1} = x_i - \frac{u(x_i)\,(x_{i-1} - x_i)}{u(x_{i-1}) - u(x_i)}$$

# PROBLEMS

## Hand Calculations

**5.1** Use the Newton-Raphson method to determine the highest root of

$$f(x) = -0.875x^2 + 1.75x + 2.625$$

Employ an initial guess of $x_i = 3.1$. Perform the computation until $\epsilon_a$ is less than $\epsilon_s = 0.01$ percent. Also perform an error check of your final answer.

**5.2** Determine the real roots of

$$f(x) = -2.1 + 6.21x - 3.9x^2 + 0.667x^3$$

(*a*) Graphically.
(*b*) Using the Newton-Raphson method to within $\epsilon_s = 0.01$ percent.

**5.3** Employ the Newton-Raphson method to determine the real roots for

$$f(x) = -23.33 + 79.35x - 88.09x^2 + 41.6x^3 - 8.68x^4 + 0.658x^5$$

using initial guesses of (*a*) $x_i = 3.5$; (*b*) $x_i = 4.0$; and (*c*) $x_i = 4.5$. Discuss and use graphical methods to explain any peculiarities in your results.

**5.4** Determine the lowest real root of

$$f(x) = 9.36 - 21.963x + 16.2965x^2 - 3.70377x^3$$

(*a*) Graphically.
(*b*) Using the secant method, to a value of $\epsilon_s$ corresponding to three significant figures.

**5.5** Locate the first positive root of

$$f(x) = 0.5x - \sin x$$

where $x$ is in radians. Use the graphical method and then use three iterations of the Newton-Raphson method with an initial guess of $x_i = 2.0$, to locate the root. Repeat the computation but with an initial guess of $x_i = 1.0$. Use the graphical method to explain your results.

**5.6** Find the positive real root of

$$f(x) = x^4 - 8.6x^3 - 35.51x^2 + 464x - 998.46$$

using the secant method. Employ initial guesses of $x_{i-1} = 7$ and $x_i = 9$ and perform four iterations. Compute $\epsilon_a$ and interpret your results.

**5.7** Perform the same computation as in Prob. 5.6 but use the Newton-Raphson method, with an initial guess of $x_i = 7$.

**5.8** Find the positive square root of 10 using three iterations of
(*a*) The Newton-Raphson method, with an initial guess of $x_i = 3$.
(*b*) The secant method, with an initial guess of $x_{i-1} = 3$ and $x_i = 3.2$.

**5.9** Determine the real root of

$$f(x) = \frac{1 - 0.6x}{x}$$

using three iterations of the secant method and initial guesses of $x_{i-1} = 1.5$ and $x_i = 2.0$. Compute the approximate error $\epsilon_a$ after the second and the third iterations.

**5.10** Determine the real root of

$$f(x) = x^3 - 100$$

with the secant method, to within $\epsilon_s = 0.1$ percent.

**5.11** Determine the highest real root of

$$x^3 - 6x^2 + 11x - 6$$

(*a*) Graphically.
(*b*) Using the bisection method (two iterations, $x_l = 2.5$ and $x_u = 3.6$).
(*c*) Using the false-position method (two iterations, $x_l = 2.5$ and $x_u = 3.6$).
(*d*) Using the Newton-Raphson method (two iterations, $x_i = 3.6$).
(*e*) Using the secant method (two iterations, $x_{i-1} = 2.5$ and $x_i = 3.6$).

**5.12** Use the Newton-Raphson method to determine all the roots of $f(x) = x^2 + 5.78x - 11.4504$ to within an $\epsilon_s = 0.001$ percent.

**5.13** Determine the lowest real root of

$$f(x) = 9.36 - 21.963x + 16.2965x^2 - 3.70377x^3$$

(*a*) Graphically
(*b*) Using the bisection method (two iterations, $x_l = 0.5$ and $x_u = 1.1$).
(*c*) Using the false-position method (two iterations, $x_l = 0.5$ and $x_u = 1.1$).
(*d*) Using the Newton-Raphson method (two iterations, $x_i = 0.5$).
(*e*) Using the secant method (two iterations, $x_{i-1} = 0.5$ and $x_i = 1.1$).

**5.14** Determine the smallest positive real root of

$$f(x) = 4x^4 - 24.8x^3 + 57.04x^2 - 56.76x + 20.57$$

(*a*) Graphically.
(*b*) Using the most efficient available method. Employ initial guesses of $x_l = x_{i-1} = 0.5$ and $x_u = x_i = 1.5$ and perform the computation to within $\epsilon_s = 15$ percent.

**5.15** Determine the roots of

$$f(x) = x^3 - 3.2x^2 - 1.92x + 9.216$$

(*a*) Graphically.
(*b*) Using the most efficient available method to within $\epsilon_s = 0.1$ percent.

**5.16** Repeat Prob. 4.12, but use the Newton-Raphson method.

**5.17** Repeat Prob. 4.12, but use the secant method.

## Computer-Related Problems

**5.18** Develop a user-friendly program for the Newton-Raphson method based on Fig. 5.4 and Sec. 5.2.3. Test it by duplicating the computation from Example 5.3.

**5.19** Use the program you developed in Prob. 5.18 to duplicate the computation from Example 5.5. Determine the root using the initial guess of $x_i = 0.5$. Perform a number of runs of 5, 10, 15, and more iterations until the true percent relative error falls below 0.1 percent. Plot true and approximate percent relative errors versus number of iterations on semilog paper. Interpret your results.

**5.20** Use the program developed in Prob. 5.18 to solve Probs. 5.1 through 5.5. For all cases, perform the computations to within a tolerance of $\epsilon_s = 0.001$ percent.

**5.21** Develop a user-friendly program for the secant method based on Fig. 5.4 and Sec. 5.3.2. Test it by duplicating the computation from Example 5.6.

**5.22** Use the program you developed in Prob. 5.21 to solve Probs. 5.6, 5.9, and 5.10. For all cases, perform the computations to within a tolerance of $\epsilon_s = 0.001$ percent.

# CHAPTER SIX

# CASE STUDIES: ROOTS OF EQUATIONS

The purpose of this chapter is to use the numerical procedures discussed in Chaps. 4 and 5 to solve actual engineering problems. Numerical techniques are important for practical applications because engineers frequently encounter problems that cannot be approached using analytical techniques. For example, simple mathematical models that can be solved analytically may not be applicable when real problems are involved. Thus, more complicated models must be employed. For these cases, it is appropriate to implement a numerical solution on a personal computer. In other situations, engineering design problems may require solutions for implicit variables in complicated equations (recall Sec. II.1.2 and Example 4.5).

The following case studies are typical of those that are routinely encountered during upper-class and graduate studies. Furthermore, they are representative of problems you will address professionally. The problems are drawn from the general area of engineering economics as well as from the four major disciplines of engineering: chemical, civil, electrical, and mechanical. These case studies also serve to illustrate the trade-offs among the various numerical techniques.

For example, *Case Study 6.1* uses all the methods with the exception of Newton-Raphson to perform an economic break-even analysis. The Newton-Raphson method is not employed because the function in the case study is difficult to differentiate. Among other things, the example demonstrates how the secant method may diverge if the initial guess is not close enough to the root.

*Case Study 6.2*, which is taken from chemical engineering, provides an excellent example of how root-location methods allow you to use realistic formulas in engineering practice. In addition, it also demonstrates how the efficiency of the Newton-Raphson technique is used to advantage when a large number of root-location computations are required.

Case Studies 6.3, 6.4, and 6.5 are engineering design problems taken from civil, electrical, and mechanical engineering. *Case Study 6.3* uses three different methods to determine the roots of a population forecasting model. *Case Study 6.4* performs a similar analysis for an electric circuit. Finally, *Case Study 6.5* deals with a vibration analysis of an automobile. Aside from investigating the efficiency of the various methods, this example has the added feature of illustrating how graphical methods provide insight into the root-location process.

## CASE STUDY 6.1 BREAK-EVEN ANALYSIS (GENERAL ENGINEERING)

Background: Good engineering practice requires that projects, products, and planning be approached in a cost-effective manner. A well-trained engineer must therefore be familiar with cost analysis. The present problem is called a "break-even problem." It is used to determine the point at which two alternative options are of equivalent value. Such choices are confronted in all fields of engineering. Although the case study is couched in personal terms, it is a prototype of other break-even problems that you may be called upon to address professionally.

You are considering the purchase of one of two personal computers: the "Lean Machine" and the "Ultimate." The estimated expenses and benefits for each computer are summarized in Table 6.1. If money can currently be borrowed at 20 percent interest ($i = 0.20$), how long must the machines be owned so that they would have equivalent worth? In other words, what is the break-even point in years?

Solution: As is common in economic problems, we have a mixture of present and future costs. For example, as depicted in Fig. 6.1, buying the Lean Machine involves an initial outlay of $3000. In addition to this one-time cost, money must be spent every year to maintain the machine. Because such costs tend to increase as the computer gets older, the maintenance costs are assumed to increase linearly with time. For example, $2000 would be required annually, by year 10, to keep the machine in working condition (Fig. 6.1). Finally, aside from these costs, you will also derive benefit from owning the computer. The annual profit and enjoyment derived from the Lean Machine are characterized by a constant annual income of $1000 each year.

In order to assess the two options, we must convert these costs into comparable measures. One way to do this is to express all the individual costs as equivalent annual payments, that is, the equivalent dollar value per year over the life span of the computer. The annual profit and enjoyment income

**TABLE 6.1 Cost and benefits for two personal computers. The negative signs indicate a cost or loss, whereas a positive sign indicates a benefit.**

| | COMPUTER | |
|---|---|---|
| | **Lean Machine** | **Ultimate** |
| Purchase cost, $ | −3000 | −10,000 |
| Increase in maintenance cost per year, $/yr/yr | −200 | −50 |
| Annual profit and enjoyment, $/yr | 1000 | 4000 |

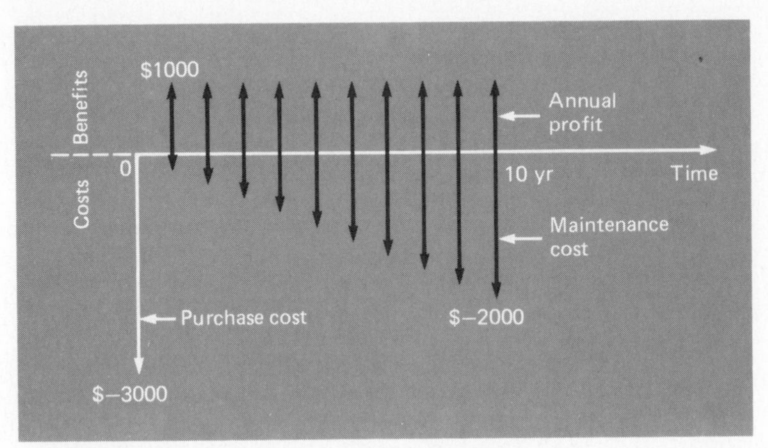


FIGURE 6.1 Cash-flow diagram for the costs and benefits of the Lean Machine computer. The abscissa is the number of years that you own this device. Cash flow is measured on the ordinate, with benefits positive and costs negative.

is already in this form. Economic formulas are available to express the purchase and maintenance costs in the same manner. For example, the initial purchase cost can be transformed into a series of uniform annual payments by the formula (Fig. 6.2*a*)

$$A_p = P\frac{i(1+i)^n}{(1+i)^n - 1} \qquad [6.1]$$

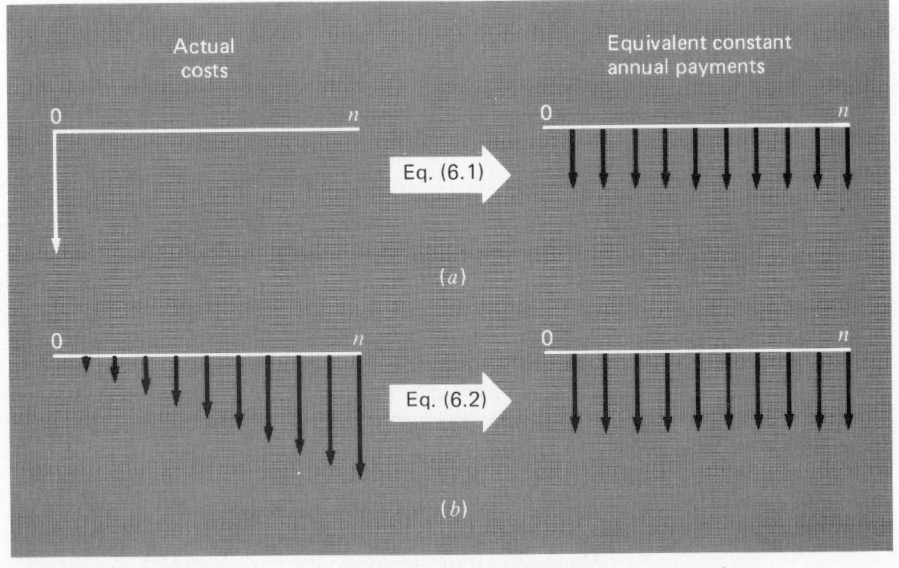


FIGURE 6.2 Graphical depiction of the use of an economic formula (*a*) to transform an initial payment to a series of equivalent annual payments using Eq. (6.1) and (*b*) to transform an arithmetic gradient series into a series of equivalent annual payments using Eq. (6.2).

where $A_p$ is the amount of the annual payment, $P$ is the purchase cost, $i$ is the interest rate, and $n$ is the number of years. For example, the initial payment for the Lean Machine is $\$-3000$, where the negative sign indicates a loss to you. If the interest rate is 20 percent ($i = 0.2$)

$$A_p = -3000\frac{0.2(1.2)^n}{1.2^n - 1}$$

For example, if the initial payments were to be spread over 10 years ($n = 10$), this formula can be used to compute that the equivalent annual payment would be $\$-715.57$ per year.

The maintenance cost is called an *arithmetic gradient series* because it increases at a constant rate. Conversion of such a series to an annual rate $A_m$ can be accomplished by the formula

$$A_m = G\left[\frac{1}{i} - \frac{n}{(1 + i)^n - 1}\right] \qquad [6.2]$$

where $G$ is the arithmetic rate of increase of maintenance. As depicted in Fig. 6.2*b*, this formula transforms the increasing maintenance cost into an equivalent series of constant annual payments.

These equations can be combined to express the value of each computer in terms of a uniform series of payments. For example, for the Lean Machine,

$$\underset{\text{Total worth}}{A_t} = \underset{-\text{purchase cost}}{-3000\frac{0.2(1.2)^n}{1.2^n - 1}} - \underset{\text{maintenance cost}}{200\left[\frac{1}{0.2} - \frac{n}{1.2^n - 1}\right]} + \underset{\text{profit}}{1000}$$

where $A_t$ designates the total annual worth. This equation can be simplified by collecting terms:

$$A_t = \frac{-600(1.2)^n}{1.2^n - 1} + \frac{200n}{1.2^n - 1} \qquad [6.3]$$

Substituting $n = 2$ into Eq. (6.3) yields the result that if you decide to discard the Lean Machine after owning it for only 2 years it will have cost \$1055 per year. If the computer is discarded after 10 years ($n = 10$), Eq. (6.3) indicates that it will have cost \$330 per year.

Similarly for the Ultimate, an equation for annual worth can be developed, as in

$$A_t = \frac{-2000(1.2)^n}{1.2^n - 1} + \frac{50n}{1.2^n - 1} + 3750 \qquad [6.4]$$

The values for Eq. (6.4) for $n = 2$ and $n = 10$ are $\$-2568$ and $\$+1461$ per year. Thus, although the Ultimate is more costly on a short-term basis, if owned long enough, it will not only be more cost-effective but will actually

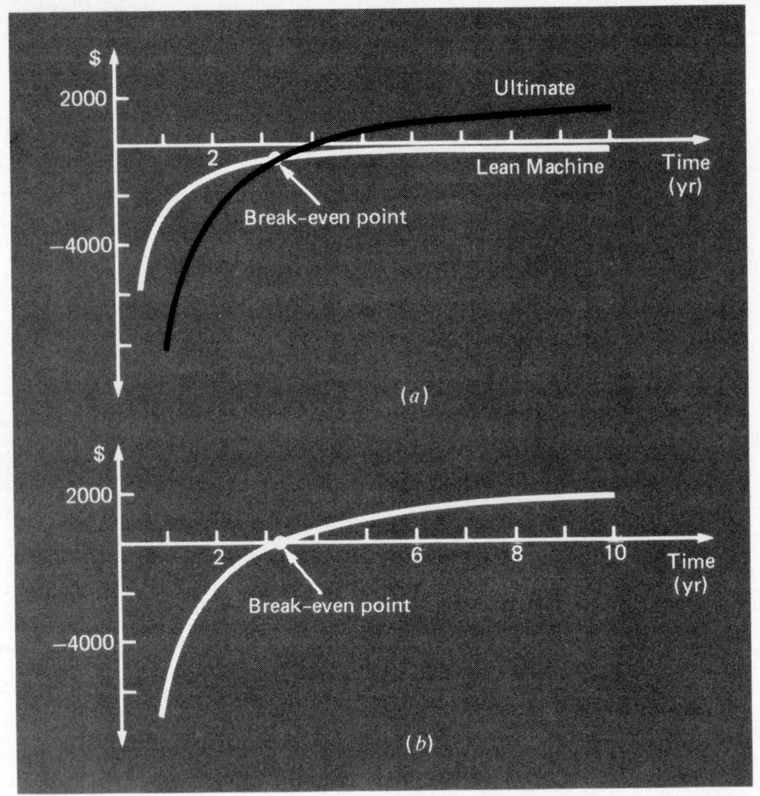


FIGURE 6.3 (*a*) Net cost curves for the Lean Machine [Eq. (6.3)] and the Ultimate [Eq. (6.4)] computers. (*b*) The break-even function [Eq. (6.5)].

earn money for you. Equations (6.3) and (6.4) are plotted for various values of $n$ in Fig. 6.3*a*.

The identification of the point at which the two machines have equivalent value designates when the Ultimate becomes the better buy. Graphically, it corresponds to the intersection of the two curves in Fig. 6.3*a*. From a mathematical perspective, the break-even point is the value of $n$ for which Eqs. (6.3) and (6.4) are equivalent, that is,

$$\frac{-600(1.2)^n}{1.2^n - 1} + \frac{200n}{1.2^n - 1} = \frac{-2000(1.2)^n}{1.2^n - 1} + \frac{50n}{1.2^n - 1} + 3750$$

By bringing all the terms of this equation to one side, the problem reduces to finding the root of

$$f(n) = \frac{-1400(1.2)^n}{1.2^n - 1} - \frac{150n}{1.2^n - 1} + 3750 = 0 \qquad [6.5]$$

Note that because of the way in which we have derived this equation, the Lean Machine is more cost-effective when $f(n) < 0$, and the Ultimate is more cost-effective when $f(n) > 0$ (Fig. 6.3*b*). The roots of Eq. (6.5) cannot be determined analytically. On the other hand, the equivalent annual payments are easy to calculate for a given $n$. Thus, as in our discussions of Sec. II.1.2 and Example 4.5, the design aspects of this problem create the need for a numerical approach.

The roots of Eq. (6.5) can be computed using some of the numerical methods described in Chaps. 4 and 5. The bracketing approaches and the secant method can be applied with minimal effort, whereas the Newton-Raphson method is awkward to use because it is time-consuming to determine $df/dn$ from Eq. (6.5).

On the basis of Fig. 6.3, we know that the root is between $n = 2$ and 10. These values provide starting values for the bisection method. Interval halving can be repeated for 18 iterations to yield a result with an $\epsilon_a$ less than 0.001 percent. The break-even point occurs at $n = 3.23$ years. This result can be checked by substituting it back into Eq. (6.5) to verify that $f(3.23) \simeq 0$.

Substituting $n = 3.23$ into either Eq. (6.3) or Eq. (6.4) yields the result that at the break-even point both machines cost about $542 per year. Beyond this point, the Ultimate becomes more cost-effective. Consequently, if you intend to own your machine for more than 3.23 years, the Ultimate is the better buy.

The method of false position can also be easily applied to this problem. A similar root and accuracy are attained after 12 iterations for the same bracketing interval of 2 to 10. On the other hand, the secant method converges to a root of −24.83 for this same bracketing interval. However, if the bracketing interval is reduced to 3 to 4, the secant method converges on $n = 3.23$ in only five iterations. Interestingly, the secant method also converges rapidly to the proper root when the initial interval is 2 to 3 and does not bracket the root. These results are typical of the trade-offs discussed subsequently in the epilogue. The best numerical method for this problem then depends on your judgment concerning the trade-offs among factors such as numerical efficiency, computer costs, and the dependability of the method.

## CASE STUDY 6.2 IDEAL AND NONIDEAL GAS LAWS (CHEMICAL ENGINEERING)

Background: The *ideal gas law* is given by

$$pV = nRT \tag{6.6}$$

where $p$ is the absolute pressure, $V$ is the volume, and $n$ is the number of moles. $R$ is the universal gas constant, and $T$ is the absolute temperature. Although this equation is widely used by engineers and scientists, it is only accurate over a limited range of pressure and temperature. Furthermore, Eq. (6.6) is more appropriate for some gases than for others.

An alternative equation of state for gases is given by

$$\left(p + \frac{a}{v^2}\right)(v - b) = RT \qquad [6.7]$$

which is known as the *van der Waals equation*. $v = V/n$ is the molal volume, and $a$ and $b$ are empirical constants that depend on the particular gas.

A chemical engineering design project requires that you accurately estimate the molal volume ($v$) of both carbon dioxide and oxygen for a number of different temperature and pressure combinations so that appropriate containment vessels can be selected. It is also of interest to examine how well each gas conforms to the ideal gas law by comparing the molal volume as calculated by Eqs. (6.6) and (6.7). The following data is provided:

$R = 0.082054$ L · atm/(mol · K)

$\left.\begin{array}{l} a = 3.592 \\ b = 0.04267 \end{array}\right\}$ carbon dioxide

$\left.\begin{array}{l} a = 1.360 \\ b = 0.03183 \end{array}\right\}$ oxygen

The design pressures of interest are 1, 10, and 100 atm for temperature combinations of 300, 500, and 700 K.

Solution: Molal volumes for both gases are calculated using the ideal gas law, with $n = 1$. For example, if $p = 1$ atm and $T = 300$ K,

$$v = \frac{V}{n} = \frac{RT}{p} = 0.082054 \frac{\text{L} \cdot \text{atm}}{\text{mol} \cdot \text{K}} \frac{300 \text{ K}}{1 \text{ atm}}$$

$$v = 24.6162 \text{ L/mol}$$

These calculations are repeated for all temperature and pressure combinations and presented in Table 6.2.

**TABLE 6.2 Computations of molal volume for Case Study 6.2**

| Temperature, K | Pressure, atm | Molal volume (ideal gas law), L/mol | Molal volume (van der Waals) carbon dioxide, L/mol | Molal volume (van der Waals) oxygen, L/mol |
|---|---|---|---|---|
| 300 | 1 | 24.6162 | 24.5126 | 24.5928 |
| | 10 | 2.4616 | 2.3545 | 2.4384 |
| | 100 | 0.2462 | 0.0795 | 0.2264 |
| 500 | 1 | 41.0270 | 40.9821 | 41.0259 |
| | 10 | 4.1027 | 4.0578 | 4.1016 |
| | 100 | 0.4103 | 0.3663 | 0.4116 |
| 700 | 1 | 57.4378 | 57.4179 | 57.4460 |
| | 10 | 5.7438 | 5.7242 | 5.7521 |
| | 100 | 0.5744 | 0.5575 | 0.5842 |

The computation of molal volume from the van der Waals equation can be accomplished using any of the numerical methods for finding roots of equations discussed in Chaps. 4 and 5, with

$$f(v) = \left(p + \frac{a}{v^2}\right)(v - b) - RT \qquad [6.8]$$

In this case, the derivative of $f(v)$ is easy to determine and the Newton-Raphson method is convenient and efficient to implement. The derivative of $f$ with respect to $v$ is given by

$$f'(v) = p - \frac{a}{v^2} + \frac{2ab}{v^3}$$

The Newton-Raphson method is described by Eq. (5.6):

$$v_{i+1} = v_i - \frac{f(v_i)}{f'(v_i)}$$

which can be used to estimate the root. For example, using the initial guess of 24.6162, the molal volume of carbon dioxide at 300 K and 1 atm is computed as 24.5126 L/mol. This result was obtained after two iterations and has an $\epsilon_a$ of less than 0.001 percent.

Similar computations for all combinations of pressure and temperature for both gases are presented in Table 6.2. It is seen that the results for the ideal gas law differ from those for van der Waals equation for both gases, depending on specific values for $p$ and $T$. Furthermore, because some of these results are significantly different, your design of the containment vessels would be quite different, depending on which equation of state were used.

In this case, a complicated equation of state was examined using the Newton-Raphson method. The results varied significantly from the ideal gas law for several cases. From a practical standpoint, the Newton-Raphson method was appropriate for this application because $f'(v)$ was easy to calculate. Thus, the rapid convergence properties of the Newton-Raphson method could be exploited.

In addition to demonstrating its power for a single computation, the present case study also illustrates how the Newton-Raphson method is especially attractive when numerous computations are required. Because of the speed of personal computers, the efficiency of various numerical methods for most roots of equations are indistinguishable for a single computation. Even a 10-s difference between the crude bisection approach and the efficient Newton-Raphson does not amount to a significant time loss when only one computation is performed. However, suppose that millions of root evaluations are required to solve a problem. In this case, the efficiency of the method could be a deciding factor in the choice of a technique.

For example, suppose that you are called upon to design an automatic computerized control system for a chemical production process. This system requires accurate estimates of molal volumes on an essentially continuous basis in order to properly manufacture the final product. Gauges are installed that provide instantaneous readings of pressure and temperature. Evaluations of $v$ must be obtained for a variety of gases that are used in the process.

For such an application, bracketing methods such as bisection or false position would probably be too time-consuming. In addition, the two initial guesses that are required for these approaches would also interject a critical delay in the procedure. This shortcoming is relevant to the secant method, which also needs two initial estimates.

In contrast, the Newton-Raphson method requires only one guess for the root. The ideal gas law could be employed to obtain this guess at the initiation of the process. Then, assuming that the time frame is short enough so that pressure and temperature do not vary wildly between computations, the previous root solution would provide a good guess for the next application. Thus, the close guess that is often a prerequisite for convergence of the Newton-Raphson method would automatically be available. All of the above considerations would greatly favor the Newton-Raphson technique for such problems.

## CASE STUDY 6.3 POPULATION-GROWTH DYNAMICS (CIVIL ENGINEERING)

Background: Population-growth dynamics are important in all engineering planning studies. Construction schedules and resource allocations for large-scale projects such as water supply and transportation systems depend on accurate forecasts of population trends. In addition, trends of nonhuman populations such as microbes are important in many engineering processes such as waste treatment and the manufacture of fermentation and pharmaceutical products.

Microbial population-growth models incorporate the assumption that the rate of change of the population ($p$) is proportional to the existing population at any time ($t$):

$$\frac{dp}{dt} = kp$$

The population may grow in an environment that has sufficient food such that $k$ is not a function of food concentration. (See Case Study 12.2 for an example where $k$ depends on food level.) When food is not scarce, growth will be limited by toxic by-products or space as the population size increases. With time, these factors retard the growth rate of the population and completely inhibit growth when the population reaches a maximum density of $p_{max}$. In this case, the above rate equation is modified as follows:

$$\frac{dp}{dt} = Kp(p_{max} - p)$$

where $K$ has units of liters per cell per day. This differential equation can be integrated using analytical techniques to give

$$p(t) = \frac{p_{max}}{1 + \left(\frac{p_{max}}{p_0} - 1\right) e^{-Kp_{max}t}} \qquad [6.9]$$

where $p(t = 0) = p_0$. Equation (6.9) is known as the *logistic growth model.* This model gives an S-shaped curve for $p(t)$, as shown in Fig. 6.4. Thus, the model simulates slow initial growth followed by a rapid growth period and finally limited growth at high population density.

As an example of the application of this model in the environmental area of civil engineering, consider the growth of a bacterial population in a lake. The growth follows Eq. (6.9). In the spring of the year at $t = 0$, the population is small, $p(t = 0) = 10$ cells per liter. It is known that the population reaches a density of 15,000 cells per liter when $t = 60$ days and that the growth rate $K$ is $2 \times 10^{-6}$ liter per cell per day. It is required to calculate the bacterial population density on Labor Day when $t = 90$ days. If the population exceeds 40,000 cells per liter, water-quality standards will require implementation of some abatement procedure to protect swimmers.

Solution: Substituting the known information into Eq. (6.9) gives

$$15{,}000 = \frac{p_{max}}{1 + \left(\frac{p_{max}}{10} - 1\right) e^{-2\times10^{-6}(p_{max})(60)}} \qquad [6.10]$$

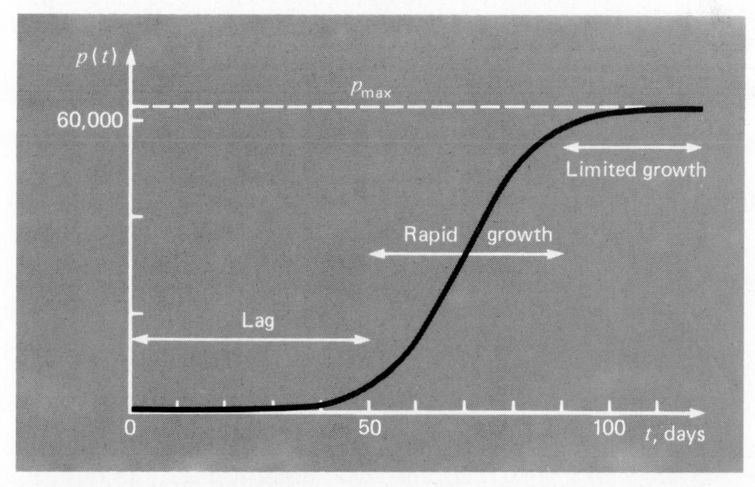


FIGURE 6.4 A logistic growth model for a population. The model simulates slow initial growth, then rapid acceleration of growth, followed by a leveling off at high population density.

which has only one unknown, $p_{max}$. If Eq. (6.10) could be solved for $p_{max}$ then $p(t = 90)$ could be easily determined from Eq. (6.9). However, because $p_{max}$ is implicit, it cannot be obtained directly from Eq. (6.10). Therefore, a numerical method from Chap. 4 or 5 must be used. The Newton-Raphson technique will not be used because the derivative of Eq. (6.10) is difficult to determine. However the bisection, false-position, and secant methods can be readily applied. Given inital guesses of 60,000 and 70,000 cells per liter results in the following estimates of $p_{max}$, with a relative error criterion of 0.01 percent:

| False position | Result | Iterations |
|---|---|---|
| Bisection | 63,198 | 11 |
| False position | 63,199 | 5 |
| Secant | 63,200 | 4 |

Note that the false-position and secant methods converge in about half the iterations required for the bisection method.

Now, from Eq. (6.9), with $p_{max} = 63{,}200$,

$$p(90) = \frac{63{,}200}{1 + \left(\dfrac{63{,}200}{10} - 1\right) e^{-2\times 10^{-6}(63{,}200)(90)}} = 58{,}930 \text{ cells per liter}$$

Such a population level exceeds the water-quality standard of 40,000 cells per liter, and therefore some corrective action must be pursued on Labor Day.

This example illustrates the relative computational efficiency of three different methods of finding roots of equations in a civil-engineering design problem. However, as noted previously, the general approach has broad application in all fields of engineering concerned with the population trends of organisms, including humans.

## CASE STUDY 6.4 DESIGN OF AN ELECTRIC CIRCUIT (ELECTRICAL ENGINEERING)

Background: Electrical engineers often use Kirchhoff's laws to study the steady-state (not time-varying) behavior of electric circuits. Such steady-state behavior will be examined in Case Study 9.4. Another important class of problems are transient in nature and involve circuits where sudden temporal changes take place. Such a situation occurs following the closing of the switch in Fig. 6.5. In this case, there will be a period of adjustment following the closing of the switch as a new steady state is reached. The length of this adjustment period is closely related to the charge-storing properties of the capacitor and energy storage by the inductor. Energy storage may oscillate

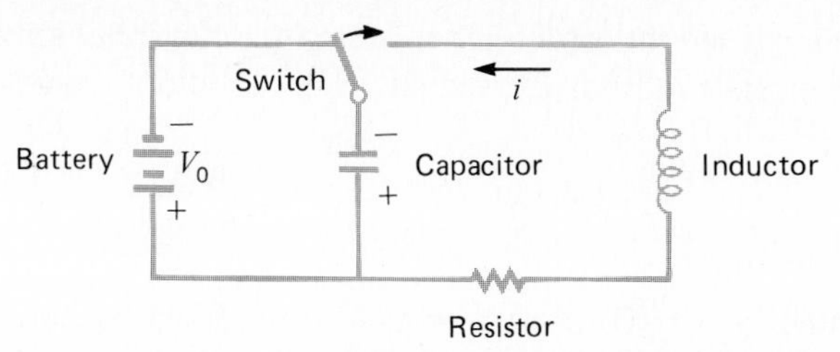


FIGURE 6.5 An electric circuit. When the switch is closed, the current will undergo a series of oscillations until a new steady state is reached.

between these two elements during a transient period. However, resistance in the circuit will dissipate the magnitude of the oscillations.

The flow of current through the resistor causes a voltage drop ($V_R$) given by

$$V_R = iR$$

where $i$ is the current and $R$ is the resistance of the resistor. When $R$ and $i$ have units of ohms and amperes, then $V_R$ has units of volts.

Similarly, an inductor resists changes in current, such that the voltage drop across it ($V_L$) is

$$V_L = L\frac{di}{dt}$$

where $L$ is the inductance. When $L$ and $i$ have units of henrys and amperes, $V_L$ has units of volts and $t$ has units of seconds.

The voltage drop across the capacitor ($V_C$) depends on the charge ($q$) on it:

$$V_C = \frac{q}{C}$$

where $C$ is the capacitance. When the charge is expressed in units of coulombs, the unit of $C$ is the farad.

Kirchhoff's second law states that the algebraic sum of voltage drops around a closed circuit is zero. After the switch is closed we have

$$L\frac{di}{dt} + Ri + \frac{q}{C} = 0$$

However, the current is related to the charge according to

$$i = \frac{dq}{dt}$$

Therefore,

$$L\frac{d^2q}{dt^2} + R\frac{dq}{dt} + \frac{q}{C} = 0$$

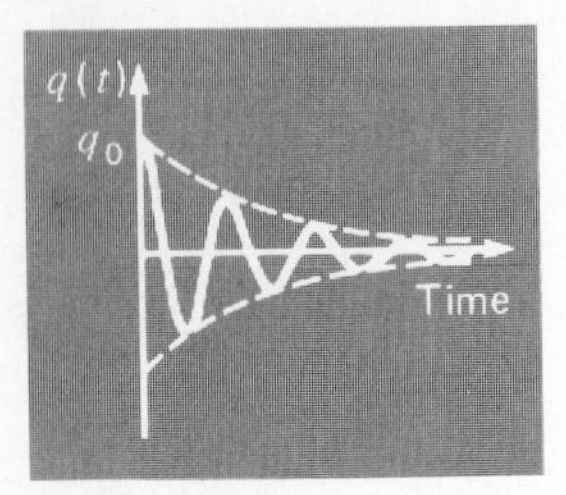


FIGURE 6.6
The charge on a capacitor as a function of time following the closing of the switch in Fig. 6.5.

This is a second-order linear ordinary differential equation that can be solved using the methods of calculus. This solution is given by

$$q(t) = q_0 e^{-Rt/2L} \cos\left(\sqrt{\frac{1}{LC} - \left(\frac{R}{2L}\right)^2}\, t\right) \qquad [6.11]$$

where at $t = 0$, $q = q_0 = V_0 C$, and $V_0$ is the voltage from the charging battery. Equation (6.11) describes the time variation of the charge on the capacitor. The solution $q(t)$ is plotted in Fig. 6.6.

A typical electrical-engineering design problem might involve determining the proper resistor to dissipate energy at a specified rate, with known values for $L$ and $C$. For the present case study, assume the charge must be dissipated to 1 percent of its original value ($q/q_0 = 0.01$) in $t = 0.05$ s, with $L = 5$ H and $C = 10^{-4}$ F.

Solution: It is necessary to solve Eq. (6.11) for $R$, with known values of $q$, $q_0$, $L$, and $C$. However, a numerical approximation technique must be employed because $R$ is an implicit variable in Eq. (6.11). The bisection method will be used for this purpose. The other methods discussed in Chaps. 4 and 5 are also appropriate, although the Newton-Raphson method would be cumbersome to use because the derivative of Eq. (6.11) is quite complicated. Rearranging Eq. (6.11),

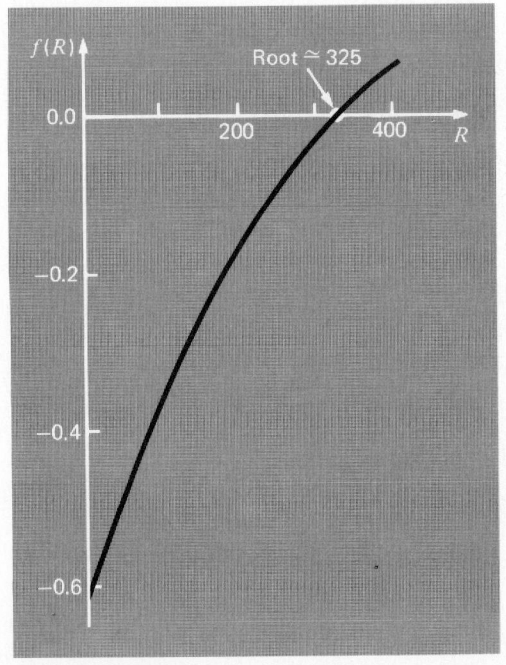


FIGURE 6.7 Plot of Eq. (6.12) used to obtain initial guesses for $R$ that bracket the root.

$$f(R) = e^{-Rt/2L} \cos\left(\sqrt{\frac{1}{LC} - \left(\frac{R}{2L}\right)^2}\, t\right) - \frac{q}{q_0}$$

or, using the numerical values given,

$$f(R) = e^{-0.005R} \cos(\sqrt{2000 - 0.01R^2}\; 0.05) - 0.01 \qquad [6.12]$$

Examination of this equation suggests that a reasonable initial range for $R$ is 0 to 400 Ω (because $2000 - 0.01R^2$ must be greater than zero). Figure 6.7, a plot of Eq. (6.12), confirms this. Twenty-one iterations of the bisection method give $R = 328.1515$ Ω, with an error of less than 0.0001 percent.

Thus, you can specify a resistor with this rating for the circuit shown in Fig. 6.5 and expect to achieve a dissipation performance that is consistent with the requirements of the problem. This design problem could not be solved efficiently without using the numerical methods in Chaps. 4 and 5.

## CASE STUDY 6.5 VIBRATION ANALYSIS (MECHANICAL ENGINEERING)

Background: Differential equations are often used to model the behavior of engineering systems. A class of such models broadly applicable to most fields of engineering is harmonic oscillators. Some basic examples of harmonic oscillators are a simple pendulum, a mass on a spring, and an inductance-capacitance electric circuit (Fig. 6.8). Although these are very different physical systems, their oscillations can all be described by the same mathematical model. Thus, although the present problem deals with the design of an automobile shock absorber, the general approach is applicable to a variety of other problems in all fields of engineering.

As depicted in Fig. 6.9, a car of mass $m$ is supported by springs. Shock absorbers offer resistance to motion of the car that is proportional to the vertical speed (up-and-down motion) of the car. Disturbance of the car from equilibrium causes the system to move with an oscillating motion $x(t)$. At any instant the net forces acting on $m$ are the resistance of the spring and the damping force of the shock absorber. The resistance of the spring is proportional to a spring constant ($k$) and the distance from equilibrium ($x$):

$$\text{Spring force} = -kx \qquad [6.13]$$

where the negative sign indicates that the restoring force acts to return the car toward the position of equilibrium. The damping force of the shock absorber is given by

$$\text{Damping force} = -c\frac{dx}{dt}$$

where $c$ is a damping coefficient and $dx/dt$ is the vertical velocity. The

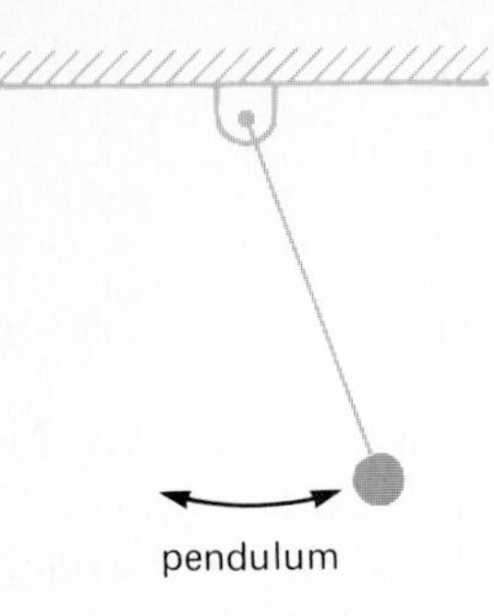


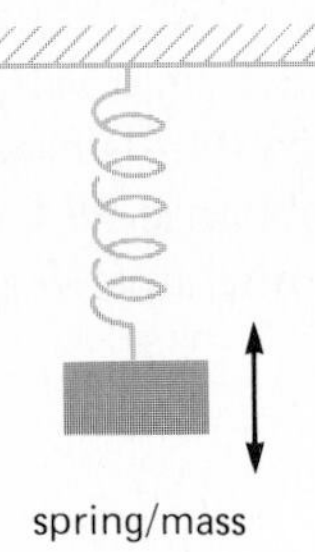


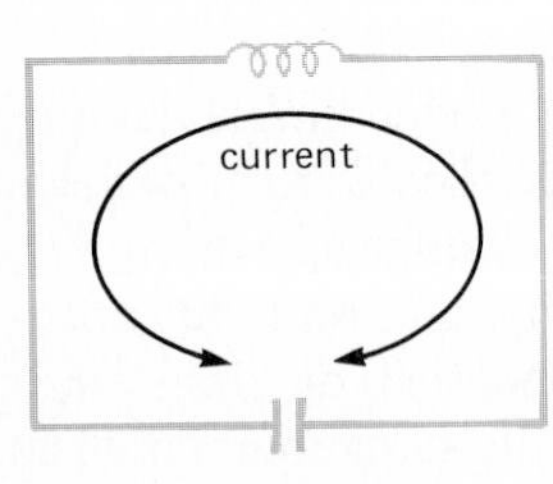


FIGURE 6.8
Three examples of simple harmonic oscillators. The two-way arrows illustrate the oscillations for each system.

negative sign indicates that the damping force acts in the opposite direction against the velocity.

The equations of motion for the system are given by Newton's second law ($F = ma$), which for the present problem is expressed as

$$m \frac{d^2x}{dt^2} = -c\frac{dx}{dt} + (-kx)$$

Mass × acceleration = damping force + spring force

or

$$\frac{d^2x}{dt^2} + \frac{c}{m}\frac{dx}{dt} + \frac{k}{m}x = 0$$

This is a second-order linear differential equation that can be solved using the methods of calculus. For example, if the car hits a hole in the road at $t = 0$, such that it is displaced from equilibrium with $x = x_0$ and $dx/dt = 0$, then,

$$x(t) = e^{-nt}\left(x_0 \cos pt + x_0 \frac{n}{p} \sin pt\right) \quad [6.14]$$

where $n = c/(2m)$, $p = \sqrt{k/m - c^2/(4m^2)}$ and $k/m > c^2/(4m^2)$. Equation (6.14) gives the vertical position of the car as a function of time. The parameter values are $c = 1.4 \times 10^7$ g/s, $m = 1.2 \times 10^6$ g, and $k = 1.25 \times 10^9$ g/s$^2$. If $x_0 = 0.3$, mechanical engineering design considerations require that estimates be provided for the first three times the car passes through the equilibrium point.

Solution: This design problem must be solved using numerical methods of Chaps. 4 and 5. The bracketing or secant methods are preferable because the derivative of Eq. (6.14) is complicated.

Estimates of the initial guesses are easily obtained by reference to Fig. 6.10. This case study illustrates how graphical methods often provide information that is essential for the successful application of the numerical techniques. The plot indicates that this problem is complicated by the existence

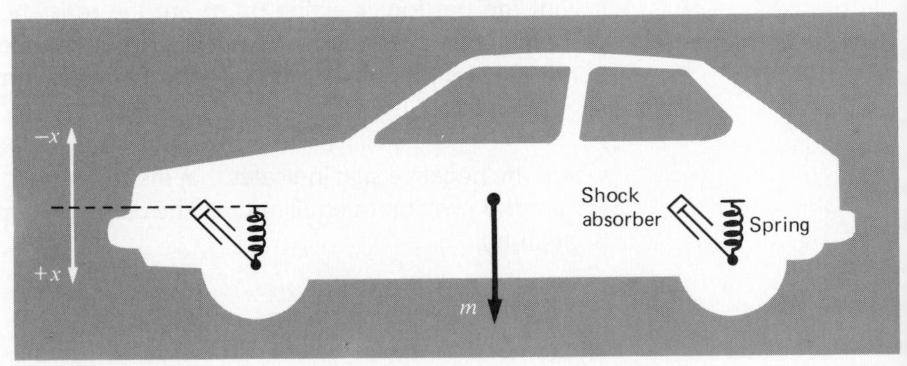


FIGURE 6.9 A car of mass $m$.

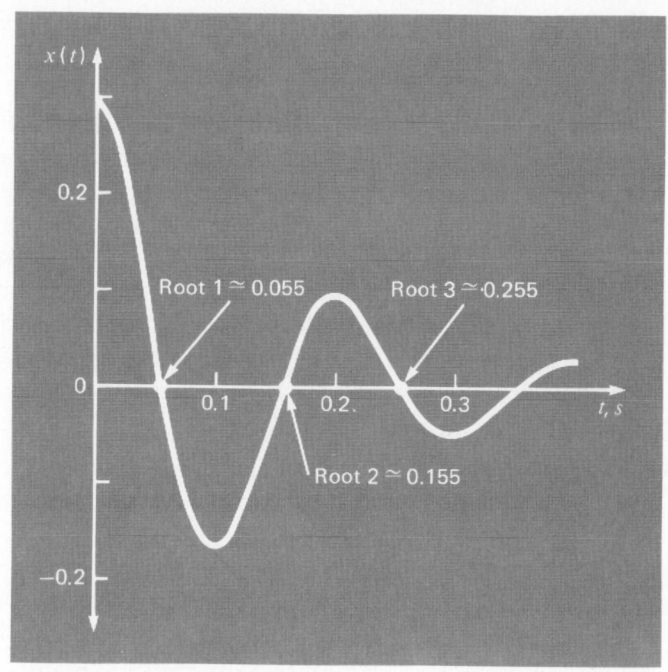


FIGURE 6.10 Plot of the position versus time for a shock absorber after the wheel of a car hits a hole in the road.

of several roots. Thus, in this case, rather narrow bracketing intervals must be used to avoid overlap.

Table 6.3 lists the results of using the bisection, false-position, and secant methods, with a stopping criterion of 0.1 percent. All the techniques converge quickly. As expected, the false-position and secant methods are more efficient than bisection.

**TABLE 6.3 Results of using the bisection, false-position, and secant methods to locate the first three roots for vibrations of a shock absorber. A stopping criterion of 0.1 percent was used to obtain these results. Note that the exact values of the roots are 0.0552095329, 0.15417813, and 0.253146726.**

| Method | Lower guess | Upper guess | Root estimate | Number of iterations | PERCENT RELATIVE ERROR Approximate | True |
|---|---|---|---|---|---|---|
| Bisection | 0.0 | 0.1 | 0.0552246 | 11 | 0.088 | 0.027 |
| | 0.1 | 0.2 | 0.1541992 | 10 | 0.063 | 0.014 |
| | 0.2 | 0.3 | 0.2533203 | 9 | 0.077 | 0.069 |
| False position | 0.0 | 0.1 | 0.0552095 | 5 | 0.002 | 0.0001 |
| | 0.1 | 0.2 | 0.1541790 | 4 | 0.069 | 0.0006 |
| | 0.2 | 0.3 | 0.2531475 | 4 | 0.043 | 0.0003 |
| Secant | 0.0 | 0.1 | 0.0552095 | 5 | 0.038 | 0.0001 |
| | 0.1 | 0.2 | 0.1541780 | 5 | 0.020 | 0.0001 |
| | 0.2 | 0.3 | 0.2531465 | 5 | 0.017 | 0.0001 |

Notice how, for all methods, the approximate percent relative errors are greater than the true errors. Thus, the results are *at least* as accurate as the stopping criterion of 0.1 percent. However, also observe that the false-position and secant methods are very conservative in this regard. Recall in our discussion in Sec. 4.3 that the termination criterion usually constitutes an approximation of the discrepancy of the previous iteration. Thus, for rapidly convergent approaches such as the false-position and secant methods, the improvement in accuracy between successive iterations is so great that $\epsilon_t$ will usually be much less than $\epsilon_a$. The practical significance of this behavior is of little importance when determining a single root. However, if numerous root locations are involved, rapid convergence becomes a valuable property to be considered when choosing a particular technique.

## PROBLEMS

### General Engineering

**6.1** Using your own software, reproduce the computation performed in Case Study 6.1.

**6.2** Perform the same computation as in Case Study 6.1, but use an interest rate of 17 percent ($i = 0.17$). If possible, use your computer software to determine the break-even point. Otherwise, use any of the numerical methods discussed in Chaps. 4 and 5 to perform the computation. Justify your choice of technique.

**6.3** For Case Study 6.1, determine the number of years the Ultimate computer must be owned for it to earn money for you. That is, compute the value of $n$ at which $A_t$ for Eq. (6.4) becomes positive.

**6.4** Using an approach similar to Case Study 6.1, the following equation can be developed for determining the total annual worth of a personal computer:

$$A_t = \frac{-3000\,(1.2)^n}{1.2^n - 1} + \frac{175n}{1.2^n - 1} + 5000$$

Find the value of $n$ such that $A_t$ is zero.

**6.5** You are interested in buying an automobile and have narrowed the choices down to two options. Just as in Case Study 6.1, the net annual worth of owning either car is a composite of purchase cost, maintenance cost, and profit:

| | **Luxury model** | **Economy model** |
|---|---|---|
| Purchase cost, $ | −15,000 | −5000 |
| Increase in maintenance, $/yr/yr | −400 | −200 |
| Annual profit and enjoyment, $ | 7500 | 3000 |

If the interest rate is 12.5 percent ($i = 0.125$), compute the break-even point ($n$) for the cars.

**6.6** You buy a \$20,000 piece of equipment for nothing down and \$5000 per year for 5 years. What interest rate are you paying? The formula relating present worth ($P$), annual payments ($A$), number of years ($n$), and interest rate ($i$) is

$$A = P\frac{i(1+i)^n}{(1+i)^n - 1}$$

**6.7** Because many engineering-economics tables were developed years ago, they are not designed to handle the large interest rates that are prevalent today. In addition, they are often not designed to handle fractional interest rates. As in the following problem, numerical methods can be used to determine economic estimates for such situations.

A new entertainment complex is estimated to cost \$10 million and to produce a net annual revenue of \$2 million. If the debt is to be paid off in 10 years, at what interest rate must the funds be borrowed? Present cost ($P$), annual payments ($A$), and interest rate ($i$) are related to each other by the following economic formula:

$$\frac{P}{A} = \frac{(1+i)^n - 1}{i(1+i)^n}$$

where $n$ is the number of annual payments. For the present problem,

$$\frac{P}{A} = \frac{10{,}000{,}000}{2{,}000{,}000} = 5$$

Therefore, the equation becomes

$$5 = \frac{(1+i)^{10} - 1}{i(1+i)^{10}}$$

The interest rate that satisfies this equation can be determined by finding the root of

$$f(i) = \frac{(1+i)^{10} - 1}{i(1+i)^{10}} - 5$$

($a$) Sketch $f(i)$ versus $i$ to make an initial graphical guess at the root.
($b$) Solve for $i$ using the bisection method (count iterations).
($c$) Solve for $i$ using the false-position method (count iterations).
In both ($b$) and ($c$), use initial guesses of $i = 0.1$ and $0.2$. Attain an error level of 2 percent for both cases.

## Chemical Engineering

**6.8** Using your own software, reproduce the computation performed in Case Study 6.2.

**6.9** Perform the same computation as in Case Study 6.2, but for ethyl alcohol ($a = 12.02$ and $b = 0.08407$) at a temperature of 350 K and $p$ of 1.5 atm.

Compare your results with the ideal gas law. If possible, use your computer software to determine the molal volume. Otherwise, use any of the numerical methods discussed in Chaps. 4 and 5 to perform the computation. Justify your choice of technique.

**6.10** Repeat Prob. 6.9, but use nitrous oxide ($a = 3.782$ and $b = 0.04415$) at a temperature of 450 K; $p = 2$ atm.

**6.11** The temperature (in kelvin) of a system varies over the course of a day according to

$$T = 400 + 200 \cos \frac{2\pi t}{1440}$$

where $t$ is expressed in minutes. Pressure is lost from the system according to $p = e^{-t/1440}$. Develop a computer program to calculate the molal volume of oxygen at minute intervals over the course of the day. Plot the results. If you have computer graphics capabilities, plot all the data. If not, plot the results at 60-min intervals. Background for this problem can be found in Case Study 6.2.

**6.12** In chemical engineering, plug flow reactors (that is, those in which fluid flows from one end to the other with minimal mixing along the longitudinal axis) are often used to convert reactants into products. It has been determined that the efficiency of the conversion can sometimes be improved by recycling a portion of the product stream so that it returns to the entrance for an additional pass through the reactor (Fig. P6.12). The recycle rate is defined as

$$R = \frac{\text{volume of fluid returned to entrance}}{\text{volume leaving the system}}$$

Suppose that we are processing a chemical A in order to generate a product B. For the case where A forms B according to an autocatalytic reaction (that is, in which one of the products acts as a catalyst or stimulus for the reaction), or

$$\mathrm{A} + \mathrm{B} \longrightarrow \mathrm{B} + \mathrm{B}$$

it can be shown that an optimal recycle rate must satisfy

$$\ln \frac{1 + R(1 - X_{Af})}{R(1 - X_{Af})} = \frac{R + 1}{R[1 + R(1 - X_{Af})]}$$

where $X_{Af}$ is the fraction of the reactant A that is converted to the product B. The optimal recycle rate corresponds to the minimum-sized reactor needed to attain the desired level of conversion.

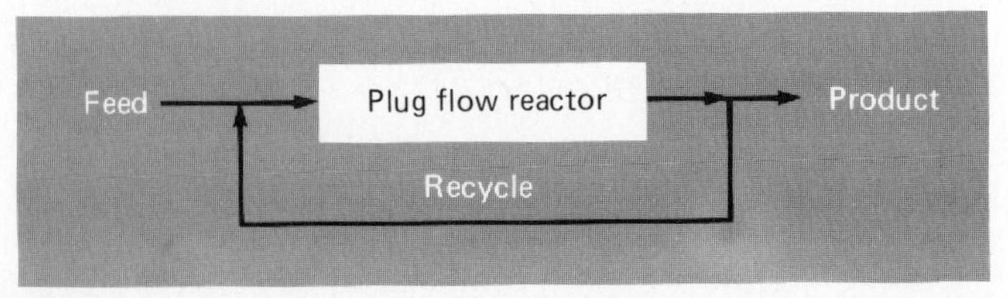


FIGURE P6.12 Schematic representation of a plug flow reactor with recycle.

Use the bisection method to determine the recycle ratios needed to minimize reactor size for fractional conversions of
(a) $X_{Af} = 0.99$
(b) $X_{Af} = 0.995$
(c) $X_{Af} = 0.999$

**6.13** In a chemical engineering process, water vapor ($H_2O$) is heated to sufficiently high temperatures that a significant portion of the water dissociates, or splits apart, to form oxygen ($O_2$) and hydrogen ($H_2$):

$$H_2O \rightleftharpoons H_2 + \frac{1}{2} O_2$$

If it is assumed that this is the only reaction involved, the mole fraction ($x$) of $H_2O$ that dissociates can be represented by

$$k_p = \frac{x}{1 - x} \sqrt{\frac{2p_t}{2 + x}} \qquad \text{[P6.13]}$$

where $k_p$ is the reaction's equilibrium constant and $p_t$ is the total pressure of the mixture. If $p_t = 2$ atm and $k_p = 0.04568$, determine the value of $x$ that satisfies Eq. (P6.13).

## Civil Engineering

**6.14** Using your own software, reproduce the computation of Case Study 6.3.

**6.15** Perform the same computation as in Case Study 6.3, but use a growth rate of $1.5 \times 10^6$ liters per cell per day.

**6.16** Perform the same computation as in Case Study 6.3 but use an initial bacteria concentration of $p_0 = 20$ cells per liter.

**6.17** The concentration of pollutant bacteria $C$ in a lake decrease according to

$$C = 80e^{-2t} + 20e^{-0.1t}$$

Determine the time required for the bacteria to be reduced to 10 using (a) the graphical method and (b) the Newton-Raphson method.

**6.18** Many fields of engineering require accurate population estimates. For example, transportation engineers might find it necessary to determine separately the population growth trends of a city and adjacent suburb. The population of the urban area is declining with time according to

$$P_u(t) = P_{u,\max}e^{-k_u t} + P_{u,\min}$$

while the suburban population is growing, as in

$$P_s(t) = \frac{P_{s,\max}}{1 + \left(\frac{P_{s,\max}}{P_o} - 1\right)e^{-k_s t}}$$

where $P_{u,\max}$, $k_u$, $P_{u,\min}$ $P_{s,\max}$, $P_o$, and $k_s$ are empirically derived parameters.

Determine the time and corresponding values of $P_u(t)$ and $P_s(t)$ when the populations are equal. The parameter values are $P_{u,\max} = 60000$; $k_u = 0.04$ yr$^{-1}$; $P_{u,\min} = 120000$; $P_{s,\max} = 300000$; $P_o = 5000$; and $k_s = 0.06$ yr$^{-1}$. To obtain your solutions, use (*a*) the graphical and (*b*) the false-position methods.

**6.19** The displacement of a structure is defined by the following equation for a damped oscillation:

$$y = 10e^{-kt} \cos \omega t$$

where $k = 0.5$ and $\omega = 2$.
(*a*) Use the graphical method to make an initial estimate of the time required for the displacement to decrease to 4.
(*b*) Use the Newton-Raphson method to determine the root to $\epsilon_s = 0.01$ percent.
(*c*) Use the secant method to determine the root to $\epsilon_s = 0.01$ percent.

**6.20** Figure P6.20 shows an open channel of constant dimensions with a rectangular cross-sectional area $A$. Under uniform flow conditions, the following relationship, based on Manning's equation, holds

$$Q = \frac{y_n B}{n}\left(\frac{y_n B}{B + 2y_n}\right)^{2/3} S^{1/2} \qquad \text{[P6.7]}$$

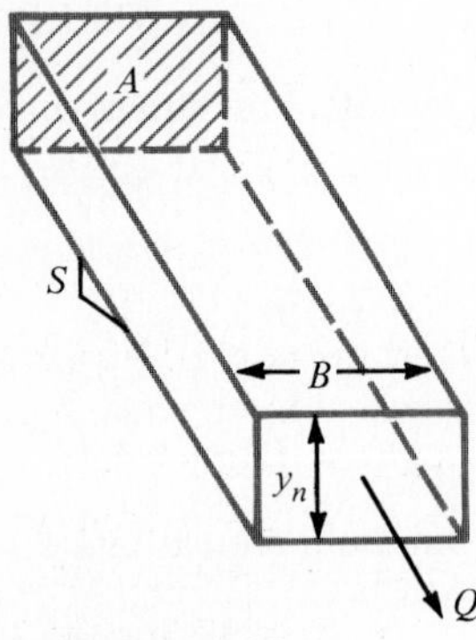

FIGURE P6.20

where $Q$ is flow, $y_n$ is normal depth, $B$ is the width of the channel, $n$ is a roughness coefficient used to parameterize the frictional effects of the channel material, and $S$ is the slope of the channel. This equation is used by fluid and water-resource engineers to determine the normal depth. If this value is less than the critical depth,

$$y_c = \left(\frac{Q^2}{B^2 g}\right)^{1/3}$$

where $g$ is acceleration due to gravity (980 cm/s$^2$), then flow will be subcritical.
Use the graphical and bisection methods to determine $y_n$ if $Q = 14.15$ m$^3$/s; $B = 4.572$ m; $n = 0.017$; and $S = 0.0015$. Compute whether flow is sub- or supercritical.

## Electrical Engineering

**6.21** Using your own software, reproduce the computation of Case Study 6.4.

**6.22** Perform the same computation as in Case Study 6.4, but assume that the charge must be dissipated to 2 percent of its original value in 0.04 s.

**6.23** Perform the same computation as in Case Study 6.4, but determine the time required for the circuit to dissipate to 10 percent of its original value, given $R = 300\ \Omega$, $C = 10^{-4}$ F, and $L = 4$ H.

**6.24** Perform the same computation as in Case Study 6.4, but determine the value of $L$ required for the circuit to dissipate to 1 percent of its original value in $t = 0.05$ s, given $R = 300\ \Omega$ and $C = 10^{-4}$ F.

**6.25** An oscillating current in an electric circuit is described by

$$I = 10e^{-t} \sin(2\pi t)$$

where $t$ is in seconds. Determine all values of $t$ such that $I = 2$.

## Mechanical Engineering

**6.26** Using your own software, reproduce the computation performed in Case Study 6.5.

**6.27** Perform the same computation as in Case Study 6.5, but use $c = 1.5 \times 10^7$ g/s, $k = 1.5 \times 10^9$ g/s$^2$, and $m = 2 \times 10^6$ g.

**6.28** Perform the same computation as in Case Study 6.5, but determine the value of $k$ so that the first root occurs at $t = 0.08$ s.

**6.29** Perform the same computation as in Case Study 6.5, but determine the value of $m$ so that the first root occurs at $t = 0.04$ s.

**6.30** Perform the same computation as in Case Study 6.5, but determine the value of $c$ so that the second root occurs at $t = 0.2$ s.

## Miscellaneous

**6.31** Read all the case studies in Chap. 6. On the basis of your reading and experience, make up your own case study for any one of the fields of engineering. This may involve modifying or reexpressing one of our case studies. However, it can also be totally original. As with our examples, it must be drawn from an engineering problem context and must demonstrate the use of the numerical methods for solving roots of equations. Write up your results using our case studies as models.

# EPILOGUE: PART II

## II.4 TRADE-OFFS

Table II.3 provides a summary of the trade-offs involved in solving for roots of algebraic and transcendental equations. Although graphical methods are time-consuming, they provide insight into the behavior of the function and are useful in identifying initial guesses and potential problems such as multiple roots. Therefore, if time permits, a quick sketch (or better yet, a computerized graph) can yield valuable information regarding the behavior of the function.

The numerical methods themselves are divided into two general categories: bracketing and open methods. The former require two initial guesses that are on either side of a root. This "bracketing" is maintained as the solution proceeds, and thus, these techniques are always convergent. However, a price is paid for this property in that the rate of convergence is relatively slow. Of the bracketing techniques, the false-position method is usually the method of preference because for most problems it converges much faster than bisection.

Open techniques differ from bracketing methods in that they use information at a single point (or two values that need not bracket the root) to extrapolate to a new root estimate. This property is a double-edged sword. Although it leads to quicker convergence, it also allows the possibility that the solution may diverge. In general, the convergence of open techniques is partially dependent on the quality of the initial guess. The closer it is to the true root, the more likely the methods will converge.

Of the open techniques, the standard Newton-Raphson method is often used because of its property of quadratic convergence. However, its major shortcoming is that it requires that the derivative of the function be obtained analytically. For some functions this is impractical. In these cases, the secant method, which employs a finite-difference representation of the derivative, provides a viable alternative. Because of the finite-difference ap-

**TABLE II.3 Comparison of the characteristics of alternative methods for finding roots of algebraic and transcendental equations. The comparisons are based on general experience and do not account for the behavior of special functions.**

| Method | Initial guesses | Relative rate of convergence | Stability | Accuracy | Breadth of application | Programming effort | Comments |
|---|---|---|---|---|---|---|---|
| Direct | — | — | — | — | Very limited | | |
| Graphical | — | — | — | Poor | General | — | May take more time than the numerical method |
| Bisection | 2 | Slow | Always converges | Good | General | Easy | |
| False position | 2 | Medium | Always converges | Good | General | Easy | |
| One-point iteration | 1 | Slow | May not converge | Good | General | Easy | |
| Newton-Raphson | 1 | Fast | May not converge | Good | Limited if $f'(x) = 0$ | Easy | Requires evaluation of $f'(x)$ |
| Modified Newton-Raphson | 1 | Fast for multiple roots; medium for simple roots | May not converge | Good | Specifically designed for multiple roots | Easy | Requires evaluation of $f''(x)$ and $f'(x)$ |
| Secant | 2 | Medium to fast | May not converge | Good | General | Easy | Initial guesses do not have to bracket the root |

proximation, the rate of convergence of the secant method is initially slower than for the Newton-Raphson method. However, as the root estimate is refined, the difference approximation becomes a better representation of the true derivative, and convergence accelerates rapidly. The modified Newton-Raphson technique can be used to attain rapid convergence for multiple roots. However, this technique requires an analytical expression for both the first and second derivative.

All the numerical methods are easy to program on personal computers and require minimal time to determine a single root. On this basis, you might conclude that simple methods such as bisection would be good enough for practical purposes. This would be true if you were exclusively interested in determining the root of only one equation. However, there are many cases in engineering where numerous root locations are required and where speed becomes important. For these cases, slow methods are very time-consuming and, hence, costly. On the other hand, the fast open methods may diverge, and the accompanying delays can also be costly. Some computer algorithms attempt to capitalize on the strong points of both classes of techniques by initially employing a bracketing method to approach the root, then switching to an open method to rapidly refine the estimate. Whether a single approach or a combination is used, the trade-offs between convergence and speed are at the heart of the choice of a root-location technique.

## II.5 IMPORTANT RELATIONSHIPS AND FORMULAS

Table II.4 summarizes important information that was presented in Part II. This table can be consulted to quickly access important relationships and formulas.

## II.6 ADVANCED METHODS AND ADDITIONAL REFERENCES

The methods in this text have been limited to determining a single real root of an algebraic or transcendental equation based on foreknowledge of its approximate location. Other techniques are available for determining complex roots and all roots of a polynomial. Good general references on the subject are Ralston and Rabinowitz (1978) and Carnahan, Luther, and Wilkes (1969). James, Smith, and Wolford (1977) and Gerald and Wheatley (1984) provide summaries and computer codes for some of the methods.

**TABLE II.4 Summary of important information presented in Part II.**

| Method | Formulation | Graphical interpretation | Errors and stopping criteria |
|---|---|---|---|
| **Bracketing methods:** | | | |
| Bisection | $x_r = \dfrac{x_l + x_u}{2}$<br>If $f(x_l)f(x_r) < 0$, $x_u = x_r$<br>If $f(x_l)f(x_r) > 0$, $x_l = x_r$ | 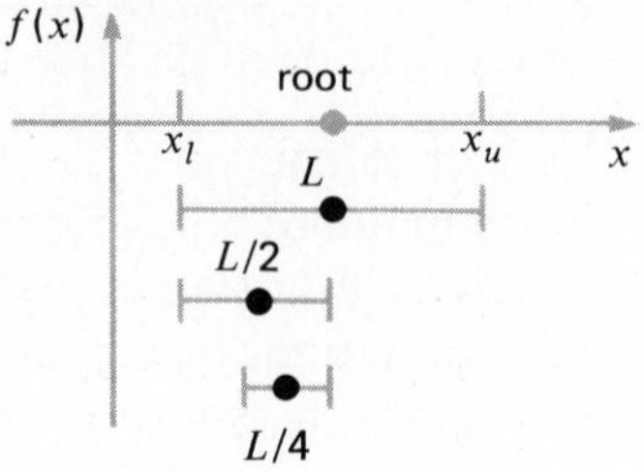 | Stopping criterion:<br>$\left\lvert \dfrac{x_r^{\text{new}} - x_r^{\text{old}}}{x_r^{\text{new}}} \right\rvert 100\% \le \epsilon_s$ |
| False position | $x_r = x_u - \dfrac{f(x_u)(x_l - x_u)}{f(x_l) - f(x_u)}$<br>If $f(x_l)f(x_r) < 0$, $x_u = x_r$<br>$f(x_l)f(x_r) > 0$, $x_l = x_r$ | 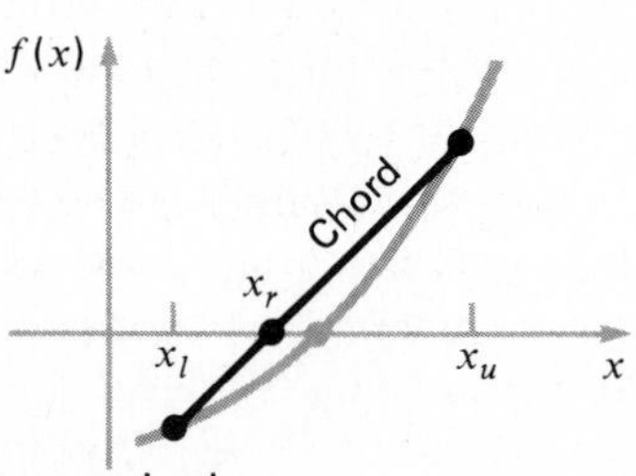 | Stopping criterion:<br>$\left\lvert \dfrac{x_r^{\text{new}} - x_r^{\text{old}}}{x_r^{\text{new}}} \right\rvert 100\% \le \epsilon_s$ |
| **Open methods:** | | | |
| Newton-Raphson | $x_{i+1} = x_i - \dfrac{f(x_i)}{f'(x_i)}$ | 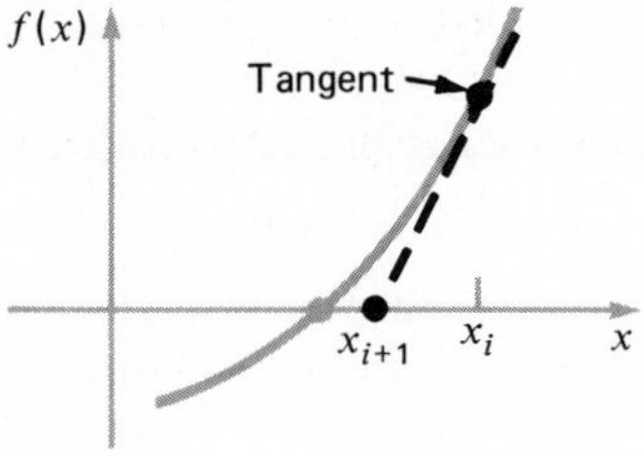 | Stopping criterion:<br>$\left\lvert \dfrac{x_{i+1} - x_i}{x_{i+1}} \right\rvert 100\% \le \epsilon_s$<br>Error: $E_{i+1} = 0(E_i^2)$ |
| Secant | $x_{i+1} = x_i - \dfrac{f(x_i)(x_{i-1} - x_i)}{f(x_{i-1}) - f(x_i)}$ | 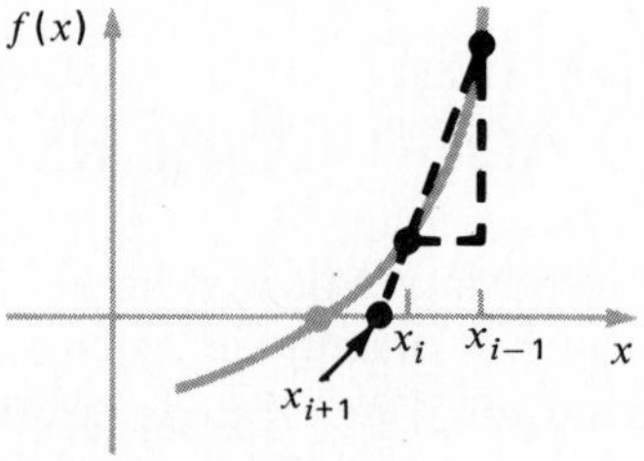 | Stopping criterion:<br>$\left\lvert \dfrac{x_{i+1} - x_i}{x_{i+1}} \right\rvert 100\% \le \epsilon_s$ |

As for specific techniques, the *Newton-Raphson method* can be employed in certain cases to locate complex roots on the basis of an initial complex guess. Because most computers cannot perform complex arithmetic, the technique is somewhat limited. However, Stark (1970) illustrates a way to circumvent this dilemma.

*Muller's method* is similar to the false-position method but uses quadratic rather than linear interpolation to locate the root. The approach can be employed to determine complex as well as real roots (Muller, 1956; Gerald and Wheatley, 1984; and Rice, 1983).

Several techniques are available to determine all the roots of polynomials. *Bairstow's method* requires good initial guesses for efficient root location (Gerald and Wheatley, 1984, and James, Smith, and Wolford, 1977). *Graeffe's method* (Scarborough, 1966, and James, Smith, and Wolford, 1977) and the *quotient-difference (QD) algorithm* (Henrici, 1964, and Gerald and Wheatley, 1984) determine all roots without initial guesses. Ralston and Rabinowitz (1978) and Carnahan, Luther, and Wilkes (1969) also contain discussions of the above methods as well as of other techniques for locating roots of polynomials.

In summary, the foregoing is intended to provide you with avenues for deeper exploration of the subject. Additionally, all of the above references provide descriptions of the basic techniques covered in Part II. We urge you to consult these alternative sources to broaden your understanding of numerical methods for root location.*

*Books are referenced only by author here; a complete bibliography will be found at the back of this text.

# PART THREE

# SYSTEMS OF LINEAR ALGEBRAIC EQUATIONS

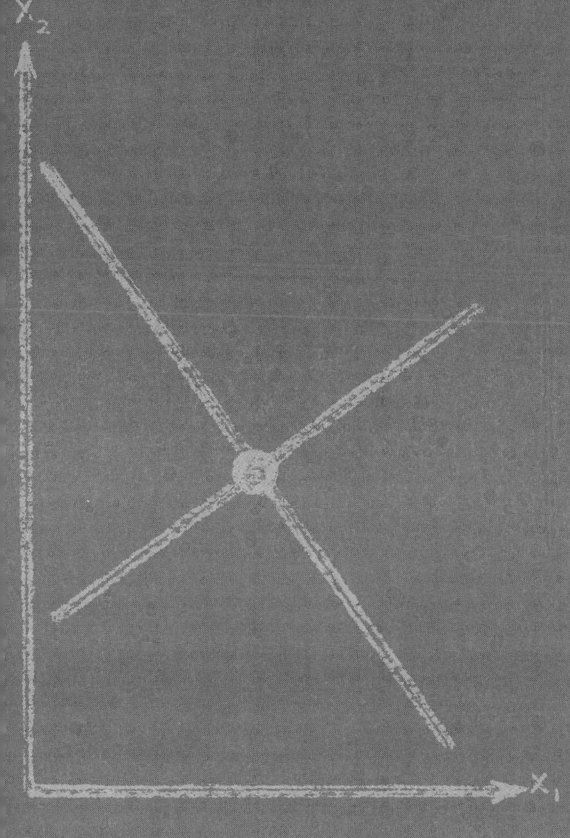


## III.1 MOTIVATION

In Part II, we determined the value $x$ that satisfied a single equation, $f(x) = 0$. Now, we deal with the case of determining the values $x_1, x_2, \ldots, x_n$, that simultaneously satisfy a set of equations:

$$
\begin{aligned}
f_1(x_1, x_2, \ldots, x_n) &= 0 \\
f_2(x_1, x_2, \ldots, x_n) &= 0 \\
&\vdots \\
f_n(x_1, x_2, \ldots, x_n) &= 0
\end{aligned}
$$

Such systems can be either linear or nonlinear. In Part III, we deal with *linear algebraic equations* that are of the general form:

$$
\begin{aligned}
a_{11}x_1 + a_{12}x_2 + \cdots + a_{1n}x_n &= c_1 \\
a_{21}x_1 + a_{22}x_2 + \cdots + a_{2n}x_n &= c_2 \\
\vdots \qquad \vdots \qquad\qquad \vdots \quad &\quad \vdots \\
a_{n1}x_1 + a_{n2}x_2 + \cdots + a_{nn}x_n &= c_n
\end{aligned}
\qquad \text{[III.1]}
$$

where the $a$'s are constant coefficients, the $c$'s are constants, and $n$ is the number of equations. All other equations are nonlinear. Nonlinear systems will be discussed briefly at the end of Part III.

### III.1.1 Precomputer Methods for Solving Systems of Equations

For small numbers of equations ($n \leq 3$), linear (and sometimes nonlinear) equations can be solved readily by simple techniques. Some of these methods will be reviewed at the beginning of Chap. 7. However, for four or more equations, solutions become arduous and computers must be utilized. Historically, the inability to solve all but the smallest sets of equations by hand has limited the scope of problems addressed in many engineering applications.

Before computers, techniques to solve systems of linear algebraic equations were time-consuming and awkward. These approaches placed a constraint on creativity because the methods were often difficult to implement and understand. Consequently, the techniques were sometimes overemphasized at the expense of other aspects of the problem-solving process such as formulation and interpretation (recall Fig. I.1 and accompanying discussion).

The advent of easily accessible personal computers makes it possible and practical for you to solve large sets of simultaneous linear algebraic equations. Thus, you can approach more complex and realistic examples and problems. Furthermore, you will have more time to test your creative skills because you will be able to place more emphasis on problem formulation and solution interpretation.

### III.1.2 Linear Algebraic Equations and Engineering Practice

Many of the fundamental equations of engineering are based on conservation laws (recall Table II.1). Some familiar quantities that conform to such laws are mass, force, energy, and momentum. In mathematical terms, these principles lead to balance or continuity equations that relate system *behavior* as represented by the *levels* or *response* of the quantity being modeled to the *properties* or *characteristics* of the system and the external *stimuli* acting on the system.

As an example, the conservation of mass can be used to formulate a mass balance for a series of chemical reactors (Fig. III.1*a*). For this case, the quantity being modeled is the mass of the chemical in each reactor. The system properties are the reaction characteristics of the chemical and the reactor sizes and flow rates. The external stimuli are the feed rates of the chemical into the system.

In the previous part of the book, you saw how single-component systems result in a single equation that can be solved using root-location techniques. Multicomponent systems result in a coupled set of mathematical equations that must be solved simultaneously. The equations are coupled because the individual parts of the system are influenced by other parts. For example, in Fig. III.1*a*, reactor 4 receives chemical inputs from reactors 2 and 3. Consequently, its response is dependent on the quantity of chemical in these other reactors.

When these dependencies are expressed mathematically, the resulting equations are often of the linear algebraic form of Eq. (III.1). The $x$'s are usually measures of the magnitudes and responses of the individual components. Using Fig. III.1*a* as an example, $x_1$ might quantify the amount of mass in the first reactor, $x_2$ might quantify the amount in the

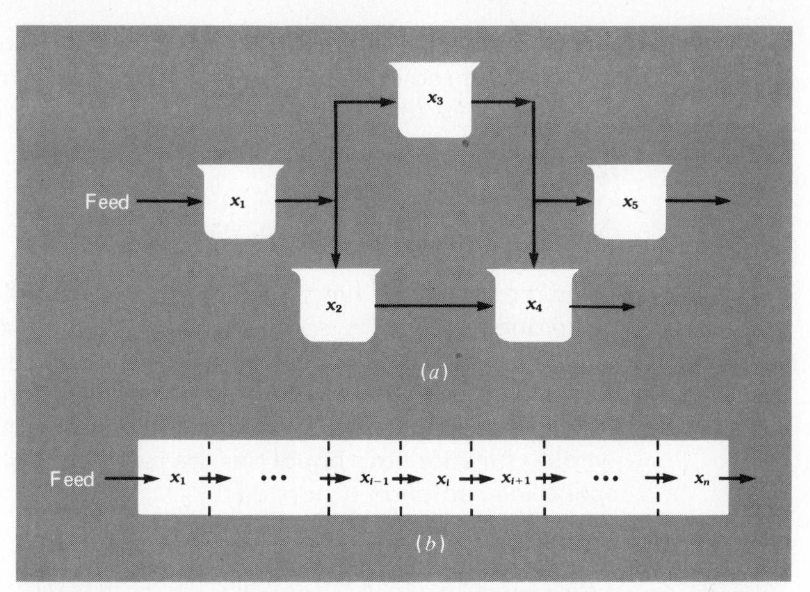


FIGURE III.1 Two types of systems that can be modeled using systems of linear algebraic equations: (*a*) macrovariable system that involves coupled finite components and (*b*) microvariable system that involves a continuum.

second, and so forth. The *a*'s typically represent the properties and characteristics that bear on the interactions between components. For instance, the *a*'s for Fig. III.1*a* might be reflective of the flow rates of mass between the reactors. Finally, the *c*'s usually represent the external stimuli acting on the system, such as the feed rate in Fig. III.1*a*. The case studies in Chap. 9 provide other examples of such equations derived from engineering practice.

Multicomponent problems of the above types arise from both *lumped* (macro-) or *continuous* (micro-) variable mathematical models (Fig. III.1). Lumped variable problems involve coupled finite components such as trusses (Case Study 9.3), reactors (Fig. III.1*a*), and electric circuits (Case Study 9.4). These types of problems use models that provide gross behavior of a system with little or no spatial detail.

Conversely, microscaled problems attempt to describe spatial detail of systems on a continuous or semicontinuous basis. The distribution of chemical along the length of an elongated, rectangular reactor (Fig. III.1*b*) is an example of a continuous variable model. Differential equations derived from the conservation laws specify the distribution of the dependent variable for such systems (Case Study 9.2). These differential equations can be solved numerically by converting them to an equivalent system of simultaneous algebraic equations. The solution of such sets of equations represents a major engineering application area for the methods in the following chapters. These equations are coupled because the variables at one location are dependent on the variables in adjoining regions. For example, the concentration at the middle of the reactor is a function of the concentration in adjoining regions. Similar examples could be developed for the spatial distribution of temperature or momentum.

Aside from physical systems, simultaneous linear algebraic equations also arise in a variety of mathematical-problem contexts. These result when mathematical functions are required to satisfy several conditions simultaneously. Each condition results in an equation that contains known coefficients and unknown variables. The techniques discussed in this part can be used to solve for the coefficients when the equations are linear and algebraic. Some widely used numerical techniques that employ simultaneous equations are regression analysis (Chap. 10) and spline interpolation (Chap. 11).

## III.2 MATHEMATICAL BACKGROUND

All parts of this book require some mathematical background. For Part III, matrix notation and algebra are useful because they provide a concise way to represent and manipulate systems of linear algebraic equations. If you are already familiar with matrices, feel free to skip to Sec. III.3. For those who are unfamiliar or require a review, the following material provides a brief introduction to the subject.

### III.2.1 Matrix Notation

A *matrix* consists of a rectangular array of elements represented by a single symbol. As depicted in Fig. III.2, $[A]$ is the shorthand notation for the matrix and $a_{ij}$ designates an individual *element* of the matrix.

A horizontal set of elements is called a *row* and a vertical set is called a *column*. The first subscript $i$ always designates the number of the row in which the element lies. The second subscript $j$ designates the column. For example, element $a_{23}$ is in row 2 and column 3.

The matrix in Fig. III.2 has $m$ rows and $n$ columns and is said to have a *dimension* of $m$ by $n$ (or $m \times n$). It is referred to as an $m$-by-$n$ matrix.

Matrices with row dimension $m = 1$, such as

$$[B] = [b_1 b_2 \; . \; . \; . \; b_n]$$

are called *row vectors*. Note that for simplicity, the first subscript of each element is dropped.

Matrices with column dimension $n = 1$, such as

Column 3

$$[A] = \begin{bmatrix} a_{11} & a_{12} & a_{13} & \cdots & a_{1n} \\ a_{21} & a_{22} & \boxed{a_{23}} & \cdots & a_{2n} \\ \cdot & \cdot & & & \cdot \\ \cdot & \cdot & & & \cdot \\ \cdot & \cdot & & & \cdot \\ a_{m1} & a_{m2} & a_{m3} & \cdots & a_{mn} \end{bmatrix} \quad \text{Row 2}$$

FIGURE III.2 A matrix.

$$[C] = \begin{bmatrix} c_1 \\ c_2 \\ \cdot \\ \cdot \\ \cdot \\ c_m \end{bmatrix}$$

are referred to as *column vectors.* For simplicity, the second subscript is dropped.

Matrices where $m = n$ are called *square matrices.* For example, a 4-by-4 matrix is

$$[A] = \begin{bmatrix} a_{11} & a_{12} & a_{13} & a_{14} \\ a_{21} & a_{22} & a_{23} & a_{24} \\ a_{31} & a_{32} & a_{33} & a_{34} \\ a_{41} & a_{42} & a_{43} & a_{44} \end{bmatrix}$$

The diagonal consisting of the elements $a_{11}$, $a_{22}$, $a_{33}$, and $a_{44}$ is termed the *main diagonal* of the matrix.

Square matrices are particularly important when solving sets of simultaneous linear equations. For such systems, the number of equations (corresponding to rows) and the number of unknowns (corresponding to columns) must be equal in order for a unique solution to be possible. Consequently, square matrices of coefficients are encountered when dealing with such systems. Some special types of square matrices are described in Box III.1.

BOX III.1 Special Types of Square Matrices

There are a number of special forms of square matrices that are important and should be noted:

A *symmetric matrix* is one where $a_{ij} = a_{ji}$ for all $i$'s and $j$'s. For example,

$$[A] = \begin{bmatrix} 5 & 1 & 2 \\ 1 & 3 & 7 \\ 2 & 7 & 8 \end{bmatrix}$$

is a 3-by-3 symmetric matrix.

A *diagonal matrix* is a square matrix where all elements off the main diagonal are equal to zero, as in

$$[A] = \begin{bmatrix} a_{11} & & & \\ & a_{22} & & \\ & & a_{33} & \\ & & & a_{44} \end{bmatrix}$$

Note that where large blocks of elements are zero, they are left blank.

An *identity matrix* is a diagonal matrix where all elements on the main diagonal are equal to 1, as in

$$[I] = \begin{bmatrix} 1 & & & \\ & 1 & & \\ & & 1 & \\ & & & 1 \end{bmatrix}$$

The symbol $[I]$ is used to denote the identity matrix. The identity matrix has properties similar to unity.

An *upper triangular matrix* is one where all the elements below the main diagonal are zero, as in

$$[A] = \begin{bmatrix} a_{11} & a_{12} & a_{13} & a_{14} \\ & a_{22} & a_{23} & a_{24} \\ & & a_{33} & a_{34} \\ & & & a_{44} \end{bmatrix}$$

A *lower triangular matrix* is one where all elements above the main diagonal are zero, as in

$$[A] = \begin{bmatrix} a_{11} & & & \\ a_{21} & a_{22} & & \\ a_{31} & a_{32} & a_{33} & \\ a_{41} & a_{42} & a_{43} & a_{44} \end{bmatrix}$$

A *banded matrix* has all elements equal to zero, with the exception of a band centered on the main diagonal:

$$[A] = \begin{bmatrix} a_{11} & a_{12} & & \\ a_{21} & a_{22} & a_{23} & \\ & a_{32} & a_{33} & a_{34} \\ & & a_{43} & a_{44} \end{bmatrix}$$

The above matrix has a bandwidth of 3 and is given a special name—the *tridiagonal matrix.*

## III.2.2 Matrix Operating Rules

Now that we have specified what we mean by a matrix, we can define some operating rules that govern its use. Two $m$-by-$n$ matrices are equal if, and only if, every element in the first is equal to every element in the second; that is $[A] = [B]$ if $a_{ij} = b_{ij}$ for all $i$ and $j$.

*Addition* of two matrices, say $[A]$ and $[B]$, is accomplished by adding corresponding terms in each matrix. The elements of the resulting matrix $[C]$ are computed as

$$c_{ij} = a_{ij} + b_{ij}$$

for $i = 1, 2, \ldots, m$ and $j = 1, 2, \ldots, n$.

Similarly, the *subtraction* of two matrices, say $[E]$ minus $[F]$, is obtained by subtracting corresponding terms, as in

$$d_{ij} = e_{ij} - f_{ij}$$

for $i = 1, 2, \ldots, m$ and $j = 1, 2, \ldots, n$. It follows directly from the above definition that addition and subtraction can only be performed between matrices having the same dimensions.

Both addition and subtraction are *commutative*:

$$[A] + [B] = [B] + [A]$$

and

$$[E] - [F] = -[F] + [E]$$

Addition and subtraction are also *associative*, that is,

$$[A] + ([B] + [C]) = ([A] + [B]) + [C]$$

The *multiplication* of a matrix $[A]$ by a *scalar* $g$ is obtained by multiplying every element of $[A]$ by $g$, as in

$$[B] = g[A] = \begin{bmatrix} ga_{11} & ga_{12} & \cdots & ga_{1n} \\ ga_{21} & ga_{22} & \cdots & ga_{2n} \\ \cdot & \cdot & & \cdot \\ \cdot & \cdot & & \cdot \\ \cdot & \cdot & & \cdot \\ ga_{m1} & ga_{m2} & \cdots & ga_{mn} \end{bmatrix}$$

The *product* of two matrices is represented as $[C] = [A][B]$, where the elements of $[C]$ are defined as (see Box III.2 for a simple way to conceptualize matrix multiplication)

$$c_{ij} = \sum_{k=1}^{n} a_{ik} b_{kj} \qquad \text{[III.2]}$$

Where $n$ = the column dimension of $[A]$ and the row dimension of $[B]$.

---

BOX III.2: A Simple Method for Multiplying Two Matrices

Although Eq. (III.2) is well-suited for implementation on a computer, it is not the simplest means for visualizing the mechanics of multiplying two matrices. What follows gives more tangible expression to the concept. Suppose that we want to multiply $[A]$ by $[B]$ to yield $[C]$:

$$[C] = [A][B] = \begin{bmatrix} 3 & 1 \\ 8 & 6 \\ 0 & 4 \end{bmatrix} \begin{bmatrix} 5 & 9 \\ 7 & 2 \end{bmatrix}$$

A simple way to represent the computation of $[C]$ is to raise $[B]$, as in

$$\begin{matrix} & \begin{bmatrix} 5 & 9 \\ 7 & 2 \end{bmatrix} & \leftarrow [B] \\ [A] \rightarrow \begin{bmatrix} 3 & 1 \\ 8 & 6 \\ 0 & 4 \end{bmatrix} & \begin{bmatrix} & & \\ & ? & \\ & & \end{bmatrix} & \leftarrow [C] \end{matrix}$$

Now the answer $[C]$ can be computed in the space vacated by $[B]$. This format has utility because it aligns the appropriate rows and columns that are to be multiplied. For example, according to Eq. (III.2) the element $c_{1,1}$ is obtained by multiplying the first row of $[A]$ by the first column of $[B]$. This amounts to adding the product of $a_{1,1}$ and $b_{1,1}$ to the product of $a_{1,2}$ and $b_{2,1}$, as in

$$\begin{matrix} & \begin{bmatrix} \mathbf{5} & 9 \\ \mathbf{7} & 2 \end{bmatrix} & \leftarrow [B] \\ [A] \rightarrow \begin{bmatrix} \mathbf{3} & \mathbf{1} \\ 8 & 6 \\ 0 & 4 \end{bmatrix} & \begin{bmatrix} \mathbf{3 \times 5 + 1 \times 7 = 22} & \\ & \\ & \end{bmatrix} & \leftarrow [C] \end{matrix}$$

Thus, $c_{1,1}$ is equal to 22. Element $c_{2,1}$ can be computed in a similar fashion, as in

$$\begin{matrix} & \begin{bmatrix} \mathbf{5} & 9 \\ \mathbf{7} & 2 \end{bmatrix} & \leftarrow [B] \\ [A] \rightarrow \begin{bmatrix} 3 & 1 \\ \mathbf{8} & \mathbf{6} \\ 0 & 4 \end{bmatrix} & \begin{bmatrix} 22 & \\ \mathbf{8 \times 5 + 6 \times 7 = 82} & \\ & \end{bmatrix} & \leftarrow [C] \end{matrix}$$

The computation can be continued in this way, following the alignment of the rows and columns, to yield the result:

$$[C] = \begin{bmatrix} 22 & 29 \\ 82 & 84 \\ 28 & 8 \end{bmatrix}$$

Note how this simple method makes it clear why it is impossible to multiply if the number of columns of the first matrix does not equal the number of rows in the second matrix. Also, note how it demonstrates that the order of multiplication matters. Problem 7.3 also illustrates these points.

That is, the $c_{ij}$ element is obtained by adding the product of individual elements from the $i$th row of the first matrix, in this case [$A$], by the $j$th column of the second matrix [$B$]. According to this definition, multiplication of two matrices can only be performed if the first matrix has as many columns as the number of rows in the second matrix. Thus, if [$A$] is an $m$-by-$n$ matrix, [$B$] could be an $n$-by-$p$ matrix. For this case, the resulting [$C$] matrix would have dimension of $m$ by $p$. However, if [$B$] were a $p$-by-$n$ matrix, the multiplication could not be performed. Figure III.3 provides an easy way to check whether two matrices can be multiplied.

If the dimensions of the matrices are suitable, matrix multiplication is *associative*:

$$([A][B])\,[C] = [A]\,([B][C])$$

and *distributive*:

$$[A]\,([B] + [C]) = [A][B] + [A][C]$$

or

$$([A] + [B])\,[C] = [A][C] + [B][C]$$

However, multiplication is not generally *commutative*:

$$[A][B] \neq [B][A]$$

That is, *the order of multiplication is important.*

Although multiplication is possible, matrix division is not a defined operation. However, if a matrix [$A$] is square, there is another matrix $[A]^{-1}$, called the *inverse* of [$A$], for which

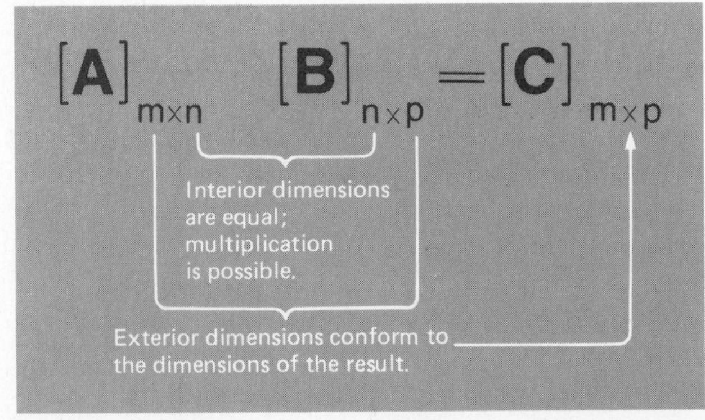


FIGURE III.3 A simple way to check whether matrix multiplication is possible.

$$[A][A]^{-1} = [A]^{-1}[A] = [I] \tag{III.3}$$

Thus, the multiplication of a matrix by the inverse is analogous to division, in the sense that a number divided by itself is equal to 1. That is, multiplication of a matrix by its inverse leads to the identity matrix (recall Box III.1).

The inverse of a two-dimensional square matrix can be represented simply by

$$[A]^{-1} = \frac{1}{a_{11}\,a_{22} - a_{12}\,a_{21}} \begin{bmatrix} a_{22} & -a_{12} \\ -a_{21} & a_{11} \end{bmatrix} \tag{III.4}$$

Higher dimensional matrices are much more involved. Section 8.2 will be devoted to a technique for computing the inverse for such systems.

The final matrix manipulations that will have utility in our discussion are transposition and augmentation of matrices. The *transpose* of a matrix involves transforming its rows into columns and its columns into rows. For the matrix

$$[A] = \begin{bmatrix} a_{11} & a_{12} & \cdots & a_{1n} \\ a_{21} & a_{22} & \cdots & a_{2n} \\ \cdot & \cdot & & \cdot \\ \cdot & \cdot & & \cdot \\ \cdot & \cdot & & \cdot \\ a_{m1} & a_{m2} & \cdots & a_{mn} \end{bmatrix}$$

the transpose, designated $[A]^T$, is defined as

$$[A]^T = \begin{bmatrix} a_{11} & a_{21} & \cdots & a_{m1} \\ a_{12} & a_{22} & \cdots & a_{m2} \\ \cdot & \cdot & & \cdot \\ \cdot & \cdot & & \cdot \\ \cdot & \cdot & & \cdot \\ a_{1n} & a_{2n} & \cdots & a_{mn} \end{bmatrix}$$

In other words, the element $a_{ij}$ of the transpose is equal to the $a_{ji}$ element of the original matrix, or $a_{ij} = a_{ji}$. The transpose has a variety of functions in matrix algebra. One simple advantage is that it allows a column vector to be written as a row. For example, if

$$[C] = \begin{bmatrix} c_{11} \\ c_{21} \\ c_{31} \\ c_{41} \end{bmatrix}$$

then

$$[C]^T = [c_{11}\ c_{21}\ c_{31}\ c_{41}]$$

where the superscript T designates the transpose. For example, this can save space when writing a column vector in a manuscript. In addition, the transpose has a number of mathematical applications.

A matrix is *augmented* by the addition of a column (or columns) to the original matrix. For example, suppose we have a matrix of coefficients:

$$[A] = \begin{bmatrix} a_{11} & a_{12} & a_{13} \\ a_{21} & a_{22} & a_{23} \\ a_{31} & a_{32} & a_{33} \end{bmatrix}$$

We might wish to augment this matrix $[A]$ with an identity matrix (recall Box III.1) to yield a 3-by-6 dimensional matrix:

$$\left[\begin{array}{ccc|ccc} a_{11} & a_{12} & a_{13} & 1 & 0 & 0 \\ a_{21} & a_{22} & a_{23} & 0 & 1 & 0 \\ a_{31} & a_{32} & a_{33} & 0 & 0 & 1 \end{array}\right]$$

Such an expression has utility when we must perform a set of identical operations on two matrices. Thus, we can perform the operations on the single augmented matrix rather than on the two individual matrices.

### III.2.3 Representing Simultaneous Linear Algebraic Equations in Matrix Form

It should be clear that matrices provide a concise notation for representing simultaneous linear equations. For example, Eq. (III.1) can be expressed as

$$[A][X] = [C] \qquad \text{[III.5]}$$

where $[A]$ is the $n$-by-$n$ square matrix of coefficients:

$$[A] = \begin{bmatrix} a_{11} & a_{12} & \cdots & a_{1n} \\ a_{21} & a_{22} & \cdots & a_{2n} \\ \cdot & \cdot & & \cdot \\ \cdot & \cdot & & \cdot \\ \cdot & \cdot & & \cdot \\ a_{n1} & a_{n2} & \cdots & a_{nn} \end{bmatrix}$$

$[C]$ is the $n$-by-1 column vector of constants:

$$[C]^T = [c_1 \; c_2 \; c_3 \; . \; . \; . \; c_n]$$

and $[X]$ is the $n$-by-1 column vector of unknowns:

$$[X]^T = [x_1 \; x_2 \; x_3 \; . \; . \; . \; x_n]$$

Recall the definition of matrix multiplication [Eq. (III.2) or Box III.2] to convince yourself that Eqs. (III.1) and (III.5) are equivalent. Also, realize that Eq. (III.5) is a valid matrix multiplication because the number of columns ($n$) of the first matrix ($[A]$) is equal to the number of rows ($n$) of the second matrix ($[X]$).

This part of the book is devoted to solving Eq. (III.5) for $[X]$. A formal way to obtain a solution using matrix algebra is to multiply each side of the equation by the inverse of $[A]$ to yield

$$[A]^{-1}[A][X] = [A]^{-1}[C]$$

Because $[A]^{-1}[A]$ equals the identity matrix, the equation becomes

$$[X] = [A]^{-1}[C] \tag{III.6}$$

Therefore, the equation has been solved for $[X]$. This is another example of how the inverse plays a role that is similar to division in matrix algebra.

Finally, we will sometimes find it useful to augment $[A]$ with $[C]$. For example, if $n = 3$ this results in a 3-by-4 dimensional matrix:

$$\left[\begin{array}{ccc|c} a_{11} & a_{12} & a_{13} & c_1 \\ a_{21} & a_{22} & a_{23} & c_2 \\ a_{31} & a_{32} & a_{33} & c_3 \end{array}\right] \tag{III.7}$$

Expressing the equations in this form has utility, because several of the techniques for solving linear systems perform identical operations to a row of coefficients and the corresponding right-hand-side constant. As expressed in Eq. (III.7), we can perform the manipulation once on an individual row of the augmented matrix rather than separately on the coefficient matrix and the right-hand-side vector.

## III.3 ORIENTATION

Before proceeding to the numerical methods, some further orientation might be helpful. The following is intended as an overview of the material discussed in Part III. In addition, we have formulated some objectives to help focus your efforts when studying the material.

### III.3.1 Scope and Preview

Figure III.4 provides an overview for Part III. *Chapter* 7 is devoted to the most fundamental technique for solving linear algebraic systems: Gauss elimination. Before launching into a detailed discussion of this technique, a preliminary section deals with simple methods for solving small systems. These approaches are presented to provide you with visual insight and because one of the methods—the elimination of unknowns—represents the basis for Gauss elimination.

After the preliminary material, "naive" Gauss elimination is discussed.

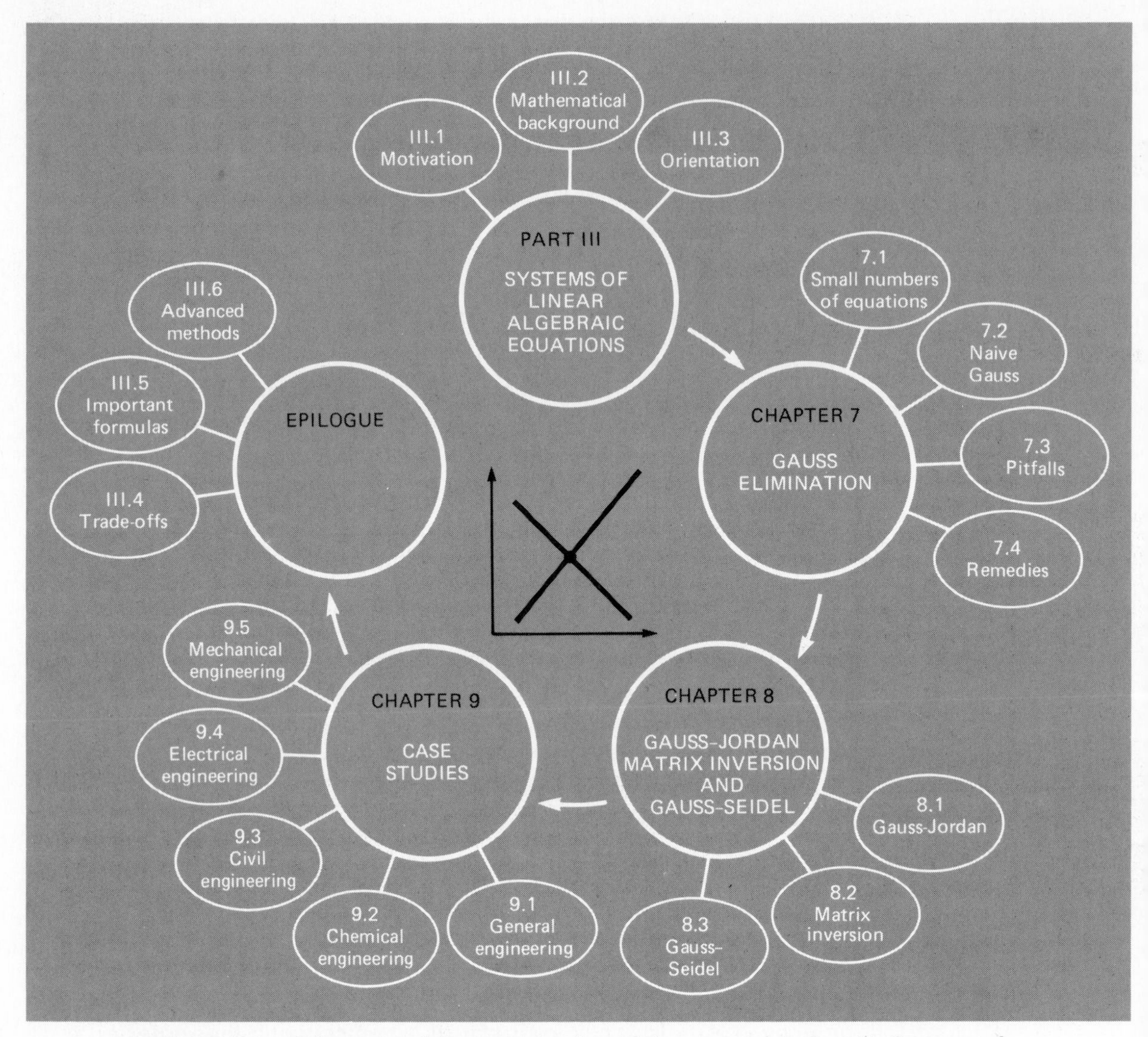


FIGURE III.4 Schematic of the organization of the material in Part III: Systems of linear algebraic equations.

We start with this "stripped-down" version because it allows the fundamental technique to be elaborated without complicating details. Then, in subsequent sections, we discuss potential problems of the naive approach and present a number of modifications to minimize and circumvent these problems. At the end of the chapter, a box is devoted to a very efficient form of Gauss elimination that is available for tridiagonal systems.

*Chapter 8* begins with a discussion of the Gauss-Jordan method. Although this technique is quite similar to that of Gauss elimination, it is presented because it allows the computation of the matrix inverse which has tremendous utility in engineering practice.

Both the Gauss elimination and the Gauss-Jordan methods are exact techniques. Chapter 8 concludes with an alternative type of approach called the Gauss-Seidel method. This technique is similar in spirit to the approximate methods for roots of equations that were discussed in Chap. 5. That is, the technique involves guessing a solution and then iterating to obtain a refined estimate.

*Chapter 9* demonstrates how the methods can actually be applied for problem solving. As with other parts of the book, case studies are drawn from all fields of engineering.

Finally, an epilogue is included at the end of Part III. This review includes discussion of trade-offs that are relevant to implementation of the methods in engineering practice. This section also summarizes the important formulas and advanced methods related to linear algebraic equations. As such, it can be used before exams or as a refresher after you have graduated and must return to linear algebraic equations as a professional.

Automatic computation capability is integrated into Part III in a number of ways. First, user-friendly NUMERICOMP software for the Gauss elimination method is available on an optional basis for the Apple-II and the IBM-PC. But computer codes for Gauss elimination, using both FORTRAN and BASIC, are also given directly in the text. This provides you with the opportunity to copy and enhance the code for implementation on your own personal or mainframe computer. Flowcharts or algorithms are included for most of the other methods described in the text. This material can form the basis for a comprehensive software package that you can develop and apply to a number of engineering problems.

### III.3.2 Goals and Objectives

*Study objectives.* After completing Part III, you should be able to solve most problems involving linear algebraic equations and appreciate the application of these equations in many fields of engineering. You should strive to master several techniques and assess their reliability. You should understand the trade-offs involved in selecting the "best" method (or methods) for any particular problem. In addition to these general objectives, the specific concepts listed in Table III.1 should be assimilated and mastered.

*Computer objectives.* Your most fundamental computer objective is to be able to use a program to successfully solve systems of linear algebraic equations. You should understand how to use the NUMERICOMP software or, alternatively, how to copy and use the computer program for naive Gauss elimination given in this book. These programs will allow you to adequately handle many practical problems that involve several simultaneous linear algebraic equations. As you progress to problems that contain more equations you can utilize the other programs, flowcharts, and algorithms provided in Part III. Eventually you may want to incorporate partial pivoting, determinant calculation, and condition evaluation into your programs. You may also want to have your own software for the Gauss-Jordan and Gauss-Seidel methods.

**TABLE III.1 Specific study objectives for Part III**

1. Understand the graphical interpretation of ill-conditioned systems and how it relates to the determinant
2. Understand the reason for the label "naive" Gauss elimination
3. Be familiar with terminology: forward elimination, back-substitution, normalization, pivot equation, and pivot coefficient
4. Understand the problems of division by zero, round-off error, and ill conditioning
5. Know how to evaluate system condition
6. Know how to compute the determinant using Gauss elimination
7. Understand the advantages of pivoting; realize the difference between partial and complete pivoting
8. Know how to apply error-correction techniques to improve solutions
9. Understand why banded systems are relatively efficient to solve
10. Know the fundamental difference between Gauss elimination and the Gauss-Jordan method
11. Understand how the Gauss-Jordan method is used to compute the matrix inverse
12. Know how to interpret the elements of the matrix inverse in evaluating stimulus-response computations in engineering
13. Realize how to use the matrix inverse to evaluate system condition
14. Understand why the Gauss-Seidel method is particularly well-suited for large systems of equations
15. Know how to assess diagonal dominance of a system of equations and how it relates to whether the system can be solved with the Gauss-Seidel method
16. Understand the rationale behind relaxation; know where underrelaxation and overrelaxation are appropriate

# CHAPTER SEVEN

# GAUSS ELIMINATION

The present chapter deals with linear simultaneous algebraic equations which can be represented generally as

$$
\begin{aligned}
a_{11}x_1 + a_{12}x_2 + \cdots + a_{1n}x_n &= c_1 \\
a_{21}x_1 + a_{22}x_2 + \cdots + a_{2n}x_n &= c_2 \\
\cdot \quad \cdot \quad \quad \cdot \quad &\cdot \\
\cdot \quad \cdot \quad \quad \cdot \quad &\cdot \\
\cdot \quad \cdot \quad \quad \cdot \quad &\cdot \\
a_{n1}x_1 + a_{n2}x_2 + \cdots + a_{nn}x_n &= c_n
\end{aligned}
\qquad [7.1]
$$

where the $a$'s are constant coefficients and the $c$'s are constants.

The technique described in this chapter is called *Gauss elimination* because it involves combining equations in order to eliminate unknowns. Although it is one of the earliest methods for solving simultaneous equations, it remains among the most important algorithms in use today. In particular, it is easy to program and apply using personal computers.

## 7.1 SOLVING SMALL NUMBERS OF EQUATIONS

Before proceeding to the computer methods, we will describe several methods that are appropriate for solving small ($n \leq 3$) sets of simultaneous equations and that do not require a computer. These are the graphical method, Cramer's rule, and the elimination of unknowns.

### 7.1.1 The Graphical Method

A graphical solution is obtainable for two equations by plotting them on cartesian coordinates with one axis corresponding to $x_1$ and the other to $x_2$. Because we are dealing with linear systems, each equation is a straight line. This can be easily illustrated for the general equations

$$a_{11}x_1 + a_{12}x_2 = c_1$$

$$a_{21}x_1 + a_{22}x_2 = c_2$$

Both equations can be solved for $x_2$:

$$x_2 = -\left(\frac{a_{11}}{a_{12}}\right)x_1 + \frac{c_1}{a_{12}}$$

$$x_2 = -\left(\frac{a_{21}}{a_{22}}\right)x_1 + \frac{c_2}{a_{22}}$$

Thus, the equations are now in the form of straight lines; that is, $x_2 =$ (slope) $x_1$ + intercept. These lines can be graphed on cartesian coordinates with $x_2$ as the ordinate and $x_1$ as the abscissa. The values of $x_1$ and $x_2$ at the intersection of the lines represent the solution.

## EXAMPLE 7.1
### The Graphical Method for Two Equations

Problem Statement: Use the graphical method to solve

$$3x_1 + 2x_2 = 18 \quad \text{[E7.1.1]}$$

$$-x_1 + 2x_2 = 2 \quad \text{[E7.1.2]}$$

Solution: Let $x_1$ be the abscissa. Solve Eq. (E7.1.1) for $x_2$:

$$x_2 = (-3/2)x_1 + 9$$

which, when plotted on Fig. 7.1, is a straight line with an intercept of 9 and a slope of $-3/2$.

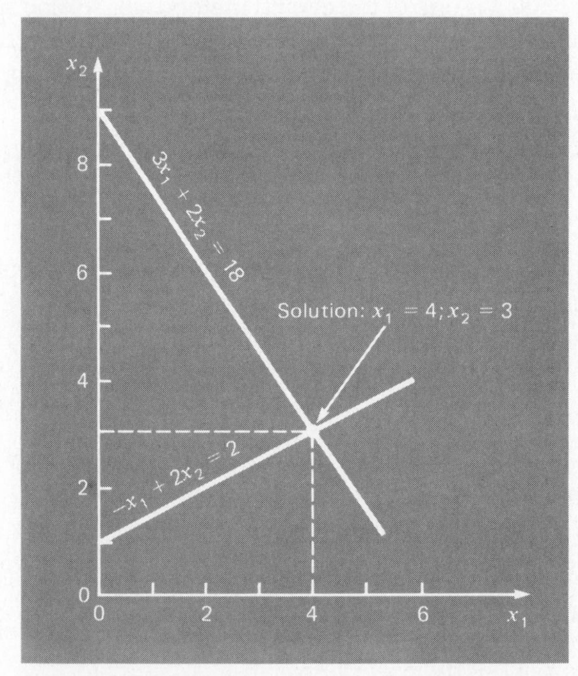


FIGURE 7.1 Graphical solution of a set of two simultaneous linear algebraic equations. The intersection of the lines represents the solution.

Equation (E7.1.2) can also be solved for $x_2$:

$$x_2 = (1/2)x_1 + 1$$

which is also plotted on Fig. 7.1. The solution is the intersection of the two lines at $x_1 = 4$ and $x_2 = 3$. This result can be checked by substituting these values into the original equations to yield

$$3(4) + 2(3) = 18$$

$$-4 + 2(3) = 2$$

Thus, the results are equivalent to the right-hand sides of the original equations.

For three simultaneous equations, each equation would be represented by a plane in a three-dimensional coordinate system. The point where the three planes intersect would represent the solution. Beyond three equations, graphical methods break down and, consequently, have little practical value for solving simultaneous equations. However, they sometimes prove useful in visualizing properties of the solutions. For example, Fig. 7.2 depicts three cases which can pose problems when solving sets of linear equations. Figure 7.2*a* shows the case where the two equations represent parallel lines. For such situations, there is no solution because the lines never cross. Figure 7.2*b* depicts the case where the two lines are coincident. For such situations there is an infinite number of solutions. Both types of systems are said to be *singular.* In addition, systems that are very close to being singular (Fig. 7.2*c*) can also cause problems. These systems are said to be *ill-conditioned.* Graphically, this corresponds to the fact that it is difficult to visualize the exact point at which the lines intersect. As illustrated in subsequent sections, ill-conditioned systems will also pose problems when they are encountered during the numerical solution of linear equations.

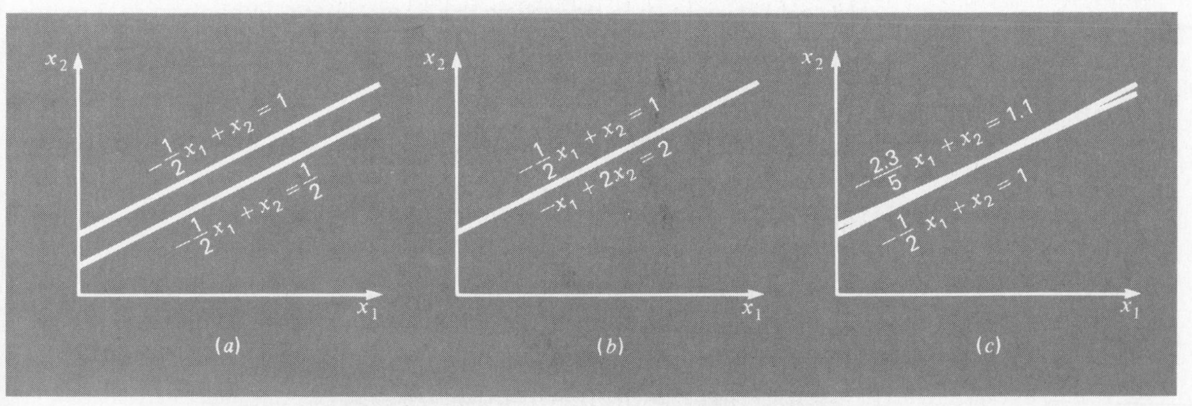


FIGURE 7.2 Graphical depiction of ill-conditioned systems: (*a*) no solution, (*b*) infinite solutions, and (*c*) an ill-conditioned system where the slopes are so close that the point of intersection is difficult to detect visually.

### 7.1.2 Determinants and Cramer's Rule

Cramer's rule is another solution technique that is best suited to small numbers of equations. Before describing this method, we will briefly introduce the concept of the determinant, which is used to implement Cramer's rule. In addition, the determinant has utility in evaluating ill conditioning of a matrix.

**Determinants.** The determinant can be illustrated for a set of three equations:

$$a_{11}x_1 + a_{12}x_2 + a_{13}x_3 = c_1$$

$$a_{21}x_1 + a_{22}x_2 + a_{23}x_3 = c_2$$

$$a_{31}x_1 + a_{32}x_2 + a_{33}x_3 = c_3$$

or, in matrix form,

$$[A][X] = [C]$$

where $[A]$ is the coefficient matrix:

$$[A] = \begin{bmatrix} a_{11} & a_{12} & a_{13} \\ a_{21} & a_{22} & a_{23} \\ a_{31} & a_{32} & a_{33} \end{bmatrix}$$

The *determinant* $D$ of this system is formed from the coefficients of the equation, as in

$$D = \begin{vmatrix} a_{11} & a_{12} & a_{13} \\ a_{21} & a_{22} & a_{23} \\ a_{31} & a_{32} & a_{33} \end{vmatrix} \quad [7.2]$$

Although the determinant $D$ and the coefficient matrix $[A]$ are composed of the same elements, they are completely different mathematical concepts. That is why they are distinguished visually by using brackets to enclose the matrix and straight lines to enclose the determinant. In contrast to a matrix, the determinant is a single number. For example, the value of the second-order determinant

$$D = \begin{vmatrix} a_{11} & a_{12} \\ a_{21} & a_{22} \end{vmatrix}$$

is calculated by

$$D = a_{11}a_{22} - a_{12}a_{21} \quad [7.3]$$

For the third-order case [Eq. (7.2)], a single numerical value for the determinant can be computed as

$$D = a_{11}\begin{vmatrix} a_{22} & a_{23} \\ a_{32} & a_{33} \end{vmatrix} - a_{12}\begin{vmatrix} a_{21} & a_{23} \\ a_{31} & a_{33} \end{vmatrix} + a_{13}\begin{vmatrix} a_{21} & a_{22} \\ a_{31} & a_{32} \end{vmatrix} \quad [7.4]$$

where the 2-by-2 determinants are called *minors*.

### EXAMPLE 7.2
Determinants

Problem Statement: Compute values for the determinants of the systems represented in Figs. 7.1 and 7.2.

Solution: For Fig. 7.1:

$$D = \begin{vmatrix} 3 & 2 \\ -1 & 2 \end{vmatrix} = 3(2) - 2(-1) = 8$$

For Fig. 7.2*a*:

$$D = \begin{vmatrix} -1/2 & 1 \\ -1/2 & 1 \end{vmatrix} = \frac{-1}{2}(1) - 1\left(\frac{-1}{2}\right) = 0$$

For Fig. 7.2*b*:

$$D = \begin{vmatrix} -1/2 & 1 \\ -1 & 2 \end{vmatrix} = \frac{-1}{2}(2) - 1(-1) = 0$$

For Fig. 7.2*c*

$$D = \begin{vmatrix} -1/2 & 1 \\ -2.3/5 & 1 \end{vmatrix} = \frac{-1}{2}(1) - 1\left(\frac{-2.3}{5}\right) = -0.04$$

In the foregoing example, the singular systems had zero determinants. Additionally, the results suggest that the system which is almost singular (Fig. 7.2*c*) has a determinant that is close to zero. These ideas will be pursued further in our subsequent discussion of ill conditioning (Sec. 7.3.3).

**Cramer's Rule.** This rule states that each unknown in a system of linear algebraic equations may be expressed as a fraction of two determinants with denominator $D$ and with the numerator obtained from $D$ by replacing the column of coefficients of the unknown in question by the constants $c_1$, $c_2$, . . . , $c_n$. For example, $x_1$ would be computed as

$$x_1 = \frac{\begin{vmatrix} c_1 & a_{12} & a_{13} \\ c_2 & a_{22} & a_{23} \\ c_3 & a_{32} & a_{33} \end{vmatrix}}{D} \qquad [7.5]$$

### EXAMPLE 7.3
Cramer's Rule

Problem Statement: Use Cramer's rule to solve

$$0.3x_1 + 0.52x_2 + x_3 = -0.01$$

$$0.5x_1 + x_2 + 1.9x_3 = 0.67$$

$$0.1x_1 + 0.3x_2 + 0.5x_3 = -0.44$$

Solution: The determinant $D$ can be written as [Eq. (7.2)]

$$D = \begin{vmatrix} 0.3 & 0.52 & 1 \\ 0.5 & 1 & 1.9 \\ 0.1 & 0.3 & 0.5 \end{vmatrix}$$

The minors are

$$A_1 = \begin{vmatrix} 1 & 1.9 \\ 0.3 & 0.5 \end{vmatrix} = 1(0.5) - 1.9(0.3) = -0.07$$

$$A_2 = \begin{vmatrix} 0.5 & 1.9 \\ 0.1 & 0.5 \end{vmatrix} = 0.5(0.5) - 1.9(0.1) = 0.06$$

$$A_3 = \begin{vmatrix} 0.5 & 1 \\ 0.1 & 0.3 \end{vmatrix} = 0.5(0.3) - 1(0.1) = 0.05$$

These can be used to evaluate the determinant, as in [Eq. (7.4)]

$$D = 0.3(-0.07) - 0.52(0.06) + 1(0.05) = -0.0022$$

Applying Eq. (7.5), the solution is

$$x_1 = \frac{\begin{vmatrix} -0.01 & 0.52 & 1 \\ 0.67 & 1 & 1.9 \\ -0.44 & 0.3 & 0.5 \end{vmatrix}}{-0.0022} = \frac{0.03278}{-0.0022} = -14.9$$

$$x_2 = \frac{\begin{vmatrix} 0.3 & -0.01 & 1 \\ 0.5 & 0.67 & 1.9 \\ 0.1 & -0.44 & 0.5 \end{vmatrix}}{-0.0022} = \frac{0.0649}{-0.0022} = -29.5$$

$$x_3 = \frac{\begin{vmatrix} 0.3 & 0.52 & -0.01 \\ 0.5 & 1 & 0.67 \\ 0.1 & 0.3 & -0.44 \end{vmatrix}}{-0.0022} = \frac{-0.04356}{-0.0022} = 19.8$$

For more than three equations, Cramer's rule becomes impractical because, as the number of equations increases, the determinants are time-consuming to evaluate by hand. Consequently, more efficient alternatives are used. Some of these alternatives are based on the last noncomputer solution technique covered in this book—the elimination of unknowns.

### 7.1.3 The Elimination of Unknowns

The elimination of unknowns by combining equations is an algebraic approach that can be illustrated for a set of two equations:

$$a_{11}x_1 + a_{12}x_2 = c_1 \qquad [7.6]$$

$$a_{21}x_1 + a_{22}x_2 = c_2 \qquad [7.7]$$

The basic strategy is to multiply the equations by constants in order that one of the unknowns will be eliminated when the two equations are combined. The result is a single equation that can be solved for the remaining unknown. This value can then be substituted into either of the original equations to compute the other variable.

For example, Eq. (7.6) might be multiplied by $a_{21}$ and Eq. (7.7) by $a_{11}$ to give

$$a_{11}a_{21}x_1 + a_{12}a_{21}x_2 = c_1a_{21} \qquad [7.8]$$

$$a_{21}a_{11}x_1 + a_{22}a_{11}x_2 = c_2a_{11} \qquad [7.9]$$

Subtracting Eq. (7.8) from Eq. (7.9) will, therefore, eliminate the $x_1$ term from the equations to yield

$$a_{22}a_{11}x_2 - a_{12}a_{21}x_2 = c_2a_{11} - c_1a_{21}$$

which can be solved for

$$x_2 = \frac{a_{11}c_2 - a_{21}c_1}{a_{11}a_{22} - a_{12}a_{21}} \qquad [7.10]$$

Equation (7.10) can then be substituted into Eq. (7.6), which can be solved for

$$x_1 = \frac{a_{22}c_1 - a_{12}c_2}{a_{11}a_{22} - a_{12}a_{21}} \qquad [7.11]$$

Notice that Eqs. (7.10) and (7.11) follow directly from Cramer's rule, which states

$$x_1 = \frac{\begin{vmatrix} c_1 & a_{12} \\ c_2 & a_{22} \end{vmatrix}}{\begin{vmatrix} a_{11} & a_{12} \\ a_{21} & a_{22} \end{vmatrix}} = \frac{c_1a_{22} - a_{12}c_2}{a_{11}a_{22} - a_{12}a_{21}}$$

and

$$x_2 = \frac{\begin{vmatrix} a_{11} & c_1 \\ a_{21} & c_2 \end{vmatrix}}{\begin{vmatrix} a_{11} & a_{12} \\ a_{21} & a_{22} \end{vmatrix}} = \frac{a_{11}c_2 - c_1a_{21}}{a_{11}a_{22} - a_{12}a_{21}}$$

### EXAMPLE 7.4
Elimination of Unknowns

Problem Statement: Use the elimination of unknowns to solve (recall Example 7.1),

$$3x_1 + 2x_2 = 18$$
$$-x_1 + 2x_2 = 2$$

Solution: Using Eqs. (7.11) and (7.10)

$$x_1 = \frac{2(18) - 2(2)}{3(2) - 2(-1)} = 4$$

$$x_2 = \frac{3(2) - (-1)18}{3(2) - 2(-1)} = 3$$

which is consistent with our graphical solution (Fig. 7.1).

The elimination of unknowns can be extended to systems with more than two or three equations. However, the numerous calculations that are required for larger systems makes the method extremely tedious to implement by hand. However, as described in the next section, the technique can be formalized and readily programmed for the personal computer.

## 7.2 NAIVE GAUSS ELIMINATION

In the previous section, the elimination of unknowns was used to solve a pair of simultaneous equations. The procedure consisted of two steps:

1. The equations were manipulated in order to eliminate one of the unknowns from the equations. The result of this *elimination* step was that we had one equation with one unknown.
2. Consequently, this equation could be solved directly and the result *back-substituted* into one of the original equations in order to solve for the remaining unknown.

This basic approach can be extended to large sets of equations by developing a systematic scheme to eliminate unknowns and to back-substitute. *Gauss elimination* is the most common of these schemes.

### 7.2.1 Algorithm for Naive Gauss Elimination

The present section includes the systematic techniques for forward elimination and back-substitution that comprise Gauss elimination. Although these techniques are ideally suited for implementation on personal computers, some modifications will be required in order to obtain a reliable algorithm. In particular, the computer program must avoid division by zero. The following method is called *"naive" Gauss elimination* because it does not avoid this contingency. Subsequent sections will deal with the additional features required for an effective computer program.

The approach is designed to solve a general set of $n$ equations:

$$a_{11}x_1 + a_{12}x_2 + a_{13}x_3 + \cdots + a_{1n}x_n = c_1 \qquad [7.12a]$$

$$a_{21}x_1 + a_{22}x_2 + a_{23}x_3 + \cdots + a_{2n}x_n = c_2 \qquad [7.12b]$$

$$\vdots$$

$$a_{n1}x_1 + a_{n2}x_2 + a_{n3}x_3 + \cdots + a_{nn}x_n = c_n \qquad [7.12c]$$

As was the case with the solution of two equations, the technique for $n$ equations consists of two phases: elimination of unknowns and solution through back-substitution.

**Forward Elimination of Unknowns.** The first phase is designed to reduce the set of equations to an upper triangular system (Fig. 7.3). The initial step in the procedure is to divide the first equation [Eq. (7.12$a$)] by the coefficient of the first unknown, $a_{11}$:

$$x_1 + \frac{a_{12}}{a_{11}}x_2 + \cdots + \frac{a_{1n}}{a_{11}}x_n = \frac{c_1}{a_{11}}$$

This is called *normalization* and is intended to make the first coefficient of the normalized equation equal to 1.

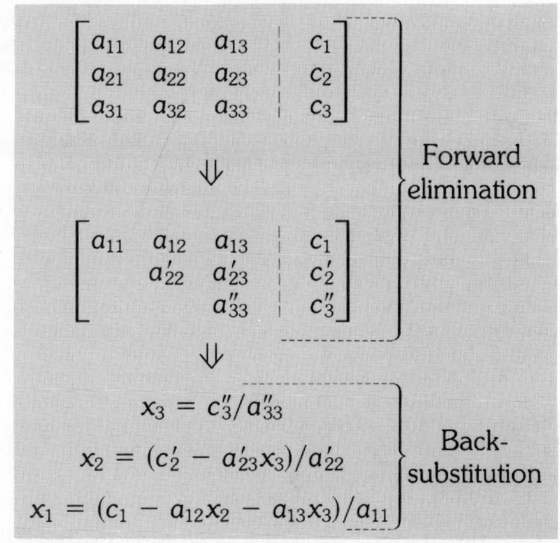

$$\left[\begin{array}{ccc|c} a_{11} & a_{12} & a_{13} & c_1 \\ a_{21} & a_{22} & a_{23} & c_2 \\ a_{31} & a_{32} & a_{33} & c_3 \end{array}\right]$$

$$\Downarrow \quad \text{Forward elimination}$$

$$\left[\begin{array}{ccc|c} a_{11} & a_{12} & a_{13} & c_1 \\ & a'_{22} & a'_{23} & c'_2 \\ & & a''_{33} & c''_3 \end{array}\right]$$

$$\Downarrow$$

$$x_3 = c''_3/a''_{33}$$

$$x_2 = (c'_2 - a'_{23}x_3)/a'_{22} \quad \text{Back-substitution}$$

$$x_1 = (c_1 - a_{12}x_2 - a_{13}x_3)/a_{11}$$

FIGURE 7.3 Graphical depiction of the two phases of naive Gauss elimination. Forward elimination reduces the coefficient matrix to an upper triangular form. Then, back-substitution can be used to solve for the unknowns.

Next, multiply the normalized equation by the first coefficient of the second equation [Eq. (7.12*b*)], $a_{21}$:

$$a_{21}x_1 + \left(a_{21}\frac{a_{12}}{a_{11}}\right)x_2 + \cdots + \left(a_{21}\frac{a_{1n}}{a_{11}}\right)x_n = a_{21}\frac{c_1}{a_{11}} \quad [7.13]$$

Note that the first term in the first equation is now identical to the first term in the second equation. Consequently, the first unknown can be eliminated from the second equation by subtracting Eq. (7.13) from Eq. (7.12*b*) to yield

$$\left(a_{22} - a_{21}\frac{a_{12}}{a_{11}}\right)x_2 + \cdots + \left(a_{2n} - a_{21}\frac{a_{1n}}{a_{11}}\right)x_n = c_2 - a_{21}\frac{c_1}{a_{11}}$$

or

$$a'_{22}x_2 + \cdots + a'_{2n}x_n = c'_2$$

where the prime indicates that the elements have been changed from their original values.

The procedure is then repeated in order to eliminate the first unknown from the remaining equations. For instance, the normalized equation is multiplied by $a_{31}$ and the result subtracted from the third equation to yield

$$a'_{32}x_2 + a'_{33}x_3 + \cdots + a'_{3n}x_n = c'_3$$

Repeating the procedure for the remaining equations results in the following modified system:

$$a_{11}x_1 + a_{12}x_2 + a_{13}x_3 + \cdots + a_{1n}x_n = c_1 \quad [7.14a]$$

$$a'_{22}x_2 + a'_{23}x_3 + \cdots + a'_{2n}x_n = c'_2 \quad [7.14b]$$

$$a'_{32}x_2 + a'_{33}x_3 + \cdots + a'_{3n}x_n = c'_3 \quad [7.14c]$$

$$\vdots$$

$$a'_{n2}x_2 + a'_{n3}x_3 + \cdots + a'_{nn}x_n = c'_n \quad [7.14d]$$

For the foregoing steps, Eq. (7.12*a*) is called the *pivot equation,* and $a_{11}$ is called the *pivot coefficient.*

Now repeat the above in order to eliminate the second unknown from Eq. (7.14*c*) through (7.14*d*). To do this, use Eq. (7.14*b*) as the pivot equation and normalize by dividing by the pivot coefficient $a'_{22}$. Multiply the normalized equation by $a'_{32}$ and subtract the result from Eq. (7.14*c*) in order to eliminate the second unknown. Repeat for the remaining equations to yield

$$\begin{aligned}
a_{11}x_1 + a_{12}x_2 + a_{13}x_3 + \cdots + a_{1n}x_n &= c_1 \\
a'_{22}x_2 + a'_{23}x_3 + \cdots + a'_{2n}x_n &= c'_2 \\
a''_{33}x_3 + \cdots + a''_{3n}x_n &= c''_3 \\
\vdots \qquad\qquad & \quad \vdots \\
a''_{n3}x_3 + \cdots + a''_{nn}x_n &= c''_n
\end{aligned}$$

where the double prime indicates that the elements have been modified twice.

The procedure can be continued using the remaining pivot equations. The final manipulation in the sequence is to use the $(n-1)$th equation to eliminate the $x_{n-1}$ term from the $n$th equation. At this point, the system will have been transformed to an upper triangular system (recall Box III.1).

$$a_{11}x_1 + a_{12}x_2 + a_{13}x_3 + \cdots + a_{1n}x_n = c_1 \tag{7.15a}$$

$$a'_{22}x_2 + a'_{23}x_3 + \cdots + a'_{2n}x_n = c'_2 \tag{7.15b}$$

$$a''_{33}x_3 + \cdots + a''_{3n}x_n = c''_3 \tag{7.15c}$$

$$\vdots$$

$$a_{nn}^{(n-1)}x_n = c_n^{(n-1)} \tag{7.15d}$$

**Backward-Substitution.** Equation (7.15*d*) can now be solved for $x_n$:

$$x_n = \frac{c_n^{n-1}}{a_{nn}^{n-1}} \tag{7.16}$$

This result can be back-substituted into the $(n-1)$th equation to solve for $x_{n-1}$. The procedure, which is repeated to evaluate the remaining $x$'s, can be represented by the following formula:

$$x_i = \frac{c_i^{i-1} - \sum_{j=i+1}^{n} a_{ij}^{i-1}x_j}{a_{ii}^{i-1}} \tag{7.17}$$

for $i = n-1, n-2, \ldots, 1$.

## EXAMPLE 7.5
### Naive Gauss Elimination

Problem Statement: Use Gauss elimination to solve

$$3x_1 - 0.1x_2 - 0.2x_3 = 7.85 \tag{E7.5.1}$$

$$0.1x_1 + 7x_2 - 0.3x_3 = -19.3 \tag{E7.5.2}$$

$$0.3x_1 - 0.2x_2 + 10x_3 = 71.4 \quad \text{[E7.5.3]}$$

Carry six significant figures during the computation.

Solution: The first part of the procedure is forward elimination. Normalize Eq. (E7.5.1) by dividing it by the pivot element to yield

$$x_1 - 0.0333333x_2 - 0.0666667x_3 = 2.61667 \quad \text{[E7.5.4]}$$

Next, multiply Eq. (E7.5.4) by 0.1 and subtract the result from Eq. (E7.5.2) to give

$$7.00333x_2 - 0.293333x_3 = -19.5617 \quad \text{[E7.5.5]}$$

Then multiply Eq. (E7.5.4) by 0.3 and subtract it from Eq. (E7.5.3) to eliminate $x_1$. After these operations, the set of equations is

$$3x_1 - 0.1x_2 - 0.2x_3 = 7.85 \quad \text{[E7.5.6]}$$

$$7.00333x_2 - 0.293333x_3 = -19.5617 \quad \text{[E7.5.7]}$$

$$-0.190000x_2 + 10.0200x_3 = 70.6150 \quad \text{[E7.5.8]}$$

To complete the forward elimination, $x_2$ must be removed from Eq. (E7.5.8). To accomplish this, normalize Eq. (E7.5.7) by dividing it by 7.00333:

$$x_2 - 0.0418848x_3 = -2.79320 \quad \text{[E7.5.9]}$$

Then multiply Eq. (E7.5.9) by −0.190000 and subtract the result from Eq. (E7.5.8). This eliminates $x_2$ from the third equation and reduces the system to an upper triangular form, as in

$$3x_1 - 0.1x_2 - 0.2x_3 = 7.85$$

$$7.00333x_2 - 0.293333x_3 = -19.5617 \quad \text{[E7.5.10]}$$

$$10.0120x_3 = 70.0843 \quad \text{[E7.5.11]}$$

We can now solve these equations by back-substitution. First, Eq. (E7.5.11) can be solved for

$$x_3 = 7.00003 \quad \text{[E7.5.12]}$$

This result can be back-substituted into Eq. (E7.5.10):

$$7.00333x_2 - 0.293333(7.00003) = -19.5617$$

which can be solved for

$$x_2 = -2.50000 \quad \text{[E7.5.13]}$$

Finally, Eqs. (E7.5.12) and (E7.5.13) can be substituted into Eq. (E7.5.6):

$$3x_1 - 0.1(-2.50000) - 0.2(7.00003) = 7.85$$

which can be solved for

$$x_1 = 3.00000$$

Although there is a slight round-off error in Eq. (E7.5.12), the results are very close to the exact solution of $x_1 = 3$, $x_2 = -2.5$, and $x_3 = 7$. This can be verified by substituting the results into the original equation set

$$3(3) - 0.1(-2.5) - 0.2(7.00003) = 7.84999 \quad \simeq 7.85$$

$$0.1(3) + 7(-2.5) - 0.3(7.00003) = -19.3000 = -19.3$$

$$0.3(3) - 0.2(-2.5) + 10(7.00003) = 71.4003 \quad \simeq 71.4$$

### 7.2.2 Computer Program for Naive Gauss Elimination

Figure 7.4 presents computer codes to implement naive Gauss elimination.

**FORTRAN**

```
      DIMENSION A(15,16),X(15)
      READ(5,1)N
    1 FORMAT(I5)
      M=N+1
      DO 160 I=1,N
      DO 150 J=1,M
      READ(5,2)A(I,J)
    2 FORMAT(F10.0)
  150 CONTINUE
  160 CONTINUE
      CALL GAUSS(N,A,X)
      DO 200 I=1,N
      WRITE(6,3)X(I)
    3 FORMAT(' ',F10.3)
  200 CONTINUE
      STOP
      END

      SUBROUTINE GAUSS(N,A,X)
      DIMENSION A(15,16),X(15)
      M=N+1
      L=N-1
      DO 1140 K=1,L
      KK=K+1
      DO 1100 I=KK,N
      QT=A(I,K)/A(K,K)
      DO 1090 J=KK,M
      A(I,J)=A(I,J)-QT*A(K,J)
 1090 CONTINUE
 1100 CONTINUE
      DO 1130 I=KK,N
      A(I,K)=0.
 1130 CONTINUE
 1140 CONTINUE
      X(N)=A(N,M)/A(N,N)
      DO 1240 NN=1,L
      SUM=0.
      I=N-NN
      II=I+1
      DO 1220 J=II,N
      SUM=SUM+A(I,J)*X(J)
 1220 CONTINUE
      X(I)=(A(I,M)-SUM)/A(I,I)
 1240 CONTINUE
      RETURN
      END
```

**BASIC**

```
100  DIM A(15,16),X(15)
110  INPUT N            ---- N = number of equations
120  FOR I = 1 TO N
130  FOR J = 1 TO N + 1
140  INPUT A(I,J)       ---- A = augmented matrix of
                             coefficients and
                             right-hand-side constants
150  NEXT J
160  NEXT I
170  GOSUB 1000
180  FOR I = 1 TO N
190  PRINT X(I)         ---- X = solution vector
200  NEXT I
210  END

1000  FOR K = 1 TO N - 1                 } (Forward elimination)
1020  FOR I = K + 1 TO N
1030 QT = A(I,K) / A(K,K)
1040  FOR J = K + 1 TO N + 1
1050 A(I,J) = A(I,J) - QT * A(K,J
     )
1060  NEXT J
1070  NEXT I
1080  FOR I = K + 1 TO N
1090 A(I,K) = 0
1100  NEXT I
1110  NEXT K
1120 X(N) = A(N,N + 1) / A(N,N)          } (Back-substitution)
1130  FOR NX = 1 TO N - 1
1140 SUM = 0
1150 I = N - NX
1160  FOR J = I + 1 TO N
1170 SUM = SUM + A(I,J) * X(J)
1180  NEXT J
1190 X(I) = (A(I,N + 1) - SUM) /
     A(I,I)
1200  NEXT NX
1210  RETURN
```

FIGURE 7.4 Annotated computer programs in FORTRAN and BASIC for naive Gauss elimination.

FIGURE 7.5 Three parachutists free-falling while connected by weightless cords.

The programs consist of four parts: input of data, forward elimination, back-substitution, and output of results. Notice that the matrix of coefficients is augmented by the right-hand-side constants. This information is stored in the matrix A. Because this matrix is augmented, the fact that it is dimensioned as 15 by 16 means that the program can handle up to 15 simultaneous equations in its present form.

Notice also that we have programmed the main body of the algorithm as a subroutine. We have done this because, aside from direct solution of engineering problems, Gauss elimination also has utility as a part of other computer algorithms. In a later part of this chapter, we will develop error-correction techniques that will require a subroutine for Gauss elimination. Additionally, in Chap. 10, we will need to solve simultaneous linear algebraic equations as a part of curve-fitting techniques called multiple and polynomial regression.

The program in Fig. 7.4 is not user-friendly. In Prob. 7.16 at the end of the chapter, you will have the task of making this skeletal computer code easier to use and understand.

## EXAMPLE 7.6
### Solution of Linear Algebraic Equations Using the Computer

Problem Statement: A user-friendly computer program to implement Gauss elimination is contained on the NUMERICOMP software. We can use this software to solve a problem associated with the falling parachutist example discussed in Chap. 1. Suppose that a team of three parachutists are connected by a weightless cord while free-falling at a velocity of 5 m/s (Fig. 7.5). Calculate the tension in each section of cord and the acceleration of the team, given the following:

| Parachutist | Mass, kg | Drag coefficient, kg/s |
|---|---|---|
| 1 | 70 | 10 |
| 2 | 60 | 14 |
| 3 | 40 | 17 |

Solution: Free-body diagrams for each of the parachutists are depicted in Fig. 7.6. Summing the forces in the vertical direction and using Newton's second law gives a set of three simultaneous linear equations:

$$m_1 g - T - c_1 v \qquad = m_1 a$$

$$m_2 g + T - c_2 v - R = m_2 a$$

$$m_3 g \qquad - c_3 v + R = m_3 a$$

These equations have three unknowns: $a$, $T$, and $R$. After substituting the known values, the equations can be expressed in matrix form (as $g = 9.8$ m/s$^2$)

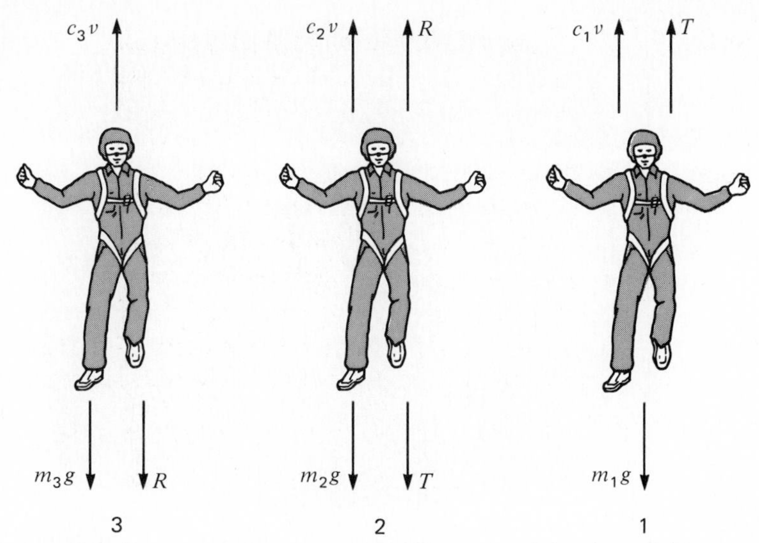


FIGURE 7.6 Free-body diagrams for each of the three falling parachutists.

$$\begin{bmatrix} 70 & 1 & 0 \\ 60 & -1 & 1 \\ 40 & 0 & -1 \end{bmatrix} \begin{bmatrix} a \\ T \\ R \end{bmatrix} \begin{matrix} = \\ = \\ = \end{matrix} \begin{bmatrix} 636 \\ 518 \\ 307 \end{bmatrix}$$

This system can be solved using your own software or the Gauss elimination option on NUMERICOMP. Figure 7.7$a$ shows the NUMERICOMP solution of $a = 8.5941$ m/s$^2$; $T = 34.4118$ N; and $R = 36.7647$ N. As de-

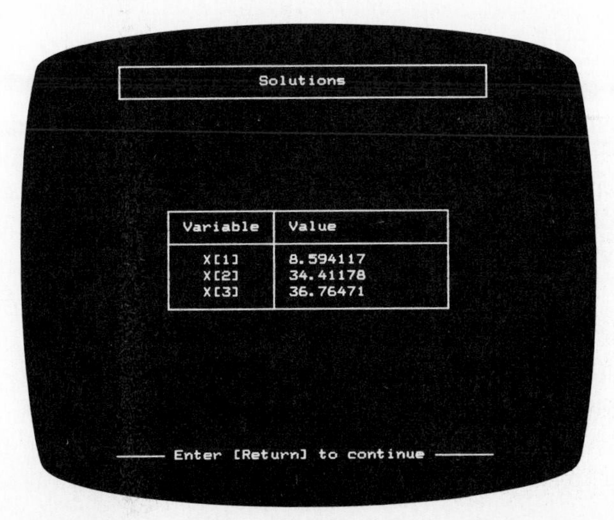


FIGURE 7.7 Computer screens showing ($a$) the NUMERICOMP solution for the three falling parachutists. Part ($b$) on the next page.

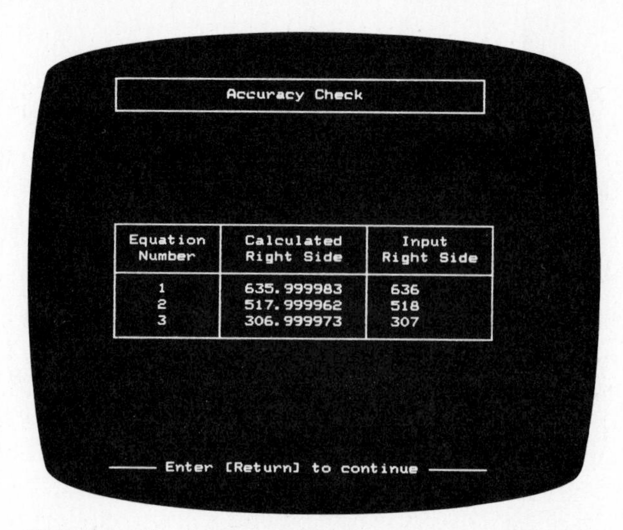


| Equation Number | Calculated Right Side | Input Right Side |
|---|---|---|
| 1 | 635.999983 | 636 |
| 2 | 517.999962 | 518 |
| 3 | 306.999973 | 307 |

FIGURE 7.7 (continued) (*b*) Accuracy check obtained by substituting the solution into the original equations to verify that the results equal the original right-hand-side constants.

picted in Fig. 7.7*b*, the software also includes an option to check the accuracy of the computation.

The foregoing results are based on a simple algorithm for the Gauss elimination method with user-friendly input and output routines. The algorithm employed is similar to that shown in Fig. 7.4. You should be able to write your own program for the Gauss elimination method. If you have our software, use it to check the adequacy of your own program.

## 7.3 PITFALLS OF ELIMINATION METHODS

Whereas there are many systems of equations that can be solved with naive Gauss elimination, there are some pitfalls that must be explored before writing a general computer program to implement the method. Although the following material relates directly to naive Gauss elimination, the information has relevance to other elimination techniques.

### 7.3.1 Division by Zero

The primary reason that the foregoing technique is called "naive" is that during both the elimination and the back-substitution phases, it is possible that a division by zero could occur. For example, if we used naive Gauss elimination to solve

$$2x_2 + 3x_3 = 8$$

$$4x_1 + 6x_2 + 7x_3 = -3$$

$$2x_1 + x_2 + 6x_3 = 5$$

the normalization of the first row would involve division by $a_{11} = 0$. Problems also can arise when a coefficient is very close to zero. The technique of pivoting has been developed to partially avoid these problems. It will be described in Sec. 7.4.2.

### 7.3.2 Round-Off Errors

Even though the solution in Example 7.5 was close to the true answer, there was a slight discrepancy in the result for $x_3$ [Eq. (E7.5.12)]. This discrepancy, which amounted to a relative error of −0.00043 percent, was due to our use of six significant figures during the computation. If we had used more significant figures, the error in the results would be reduced further. If we had used fractions instead of decimals (and consequently avoided round off altogether), the answers would have been exact. However, because personal computers carry only a limited number of significant figures ($\simeq 10$), round-off errors can occur and must be considered when evaluating the results.

#### EXAMPLE 7.7
Effect of Round-Off Error on Gauss Elimination

Problem Statement: Solve the same problem as in Example 7.5, but carry three significant figures during the computation.

Solution: Elimination of $x_1$ from the second and third equations and elimination of $x_2$ from the third equation yields the following triangular system

$$\begin{aligned} 3x_1 - 0.1x_2 - 0.2x_3 &= 7.85 \\ 7.00x_2 - 0.293x_3 &= -19.6 \\ 9.99x_3 &= 70.1 \end{aligned}$$

Compare this system with the one derived previously using six significant figures [Eqs. (E7.5.10) through (E7.5.12)]. Back-substitution can be used to solve for

$$x_1 = 3.17 \qquad |\epsilon_t| = 5.7\%$$

$$x_2 = -2.51 \qquad |\epsilon_t| = 0.4\%$$

$$x_3 = 7.02 \qquad |\epsilon_t| = 0.29\%$$

Substituting these values into the original equations yields

$$3(3.17) - 0.1(-2.51) - 0.2(7.02) = 8.36 \neq 7.85$$

$$0.1(3.17) + 7(-2.51) - 0.3(7.02) = -19.4 \neq -19.3$$

$$0.3(3.17) - 0.2(-2.51) + 10(7.02) = 71.7 \neq 71.4$$

Although the use of only three significant figures makes Example 7.7 somewhat unrealistic, the problem of round-off error is real and can become particularly important when large numbers of equations are to be solved. This is due to the fact that every result is dependent on all previous results. Consequently, an error in the early steps will tend to propagate—that is, it will cause errors in subsequent steps.

Specifying the system size where round-off error becomes significant is complicated by the fact that the type of computer and the properties of the equations are determining factors. A rough rule of thumb is that round-off error may be important when dealing with greater than 25 to 50 equations. In any event, you should always substitute your answers back into the original equations to check whether a substantial error has occurred. However, as discussed below, the magnitudes of the coefficients themselves can influence whether such an error check ensures a reliable result.

### 7.3.3 Ill-Conditioned Systems

The adequacy of the solution depends on the condition of the system. In Sec. 7.1.1, a graphical depiction of system condition was developed. In a mathematical sense, *well-conditioned systems* are those where a small change in one or more of the coefficients results in a similar small change in the solution. *Ill-conditioned systems* are those where small changes in coefficients result in large changes in the solution. An alternative interpretation of ill conditioning is that a wide range of answers can approximately satisfy the equations. Because round-off errors can induce small changes in the coefficients, these artificial changes can lead to large solution errors for ill-conditioned systems, as illustrated in the following example.

#### EXAMPLE 7.8
Ill-Conditioned Systems

Problem Statement: Solve the following system:

$$x_1 + 2x_2 = 10 \tag{E7.8.1}$$

$$1.1x_1 + 2x_2 = 10.4 \tag{E7.8.2}$$

Then, solve it again, but with the coefficient of $x_1$ in the second equation modified slightly to 1.05.

Solution: Using Eqs. (7.10) and (7.11), the solution is

$$x_1 = \frac{2(10) - 2(10.4)}{1(2) - 2(1.1)} = 4$$

$$x_2 = \frac{1(10.4) - 1.1(10)}{1(2) - 2(1.1)} = 3$$

However, with the slight change of the coefficient $a_{21}$ from 1.1 to 1.05, the result is changed dramatically to

$$x_1 = \frac{2(10) - 2(10.4)}{1(2) - 2(1.05)} = 8$$

$$x_2 = \frac{1(10.4) - 1.05(10)}{1(2) - 2(1.05)} = 1$$

Notice that the primary reason for the discrepancy between the two results is that the denominator represents the difference of two almost-equal numbers. As illustrated previously in Example 3.4, such differences are highly sensitive to slight variations in the numbers being manipulated.

At this point, you might suggest that substitution of the results into the original equations would alert you to the problem. Unfortunately, for ill-conditioned systems this is not the case. Substitution of the erroneous values of $x_1 = 8$ and $x_2 = 1$ into Eqs. (E.7.8.1) and (E7.8.2) yields

$$8 + 2(1) = 10 = 10$$

$$1.1(8) + 2(1) = 10.8 \simeq 10.4$$

Therefore, although $x_1 = 8$ and $x_2 = 1$ is not the true solution to the original problem, the error check is close enough to possibly mislead you into believing that your solutions are adequate.

As was done previously in our section on graphical methods, a visual representation of ill conditioning can be developed by plotting Eqs. (E7.8.1) and (E7.8.2) (recall Fig. 7.2). Because the slopes of the lines are almost equal, it is visually difficult to see exactly where they intersect. This visual difficulty is reflected quantitatively in the nebulous results of Example 7.8. We can mathematically characterize this situation by writing the two equations in general form:

$$a_{11}x_1 + a_{12}x_2 = c_1 \tag{7.18}$$

$$a_{21}x_1 + a_{22}x_x = c_2 \tag{7.19}$$

Dividing Eq. (7.18) by $a_{12}$ and Eq. (7.19) by $a_{22}$ and rearranging yields alternative versions that are in the format of straight lines [$x_2 =$ (slope) $x_1$ + intercept].

$$x_2 = -\frac{a_{11}}{a_{12}}x_1 + \frac{c_1}{a_{12}}$$

$$x_2 = -\frac{a_{21}}{a_{22}}x_1 + \frac{c_2}{a_{22}}$$

Consequently, if the slopes are nearly equal,

$$\frac{a_{11}}{a_{12}} \simeq \frac{a_{21}}{a_{22}}$$

or, cross-multiplying,

$$a_{11}a_{22} \simeq a_{21}a_{12}$$

which can be also expressed as

$$a_{11}a_{22} - a_{12}a_{21} \simeq 0 \qquad [7.20]$$

Now, recalling that $a_{11}a_{22} - a_{12}a_{21}$ is the determinant of a two-dimensional system [Eq. (7.3)], we arrive at the general conclusion that an ill-conditioned system is one with a determinant close to zero. In fact, if the determinant is exactly zero, the two slopes are identical, which connotes either no solution or an infinite number of solutions, as is the case for the singular systems depicted in Fig. 7.2*a* and *b*.

It is difficult to specify *how close* to zero the determinant must be to indicate ill conditioning. This is complicated by the fact that the determinant can be changed by multiplying one or more of the equations by a scale factor without changing the solution. Consequently, the determinant is a relative value that is influenced by the magnitude of the coefficients.

## EXAMPLE 7.9
Effect of Scale on the Determinant

Problem Statement: Evaluate the determinant of the following systems:

(*a*) From Example 7.1:

$$3x_1 + 2x_2 = 18 \qquad [E7.9.1]$$

$$-x_1 + 2x_2 = 2 \qquad [E7.9.2]$$

(*b*) From Example 7.8:

$$x_1 + 2x_2 = 10 \qquad [E7.9.3]$$

$$1.1x_1 + 2x_2 = 10.4 \qquad [E7.9.4]$$

(*c*) Repeat (*b*) but with the equations multiplied by 10.

Solution:
(*a*) The determinant of Eqs. (E7.9.1) and (E7.9.2), which are well-conditioned, is

$$D = 3(2) - 2(-1) = 8$$

(*b*) The determinant of Eqs. (E7.9.3) and (E7.9.4), which are ill-conditioned, is

$$D = 1(2) - 2(1.1) = -0.2$$

(*c*) The results of (*a*) and (*b*) seem to bear out the contention that ill-conditioned systems have near-zero determinants. However, suppose that the ill-conditioned system in (*b*) is multiplied by 10 to give

$$10x_1 + 20x_2 = 100$$

$$11x_1 + 20x_2 = 104$$

The multiplication of an equation by a constant has no effect on its solution. In addition, it is still ill-conditioned. This can be verified by the fact that multiplying by a constant has no effect on the graphical solution. However, the determinant is dramatically affected:

$$D = 10(20) - 20(11) = -20$$

Not only has it been raised two orders of magnitude, but it is now over twice as large as the determinant of the well-conditioned system in (*a*).

As illustrated by the previous example, the magnitude of the coefficients interjects a scale effect that complicates the relationship between system condition and determinant size. One way to partially circumvent this difficulty is to scale the equations so that the maximum element in any row is equal to 1.

## EXAMPLE 7.10
### Scaling

Problem Statement: Scale the systems of equations in Example 7.9 to a maximum value of 1 and recompute their determinants.

Solution:
(*a*) For the well-conditioned system, scaling results in

$$x_1 + 0.667x_2 = 6$$

$$-0.5x_1 + x_2 = 1$$

for which the determinant is

$$1(1) - 0.667(-0.5) = 1.333$$

(*b*) For the ill-conditioned system, scaling gives

$$0.5x_1 + x_2 = 5$$

$$0.55x_1 + x_2 = 5.2$$

for which the determinant is

$$0.5(1) - 1(0.55) = -0.05$$

(*c*) For the last case, scaling changes the system to the same form as in (*b*). Thus, the determinant is also $-0.05$, and scaling has no effect.

In a previous section (Sec. 7.1.2), we suggested that the determinant is difficult to compute for more than three simultaneous equations. Therefore, it might seem that it does not provide a practical means for evaluating system condition. However, as described in Box 7.1, there is a simple algorithm that results from Gauss elimination that can be used to evaluate the determinant.

Aside from the approach used in the previous example, there are a variety of other ways to evaluate system condition. For example, there are alternative methods for normalizing the elements (see Stark, 1970). In addition, as described in the next chapter (Sec. 8.2.2), the matrix inverse can be employed to evaluate system condition. Finally, a simple (but time-consuming) test is to modify the coefficients slightly and repeat the solution. If such modifications lead to drastically different results, the system is likely to be ill-conditioned.

### BOX 7.1 Determinant Evaluation Using Gauss Elimination

In Sec. 7.1.2, we stated that determinant evaluation by expansion of minors was impractical for large sets of equations. Thus, we concluded that Cramer's rule would only be applicable to small systems. However, as mentioned in Sec. 7.3.3, the determinant has value in assessing system condition. It would, therefore, be useful to have a practical method for computing this quantity.

Fortunately, Gauss elimination provides a simple way to do this. The method is based on the fact that the determinant of a triangular matrix can be simply computed as the product of its diagonal elements:

$$D = a_{11}a_{22}a_{33} \ldots a_{nn} \qquad [B7.1.1]$$

The validity of this formulation can be illustrated for a 3-by-3 system:

$$D = \begin{vmatrix} a_{11} & a_{12} & a_{13} \\ 0 & a_{22} & a_{23} \\ 0 & 0 & a_{33} \end{vmatrix}$$

where the determinant can be evaluated as [recall Eq. (7.4)]

$$D = a_{11}\begin{vmatrix} a_{22} & a_{23} \\ 0 & a_{33} \end{vmatrix} - a_{12}\begin{vmatrix} 0 & a_{23} \\ 0 & a_{33} \end{vmatrix} + a_{13}\begin{vmatrix} 0 & a_{22} \\ 0 & 0 \end{vmatrix}$$

or, by evaluating the minors (that is, the 2-by-2 determinants),

$$D = a_{11}a_{22}a_{33} - a_{12}(0) + a_{13}(0) = a_{11}a_{22}a_{33}$$

Recall that the forward-elimination step of Gauss elimination results in an upper triangular system. Because the value of the determinant is not changed by the forward-elimination process, the determinant can be simply evaluated at the end of this step via

$$D = a_{11}a'_{22}a''_{33} \ldots a_{nn}^{(n-1)} \qquad [B7.1.2]$$

where the superscripts signify that the elements have been modified by the elimination process. Thus, we can capitalize on the effort that has already been expended in reducing the system to triangular form and, in the bargain, come up with a simple estimate of the determinant.

There is a slight modification to the above approach when the program employs partial pivoting (Sec. 7.4.2). For this case, the determinant changes sign every time a row is pivoted. One way to represent this is to modify Eq. (B7.1.2):

$$D = a_{11}a'_{22}a''_{33} \ldots a_{nn}^{(n-1)}(-1)^p \qquad [B7.1.3]$$

where $p$ represents the number of times that rows are pivoted. This modification can be incorporated simply into a program; merely keep track of the number of pivots that take place during the course of the computation and then use Eq. (B7.1.3) to evaluate the determinant.

As you might gather from the foregoing discussion, there is no definitive, simple test for ill conditioning. For Gauss elimination, we recommend the scaled determinant as in Example 7.10. Fortunately, most linear algebraic equations derived from engineering-problem settings are naturally well-conditioned. In addition, some of the techniques outlined in the following section help to alleviate the problem.

## 7.4 TECHNIQUES FOR IMPROVING SOLUTIONS

The following techniques can be incorporated into the naive Gauss elimination algorithm to circumvent some of the pitfalls discussed in the previous section.

### 7.4.1 Use of More Significant Figures

The simplest remedy for ill conditioning is to use more significant figures in the computation (compare Examples 7.5 and 7.7). If your computer has the capability of being extended to handle larger word size, such a feature will greatly reduce the problem.

### 7.4.2 Pivoting

As mentioned at the beginning of Sec. 7.3, obvious problems occur when a pivot element is zero because the normalization step leads to division by zero. Problems may also arise when the pivot element is close to, rather than exactly equal to, zero because if the magnitude of the pivot element is small compared to the other elements, then round-off errors can be introduced.

Therefore, before each row is normalized, it is advantageous to determine the largest available coefficient. The rows can then be switched so that the largest element is the pivot element. This is called *partial pivoting.* If columns as well as rows are searched for the largest element and then switched, the procedure is called *complete pivoting.* Complete pivoting is rarely used in elementary programs because switching columns changes the order of the $x$'s and, consequently, adds significant and usually unjustified complexity to the computer program. The following example illustrates the advantages of partial pivoting. Aside from avoiding division by zero, pivoting also minimizes round-off error. As such, it also serves as a partial remedy for ill conditioning.

EXAMPLE 7.11
Partial Pivoting

Problem Statement: Use Gauss elimination to solve

$$0.0003x_1 + 3.0000x_2 = 2.0001$$

$$1.0000x_1 + 1.0000x_2 = 1.0000$$

Note that in this form the first pivot element, $a_{11} = 0.0003$, is very close to zero. Then repeat the computation, but partial pivot by reversing the order of the equations. The exact solution is $x_1 = 1/3$ and $x_2 = 2/3$.

Solution: Normalizing the first equation yields

$$x_1 + 10{,}000x_2 = 6667$$

which can be used to eliminate $x_1$ from the second equation:

$$-9999x_2 = -6666$$

which can be solved for

$$x_2 = 2/3$$

This result can be substituted back into the first equation to evaluate $x_1$:

$$x_1 = \frac{2.0001 - 3(2/3)}{0.0003} \quad \text{[E7.11.1]}$$

However, the result is very sensitive to the number of significant figures carried in the computation:

| **Significant figures** | $x_2$ | $x_1$ | **Absolute value of percent relative error for** $x_1$ |
|---|---|---|---|
| 3 | 0.667 | −3.33 | 1099 |
| 4 | 0.6667 | 0.0000 | 100 |
| 5 | 0.66667 | 0.30000 | 10 |
| 6 | 0.666667 | 0.330000 | 1 |
| 7 | 0.6666667 | 0.3330000 | 0.1 |

Note how the solution for $x_1$ is highly dependent on the number of significant figures. This is due to the fact that in Eq. (E7.11.1), we are subtracting two almost-equal numbers (recall Example 3.4). On the other hand, if the equations are solved in reverse order, the row with the largest pivot element is normalized. The equations are

$$1.0000x_1 + 1.0000x_2 = 1.0000$$

$$0.0003x_1 + 3.0000x_2 = 2.0001$$

Normalization and elimination yields $x_2 = 2/3$. For different numbers of significant figures, $x_1$ can be computed from the first equation, as in

$$x_1 = \frac{1 - (2/3)}{1} \quad \text{[E7.11.2]}$$

This case is much less sensitive to the number of significant figures in the computation:

| Significant figures | $x_2$ | $x_1$ | Absolute value of percent relative error for $x_1$ |
|---|---|---|---|
| 3 | 0.667 | 0.333 | 0.1 |
| 4 | 0.6667 | 0.3333 | 0.01 |
| 5 | 0.66667 | 0.33333 | 0.001 |
| 6 | 0.666667 | 0.333333 | 0.0001 |
| 7 | 0.6666667 | 0.3333333 | 0.00001 |

Thus, a pivot strategy is much more satisfactory.

General-purpose computer programs such as the NUMERICOMP software accompanying this book often include a pivot strategy. Figure 7.8 provides one algorithm for implementing such a strategy. This code can be integrated with Fig. 7.4 to incorporate partial pivoting into your Gauss elimination software.

### 7.4.3 Scaling

In Sec. 7.3.3, we proposed that scaling had value in standardizing the size of the determinant. Beyond this application, it has utility in minimizing round-off errors for those cases where some of the equations in a system have much larger coefficients than others. Such situations are frequently encountered in engineering practice when widely different units are used in the development of simultaneous equations. For instance, in electric-circuit problems, the unknown voltages can be expressed in units ranging from microvolts to kilovolts. Similar examples can arise in all fields of engineering. As long as

**FORTRAN**

```
      SUBROUTINE PARPIV(A)
      DIMENSION A(15,16)
      COMMON N,K
      JJ=K
      KK=K+1
      N1=N+1
      B=ABS(A(K,K))
      DO 3080 I=KK,N
      BP=ABS(A(I,K))
      IF(B-BP.GE.0.)GO TO 3080
      B=BP
      JJ=I
3080  CONTINUE
3090  IF(JJ-K.EQ.0.)GO TO 3150
      DO 3140 J=K,N1
      TE=A(JJ,J)
      A(JJ,J)=A(K,J)
      A(K,J)=TE
3140  CONTINUE
3150  CONTINUE
      RETURN
      END
```

**BASIC**

```
3000 JJ = K                          K = present pivot row
3020 B =  ABS (A(K,K))               (Store absolute value of present pivot element)
3030  FOR I = K + 1 TO N
3040 BP =  ABS (A(I,K))
3050  IF B - BP > = 0 THEN 3080      (Loop to compare other column
3060 B = BP                           elements with pivot element)
3070 JJ = I
3080  NEXT I
3090  IF JJ - K = 0 THEN 3150        (If present pivot element
3100  FOR J = K TO N + 1              is biggest, return
3110 TE = A(JJ,J)                     to main program)
3120 A(JJ,J) = A(K,J)
3130 A(K,J) = TE                     (If not, this loop
3140  NEXT J                          switches rows)
3150  RETURN
                                     (Return to main program
                                      to continue elimination)
```

FIGURE 7.8 Annotated computer code in FORTRAN and BASIC to implement partial pivoting.

each equation is consistent, the system will be technically correct and solvable. However, the use of widely differing units can lead to coefficients of widely differing magnitudes. This, in turn, can have an impact on round-off error as it affects pivoting, as illustrated by the following example.

## EXAMPLE 7.12
### Effect of Scaling on Pivoting and Round Off

Problem Statement:
(*a*) Solve the following set of equations using Gauss elimination and a pivoting strategy:

$$2x_1 + 100{,}000x_2 = 100{,}000$$

$$x_1 + \qquad x_2 = 2$$

(*b*) Repeat the solution after scaling the equations so that the maximum coefficient in each row is 1. For both cases, retain only three significant figures. Note that the correct answers are $x_1 = 1.00002$ and $x_2 = 0.99998$ or, for three significant figures, $x_1 = x_2 = 1.00$.

Solution:
(*a*) Without scaling, forward elimination is applied to give

$$2x_1 + 100{,}000x_2 = 100{,}000$$

$$-50{,}000x_2 = -50{,}000$$

which can be solved by back-substitution for

$$x_2 = 1.00$$

$$x_1 = 0.00$$

Although $x_2$ is correct, $x_1$ is 100 percent in error because of round-off.
(*b*) Scaling transforms the original equations to

$$0.00002x_1 + x_2 = 1$$

$$x_1 + x_2 = 2$$

Therefore, the rows should be pivoted to put the greatest value on the diagonal.

$$x_1 + x_2 = 2$$

$$0.00002x_1 + x_2 = 1$$

Forward elimination yields

$$x_1 + x_2 = 2$$

$$x_1 = 1.00$$

which can be solved for

$$x_1 = x_2 = 1$$

Thus, scaling leads to the correct answer.

As in the previous example, scaling has utility here in minimizing round off. However, it should be noted that scaling itself also leads to round off. For example, given the equation

$$2x_1 + 300{,}000x_2 = 1$$

and using three significant figures, scaling leads to

$$0.00000667x_1 + x_2 = 0.00000333$$

Thus, scaling introduces a round-off error to the first coefficient. For this reason, it is sometimes suggested that scaling should only be employed in cases where it is needed, that is, when the system involves coefficients of widely differing magnitude.

### 7.4.4 Error Corrections

In some cases, partial pivoting and scaling are not sufficient to ensure precise results. For instance, recall Example 7.7, which had the largest elements on the diagonal but because of round off still had error in the final solution. Such errors can usually be reduced by the following procedure. Consider a system of linear equations of the form

$$\begin{array}{cccccccc} a_{11}x_1 & + & a_{12}x_2 & + \cdots + & a_{1n}x_n & = & c_1 \\ a_{21}x_1 & + & a_{22}x_2 & + \cdots + & a_{2n}x_n & = & c_2 \\ \cdot & & \cdot & & \cdot & & \cdot \\ \cdot & & \cdot & & \cdot & & \cdot \\ \cdot & & \cdot & & \cdot & & \cdot \\ a_{n1}x_1 & + & a_{n2}x_2 & + \cdots + & a_{nn}x_n & = & c_n \end{array} \qquad [7.21]$$

Suppose an approximate solution vector is given by $\tilde{x}_1, \tilde{x}_2, \ldots, \tilde{x}_n$. These results are substituted into Eq. (7.21), giving

$$\begin{array}{cccccccc} a_{11}\tilde{x}_1 & + & a_{12}\tilde{x}_2 & + \cdots + & a_{1n}\tilde{x}_n & = & \tilde{c}_1 \\ a_{21}\tilde{x}_1 & + & a_{22}\tilde{x}_2 & + \cdots + & a_{2n}\tilde{x}_n & = & \tilde{c}_2 \\ \cdot & & \cdot & & \cdot & & \cdot \\ \cdot & & \cdot & & \cdot & & \cdot \\ \cdot & & \cdot & & \cdot & & \cdot \\ a_{n1}\tilde{x}_1 & + & a_{n2}\tilde{x}_2 & + \cdots + & a_{nn}\tilde{x}_n & = & \tilde{c}_n \end{array} \qquad [7.22]$$

Now suppose that the exact solutions $x_1, x_2, \ldots, x_n$ are expressed as a function of $\tilde{x}_1, \tilde{x}_2, \ldots, \tilde{x}_n$ and correction factors $\Delta x_1, \Delta x_2, \ldots, \Delta x_n$, where

$$\begin{aligned} x_1 &= \tilde{x}_1 + \Delta x_1 \\ x_2 &= \tilde{x}_2 + \Delta x_2 \\ &\vdots \\ x_n &= \tilde{x}_n + \Delta x_n \end{aligned} \tag{7.23}$$

If these results are substituted into Eq. (7.21) the following system results:

$$\begin{aligned} a_{11}(\tilde{x}_1 + \Delta x_1) + a_{12}(\tilde{x}_2 + \Delta x_2) + \cdots + a_{1n}(\tilde{x}_n + \Delta x_n) &= c_1 \\ a_{21}(\tilde{x}_1 + \Delta x_1) + a_{22}(\tilde{x}_2 + \Delta x_2) + \cdots + a_{2n}(\tilde{x}_n + \Delta x_n) &= c_2 \\ \vdots \qquad\qquad \vdots \qquad\qquad\qquad \vdots \qquad & \quad \vdots \\ a_{n1}(\tilde{x}_1 + \Delta x_1) + a_{n2}(\tilde{x}_2 + \Delta x_2) + \cdots + a_{nn}(\tilde{x}_n + \Delta x_n) &= c_n \end{aligned} \tag{7.24}$$

Now Eq. (7.22) can be subtracted from Eq. (7.24) to yield

$$\begin{aligned} a_{11}\Delta x_1 + a_{12}\Delta x_2 + \cdots + a_{1n}\Delta x_n &= c_1 - \tilde{c}_1 = E_1 \\ a_{21}\Delta x_1 + a_{22}\Delta x_2 + \cdots + a_{2n}\Delta x_n &= c_2 - \tilde{c}_2 = E_2 \\ \vdots \qquad\qquad \vdots \qquad\qquad \vdots \qquad & \quad \vdots \\ a_{n1}\Delta x_1 + a_{n2}\Delta x_2 + \cdots + a_{nn}\Delta x_n &= c_n - \tilde{c}_n = E_n \end{aligned} \tag{7.25}$$

This system itself is a set of simultaneous linear equations that can be solved to obtain the correction factors. The factors can then be applied to improve the solution as specified by Eq. (7.23).

### EXAMPLE 7.13
### Use of Error Equations to Correct Round-Off Error

Problem Statement: Recall that in Example 7.7 we used Gauss elimination and three significant figures to solve

$$\begin{aligned} 3x_1 - 0.1x_2 - 0.2x_3 &= 7.85 \\ 0.1x_1 + 7x_2 - 0.3x_3 &= -19.3. \\ 0.3x_1 - 0.2x_2 + 10x_3 &= 71.4 \end{aligned}$$

Because of the limited number of significant figures, the solution differed from the true solution ($x_1 = 3$, $x_2 = -2.5$, $x_3 = 7$), as in

$x_1 = 3.17 \qquad \epsilon_t = 5.7\%$

$x_2 = -2.51 \qquad \epsilon_t = 0.4\%$

$x_3 = 7.02 \qquad \epsilon_t = 0.29\%$

Use error equations to refine these estimates.

Solution: Substituting the solutions into the original set of equations gives the right-hand vector:

$$[\tilde{C}]^T = [8.36 \quad -19.4 \quad 71.7]$$

which is not equal to the actual right-hand-side vector. Therefore, an error vector can be developed:

$$[E] = \begin{bmatrix} c_1 - \tilde{c}_1 \\ c_2 - \tilde{c}_2 \\ c_3 - \tilde{c}_3 \end{bmatrix} = \begin{bmatrix} 7.85 & - 8.36 \\ -19.3 & - (-19.4) \\ 71.4 & - 71.7 \end{bmatrix} = \begin{bmatrix} -0.51 \\ 0.1 \\ -0.3 \end{bmatrix}$$

A set of error equations [Eq. (7.25)] can now be generated:

$$3\Delta x_1 - 0.1\Delta x_2 - 0.2\Delta x_3 = -0.51$$

$$0.1\Delta x_1 + 7\Delta x_2 - 0.3\Delta x_3 = 0.1$$

$$0.3\Delta x_1 - 0.2\Delta x_2 + 10\Delta x_3 = -0.3$$

which can be solved (using three significant figures so that we are consistent with the original problem) for

$$[\Delta X]^T = [-0.171 \quad 0.0157 \quad -0.0246]$$

which can be used to correct the values

$$x_1 = 3.17 - 0.171 = 3.00$$

$$x_2 = -2.51 + 0.0157 = -2.49$$

$$x_3 = 7.02 - 0.0246 = 7.00$$

which are much closer to the true solution.

**Error Equations in Computer Programs.** Error equations can be simply integrated into computer programs for Gauss elimination. An algorithm to accomplish this is delineated in Fig. 7.9. Note that in order to be most effective for correcting highly ill-conditioned systems, the $E$'s in Eq. (7.25) should be expressed in double-precision arithmetic. This is simple to do for FORTRAN but is not possible for many dialects of BASIC.

**Step 1:** Input equations.

**Step 2:** Solve equations using Gauss elimination.

**Step 3:** Substitute results into original equations in order to compute right-hand-side values.

**Step 4:** Compute the difference between the computed right-hand-side constants and the original values. If the differences are zero, discontinue the computation and print out the results. If a difference exists, continue to step 5.

**Step 5:** Solve the equations with Gauss elimination using the differences computed in step 4 as the right-hand side.

**Step 6:** Add these results to the previous solution and return to step 3.

FIGURE 7.9 Algorithm for including error corrections in Gauss elimination.

## 7.5 SUMMARY

In summary, we have devoted this chapter to Gauss elimination, the most fundamental method for solving simultaneous linear algebraic equations. Although it is one of the earliest techniques developed for this purpose, it is nevertheless still an extremely effective algorithm for obtaining solutions for many engineering problems. Aside from this practical utility, it also provided a context for our discussion of general issues such as round off, scaling, and conditioning.

Answers obtained using Gauss elimination may be checked by substituting them into the original equations. However, this does not always represent a reliable check for ill-conditioned systems. Therefore, some measure of condition, such as the determinant of the scaled system, should be computed if round-off error is suspected. Using partial pivoting and more significant figures in the computation are two options for mitigating round-off error. If the problem seems to be substantial, error corrections (Sec. 7.4.4) can sometimes be applied to refine the solution.

Other approaches and variations of Gauss elimination are available to suit particular needs. For example, as described in Box 7.2, a very efficient version of Gauss elimination can be formulated for tridiagonal systems. Chapter 8 is devoted to two alternative methods, Gauss-Jordan and Gauss-Seidel.

BOX 7.2 Banded Systems: The Tridiagonal Case

A banded matrix is a square matrix that has all elements equal to zero, with the exception of a band centered on the main diagonal (recall Box III.1). For the case where the bandwidth is 3, the matrix is given a special name—the *tridiagonal matrix.* Tridiagonal systems are frequently encountered in engineering and scientific practice. For example, they typically result from finite-difference solutions of partial differential equations. In addition, other numerical methods such as cubic splines (Sec. 11.4) require the evaluation of tridiagonal systems.

A tridiagonal system is one in which the coefficients are in a tridiagonal form, as in

$$
\begin{aligned}
e_1x_1 + f_1x_2 \quad & = g_1 \\
d_2x_1 + e_2x_2 + f_2x_3 \quad & = g_2 \\
d_3x_2 + e_3x_3 + f_3x_4 \quad & = g_3 \\
\cdot \quad \cdot \quad \cdot \quad & \quad \vdots \\
d_{n-1}x_{n-2} + e_{n-1}x_{n-1} + f_{n-1}x_n & = g_{n-1} \\
d_nx_{n-1} + e_nx_n & = g_n
\end{aligned}
$$

Notice that we have changed our notation for the coefficients of the tridiagonal system from $a$'s and $c$'s to $d$'s, $e$'s, $f$'s, and $g$'s. This was done to avoid storing large numbers of useless zeros in the matrix of $a$'s. This space-saving modification is advantageous because the resulting algorithm requires less computer memory.

As would be expected, banded systems can be solved by a technique such as Gauss elimination. However, due to the system's unique structure, the algorithm for implementing Gauss elimination can be simplified greatly, and solutions are obtained in a very efficient manner. For the tridiagonal system, the forward-elimination steps are simplified because most of the matrix elements are already zero. Then the remaining unknowns are evaluated by back-substitution. The total algorithm is expressed concisely by the following computer codes:

FORTRAN

```
      SUBROUTINE TRI(D,E,F,G)
      DIMENSION D(15),E(15),F(15),G(15)
      COMMON N,M
      DO 1040 I=1,M
      FA=D(I+1)/E(I)
      E(I+1)=E(I+1)-FA*F(I)
      G(I+1)=G(I+1)-FA*G(I)
      G(N)=G(N)/E(N)
 1040 CONTINUE
      DO 1070 I=1,M
      G(N-I)=(G(N-I)-F(N-I)*G(N-I+1))/E(N-I)
 1070 CONTINUE
      RETURN
      END
```

BASIC

```
1000  FOR I = 1 TO N - 1
1010  FA = D(I + 1) / E(I)
1020  E(I + 1) = E(I + 1) - FA * F
      (I)
1030  G(I + 1) = G(I + 1) - FA * G
      (I)
1035  G(N) = G(N) / E(N)
1040  NEXT I
1050  FOR I = 1 TO N - 1
1060  G(N - I) = (G(N - I) - F(N -
      I) * G(N - I + 1)) / E(N - I
      )
1070  NEXT I
1080  RETURN
```

N = number of equations
M = N − 1
D = vector of lower diagonal elements
E = vector of main diagonal elements
F = vector of upper diagonal elements
G = vector of right-hand-side constants
(After modification, the vector G contains the solution)

# PROBLEMS

## Hand Calculations

**7.1** Write the following set of equations in matrix form:

$$50 = 3x_2 + 8x_1$$

$$45 = 7x_1 + 6x_3$$

$$22 = 9x_3 + 4x_2$$

Write the transpose of the matrix of coefficients.

**7.2** A number of matrices are defined as

$$[A] = \begin{bmatrix} 1 & 5 & 6 \\ 2 & 1 & 3 \\ 4 & 0 & 5 \end{bmatrix} \qquad [B] = \begin{bmatrix} 4 & 3 & 1 \\ 1 & 2 & 6 \\ 1 & 0 & 4 \end{bmatrix} \qquad [C] = \begin{bmatrix} 2 \\ 6 \\ 1 \end{bmatrix}$$

$$[D] = \begin{bmatrix} 5 & 4 & 3 & 6 \\ 2 & 1 & 7 & 5 \end{bmatrix} \qquad [E] = \begin{bmatrix} 4 & 5 \\ 1 & 1 \\ 5 & 6 \end{bmatrix}$$

$$[F] = \begin{bmatrix} 2 & 0 & 1 \\ 1 & 2 & 3 \end{bmatrix} \qquad [G] = [8 \quad 6 \quad 4]$$

Answer the following questions regarding these matrices:
(*a*) What are the dimensions of the matrices?
(*b*) Identify the square, column, and row matrices.
(*c*) What are the values of the elements

$$a_{12} \qquad b_{23} \qquad d_{32} \qquad e_{22} \qquad f_{12} \qquad g_{12}$$

(*d*) Perform the operations

(1) $[A] + [B]$
(2) $[B] - [A]$
(3) $[A] + [F]$
(4) $5 \times [B]$
(5) $[A] \times [B]$
(6) $[B] \times [A]$
(7) $[A] \times [C]$
(8) $[C]^T$
(9) $[D]^T$
(10) $[I] \times [B]$

**7.3** Three matrices are defined as

$$[X] = \begin{bmatrix} 2 & 6 \\ 6 & 10 \\ 7 & 4 \end{bmatrix} \qquad [Y] = \begin{bmatrix} 6 & 0 \\ 4 & 1 \end{bmatrix} \qquad [Z] = \begin{bmatrix} 2 & 1 \\ 6 & 8 \end{bmatrix}$$

(*a*) Perform all possible multiplications that can be computed between pairs of these matrices.
(*b*) Use the method in Box III.2 to justify why the remaining pairs cannot be multiplied.
(*c*) Use the results of (*a*) to illustrate why the order of multiplication is important.

**7.4** Use the graphical method to solve

$$4x_1 - 6x_2 = -22$$

$$-x_1 + 12x_2 = 58$$

Check your results by substituting them back into the equations.

**7.5** Given the system of equations

$$0.75x_1 + x_2 = 14.25$$

$$1.1x_1 + 1.6x_2 = 22.1$$

(*a*) Solve graphically.

(*b*) On the basis of the graphical solution, what do you expect regarding the condition of the system?
(*c*) Solve by the elimination of unknowns.
(*d*) Check your answers by substituting them back into the original equations.

**7.6** For the set of equations

$$7x_1 - 3x_2 + 8x_3 = -49$$

$$x_1 - 2x_2 - 5x_3 = 5$$

$$4x_1 - 6x_2 + 10x_3 = -84$$

(*a*) Compute the determinant.
(*b*) Use Cramer's rule to solve for the $x$'s.
(*c*) Substitute your results back into the original equation to check your results.

**7.7** Given the equations

$$0.5x_1 - x_2 = -9.5$$

$$0.28x_1 - 0.5x_2 = -4.72$$

(*a*) Solve graphically.
(*b*) After scaling, compute the determinant.
(*c*) On the basis of (*a*) and (*b*) what would you expect regarding the system's condition?
(*d*) Solve by the elimination of unknowns.
(*e*) Solve again, but with $a_{11}$ modified slightly to 0.55. Interpret your results in light of the discussion of ill conditioning in Sec. 7.1.1.

**7.8** Given the system

$$-12x_1 + x_2 - 7x_3 = -80$$

$$x_1 - 6x_2 + 4x_3 = 13$$

$$-2x_1 - x_2 + 10x_3 = 92$$

(*a*) Solve by naive Gauss elimination. Show all steps of the computation.
(*b*) Substitute your results into the original equation to check your answers.

**7.9** Use Gauss elimination to solve:

$$4x_1 + 5x_2 - 6x_3 = 28$$

$$2x_1 - 7x_3 = 29$$

$$-5x_1 - 8x_2 = -64$$

Employ partial pivoting and check your answers by substituting them into the original equations.

**7.10** Use Gauss elimination to solve:

$$3x_2 - 13x_3 = -50$$

$$2x_1 - 6x_2 + x_3 = 44$$
$$4x_1 + 8x_3 = 4$$

Employ partial pivoting. Check your answers by substituting them into the original equations.

**7.11** Repeat Example 7.6 but double the drag coefficients.

**7.12** Solve the following tridiagonal system:

$$5x_1 + 4x_2 = 25$$
$$4x_1 - 3x_2 + 7x_3 = 3$$
$$x_2 - 6x_3 + 4x_4 = 17$$
$$12x_3 + 2x_4 = 36$$

**7.13** Perform the same computation as in Example 7.6, but use five parachutists with the following characteristics.

| Parachutist | Mass, kg | Drag coefficient, kg/s |
|---|---|---|
| 1 | 60 | 15 |
| 2 | 80 | 14 |
| 3 | 75 | 18 |
| 4 | 75 | 12 |
| 5 | 90 | 10 |

The parachutists have a velocity of 10 m/s.

## Computer-Related Problems

**7.14** Write a general computer program to multiply two matrices, that is, $[X] = [Y][Z]$ where $[Y]$ is $m$ by $n$ and $[Z]$ is $n$ by $p$. Test the program using

$$[Y] = \begin{bmatrix} 3 & 6 \\ 2 & 0 \\ 8 & 7 \end{bmatrix} \qquad [Z] = \begin{bmatrix} 5 & 6 & 7 & 4 \\ 3 & 0 & 1 & 2 \end{bmatrix}$$

**7.15** Write a computer program to generate the transpose of a matrix. Test it on

$$[A] = \begin{bmatrix} 5 & 0 & 7 \\ 4 & 1 & 5 \\ 2 & 6 & 1 \end{bmatrix}$$

**7.16** Reprogram Fig. 7.4 so that it is user-friendly. Among other things:
(*a*) Integrate Fig. 7.8 into the code so that the program performs partial pivoting.

(*b*) Place documentation statements throughout the program to identify what each section is intended to accomplish.
(*c*) Label the input and output.
(*d*) Scale the equations so that the maximum coefficient in each row is 1. Compute the determinant as a measure of system condition (optional).

**7.17** Test the program you developed in Prob. 7.16 by duplicating the computations from Examples 7.5 and 7.6.

**7.18** Use the program you developed in Prob. 7.16 to repeat Probs. 7.8 through 7.11.

**7.19** Repeat Probs. 7.17 and 7.18, but use the NUMERICOMP software available with the text. Also use the program to perform an error check for each problem.

**7.20** Develop a user-friendly program for tridiagonal systems based on Box 7.2.

**7.21** Test the program developed in Prob. 7.20 by solving Prob. 7.12.

**7.22** Solve Prob. 7.13 using the software developed in Prob. 7.16.

CHAPTER EIGHT

# GAUSS-JORDAN, MATRIX INVERSION, AND GAUSS-SEIDEL

In the present chapter, we describe two additional methods for solving simultaneous linear equations. The first, *Gauss-Jordan*, is very similar to Gauss elimination. Our primary motive for introducing this method to you is that the Gauss-Jordan technique provides a simple and convenient numerical method for computing the matrix inverse. The inverse has a number of valuable applications in engineering practice. It also provides a means for evaluating system condition.

The second method, *Gauss-Seidel*, is fundamentally different from Gauss elimination and the Gauss-Jordan method in that it is an *approximate, iterative* method. That is, it employs initial guesses and then iterates to obtain refined estimates of the solution. The Gauss-Seidel method is particularly well-suited for large numbers of equations. In these cases, elimination methods can be subject to round-off errors. Because the error of the Gauss-Seidel method is controlled by the number of iterations, round-off error is not an issue of concern with this method. However, there are certain instances where the Gauss-Seidel technique will not converge on the correct answer. These and other trade-offs between elimination and iterative methods will be discussed in subsequent pages.

## 8.1 THE GAUSS-JORDAN METHOD

The Gauss-Jordan method is a variation of Gauss elimination. The major difference is that when an unknown is eliminated in the Gauss-Jordan method, it is eliminated from all other equations rather than just the subsequent ones. Thus, the elimination step results in an identity matrix rather than a triangular matrix (Fig. 8.1). Consequently, it is not necessary to employ back-substitution to obtain the solution. The method is best illustrated by an example.

$$\begin{bmatrix} a_{11} & a_{12} & a_{13} & | & c_1 \\ a_{21} & a_{22} & a_{23} & | & c_2 \\ a_{31} & a_{32} & a_{33} & | & c_3 \end{bmatrix}$$

$$\Downarrow$$

$$\begin{bmatrix} 1 & 0 & 0 & | & c_1^* \\ 0 & 1 & 0 & | & c_2^* \\ 0 & 0 & 1 & | & c_3^* \end{bmatrix}$$

$$\Downarrow$$

$$\begin{aligned} x_1 \quad\quad\quad &= c_1^* \\ x_2 \quad\quad &= c_2^* \\ x_3 &= c_3^* \end{aligned}$$

FIGURE 8.1 Graphical depiction of the Gauss-Jordan method. Compare with Fig. 7.3 to elucidate the differences between this technique and Gauss elimination. The asterisks connote that the elements of the right-hand-side vector have been modified several times.

### EXAMPLE 8.1
### Gauss-Jordan Method

Problem Statement: Use the Gauss-Jordan technique to solve the same system as in Example 7.5:

$$3x_1 - 0.1x_2 - 0.2x_3 = 7.85$$

$$0.1x_1 + 7x_2 - 0.3x_3 = -19.3$$

$$0.3x_1 - 0.2x_2 + 10x_3 = 71.4$$

Solution: First, reexpress the coefficients and the right-hand side as an augmented matrix:

$$\begin{bmatrix} 3 & -0.1 & -0.2 & | & 7.85 \\ 0.1 & 7 & -0.3 & | & -19.3 \\ 0.3 & -0.2 & 10 & | & 71.4 \end{bmatrix}$$

Then normalize the first row by dividing it by the pivot element, 3, to yield

$$\begin{bmatrix} 1 & -0.0333333 & -0.0666667 & | & 2.61667 \\ 0.1 & 7 & -0.3 & | & -19.3 \\ 0.3 & -0.2 & 10 & | & 71.4 \end{bmatrix}$$

The $x_1$ term can be eliminated from the second row by subtracting 0.1 times the first row from the second row. Similarly, subtracting 0.3 times the first row from the third row will eliminate the $x_1$ term from the third row:

$$\left[\begin{array}{ccc|c} 1 & -0.0333333 & -0.0666667 & 2.61667 \\ 0 & 7.00333 & -0.293333 & -19.5617 \\ 0 & -0.190000 & 10.0200 & 70.6150 \end{array}\right]$$

Next, normalize the second row by dividing it by 7.00333:

$$\left[\begin{array}{ccc|c} 1 & -0.0333333 & -0.0666667 & 2.61667 \\ 0 & 1 & -0.0418848 & -2.79320 \\ 0 & -0.190000 & 10.0200 & 70.6150 \end{array}\right]$$

Reduction of the $x_2$ terms from the first and third equations gives

$$\left[\begin{array}{ccc|c} 1 & 0 & -0.0680629 & 2.52356 \\ 0 & 1 & -0.0418848 & -2.79320 \\ 0 & 0 & 10.0120 & 70.0843 \end{array}\right]$$

The third row is then normalized by dividing it by 10.0120:

$$\left[\begin{array}{ccc|c} 1 & 0 & -0.0680629 & 2.52356 \\ 0 & 1 & -0.0418848 & -2.79320 \\ 0 & 0 & 1 & 7.00003 \end{array}\right]$$

Finally, the $x_3$ terms can be reduced from the first and the second equations to give

$$\left[\begin{array}{ccc|c} 1 & 0 & 0 & 3.00000 \\ 0 & 1 & 0 & 2.50001 \\ 0 & 0 & 1 & 7.00003 \end{array}\right]$$

Thus, as depicted in Fig. 8.1, the coefficient matrix has been transformed to the identity matrix, and the solution is obtained in the right-hand-side vector. Notice that no back-substitution was required to obtain the solution.

All the material in Chap. 7 regarding the pitfalls and improvements in Gauss elimination also applies to the Gauss-Jordan method. For example, a similar pivoting strategy can be used to avoid division by zero and to reduce round-off error.

Although the Gauss-Jordan technique and Gauss elimination might appear almost identical, the former requires approximately 50 percent more operations. Therefore, Gauss elimination is the simple method of preference for obtaining exact solutions of simultaneous linear equations. One of the primary reasons that we have introduced the Gauss-Jordan, however, is that it provides a straightforward method for obtaining the matrix inverse, as described in Sec. 8.2.

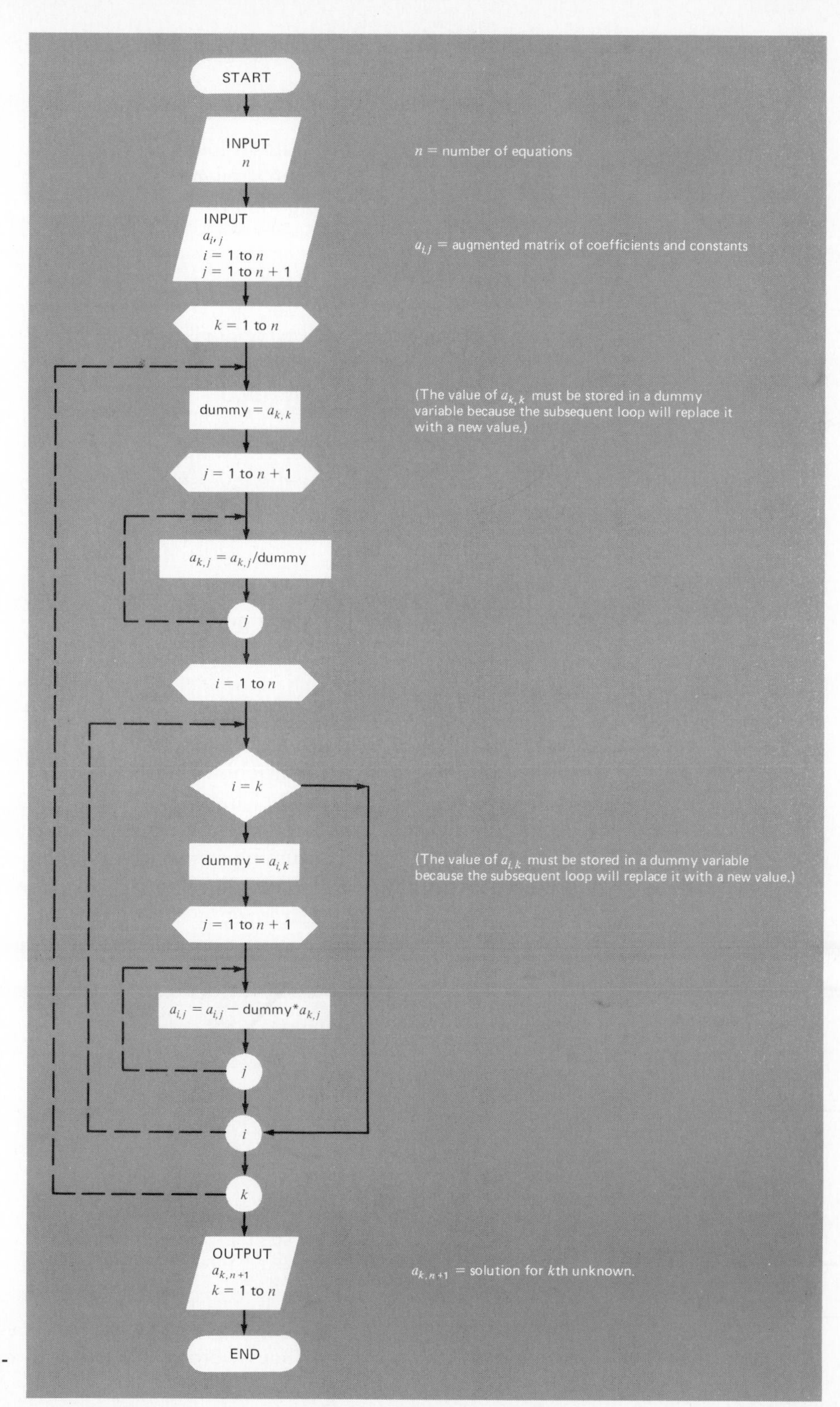


FIGURE 8.2 Flowchart for the Gauss-Jordan method, without partial pivoting.

### 8.1.1 Computer Algorithm for the Gauss-Jordan Method

A flowchart for the Gauss-Jordan method without partial pivoting is shown in Fig. 8.2. A pivoting scheme similar to the one presented in Fig. 7.10 can be incorporated into this algorithm.

## 8.2 MATRIX INVERSION

In our introduction to matrix operations (Sec. III.2.2), we introduced the notion that if a matrix $[A]$ is square, there is another matrix, $[A]^{-1}$, called the inverse of $[A]$, for which [Eq. (III.3)]

$$[A]\,[A]^{-1} = [A]^{-1}[A] = [I]$$

We also demonstrated how the inverse can be used to solve a set of simultaneous equations, as in [Eq. (III.6)]

$$[X] = [A]^{-1}\,[C] \tag{8.1}$$

One application of the inverse occurs when it is necessary to solve several systems of equations of the form

$$[A][X] = [C]$$

differing only by the right-hand-side vector $[C]$. Rather than solve each system individually, an alternative approach would be to determine the inverse of the matrix of coefficients. Then, Eq. (8.1) can be used to obtain each solution by simply multiplying $[A]^{-1}$ by the particular right-hand-side vector $[C]$. Because matrix multiplication is much quicker than inversion, we only perform the time-consuming step once and then obtain the additional solutions in an efficient manner. Also, as elaborated in Sec. 8.2.1, the elements of the inverse are extremely useful in themselves.

One straightforward way to compute the inverse is using the Gauss-Jordan method. To do this, the coefficient matrix is augmented with an

$$\overset{[A] \qquad\qquad [I]}{\left[\begin{array}{ccc|ccc} a_{11} & a_{12} & a_{13} & 1 & 0 & 0 \\ a_{21} & a_{22} & a_{23} & 0 & 1 & 0 \\ a_{31} & a_{32} & a_{33} & 0 & 0 & 1 \end{array}\right]}$$

$$\Downarrow$$

$$\underset{[I] \qquad\qquad [A]^{-1}}{\left[\begin{array}{ccc|ccc} 1 & 0 & 0 & a_{11}^{-1} & a_{12}^{-1} & a_{13}^{-1} \\ 0 & 1 & 0 & a_{21}^{-1} & a_{22}^{-1} & a_{23}^{-1} \\ 0 & 0 & 1 & a_{31}^{-1} & a_{32}^{-1} & a_{33}^{-1} \end{array}\right]}$$

FIGURE 8.3 Graphical depiction of the Gauss-Jordan method, with matrix inversion.

identity matrix (Fig. 8.3). Then the Gauss-Jordan method is applied in order to reduce the coefficient matrix to an identity matrix. When this is accomplished, the right-hand side of the augmented matrix will contain the inverse. The technique is illustrated in the following example.

## EXAMPLE 8.2
Use of the Gauss-Jordan Method to Compute the Matrix Inverse

Problem Statement: Determine the matrix inverse of the system solved previously in Example 7.5. Obtain the solution by multiplying $[A]^{-1}$ by the right-hand-side vector: $[C]^T = [7.85 \quad -19.3 \quad 71.4]$. In addition, obtain the solution for a different right-hand-side vector: $[C]^T = [20 \quad 50 \quad 15]$.

Solution: Augment the coefficient matrix with an identity matrix:

$$[A] = \left[\begin{array}{ccc|ccc} 3 & -0.1 & -0.2 & 1 & 0 & 0 \\ 0.1 & 7 & -0.3 & 0 & 1 & 0 \\ 0.3 & -0.2 & 10 & 0 & 0 & 1 \end{array}\right]$$

Using $a_{11}$ as the pivot element, normalize row 1 and use it to eliminate $x_1$ from the other rows:

$$\left[\begin{array}{ccc|ccc} 1 & -0.0333333 & -0.0666667 & 0.333333 & 0 & 0 \\ 0 & 7.00333 & -0.293333 & -0.0333333 & 1 & 0 \\ 0 & -0.190000 & 10.0200 & -0.0999999 & 0 & 1 \end{array}\right]$$

Next, $a_{22}$ can be used as the pivot element and $x_2$ eliminated from the other rows:

$$\left[\begin{array}{ccc|ccc} 1 & 0 & -0.068057 & 0.333175 & 0.004739329 & 0 \\ 0 & 1 & -0.0417061 & -0.00473933 & 0.142180 & 0 \\ 0 & 0 & 10.0121 & -0.10090 & 0.0270142 & 1 \end{array}\right]$$

Finally, $a_{33}$ is used as the pivot element and $x_3$ eliminated from the other rows:

$$\left[\begin{array}{ccc|ccc} 1 & 0 & 0 & 0.332489 & 0.00492297 & 0.00679813 \\ 0 & 1 & 0 & -0.0051644 & 0.142293 & 0.00418346 \\ 0 & 0 & 1 & -0.0100779 & 0.00269816 & 0.0998801 \end{array}\right]$$

Therefore, the inverse is

$$[A]^{-1} = \begin{bmatrix} 0.332489 & 0.00492297 & 0.00679813 \\ -0.0051644 & 0.142293 & 0.00418346 \\ -0.0100779 & 0.00269816 & 0.0998801 \end{bmatrix}$$

Now, the inverse can be multiplied by the first right-hand-side vector to determine the solution:

$$\begin{aligned} x_1 &= 7.85(0.332489) - 19.3(0.00492297) + 71.4(0.00679813) \\ &= 3.00041181 \end{aligned}$$

$$x_2 = 7.85(-0.0051644) - 19.3(0.142293) + 71.4(0.00418346)$$
$$= -2.48809640$$

$$x_3 = 7.85(-0.0100779) - 19.3(0.00269816) + 71.4(0.0998801)$$
$$= 7.00025314$$

The second solution is simply obtained by performing another multiplication, as in

$$x_1 = 20(0.332489) + 50(0.00492297) + 15(0.00679813)$$
$$= 6.99790045$$

$$x_2 = 20(-0.0051644) + 50(0.142293) + 15(0.00418346)$$
$$= 7.0741139$$

$$x_3 = 20(-0.0100779) + 50(0.00269816) + 15(0.0998801)$$
$$= 1.43155150$$

### 8.2.1 Stimulus-Response Computations

As discussed in Sec. III.1.2, many of the linear systems of equations confronted in engineering practice are derived from conservation laws. The mathematical expression of these laws is some form of balance equation to ensure that a particular property—mass, force, heat, momentum, or other—is conserved. For a force balance on a structure, the properties might be horizontal or vertical components of the forces acting on each node of the structure (see Case Study 9.3). For a mass balance, the properties might be the mass in each reactor of a chemical process. Other fields of engineering would yield similar examples.

A single balance equation can be written for each part of the system, resulting in a set of equations defining the behavior of the property for the entire system. These equations are interrelated, or coupled, in that each equation includes one or more of the variables from the other equations. For many cases, these systems are linear and, therefore, of the exact form dealt with in the present chapter.

$$[A][X] = [C] \qquad [8.2]$$

Now, for balance equations, the terms of Eq. (8.2) have a definite physical interpretation. For example, the elements of $[X]$ are the levels of the property being balanced for each part of the system. In a force balance of a structure, they represent the horizontal and vertical forces in each member. For the mass balance, they are the mass of chemical in each reactor. In either case, they represent the system's state or response, which we are trying to determine.

The right-hand-side vector $[C]$ contains those elements of the balance that are independent of behavior of the system—that is, they are constants.

As such, they often represent the external forces or stimuli that drive the system.

Finally, the matrix of coefficients $[A]$ usually contains the parameters that express how the parts of the system interact or are coupled. Consequently, Eq. (8.2) might be reexpressed as

$$[\text{Interactions}][\text{response}] = [\text{stimuli}]$$

Now, as we know from the present chapter, there are a variety of ways to solve Eq. (8.2). However, using the matrix inverse yields a particularly interesting result. The formal solution [Eq. (8.1)] can be expressed as

$$[X] = [A]^{-1}[C]$$

or (recalling our definition of matrix multiplication from Box III.2)

$$x_1 = a_{11}^{-1}c_1 + a_{12}^{-1}c_2 + a_{13}^{-1}c_3$$

$$x_2 = a_{21}^{-1}c_1 + a_{22}^{-1}c_2 + a_{23}^{-1}c_3$$

$$x_3 = a_{31}^{-1}c_1 + a_{32}^{-1}c_2 + a_{33}^{-1}c_3$$

Thus, we find that the inverted matrix itself, aside from providing a solution, has extremely useful properties. That is, each of its elements represents the response of a single part of the system to a unit stimulus of any other part of the system.

Notice that these formulations are linear and, therefore, superposition and proportionality hold. *Superposition* means that if a system is subject to several different stimuli (the $c$'s), the responses can be computed individually and the results summed to obtain a total response. *Proportionality* means that multiplying the stimuli by a quantity results in the response to that stimuli being multiplied by the same quantity. Thus, the coefficient $a_{11}^{-1}$ is a proportionality constant that gives the value of $x_1$ due to a unit level of $c_1$. This result is independent of the effects of $c_2$ and $c_3$ on $x_1$, which are reflected in the coefficients $a_{12}^{-1}$ and $a_{13}^{-1}$, respectively. Therefore, we can draw the general conclusion that the element $a_{1j}^{-1}$ of the inverted matrix represents the value of $x_1$ due to a unit quantity of $c_j$. Using the example of the structure, element $a_{ij}^{-1}$ of the matrix inverse would represent the force in member $i$ due to a unit external force at node $j$. Even for small systems, such behavior of individual stimulus-response interactions would not be intuitively obvious. As such, the matrix inverse provides a powerful technique for understanding the interrelationships of component parts of complicated systems. This power will be demonstrated in Case Study 9.3.

### 8.2.2 The Inverse and Ill Conditioning

Aside from its engineering applications, the inverse also provides a means to discern whether systems are ill-conditioned. Three methods are available for this purpose:

1. Scale the matrix of coefficients, $[A]$, so that the largest element in each row is 1. If there are elements of $[A]^{-1}$ that are several orders of magnitude greater than the elements of the original matrix, it is likely that the system is ill-conditioned.
2. Multiply the inverse by the original coefficient matrix and assess whether the result is close to the identity matrix. If not, it indicates ill conditioning.
3. Invert the inverted matrix and assess whether the result is sufficiently close to the original coefficient matrix. If not, it again indicates that the system is ill-conditioned.

### 8.2.3 Computer Algorithm for Matrix Inversion

The computer algorithm from Fig. 8.2 can be modified to calculate the matrix inverse. This involves augmenting the coefficient matrix with an identity matrix at the beginning of the program. In addition, some of the loop indices must be doubled in order that the computations are performed for all the columns of the augmented coefficient matrix.

If partial pivoting has been incorporated into the Gauss-Jordan algorithm, more substantial modifications are necessary. This is due to the fact that every time a row of the coefficient matrix is pivoted, the column of the matrix inverse must be shifted in a like manner.

Figure 8.4 illustrates this phenomenon. For example, if row 3 is pivoted or shifted between rows 1 and 2, the "meaning" or "interpretation" of column 2 of the inverted matrix is modified. Rather than indicating the effect of a unit change of $c_2$ on the $x$'s, it would indicate the effect of a unit change of $c_3$ on the $x$'s.

In addition to the above features, the computer program should be designed to calculate solutions for any number of right-hand-side vectors, as discussed at the beginning of Sec. 8.2. This can be simply accomplished by a loop placed after the computation of the matrix inverse. This loop can prompt the user for a right-hand-side vector which can then be multiplied by $[A]^{-1}$ to obtain a solution. The process can be continued until the user responds that solutions are no longer necessary.

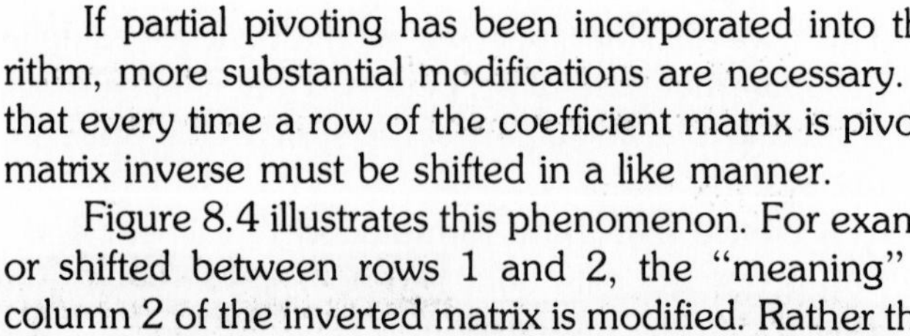

Coefficient Matrix

$$\begin{matrix} a_{11} & a_{12} & a_{13} & a_{14} \\ a_{21} & a_{22} & a_{23} & a_{24} \\ a_{31} & a_{32} & a_{33} & a_{34} \\ a_{41} & a_{42} & a_{43} & a_{44} \end{matrix}$$

Inverse

$$\begin{matrix} a_{11}^{-1} & a_{12}^{-1} & a_{13}^{-1} & a_{14}^{-1} \\ a_{21}^{-1} & a_{22}^{-1} & a_{23}^{-1} & a_{24}^{-1} \\ a_{31}^{-1} & a_{32}^{-1} & a_{33}^{-1} & a_{34}^{-1} \\ a_{41}^{-1} & a_{42}^{-1} & a_{43}^{-1} & a_{44}^{-1} \end{matrix}$$

FIGURE 8.4 Graphical depiction of how movement of a row of the coefficient matrix (that is, partial pivoting) has an impact on the interpretation of the elements of the resulting matrix inverse.

## 8.3 THE GAUSS-SEIDEL METHOD

The exact elimination methods discussed previously can be used to solve approximately 25 to 50 simultaneous linear equations. This number can sometimes be expanded if the system is well-conditioned, pivot strategy is employed, the error equations are used, or the matrix is sparse. However, because of round-off error, elimination methods sometimes prove inadequate for larger systems. For these types of problems, iterative or approximate methods can often be used to advantage.

We used similar types of techniques to obtain the roots of a single equation in Chap. 5. Those approaches consisted of guessing a value and then using a systematic method to obtain a refined estimate of the root. Because the present part of the book deals with a similar problem—obtaining the values that simultaneously satisfy a set of equations—we might suspect that such approximate methods could be useful in this context.

The reason that iterative methods are useful for round-off-prone systems is that an approximate method can be continued until it has converged within some prespecified tolerance of error. Thus, round off is no longer an issue, because you control the level of error that is acceptable.

The *Gauss-Seidel method* is the most commonly used iterative method. Assume that we are given a set of $n$ equations:

$$[A][X] = [C]$$

If the diagonal elements are all nonzero, the first equation can be solved for $x_1$, the second for $x_2$, and so on to yield

$$x_1 = \frac{c_1 - a_{12}x_2 - a_{13}x_3 - \cdots - a_{1n}x_n}{a_{11}} \quad [8.3a]$$

$$x_2 = \frac{c_2 - a_{21}x_1 - a_{23}x_3 - \cdots - a_{2n}x_n}{a_{22}} \quad [8.3b]$$

$$x_3 = \frac{c_3 - a_{31}x_1 - a_{32}x_2 - \cdots - a_{3n}x_n}{a_{33}} \quad [8.3c]$$

$$\vdots \qquad \vdots$$

$$x_n = \frac{c_n - a_{n1}x_1 - a_{n2}x_2 - \cdots - a_{n,n-1}x_{n-1}}{a_{nn}} \quad [8.3d]$$

Now, we can start the solution process by using guesses for the $x$'s. A simple way to obtain initial guesses is to assume that they are all zero. These zeros can be substituted into Eq. (8.3*a*), which can be used to calculate a new value for $x_1 = c_1/a_{11}$. Then, we substitute this new value of $x_1$, along with the previous guesses of zero for $x_3, \ldots, x_n$, into Eq. (8.3*b*) to compute a new value for $x_2$. The process is repeated for each of the equations until we use Eq. (8.3*d*) to calculate a new estimate for $x_n$. Then we return to the first equation and repeat the entire procedure until our solution converges closely enough to the true values. Convergence can be checked using the criterion [recall Eq. (3.5)]

$$\epsilon_{a,i} = \left| \frac{x_i^{j-1} - x_i^j}{x_i^j} \right| 100\% < \epsilon_s \qquad [8.4]$$

for all $i$, where $j$ and $j - 1$ are the present and previous iterations.

## EXAMPLE 8.3
### Gauss-Seidel Method

Problem Statement: Use the Gauss-Seidel method to obtain the solution of the same system used in Examples 7.5 and 8.1:

$$3x_1 - 0.1x_2 - 0.2x_3 = 7.85$$

$$0.1x_1 + 7x_2 - 0.3x_3 = -19.3$$

$$0.3x_1 - 0.2x_2 + 10x_3 = 71.4$$

Recall that the true solution is $x_1 = 3$, $x_2 = -2.5$, and $x_3 = 7$.

Solution: First, solve each of the equations for its unknown on the diagonal.

$$x_1 = \frac{7.85 + 0.1x_2 + 0.2x_3}{3} \qquad [E8.3.1]$$

$$x_2 = \frac{-19.3 - 0.1x_1 + 0.3x_3}{7} \qquad [E8.3.2]$$

$$x_3 = \frac{71.4 - 0.3x_1 + 0.2x_2}{10} \qquad [E8.3.3]$$

By assuming that $x_2$ and $x_3$ are zero, Eq. (E8.3.1) can be used to compute

$$x_1 = \frac{7.85}{3} = 2.616666667$$

This value, along with the assumed value of $x_3 = 0$, can be substituted into Eq. (E8.3.2) to calculate

$$x_2 = \frac{-19.3 - 0.1(2.616666667) + 0}{7} = -2.794523810$$

The first iteration is completed by substituting the calculated values for $x_1$ and $x_2$ into Eq. (E8.3.3) to yield

$$x_3 = \frac{71.4 - 0.3(2.616666667) + 0.2(-2.794523810)}{10}$$

$$= 7.005609524$$

For the second iteration, the same process is repeated to compute

$$x_1 = \frac{7.85 + 0.1(-2.794523810) + 0.2(7.005609524)}{3}$$

$$= 2.990556508 \qquad |\epsilon_t| = 0.31\%$$

$$x_2 = \frac{-19.3 - 0.1(2.990556508) + 0.3(7.005609524)}{7}$$

$$= -2.499624684 \qquad |\epsilon_t| = 0.015\%$$

$$x_3 = \frac{71.4 - 0.3(2.990556508) + 0.2(-2.499624684)}{10}$$

$$= 7.00029081 \qquad |\epsilon_t| = 0.0042\%$$

The method is, therefore, converging on the true solution. Additional iterations could be applied to improve the answers. However, in an actual problem, we would not know the true answer a priori. Consequently, Eq. (8.4) provides a means to estimate the error.

$$\epsilon_{a,1} = \left|\frac{2.990556508 - 2.616666667}{2.990556508}\right| 100 = 12.5\%$$

$$\epsilon_{a,2} = \left|\frac{-2.499624684 - (-2.794523810)}{-2.499624684}\right| 100 = 11.8\%$$

$$\epsilon_{a,3} = \left|\frac{7.000290811 - 7.005609524}{7.000290811}\right| 100 = 0.076\%$$

Note that, as was the case when determining roots of a single equation, formulations such as Eq. (8.4) usually provides a conservative appraisal of convergence. Thus, when they are met, they ensure that the result is known to *at least* the tolerance specified by $\epsilon_s$.

Note that for the Gauss-Seidel method, as each new *x* value is computed it is immediately used in the next equation to determine another *x* value. Thus, if the solution is converging, the best available estimates will be employed. An alternative approach, called *Jacobi iteration,* utilizes a somewhat different tactic. Rather than using the latest available *x*'s, this technique uses Eq. (8.3) to compute a set of new *x*'s on the basis of a set of old *x*'s. Thus, as new values are generated, they are not immediately used but rather are retained for the next iteration.

The difference between the Gauss-Seidel method and Jacobi iteration is depicted in Fig. 8.5. Although there are certain cases where the Jacobi method converges faster, its utilization of the best available estimates usually makes Gauss-Seidel the method of preference.

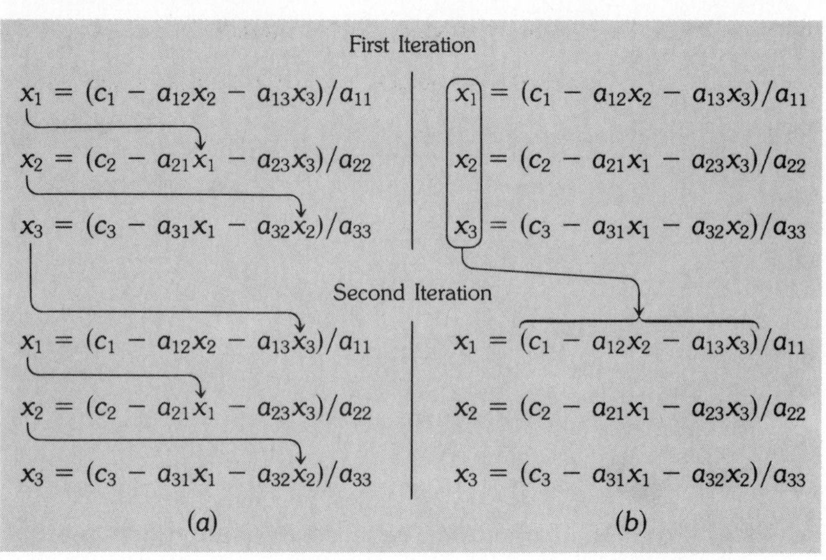


FIGURE 8.5 Graphical depiction of the difference between (*a*) the Gauss-Seidel and (*b*) the Jacobi iterative methods for solving simultaneous linear algebraic equations.

## 8.3.1 Convergence Criterion for the Gauss-Seidel Method

Note that the Gauss-Seidel method is similar in spirit to the technique of simple one-point iteration that was used in Sec. 5.1 to solve for the roots of a single equation. Recall that simple one-point iteration had two fundamental problems: (1) it was sometimes nonconvergent and (2) when it converged it often did so very slowly. The Gauss-Seidel method can also exhibit these shortcomings.

A condition for convergence is that the diagonal coefficient in each of the equations be larger than the sum of the other coefficients in the equation. A quantitative expression of this criterion is

$$|a_{i,i}| > \Sigma |a_{i,j}| \qquad [8.5]$$

where the summation is taken from $j = 1$ to $n$, excluding $j = i$. Equation (8.5) is a sufficient but not necessary criterion for convergence. That is, although the method may sometimes work if Eq. (8.5) is not met, convergence is guaranteed if the condition is satisfied. Systems where Eq. (8.5) holds are called *diagonally dominant*. Fortunately, many engineering problems of practical importance fulfill this requirement.

## 8.3.2 Improvement of Convergence Using Relaxation

*Relaxation* represents a slight modification of the Gauss-Seidel method and is designed to enhance convergence. After each new value of $x$ is computed using Eq. (8.3), that value is modified by a weighted average of the results of the previous and the present iterations:

$$x_i^{new} = \lambda x_i^{new} + (1 - \lambda)x_i^{old} \qquad [8.6]$$

where $\lambda$ is a weighting factor that is assigned a value between 0 and 2.

If $\lambda = 1$, $(1 - \lambda)$ is equal to zero and the result is unmodified. However, if $\lambda$ is set at a value between 0 and 1, the result is a weighted average of the present and the previous results. This type of modification is called *underrelaxation.* This option is typically employed to make a nonconvergent system converge.

For values of $\lambda$ from 1 to 2, extra weight is placed on the present value. In this instance, there is an implicit assumption that the new value is moving in the correct direction toward the true solution but at too slow a rate. Thus, the added weight of $\lambda$ is intended to improve the estimate by pushing it closer to the truth. Hence, this type of modification, which is called *overrelaxation*, is designed to accelerate the convergence of an already convergent system.

The choice of a proper value for $\lambda$ is highly problem-specific and is often determined via trial and error. For a single solution of a set of equations it is often unnecessary. However, if the system under study is to be solved repeatedly, the efficiency introduced by a wise choice of $\lambda$ can be extremely important. Good examples are the very large systems of partial differential equations that often arise when modeling continuous variations of variables (recall the microscale system depicted in Fig. III.1*b*). The second case study in Chap. 9 provides an example of the employment of relaxation in an engineering problem context.

### 8.3.3 Computer Algorithm for the Gauss-Seidel Method

A computer algorithm for the Gauss-Seidel method, with relaxation, is depicted in Fig. 8.6. Note that this algorithm is not guaranteed to yield adequate results if the equations are not input in a diagonally dominant form.

One way to modify the algorithm in order to partially account for this shortcoming is to search the coefficients of each equation during each iteration in order to identify the greatest coefficient. The equation in question would then be solved for the value of $x$ associated with this coefficient. For the next computation, the coefficients of the remaining $x$ values would be searched to identify the largest coefficient. The equation would be solved for the corresponding $x$ value.

By proceeding in this fashion, we would at least maximize our chances of attaining diagonal dominance. However, the scheme does not guarantee success for highly divergent systems. In addition, an algorithm to implement the scheme would not be trivial to program.

### 8.3.4 Problem Contexts for the Gauss-Seidel Method

Aside from circumventing the round-off dilemma, the Gauss-Seidel technique has a number of other advantages that make it particularly attractive in the context of certain engineering problems. For example, when the matrix

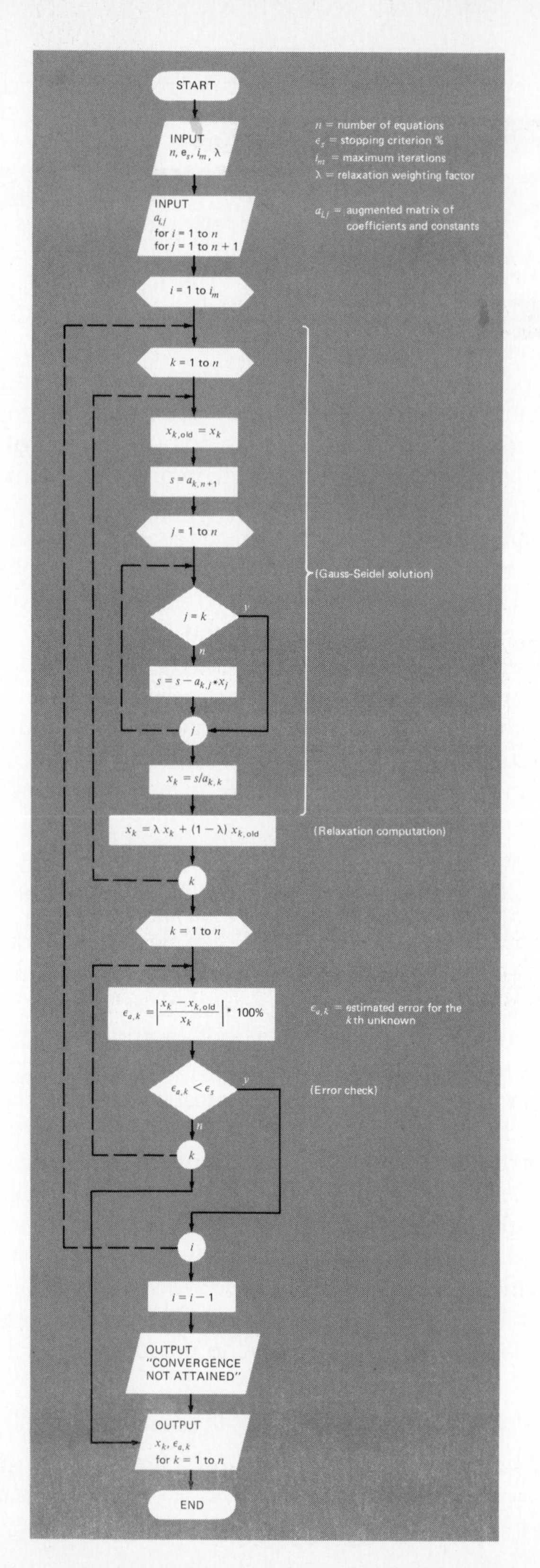


FIGURE 8.6 Flowchart for Gauss-Seidel method with relaxation.

in question is very large and very sparse (that is, most of the elements are zero), elimination methods waste large amounts of computer memory storing zeros.

In Box 7.2, we saw how this shortcoming could be circumvented if the coefficient matrix is banded. For nonbanded systems, there is no simple way to avoid large memory requirements when using elimination methods. Because all computers have a finite amount of memory, this inefficiency can place a real constraint on the size of systems for which elimination methods are practical.

Although a general algorithm such as the one in Fig. 8.6 is prone to the same constraint, the structure of the Gauss-Seidel equations [Eq. (8.3)] permits concise programs to be developed for specific systems. Because only nonzero coefficients need be included in Eq. (8.3), large savings of computer memory are possible. Although this entails more up-front investment in software development, the long-term advantages are substantial when dealing with large systems for which many simulations are to be performed. Both macrovariable and microvariable systems can result in large, sparse matrices for which the Gauss-Seidel method is appealing. In Case Study 9.2 we will elaborate a bit more on these points.

# PROBLEMS

## Hand Calculations

**8.1** Use the Gauss-Jordan method to solve Prob. 7.6.

**8.2** Determine the matrix inverse for Prob. 7.6. Check your results by multiplying $[A]$ by $[A]^{-1}$ to give the identity matrix.

**8.3** Using the Gauss-Jordan method, repeat Prob. 7.9.

**8.4** Determine the matrix inverse for Prob. 7.9. Check your results by verifying that $[A]\,[A]^{-1} = [I]$. Do not use a pivoting strategy.

**8.5** Using the Gauss-Jordan method, with partial pivoting, compute the matrix inverse for Prob. 7.10. Rearrange the inverse so that the rows and columns conform to the sequence of the original matrix prior to pivoting (see Fig. 8.4 and the discussion in Sec. 8.2.3).

**8.6** Use the Gauss-Jordan method to solve:

$$10x_1 - 3x_2 + 6x_3 = 24.5$$

$$1x_1 + 8x_2 - 2x_3 = -9$$

$$-2x_1 + 4x_2 - 9x_3 = -50$$

**8.7** Determine the matrix inverse for Prob. 8.6. Use the inverse to solve the original problem as well as to solve the additional case where the right-hand-side vector is $[C]^T = [110 \quad 55 \quad -105]$.

**8.8** Solve Prob. 8.6 using the Gauss-Seidel method, with a stopping criterion of $\epsilon_s = 10$ percent.

**8.9** Solve Prob. 7.8 using the Gauss-Seidel method, with a stopping criterion of $\epsilon_s = 10$ percent.

**8.10** Use the Gauss-Seidel method to solve ($\epsilon_s = 5$ percent):

$$\begin{aligned} x_1 + 7x_2 - 3x_3 &= -51 \\ 4x_1 - 4x_2 + 9x_3 &= 61 \\ 12x_1 - x_2 + 3x_3 &= 8 \end{aligned}$$

**8.11** Use the Gauss-Seidel method to solve ($\lambda = 0.90$ and $\epsilon_s = 5$ percent):

$$\begin{aligned} -6x_1 + \qquad 12x_3 &= 60 \\ 4x_1 - x_2 - x_3 &= -2 \\ 6x_1 + 8x_2 \qquad &= 44 \end{aligned}$$

**8.12** Given the following set of equations:

$$\begin{aligned} 4x_1 - 2x_2 - x_3 &= 39 \\ x_1 - 6x_2 + 2x_3 &= -28 \\ x_1 - 3x_2 + 12x_3 &= -86 \end{aligned}$$

solve using (*a*) Gauss elimination, (*b*) the Gauss-Jordan method, and (*c*) the Gauss-Seidel method ($\epsilon_s = 5$ percent).

**8.13** Given the following set of equations:

$$\begin{aligned} x_1 - 3x_2 + 12x_3 &= 10 \\ 5x_1 - 12x_2 + 2x_3 &= -33 \\ x_1 - 14x_2 \qquad &= -103 \end{aligned}$$

solve using (*a*) Gauss elimination, (*b*) the Gauss-Jordan method, and (*c*) the Gauss-Seidel method ($\epsilon_s = 5$ percent).

## Computer-Related Problems

**8.14** Develop a user-friendly program for the Gauss-Jordan method, based on Fig. 8.2. Incorporate a scheme similar to the one depicted in Fig. 7.10 so that partial pivoting is employed.

**8.15** Test the program developed in Prob. 8.14 by duplicating the computation from Example 8.1.

**8.16** Repeat Probs. 7.6 and 7.8 through 7.11 using the software developed in Prob. 8.14.

**8.17** Develop a user-friendly program for the Gauss-Jordan method, with matrix inversion and partial pivoting. Include the features suggested in Sec. 8.2.3 into your program.

**8.18** Repeat Probs. 8.5 and 8.7 using the software developed in Prob. 8.17.

**8.19** Develop a user-friendly computer program for the Gauss-Seidel method based on Fig. 8.6. Have the program check the convergence criterion expressed by Eq. (8.5). In addition, include relaxation as in Eq. (8.6).

**8.20** Test the program developed in Prob. 8.19 by using it to duplicate Example 8.3.

**8.21** Using the software developed in Prob. 8.19, repeat Probs. 8.8 through 8.11.

CHAPTER NINE

# CASE STUDIES: SYSTEMS OF LINEAR ALGEBRAIC EQUATIONS

The purpose of this chapter is to use the numerical procedures discussed in Chaps. 7 and 8 to solve systems of linear algebraic equations for some engineering applications. These systematic numerical techniques have practical significance because engineers frequently encounter problems involving systems of equations that are too large to solve by hand. The numerical algorithms in these applications are particularly convenient to implement on personal computers.

Among other things, we have designed the case studies to provide realistic illustrations of the characteristics and trade-offs mentioned in the theory chapters. For example, *Case Study 9.1* is a very simple illustration of how linear algebraic equations are used to simultaneously satisfy a number of independent conditions. In addition, the study is also used to illustrate the utility of the matrix inverse as an analytical tool for such problem contexts. Although the example is taken from engineering management, the basic idea has relevance to a wide variety of other technical and analytical contexts.

*Case Study 9.2*, which is framed in terms of chemical engineering, is an example of a continuous (or micro-) variable system. The case study illustrates how finite divided differences can be used to transform differential into algebraic equations. In so doing, the solution techniques developed in Chaps. 7 and 8 can be utilized to obtain solutions. Although the example pertains to the prediction of temperature in solids, the general approach can be used to simulate the continuous distribution of many other engineering variables such as velocity, force, and mass.

In contrast, Case Studies 9.3, 9.4, and 9.5 deal with lumped (or macro-) variable systems. *Case Study 9.3* places special emphasis on the use of the matrix inverse to determine the complex cause-effect interactions between forces in the members of a truss. *Case Study 9.4* is an example of the use of Kirchhoff's laws to compute the currents and voltages in a resistor circuit. Finally, *Case Study 9.5* is an illustration of how linear equations are employed to determine the dynamics of particles and rigid bodies.

## CASE STUDY 9.1 RESOURCE ALLOCATION (GENERAL ENGINEERING)

Background: All fields of engineering involve situations where the proper allocation of resources is a critical problem. Such situations arise when organizing construction schedules, product distribution, and engineering research. Although the following problem deals with product manufacturing, the general discussion has relevance to a wide range of other problems.

An industrial engineer supervises the production of four types of computers. Four kinds of resources—worker-hours, metals, plastics, and electronic components—are required for production. The amounts of these resources needed to produce each type of computer are compiled in Table 9.1. If totals of 504 h, 1970 lb, 970 lb, and 601 units of worker-hours, metal, plastic, and components, respectively, are available each day, how many of each computer can be produced per day?

Solution: The total number of each computer produced is constrained by the total resources available in each category each day. This total resource is then allocated among the various types of computers.

Let $x_1$, $x_2$, $x_3$, and $x_4$ be the total number of computers of each kind produced in a single day. We know that the total number of worker-hours available in a day is fixed at 504. Therefore, the sum of worker-hour allocations for production of each computer must be less than or equal to 504. Therefore (using the data in Table 9.1),

$$3x_1 + 4x_2 + 7x_3 + 20x_4 \le 504 \qquad [9.1]$$

Similarly for metals, plastics, and components:

$$20x_1 + 25x_2 + 40x_3 + 50x_4 \le 1970 \qquad [9.2]$$

$$10x_1 + 15x_2 + 20x_3 + 22x_4 \le 970 \qquad [9.3]$$

$$10x_1 + 8x_2 + 10x_3 + 15x_4 \le 601 \qquad [9.4]$$

Each of these equations must be satisfied simultaneously; otherwise, one or more of the resources required to produce the four types of computers will be exhausted. If the available resources represented by the right-hand side

**TABLE 9.1 Resources required to produce four types of computers**

| Computer | Worker-hours, h/computer | Metals, lb/computer | Plastics, lb/computer | Components, units/computer |
|---|---|---|---|---|
| 1 | 3 | 20 | 10 | 10 |
| 2 | 4 | 25 | 15 | 8 |
| 3 | 7 | 40 | 20 | 10 |
| 4 | 20 | 50 | 22 | 15 |

of the above equations are all depleted to zero simultaneously, we can replace the less-than or equal signs with equal signs. In this case, the total number of each computer produced can then be determined by solving the 4-by-4 system of equations using the methods of Chaps. 7 and 8.

Because this system is not diagonally dominant, the Gauss-Seidel method may be divergent. However, Gauss elimination or the Gauss-Jordan method can be applied to compute

$$x_1 = 10$$
$$x_2 = 12$$
$$x_3 = 18$$
$$x_4 = 15$$

This information can then be used to compute the total profit. For example, suppose the profit associated with each computer is given by $p_1$, $p_2$, $p_3$, and $p_4$. The total profit associated with a day's activity ($P$) is given by

$$P = p_1 x_1 + p_2 x_2 + p_3 x_3 + p_4 x_4 \tag{9.5}$$

The results of $x_1 = 10$, $x_2 = 12$, $x_3 = 18$, and $x_4 = 15$ can be substituted into Eq. (9.5) to compute a profit of (using the coefficients from Table 9.2)

$$P = 1000(10) + 700(12) + 1100(18) + 400(15) = 44{,}200$$

Thus, at the level of resources specified in the problem, a $44,200 per day profit could be generated.

Now, suppose that we had the option of increasing any one of the available resources. An objective way to assess which resource to choose would be to determine the one which would result in the greatest increase in profit. One method to make this determination would be to increase each of the four resources individually, compute the increase in profit, and then compare the results.

A simpler alternative is based on the matrix inverse, which can be computed using the Gauss-Jordan method as

$$[A]^{-1} = \begin{bmatrix} -0.0817 & 0.0396 & -0.1465 & 0.1918 \\ 0.1066 & -0.2256 & 0.4085 & 0.0107 \\ -0.1368 & 0.1728 & -0.1909 & -0.1137 \\ 0.0888 & -0.0213 & 0.0071 & 0.0089 \end{bmatrix}$$

**TABLE 9.2 Profit associated with four types of computers**

| Computer | Profit, $/computer |
|---|---|
| 1 | 1000 |
| 2 | 700 |
| 3 | 1100 |
| 4 | 400 |

Each element $a_{ij}^{-1}$ refers to the increase in computer $i$ due to a unit increase of resource $j$. For example, element $a_{12}^{-1}$ specifies that an increase of 0.0396 unit of computer 1 will result from an additional pound of metal. Notice that some of the coefficients are negative, signifying that a unit increase in some resources will actually decrease the production of certain of the computers.

Now, with this information as background, we can make a quick evaluation of the benefit to be gained in increasing each resource by multiplying the elements in each column by the unit profits in Table 9.2. For example, for the first column

$$\Delta P_1 = -0.0817(1000) + 0.1066(700) - 0.1368(1100) + 0.0888(400) = -122.04$$

where $\Delta P_j$ is the increase in profit due to a unit increase of resource $j$. Thus, a unit increase in worker-hours would actually decrease profit by \$122.04. Similar computations can be performed for the other resources to yield

$$\Delta P_2 = \$\ 63.24$$

$$\Delta P_3 = \$-67.70$$

$$\Delta P_4 = \$\ 77.78$$

Thus, an increase of components ($j = 4$) will yield the greatest profit, followed closely by an increase in the amount of metals ($j = 2$). The analysis also indicates that an increase in plastics ($j = 3$) will result in a profit decrease.

The foregoing problem is a variation of the general economic analysis technique known as *input-output modeling*. The present example differs from the classic application of this technique to quantify transfers of material between sectors of the economy. However, our use of the matrix inverse to gain insight into complex interactions for linear systems is very representative of the spirit of input-output modeling. As such, it serves to illustrate how numerical methods enhance your insight and intuition when dealing with large coupled systems.

## CASE STUDY 9.2 CALCULATION OF TEMPERATURE DISTRIBUTION (CHEMICAL ENGINEERING)

Background: Most fields of engineering deal with the distribution of temperature in solid materials. These problems are as varied as the temperature distribution in a reentry nose cone and the temperature in a river below a water-cooled power plant. The steady-state, two-dimensional spatial distribution of temperature is defined by the Laplace equation:

$$\frac{\partial^2 T}{\partial x^2} + \frac{\partial^2 T}{\partial y^2} = 0 \qquad [9.6]$$

where $T$ is temperature and $x$ and $y$ are coordinate directions. The derivatives of Eq. (9.6) can be approximated using finite differences (see Sec. 3.5.4). Figure 9.1 shows a two-dimensional grid, a useful aid in developing approximations for Eq. (9.6). Finite-divided-difference approximations for the derivatives are

$$\frac{\partial T}{\partial x} \simeq \frac{\Delta T}{\Delta x} = \frac{T_{i+1,j} - T_{i,j}}{\Delta x}$$

$$\frac{\partial^2 T}{\partial x^2} = \frac{\partial}{\partial x}\frac{\partial T}{\partial x} \simeq \frac{\dfrac{T_{i+1,j} - T_{i,j}}{\Delta x} - \dfrac{T_{i,j} - T_{i-1,j}}{\Delta x}}{\Delta x} = \frac{T_{i+1,j} - 2T_{i,j} + T_{i-1,j}}{\Delta x^2}$$

and similarly,

$$\frac{\partial^2 T}{\partial y^2} \simeq \frac{T_{i,j+1} - 2T_{i,j} + T_{i,j-1}}{\Delta y^2}$$

Then, assuming $\Delta x = \Delta y$, the Laplace equation can be approximated as

$$T_{i+1,j} + T_{i-1,j} + T_{i,j+1} + T_{i,j-1} - 4T_{i,j} = 0 \tag{9.7}$$

which is applicable at each $i$, $j$ node in Fig. 9.1. It is seen that coupled equations will result from the process of applying Eq. (9.7) to each node, because temperatures at various locations will appear in more than one equation. This results in a system of simultaneous linear algebraic equations that can be solved using the methods described in Chaps. 7 and 8.

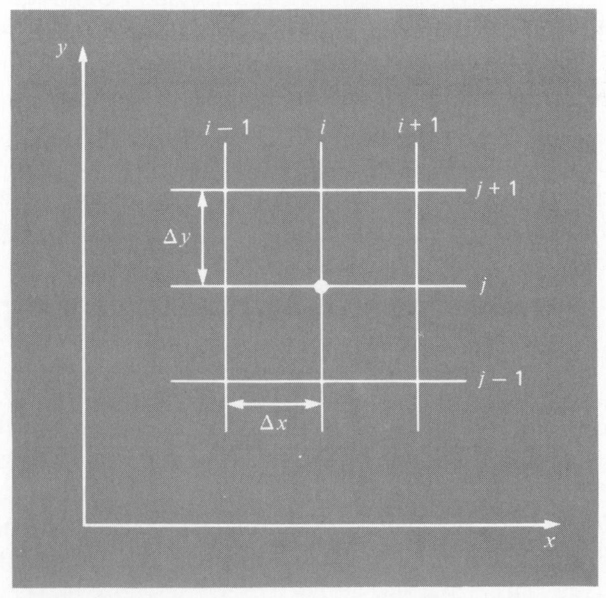


FIGURE 9.1 A two-dimensional grid used to develop finite-difference approximations of the temperature distribution on a flat plate.

Consider the flat plate shown in Fig. 9.2. The sides of the plate are held at constant temperatures of 0 and 100°C as shown. The temperature distribution inside the plate can be approximated at nine internal node points by applying Laplace's equation at each node point. This results in the following set of equations in matrix form:

$$
\begin{bmatrix}
-4 & 1 & 0 & 1 & 0 & 0 & 0 & 0 & 0 \\
1 & -4 & 1 & 0 & 1 & 0 & 0 & 0 & 0 \\
0 & 1 & -4 & 0 & 0 & 1 & 0 & 0 & 0 \\
1 & 0 & 0 & -4 & 1 & 0 & 1 & 0 & 0 \\
0 & 1 & 0 & 1 & -4 & 1 & 0 & 1 & 0 \\
0 & 0 & 1 & 0 & 1 & -4 & 0 & 0 & 1 \\
0 & 0 & 0 & 1 & 0 & 0 & -4 & 1 & 0 \\
0 & 0 & 0 & 0 & 1 & 0 & 1 & -4 & 1 \\
0 & 0 & 0 & 0 & 0 & 1 & 0 & 1 & -4
\end{bmatrix}
\begin{bmatrix}
T_{11} \\ T_{12} \\ T_{13} \\ T_{21} \\ T_{22} \\ T_{23} \\ T_{31} \\ T_{32} \\ T_{33}
\end{bmatrix}
=
\begin{bmatrix}
-100 \\ -100 \\ -200 \\ 0 \\ 0 \\ -100 \\ 0 \\ 0 \\ -100
\end{bmatrix}
$$

Solution: It is observed that the resultant system of equations is diagonal-dominant and, therefore, compatible with the Gauss-Seidel method of Chap. 8. In this case, we are guaranteed convergence because Eq. (8.5) is satisfied. Good accuracy is also assured because round-off error is not a problem for the Gauss-Seidel method. Using an $\epsilon_s = 0.05$ percent, the following results are obtained:

$$
\begin{array}{lll}
T_{11} = 49.99 & T_{21} = 28.56 & T_{31} = 14.28 \\
T_{12} = 71.42 & T_{22} = 49.99 & T_{32} = 28.57 \\
T_{13} = 85.71 & T_{23} = 71.42 & T_{33} = 50.00
\end{array}
$$

after 13 iterations. The results are displayed in Fig. 9.3.

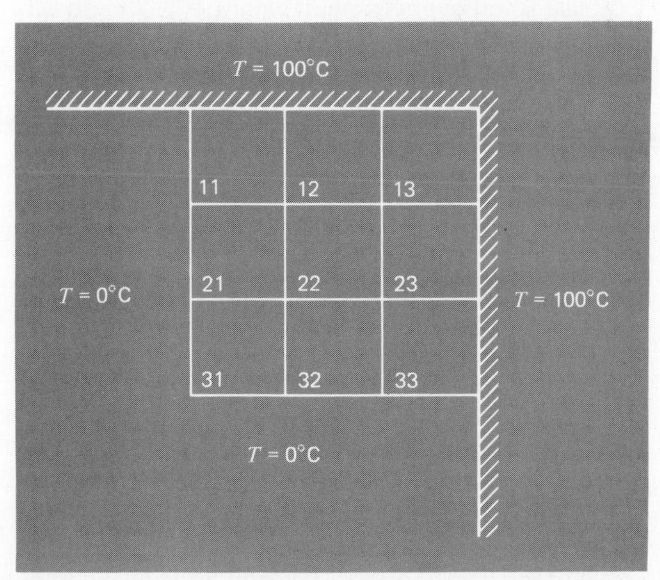


FIGURE 9.2 A flat plate with sides held at constant temperatures of 0 and 100°, as shown.

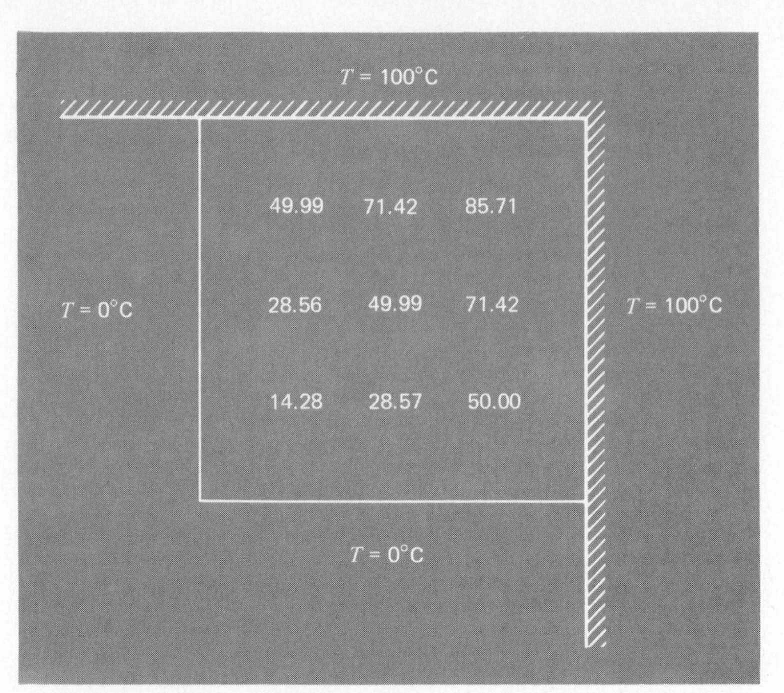


FIGURE 9.3 Temperature distribution of a flat plate, as calculated using the Gauss-Seidel method.

The simulation was performed with the standard Gauss-Seidel method. Because this is a convergent system, relaxation might serve to accelerate the rate of convergence. We therefore repeated the computation two additional times, using overrelaxation with $\lambda = 1.25$ and 1.5.

The results, as plotted in Fig. 9.4, suggest an optimal value of $\lambda$ in the vicinity of 1.25. Using techniques to be described in Chap. 11 (note, a sketch can be used to obtain a rough estimate), we fit a quadratic equation to the points in Fig. 9.4. This equation is

$$n = 96\lambda^2 - 236\lambda + 153$$

where $n$ is the number of iterations corresponding to a particular value of $\lambda$. A minimum can be determined by differentiating this equation to yield

$$\frac{dn}{d\lambda} = 192\lambda - 236$$

The minimum occurs when $dn/d\lambda$ is equal to zero, which corresponds to the point at which the slope of Fig. 9.4 flattens or bottoms out. For this case, a value of $\lambda = 1.23$ can be determined. Reperforming the computation using this relaxation coefficient results in a solution with the desired accuracy in eight iterations.

Thus, if we were to perform additional computations for this particular problem setting, we could employ a value of $\lambda \simeq 1.2$ in order to attain the

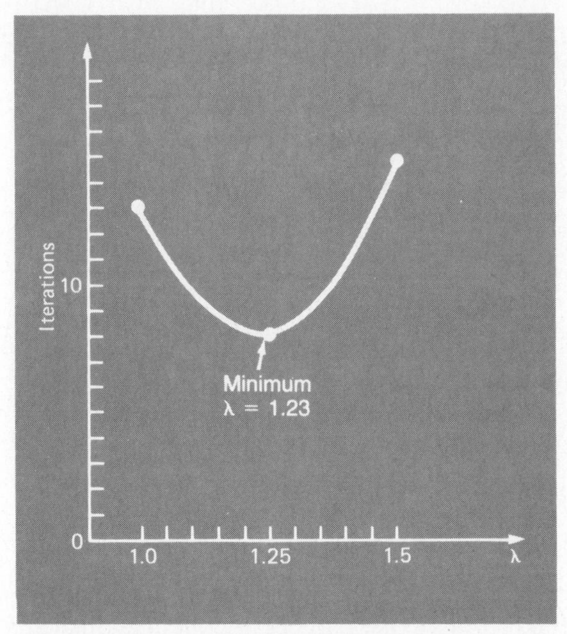


FIGURE 9.4 Plot of number of iterations versus $\lambda$, the relaxation coefficient. The three data points suggest a minimum number of iterations in the vicinity of $\lambda = 1.25$. By fitting a parabola to the data, it can be estimated that a minimum number of iterations corresponds to a value of $\lambda \simeq 1.23$.

most efficient result. For the present case, the savings in time are negligible for a single computation. However, for multiple simulations for large systems, the wise choice of $\lambda$ can possibly result in substantial savings.

This type of procedure can be extended to more complex problems involving irregular geometric shapes. Practical problems of this type require some automatic computing capability, and, except in the case of extremely large problems, the personal computer fills the role well.

## CASE STUDY 9.3 ANALYSIS OF A STATICALLY DETERMINANT TRUSS (CIVIL ENGINEERING)

Background: An important problem in structural engineering is that of finding the forces and reactions associated with a statically determinant truss. Figure 9.5 shows an example of such a truss.

The forces ($F$) represent either tension or compression on the members of the truss. External reactions ($H_2$, $V_2$, and $V_3$) are forces which characterize how the truss interacts with the supporting surface. The hinge at node 2 can transmit both horizontal and vertical forces to the surface, whereas the roller at node 3 only transmits vertical forces. It is observed that the effect of the external loading of 1000 lb is distributed among the various members of the truss.

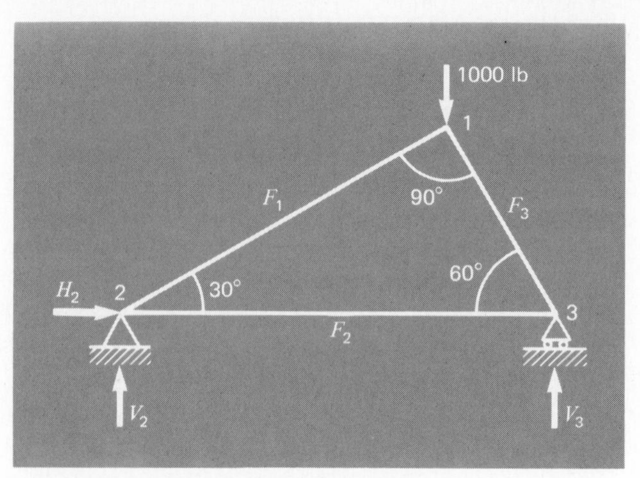


FIGURE 9.5 Forces on a statically determinant truss.

Solution: This type of structure can be described as a system of coupled linear algebraic equations. Free-body-force diagrams are shown for each node in Fig. 9.6. The sum of the forces in both horizontal and vertical directions must be zero at each node, because the system is at rest. Therefore, for node 1,

$$\Sigma F_H = 0 = -F_1 \cos 30° + F_3 \cos 60° + F_{1,h} \tag{9.8}$$

$$\Sigma F_V = 0 = -F_1 \sin 30° - F_3 \sin 60° + F_{1,v} \tag{9.9}$$

for node 2,

$$\Sigma F_H = 0 = F_2 + F_1 \cos 30° + F_{2,h} + H_2 \tag{9.10}$$

$$\Sigma F_V = 0 = F_1 \sin 30° + F_{2,v} + V_2 \tag{9.11}$$

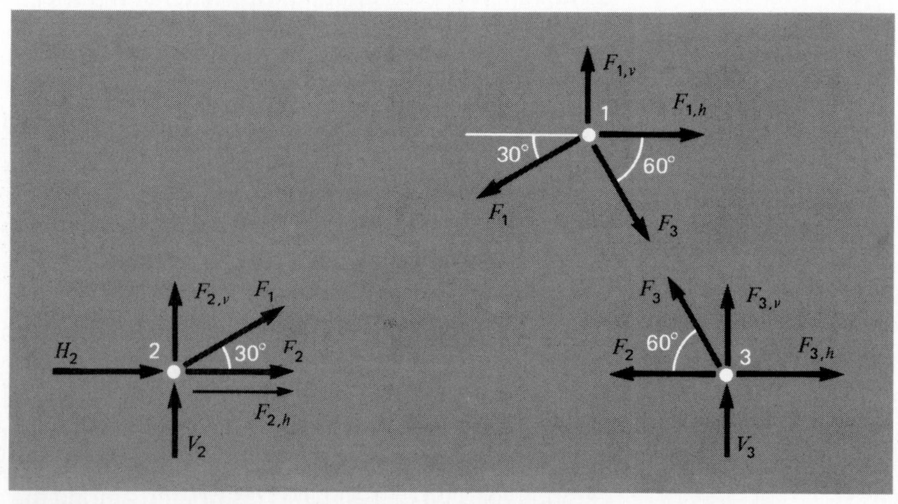


FIGURE 9.6 Free-body-force diagrams for the nodes of a statically determinant truss.

for node 3,

$$\Sigma F_H = 0 = -F_2 - F_3 \cos 60° + F_{3,h} \qquad [9.12]$$

$$\Sigma F_V = 0 = F_3 \sin 60° + F_{3,v} + V_3 \qquad [9.13]$$

where $F_{i,h}$ is the external horizontal force applied to node $i$ (where a positive force is from left to right) and $F_{i,v}$ is the external vertical force applied to node $i$ (where a positive force is upward). Thus, in this problem, the 1000-lb downward force on node 1 corresponds to $F_{1,v} = -1000$. For this case all other $F_{i,v}$'s and $F_{i,h}$'s are zero. Note that the directions of the internal forces and reactions are unknown. Proper application of Newton's laws only requires consistent assumptions regarding direction. Solutions are negative if the directions are assumed incorrectly. Also note that in this problem, the forces in all members are assumed to be in tension and act to pull adjoining nodes together. This problem can be written as the following system of six equations and six unknowns:

$$\begin{bmatrix} 0.866 & 0 & -0.5 & 0 & 0 & 0 \\ 0.5 & 0 & 0.866 & 0 & 0 & 0 \\ -0.866 & -1 & 0 & -1 & 0 & 0 \\ -0.5 & 0 & 0 & 0 & -1 & 0 \\ 0 & 1 & 0.5 & 0 & 0 & 0 \\ 0 & 0 & -0.866 & 0 & 0 & -1 \end{bmatrix} \begin{bmatrix} F_1 \\ F_2 \\ F_3 \\ H_2 \\ V_2 \\ V_3 \end{bmatrix} = \begin{bmatrix} 0 \\ -1000 \\ 0 \\ 0 \\ 0 \\ 0 \end{bmatrix} \qquad [9.14]$$

Notice that as formulated in Eq. (9.14), partial pivoting is required to avoid division by zero diagonal elements. Employing a pivot strategy, the system can be solved using the elimination techniques discussed in Chaps. 7 and 8. However, because this problem is an ideal case study for demonstrating the utility of the matrix inverse, the Gauss-Jordan method is used to compute:

$$\begin{array}{lll} F_1 = -500 & F_2 = 433 & F_3 = -866 \\ H_2 = 0 & V_2 = 250 & V_3 = 750 \end{array}$$

and the matrix inverse is

$$[A]^{-1} = \begin{bmatrix} 0.866 & 0.5 & 0 & 0 & 0 & 0 \\ 0.25 & -0.433 & 0 & 0 & 1 & 0 \\ -0.5 & 0.866 & 0 & 0 & 0 & 0 \\ -1 & 0 & -1 & 0 & -1 & 0 \\ -0.433 & -0.25 & 0 & -1 & 0 & 0 \\ 0.433 & -0.75 & 0 & 0 & 0 & -1 \end{bmatrix}$$

Now, realize that the right-hand-side vector represents the externally applied horizontal and vertical forces on each node, as in

$$[\text{Right-hand-side vector}]^T = [F_{1,h}\ F_{1,v}\ F_{2,h}\ F_{2,v}\ F_{3,h}\ F_{3,v}] \qquad [9.15]$$

Because the external forces have no effect on the matrix of coefficients, the Gauss-Jordan method need not be implemented over and over again to study the effect of different external forces on the truss. Rather, all that we have to do is multiply the matrix inverse by each right-hand-side vector to efficiently obtain alternative solutions. For example, we might want to study the effect of horizontal forces induced by a wind blowing from left to right. If the wind force can be idealized as two point forces of 1000 lb on nodes 1 and 2 (Fig. 9.7) the right-hand-side vector is

$$[\text{Right-hand-side vector}]^T = [1000 \quad 0 \quad 1000 \quad 0 \quad 0 \quad 0]$$

which can be multiplied by the matrix inverse to yield

$$\begin{array}{lll} F_1 = 866 & F_2 = 250 & F_3 = -500 \\ H_2 = -2000 & V_2 = -433 & V_3 = 433 \end{array}$$

For a wind from the right, $F_{1,h} = -1000$, $F_{3,h} = -1000$, and all other external forces are zero, with the result that

$$\begin{array}{lll} F_1 = -866 & F_2 = -1250 & F_3 = 500 \\ H_2 = 2000 & V_2 = 433 & V_3 = -433 \end{array}$$

The results indicate that the winds have markedly different effects on the structure. Both cases are depicted in Fig. 9.7.

The individual elements of the inverted matrix also have direct utility in elucidating stimulus-response interactions for the structure. Each element represents the change of one of the unknown variables to a unit change of one of the external stimuli. For example, element $a_{32}^{-1}$ indicates that the third unknown ($F_3$) will change 0.866 due to a unit change of the second external stimuli ($F_{1,v}$). Thus, if the vertical load at the first node were increased by one, $F_3$ would increase by 0.866. The fact that elements are zero indicates that certain unknowns are unaffected by some of the external stimuli. For instance $a_{13}^{-1} = 0$ means that $F_1$ is unaffected by changes in $F_{2,h}$. This ability to isolate interactions has a number of engineering applications including the

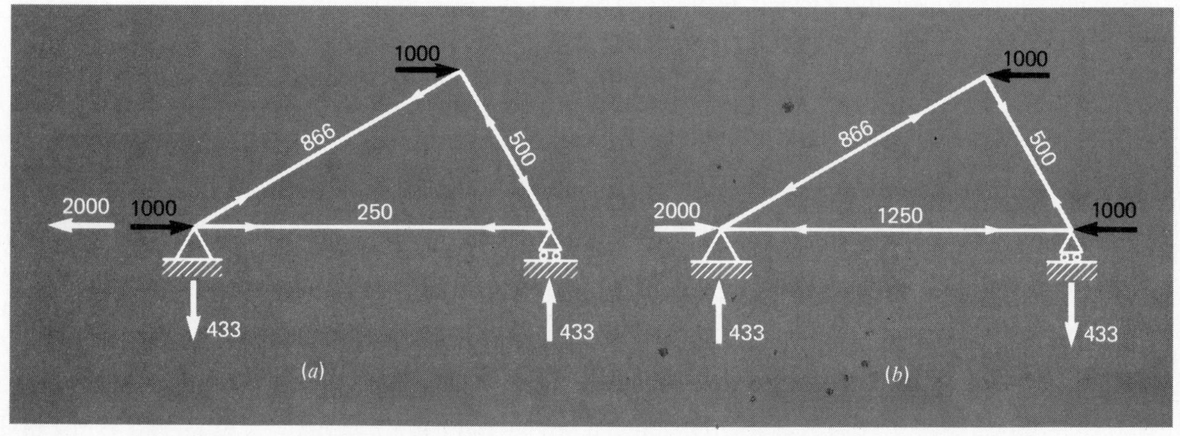


FIGURE 9.7 Two test cases showing (*a*) winds from the left and (*b*) winds from the right.

identification of those components that are most sensitive to external stimuli and, as a consequence, most prone to failure.

The foregoing approach becomes particularly useful when applied to large complex structures. In engineering practice it may be necessary to solve trusses with hundreds or even thousands of structural members. Linear equations provide one powerful approach for gaining insight into the behavior of these structures.

## CASE STUDY 9.4 CURRENTS AND VOLTAGES IN RESISTOR CIRCUITS (ELECTRICAL ENGINEERING)

Background: A common problem in electrical engineering involves determining the currents and voltages at various locations in complex resistor circuits. These problems are solved using Kirchhoff's current and Ohm's laws. The current law states that the algebraic sum of all currents entering a node must be zero (Fig. 9.8*a*), or

$$\Sigma i_k = 0 \qquad [9.16]$$

where all current entering the node is assumed positive in sign.

Ohm's law states that the current through a resistor is related to the voltage change and the value of the resistor (Fig. 9.8*b*),

$$i_{ij} = \frac{V_i - V_j}{R_{ij}} \qquad [9.17]$$

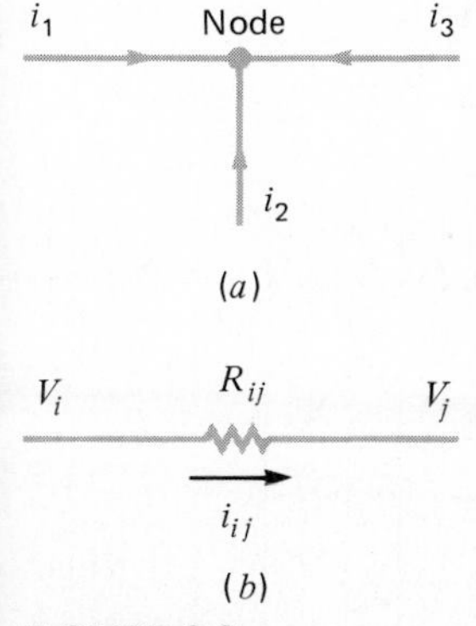

FIGURE 9.8
Schematic representations of (*a*) Kirchhoff's current and (*b*) Ohm's laws.

Solution: Problems of this type result in systems of simultaneous linear algebraic equations because the various loops within a circuit are coupled to the others. For example, consider the circuit shown in Fig. 9.9. The currents associated with this circuit are unknown both in magnitude and direction. This presents no great difficulty because one simply assumes a direction for each current. If the resultant solution from Kirchhoff's laws is negative, then the assumed direction is incorrect. For example, Fig. 9.10 shows some assumed currents.

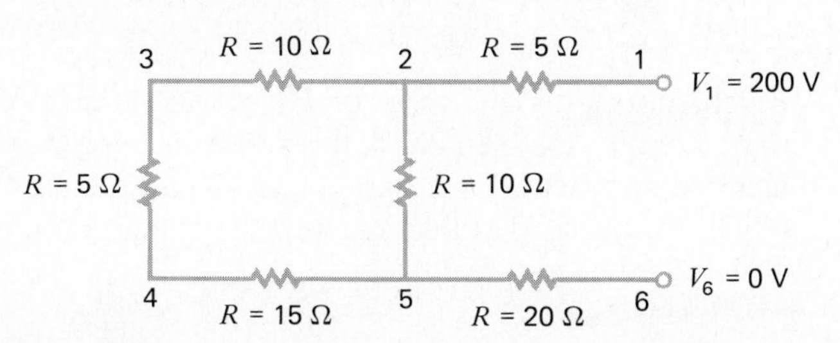


FIGURE 9.9 A resistor circuit to be solved using simultaneous linear algebraic equations.

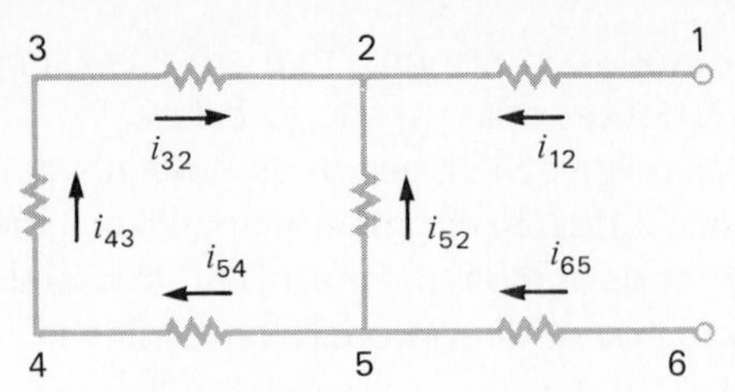


FIGURE 9.10 Assumed currents.

Given these assumptions, four node equations for currents are written as

$$i_{12} + i_{52} + i_{32} = 0$$
$$i_{65} - i_{52} - i_{54} = 0$$
$$i_{43} - i_{32} = 0$$
$$i_{54} - i_{43} = 0$$

and the six voltage equations are

$$i_{32} = \frac{V_3 - V_2}{10} \qquad i_{43} = \frac{V_4 - V_3}{5} \qquad i_{65} = \frac{0 - V_5}{20}$$
$$i_{12} = \frac{200 - V_2}{5} \qquad i_{54} = \frac{V_5 - V_4}{15} \qquad i_{52} = \frac{V_5 - V_2}{10}$$

where the current flows from the high to low voltage. These equations are equivalent to the following matrix notation:

$$\begin{bmatrix} 1 & 1 & 0 & 0 & 0 & 1 & 0 & 0 & 0 & 0 \\ 0 & 0 & 0 & -1 & 1 & -1 & 0 & 0 & 0 & 0 \\ 0 & -1 & 1 & 0 & 0 & 0 & 0 & 0 & 0 & 0 \\ 0 & 0 & -1 & 1 & 0 & 0 & 0 & 0 & 0 & 0 \\ 0 & 10 & 0 & 0 & 0 & 0 & 1 & -1 & 0 & 0 \\ 5 & 0 & 0 & 0 & 0 & 0 & 1 & 0 & 0 & 0 \\ 0 & 0 & 5 & 0 & 0 & 0 & 0 & 1 & -1 & 0 \\ 0 & 0 & 0 & 15 & 0 & 0 & 0 & 0 & 1 & -1 \\ 0 & 0 & 0 & 0 & 20 & 0 & 0 & 0 & 0 & 1 \\ 0 & 0 & 0 & 0 & 0 & 10 & 1 & 0 & 0 & -1 \end{bmatrix} \begin{bmatrix} i_{12} \\ i_{32} \\ i_{43} \\ i_{54} \\ i_{65} \\ i_{52} \\ V_2 \\ V_3 \\ V_4 \\ V_5 \end{bmatrix} = \begin{bmatrix} 0 \\ 0 \\ 0 \\ 0 \\ 0 \\ 200 \\ 0 \\ 0 \\ 0 \\ 0 \end{bmatrix}$$

which represents a system of 10 equations and 10 unknowns. Although impractical to solve by hand, this system is easily solved using any automatic elimination method such as Gauss elimination or the Gauss-Jordan technique. Proceeding in this manner, the solution is

$$\begin{array}{lll} i_{12} = 6.1538 & i_{65} = -6.1538 & V_4 = 146.15 \\ i_{32} = -1.5385 & i_{52} = -4.6154 & V_5 = 123.08 \\ i_{43} = -1.5385 & V_2 = 169.23 & \\ i_{54} = -1.5385 & V_3 = 153.85 & \end{array}$$

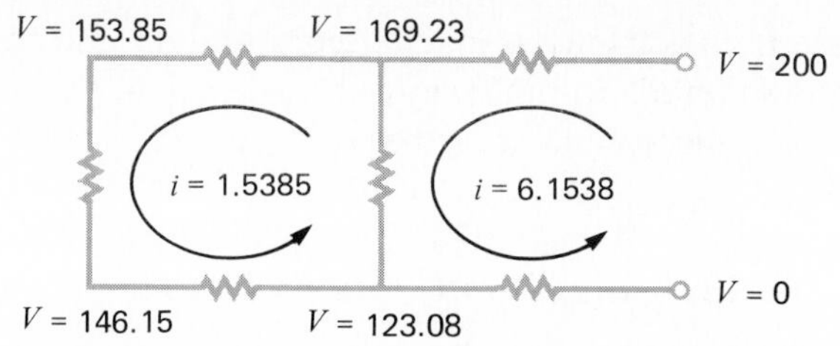


FIGURE 9.11 The solution for currents and voltages obtained using an elimination method.

Thus, with proper interpretation of the signs of the result, the circuit currents and voltages are as shown in Fig. 9.11. The advantages of using numerical algorithms and personal computers for problems of this type should be evident.

## CASE STUDY 9.5 DYNAMICS OF PARTICLES AND RIGID BODIES (MECHANICAL ENGINEERING)

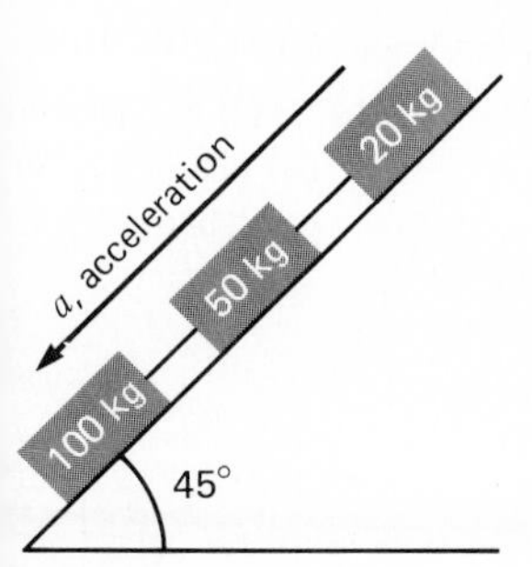


FIGURE 9.12 Three blocks connected by a weightless rope over an inclined plane.

Background: The dynamic motion of particles and rigid bodies plays an important role in many mechanical and other engineering problems. This motion can be described using Newton's laws. The application of Newton's laws to a single particle results in two equations. However, if some particles in a system affect others then a large number of simultaneous equations may arise.

For example, consider the system shown in Fig. 9.12. Three blocks are attached by a weightless cord and rest on a smooth plane inclined at 45° with the horizontal. The coefficient of friction between the plane and the 100-kg mass is 0.25, and between the 50- and 20-kg mass it is 0.375.

Solution: Free-body-force diagrams for the three blocks are given in Fig. 9.13. The forces have units of newtons (kilograms per meter per second

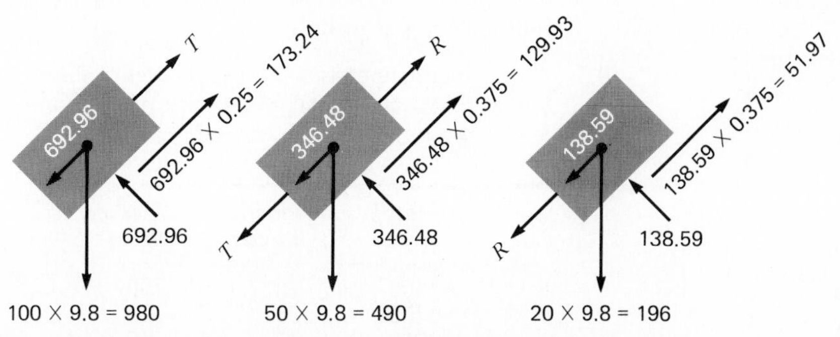


FIGURE 9.13 Free-body-force diagrams for the blocks on an inclined plane.

squared), $m$ is the mass in kilograms, and $a$ is the acceleration in meters per second squared. Summing forces in a direction parallel to the plane and using Newton's second law ($F = ma$),

$$692.96 - 173.24 - T = 100a$$

$$346.48 - 129.93 + T - R = 50a$$

$$138.59 - 51.97 + R = 20a$$

or, in matrix form,

$$\begin{bmatrix} 100 & 1 & 0 \\ 50 & -1 & 1 \\ 20 & 0 & -1 \end{bmatrix} \begin{bmatrix} a \\ T \\ R \end{bmatrix} = \begin{bmatrix} 519.72 \\ 216.55 \\ 86.62 \end{bmatrix}$$

Solving, using Gauss elimination, gives

$$a = 4.8405 \text{ m/s}^2$$

$$T = 36.6671 \text{ N}$$

$$R = 10.1906 \text{ N}$$

Expressing the equations of motion in matrix form is a general and versatile approach for problems of this type. Although the problem solved here was easy, the case study serves to illustrate the general approach and to inspire, it is hoped, application to more difficult problems. When coupled with a numerical algorithm and a personal computer, it is an extremely powerful tool that can be used to solve a variety of complex problems.

# PROBLEMS

## General Engineering

**9.1** Reproduce the computation performed in Case Study 9.1 using your own software.

**9.2** Perform the same computation as in Case Study 9.1, but change the totals of worker-hours, metal, plastics, and components to 856 h, 3050 lb, 1450 lb, and 948 units respectively.

**9.3** An engineer supervises the production of three types of automobiles. Three kinds of material—metal, plastic, and rubber—are required for production. The amounts needed to produce each car are

| *Car* | *Metal, lb/car* | *Plastic, lb/car* | *Rubber, lb/car* |
|---|---|---|---|
| 1 | 1500 | 25 | 100 |
| 2 | 1700 | 33 | 120 |
| 3 | 1900 | 42 | 160 |

If totals of 106, 2.17, and 8.2 tons of metal, plastic, and rubber, respectively, are available each day, how many automobiles can be produced per day?

**9.4** An engineer requires 4800, 5810, and 5690 $yd^3$ of sand, fine gravel, and coarse gravel, respectively, for a building project. There are three pits from which these materials can be obtained. The composition of these pits is

| | *Sand,* % | *Fine gravel,* % | *Coarse gravel,* % |
|---|---|---|---|
| Pit 1 | 52 | 30 | 18 |
| Pit 2 | 20 | 50 | 30 |
| Pit 3 | 25 | 20 | 55 |

How many cubic yards must be hauled from each pit in order to meet the engineer's needs?

## Chemical Engineering

**9.5** Reproduce the computation performed in Case Study 9.2 using your own software.

**9.6** Perform the same computation as in Case Study 9.2, but change the temperature of the wall to 200°C.

**9.7** Using the same approach as in Case Study 9.2, compute the temperature distribution in a rod that is heated at both ends, as depicted in Fig. P9.7. Apply the one-dimensional form of Eq. (9.6):

$$\frac{d^2T}{dx^2} = 0$$

where $x$ is distance along the rod. Make a plot of $T$ versus $x$.

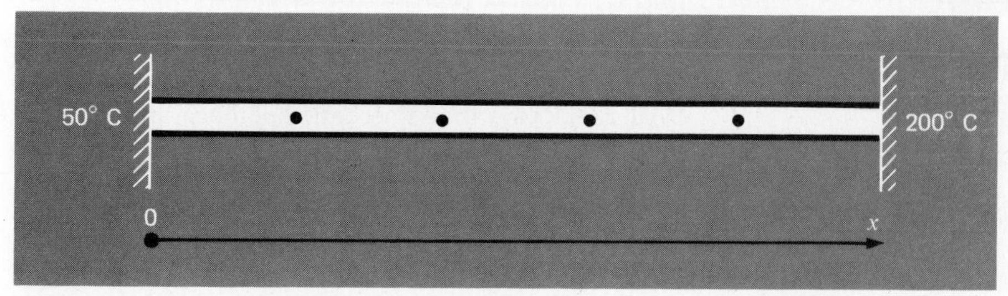


FIGURE P9.7 A one-dimensional insulated rod held at a constant temperature at each end. The points designate the locations where you should apply the one-dimensional form of Eq. (9.6) in order to compute the temperature distribution along the rod.

**9.8** Repeat Prob. 9.7 but now include a heat loss in the equation:

$$\frac{d^2T}{dx^2} - rT = 0$$

where $r$, the heat loss coefficient, equals 0.01 $\text{cm}^{-2}$ and the length of the rod is 10 cm. Make a plot of $T$ versus $x$.

**9.9** Figure P9.9 shows three reactors linked by pipes. As indicated, the rate of transfer of chemicals through each pipe is equal to a flow rate ($Q$, with units of cubic meters per second) multiplied by the concentration of the reactor from which the flow originates ($c$, with units of milligrams per cubic meter). If the system is at a steady state, the transfer into each reactor will balance the transfer out. For example, for reactor 1, (in) = (out), or

$$500 + Q_{21}c_2 = Q_{12}c_1 + Q_{13}c_1$$

Or using the flow rates as specified in Fig. P9.9,

$$500 + 20c_2 = 80c_1 + 40c_1$$

where 500 is a direct input (milligrams per second). Develop comparable mass balance equations for the other two reactors and solve the three simultaneous linear algebraic equations for the concentrations of the reactors.

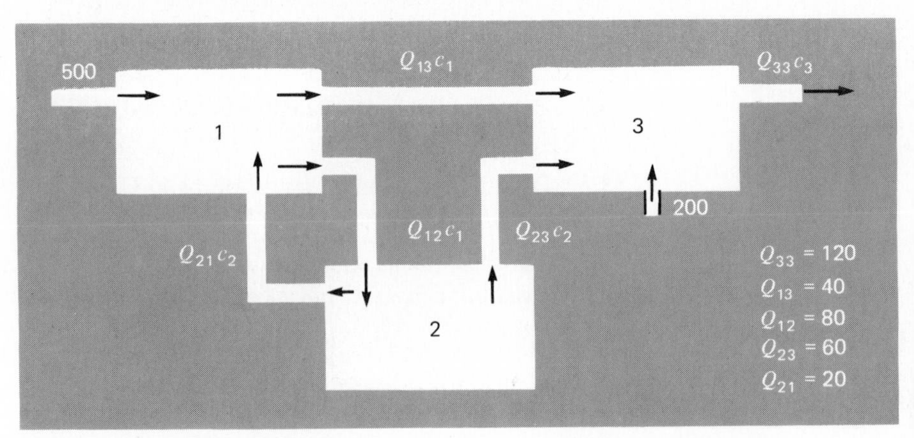


FIGURE P9.9 Three reactors linked by pipes. The rate of mass transfer through each pipe is equal to the product of flow $Q$ and concentration $c$ of the reactor from which the flow originates.

**9.10** Employing the same basic approach as in Prob. 9.9, determine the concentration of chloride in each of the Great Lakes using the information shown in Fig. P9.10.

## Civil Engineering

**9.11** Reproduce the computation performed in Case Study 9.3 using your own software.

**9.12** Perform the same computation as in Case Study 9.3, but change the angle at node 2 to 40° and at node 3 to 55°.

**9.13** Perform the same computation as in Case Study 9.3, but for the truss depicted in Fig. P9.13.

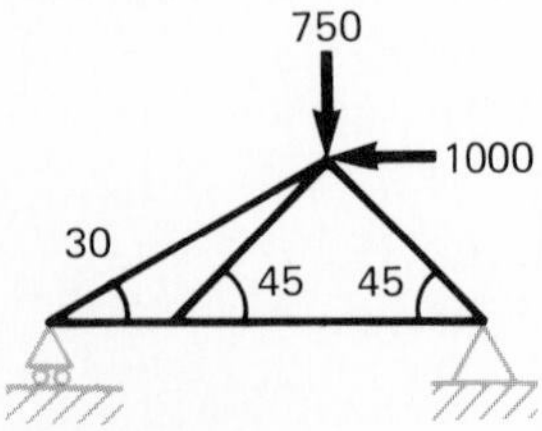

750 1000 30 45 45

FIGURE P9.13

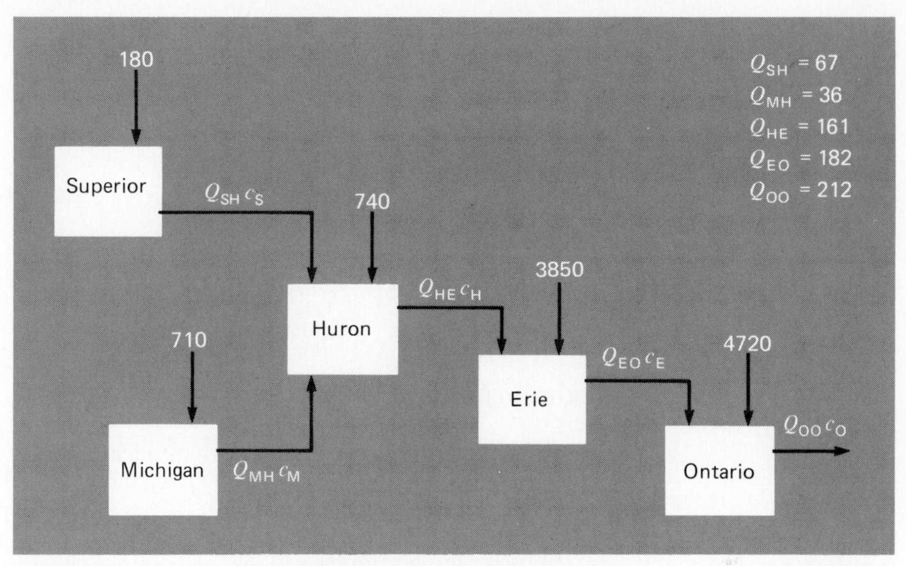


FIGURE P9.10 A chloride balance for the Great Lakes. Numbered arrows are direct inputs.

**9.14** Perform the same computation as in Case Study 9.3, but for the truss depicted in Fig. P9.14.

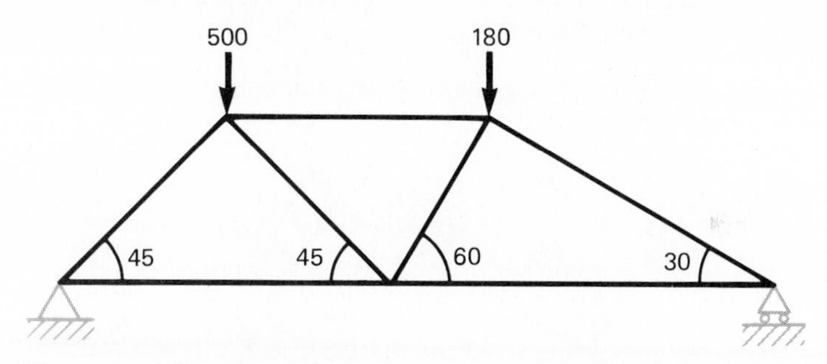


FIGURE P9.14

## Electrical Engineering

**9.15** Reproduce the computation performed in Case Study 9.4 using your own software.

**9.16** Perform the same computation as in Case Study 9.4, but change the resistance between nodes 3 and 4 to 15 Ω and change $V_6$ to 50 V.

**9.17** Perform the same computation as in Case Study 9.4, but for the circuit depicted in Fig. P9.17.

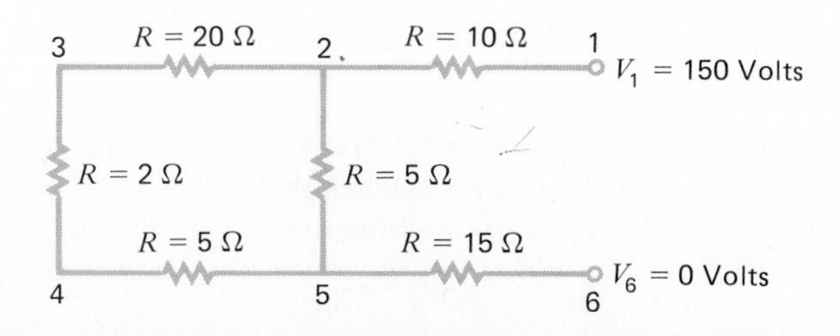


FIGURE P9.17

**9.18** Perform the same computation as in Case Study 9.4, but for the circuit depicted in Fig. P9.18.

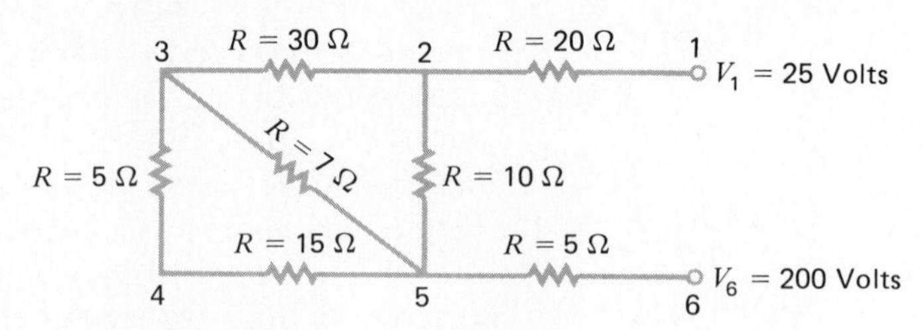


FIGURE P9.18

## Mechanical Engineering

**9.19** Reproduce the computation performed in Case Study 9.5 using your own software.

**9.20** Perform the same computation as in Case Study 9.5, but change the angle to 55° with the horizontal.

**9.21** Perform the same computation as in Case Study 9.5, but change the coefficient of friction for the 100-kg mass to 0.5 and for the 50- and 25-kg masses to 0.25.

**9.22** Perform the same computation as in Case Study 9.5, but change the masses from 100, 50, and 20 kg to 45, 20, and 80 kg.

**9.23** Perform the same computation as in Case Study 9.5, but for the system depicted in Fig. P9.23.

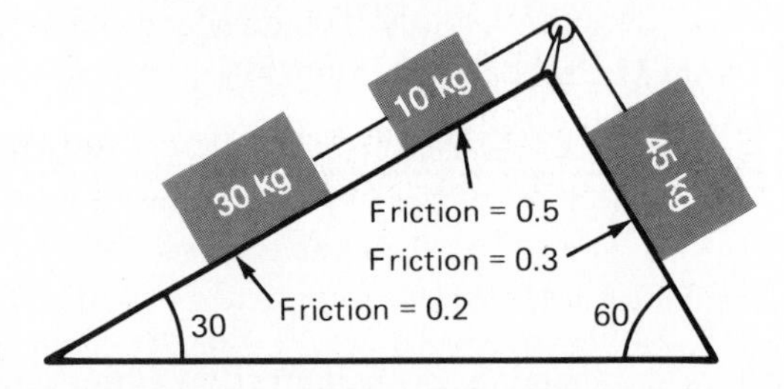


FIGURE P9.23

**9.24** Perform the same computation as in Case Study 9.5, but for the system depicted in Fig. P9.24 (angles are 45°).

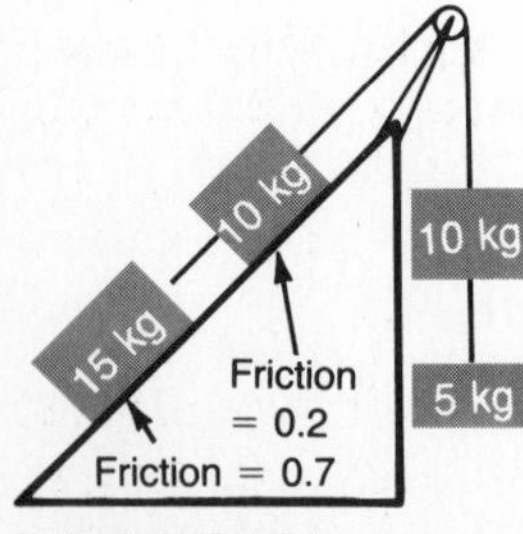


FIGURE P9.24

## Miscellaneous

**9.25** Read all the case studies in Chap. 9. On the basis of your reading and experience make up your own case study for any one of the fields of engineering. This may involve modifying or reexpressing one of our case studies. However, it can also be totally original. As with our examples, it must be drawn from an engineering problem context and must demonstrate the use of numerical methods for solving linear algebraic equations. Write up your results using our case studies as models.

# EPILOGUE: PART III

## III.4 TRADE-OFFS

Table III.2 provides a summary of the trade-offs involved in solving simultaneous linear algebraic equations. Three methods—graphical, Cramer's rule, and algebraic manipulation—are limited to small ($\leq 3$) numbers of equations and thus have little utility for practical problem solving. However, these techniques are useful didactic tools for understanding the behavior of linear systems in general.

The numerical methods themselves are divided into two general categories: exact and approximate methods. As the name implies, the former are intended to yield exact answers. However, because they are affected by round-off errors they sometimes yield imprecise results. The magnitude of the round-off error varies from system to system and is dependent on a number of factors.These include the system's dimensions, its condition, and whether the matrix of coefficients is sparse or full. In addition, computer precision will affect round-off error. In general, exact methods are usually the method of choice for small equation sets (that is, those less than 50).

Two exact methods are commonly used—the Gauss elimination and the Gauss-Jordan techniques. It is recommended that a pivoting strategy be employed in any computer program implementing these methods. The inclusion of such a strategy minimizes round-off error and avoids problems such as division by zero. All other things being equal, Gauss elimination is preferable to the Gauss-Jordan method because the former is about 50 percent quicker. However, the Gauss-Jordan technique has utility because it can be modified slightly so that the matrix inverse is obtained as an ancillary benefit of the computation.

Although elimination methods have great utility, their use of the entire matrix of coefficients can be somewhat limiting when dealing with very large, sparse systems. This is due to the fact that large portions of computer memory would be devoted to

**TABLE III.2 Comparison of the characteristics of alternative methods for finding solutions of simultaneous linear algebraic equations.**

| Method | Approximate maximum number of equations* | Stability | Precision | Breadth of application | Programming effort | Comments |
|---|---|---|---|---|---|---|
| Graphical | 2 | – | Poor | Limited | – | May take more time than the numerical method |
| Cramer's rule | 3 | – | Affected by round-off error | Limited | – | Excessive computational effort required for more than three equations |
| Algebraic manipulation (elimination of unknowns) | 2 | – | Affected by round-off error | Limited | | |
| Gauss elimination (with partial pivoting) | 50 | – | Affected by round-off error | General | Moderate | |
| Gauss-Jordan (with partial pivoting) | 50 | – | Affected by round-off error | General | Moderate | Allows computations of matrix inverse |
| Gauss-Seidel | 1000† | May not converge if not diagonally dominant | Excellent | Appropriate only for diagonally dominant systems | Easy | |

*Does not account for conditioning or sparseness or computer hardware.
†Upper limit depends on computer system and degree of sparseness of the equations.

storage of meaningless zeros. For banded systems, techniques are available to implement elimination methods without having to store the entire coefficient matrix. Box 7.2 describes a simple algorithm of this sort for a particular type of banded system—the tridiagonal case.

The approximate technique described in this book is called the Gauss-Seidel method. It differs from the exact techniques in that it employs an iterative scheme to obtain progressively closer estimates of the solution. Thus, the effect of round off is a moot point with the Gauss-Seidel method, because the iterations can be continued as long as is necessary to obtain the desired precision. In addition, versions of the Gauss-Seidel method can be developed to efficiently utilize computer storage requirements for sparse systems. Consequently, the Gauss-Seidel technique is usually the method of preference for large systems of equations ($> 100$), where round-off errors and storage requirements would pose significant problems for the exact techniques.

The disadvantage of the Gauss-Seidel method is that it does not always converge or sometimes converges slowly on the true solution. It is only strictly reliable for those systems which are diagonally dominant. However, relaxation methods are available that sometimes offset these disadvantages. In addition, because many sets of linear algebraic equations originating from physical systems exhibit diagonal dominance, the Gauss-Seidel method has great utility for engineering problem solving.

In summary, a variety of factors will bear on your choice of a technique for a particular problem involving linear algebraic equations. However, as outlined above, the size and sparseness of the system are particularly important factors in determining your choice.

## III.5 IMPORTANT RELATIONSHIPS AND FORMULAS

Every part of this book includes a section that summarizes important formulas. Although Part III does not really deal with single formulas, we have used Table III.3 to summarize the algorithms that were covered. The table provides an overview that should be helpful for review and in elucidating the major differences between the methods.

## III.6 ADVANCED METHODS AND ADDITIONAL REFERENCES

The methods in this text have been limited to some of the simpler techniques for solving simultaneous linear equations. Other techniques

**TABLE III.3 Summary of important information presented in Part III.**

| Method | Procedure | Potential problems and remedies |
|---|---|---|
| Gauss elimination | $\left[\begin{array}{ccc\|c} a_{11} & a_{12} & a_{13} & c_1 \\ a_{21} & a_{22} & a_{23} & c_2 \\ a_{31} & a_{32} & a_{33} & c_3 \end{array}\right] \Rightarrow \left[\begin{array}{ccc\|c} a_{11} & a_{12} & a_{13} & c_1 \\ & a'_{22} & a'_{23} & c'_2 \\ & & a''_{33} & c''_3 \end{array}\right] \Rightarrow \begin{array}{l} x_3 = c''_3/a''_{33} \\ x_2 = (c'_2 - a'_{23}x_3)/a'_{22} \\ x_1 = (c_1 - a_{12}x_1 - a_{13}x_3)/a_{11} \end{array}$ | Problems:<br>Ill-conditioning<br>Round-off<br>Division by zero<br>Remedies:<br>Higher precision<br>Partial pivoting<br>Error equations |
| Gauss-Jordan method | $\left[\begin{array}{ccc\|c} a_{11} & a_{12} & a_{13} & c_1 \\ a_{21} & a_{22} & a_{23} & c_2 \\ a_{31} & a_{32} & a_{33} & c_3 \end{array}\right] \Rightarrow \left[\begin{array}{ccc\|c} 1 & 0 & 0 & c^*_1 \\ 0 & 1 & 0 & c^*_2 \\ 0 & 0 & 1 & c^*_3 \end{array}\right] \Rightarrow \begin{array}{l} x_1 = c^*_1 \\ x_2 = c^*_2 \\ x_3 = c^*_3 \end{array}$ | Problems:<br>Ill-conditioning<br>Round-off<br>Division by zero<br>Remedies:<br>Higher precision<br>Partial pivoting<br>Error equations |
| Matrix inversion | $\left[\begin{array}{cccccc} a_{11} & a_{12} & a_{13} & 1 & 0 & 0 \\ a_{21} & a_{22} & a_{23} & 0 & 1 & 0 \\ a_{31} & a_{32} & a_{33} & 0 & 0 & 1 \end{array}\right] \Rightarrow \left[\begin{array}{cccccc} 1 & 0 & 0 & a^{-1}_{11} & a^{-1}_{12} & a^{-1}_{13} \\ 0 & 1 & 0 & a^{-1}_{21} & a^{-1}_{22} & a^{-1}_{23} \\ 0 & 0 & 1 & a^{-1}_{31} & a^{-1}_{32} & a^{-1}_{33} \end{array}\right]$ | Problems:<br>Ill-conditioning<br>Round-off<br>Division by zero<br>Remedies:<br>Higher precision<br>Partial pivoting |
| Gauss-Seidel method | $\left.\begin{array}{l} x^i_1 = (c_1 - a_{12}x^{i-1}_2 - a_{13}x^{i-1}_3)/a_{11} \\ x^i_2 = (c_2 - a_{21}x^i_1 - a_{23}x^{i-1}_3)/a_{22} \\ x^i_3 = (c_3 - a_{31}x^i_1 - a_{32}x^i_2)/a_{33} \end{array}\right\}$ continue iteratively until $\left\| \frac{x^i_i - x^{i-1}_i}{x^i_i} \right\| 100\% < \epsilon_s$ for all $x_i$'s | Problems:<br>Divergent or converges slowly<br>Remedies:<br>Diagonal dominance<br>Relaxation |

are available for use in the same problem context as well as in some related subjects such as eigenvalues and nonlinear simultaneous equations.

*LU decomposition* (also called *Cholesky's method* or *Crout's method*) is a particularly efficient technique for solving the same sorts of problems we have been dealing with in Part III. Good descriptions and computer algorithms for this method are found in James, Smith, and Wolford (1977) and Gerald and Wheatley (1984).

A variety of techniques are available for determining *eigenvalues*. James, Smith, and Wolford (1977), Gerald and Wheatley (1984), and Hornbeck (1975) provide introductions to the subject. In-depth treatments are contained in Ralston and Rabinowitz (1978), Householder (1964), and Wilkonson (1965).

Simultaneous nonlinear equations can sometimes be solved using the *Gauss-Seidel method*. In addition, a multidimensional version of the *Newton-Raphson method* offers a more efficient, although somewhat more complicated, approach. Carnahan, Luther, and Wilkes (1969), Gerald and Wheatley (1984), and James, Smith, and Wolford (1977) discuss the methods. Ortega and Rheinboldt (1970) offer a comprehensive work on the subject.

In summary, the foregoing is intended to provide you with avenues for further study of the subject and related areas. All of the above references provide descriptions of the basic techniques in Part III. In addition, Ralston and Rabinowitz (1978) provides an in-depth survey, and Stark (1970) includes discussion of topics such as ill conditioning. We urge you to consult these alternative sources to supplement the material in our book and enrich your understanding of simultaneous linear algebraic equations.*

*Books are referenced only by author here; a complete bibliography will be found at the back of this text.

# PART FOUR

# CURVE FITTING

## IV.1 MOTIVATION

Data is often given for discrete values along a continuum. However, you may require estimates at points between the discrete values. The present part of this book describes techniques to fit curves to such data in order to obtain intermediate estimates. In addition, you may require a simplified version of a complicated function. One way to do this is to compute values of the function at a number of discrete values along the range of interest. Then a simpler function may be derived to fit these values. Both of these applications are known as *curve fitting.*

There are two general approaches for curve fitting that are distinguished from each other on the basis of the amount of error associated with the data. First, where the data exhibits a significant degree of error or "noise," the strategy is to derive a single curve that represents the general trend of the data. Because any individual data point may be incorrect, we make no effort to intersect every point. Rather, the curve is designed to follow the pattern of the points taken as a group. One approach of this nature is called *least-squares regression* (Fig. IV.1*a*)

Second, where the data is known to be very precise, the basic approach is to fit a curve or a series of curves that pass directly through each of the points. Such data usually originates from tables. Examples are values for the density of water or for the heat capacity of gases as a function of temperature. The estimation of values between well-known discrete points is called *interpolation* (Fig. IV.1*b* and *c*).

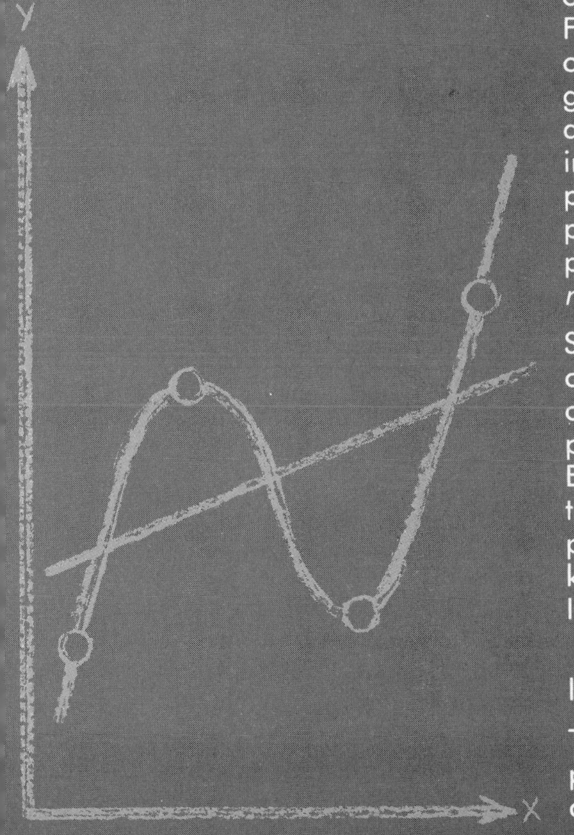


### IV.1.1 Precomputer Methods for Curve Fitting

The simplest method for fitting a curve to data is to plot the points and then sketch a line that visually conforms to the data. Although this is a valid op-

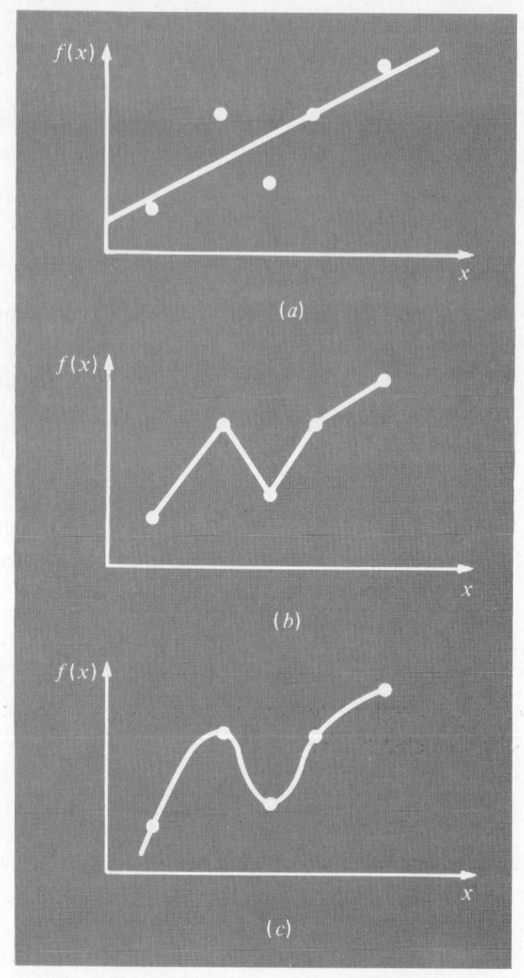


FIGURE IV.1 Three attempts to fit a "best" curve through five data points. (*a*) Least-squares regression, (*b*) linear interpolation, and (*c*) curvilinear interpolation.

tion when quick estimates are required, the results are dependent on the subjective viewpoint of the person sketching the curve.

For example, Fig. IV.1 shows sketches developed from the same set of data by three students. The first did not attempt to connect the points but rather characterized the general upward trend of the data with a straight line (Fig. IV.1*a*). The second student used straight-line segments or linear interpolation to connect the points (Fig. IV.1*b*). This is a very common practice in engineering. If the values are truly close to being linear or are spaced closely, such an approximation provides estimates that are adequate for many engineering calculations. However, where the underlying relationship is highly curvilinear or the data is widely spaced, significant errors can be introduced by such linear interpolation. The third student used curves to try to capture the meanderings suggested by the data (Fig. IV.1*c*). A fourth or fifth student would likely develop alternative fits. Obviously, our goal here is to develop systematic and objective methods for the purpose of deriving such curves.

### IV.1.2 Curve Fitting and Engineering Practice

Your first exposure to curve fitting may have been to determine intermediate values from tabulated data—for instance, from interest tables for engineering economics or from steam tables for thermodynamics. Throughout the remainder of your career, you will have frequent occasion to estimate intermediate values from such tables.

Although many of the widely used engineering properties have been tabulated, there are a great many more that are not available in this convenient form. Special cases and new problem contexts often require that you measure your own data and develop your own predictive relationships. Two types of applications are generally encountered when fitting experimental data: trend analysis and hypothesis testing.

*Trend analysis* represents the process of using the pattern of the data to make predictions. For cases where the data is measured with high precision, you might utilize interpolating polynomials. Imprecise data is usually analyzed with least-squares regression.

Trend analysis may be used to predict or forecast values of the dependent variable. This can involve *extrapolation* beyond the limits of the observed data or *interpolation* within the range of the data. All fields of engineering commonly involve problems of this type.

A second engineering application of experimental curve fitting is *hypothesis testing.* Here an existing mathematical model is compared with measured data. If the model coefficients are unknown it may be necessary to determine values that best fit the observed data. On the other hand, if estimates of the model coefficients are already available it may be appropriate to compare predicted values of the model with observed values to test the adequacy of the model. Often, alternative models are compared and the "best" selected on the basis of empirical observations.

In addition to the above engineering applications, curve fitting is important in other numerical methods such as integration and the approximate solution of differential equations. Finally, curve-fitting techniques can be used to derive simple functions to approximate complicated functions.

# IV.2 MATHEMATICAL BACKGROUND

The prerequisite mathematical background for interpolation is found in the material on Taylor series expansions and finite divided differences introduced in Chap. 3. Least-squares regression requires additional information from the field of statistics. If you are familiar with the concepts of the mean, standard deviation, residual sum of the squares, and normal distribution, feel free to skip the following pages and proceed directly to Sec. IV.3. If you are unfamiliar with these concepts or are in need of a review, the following material is designed as a brief introduction to these topics.

## IV.2.1 Simple Statistics

Suppose that in the course of an engineering study, several measurements were made of a particular quantity. For example, Table IV.1 contains 24 readings of the coefficient of thermal expansion of a structural steel. Taken at face value, the data provides a limited amount of information—that is, that the values range from a minimum of 6.395 to a maximum of 6.775. Additional insight can be gained by summarizing the data in one or more well-chosen statistics that convey as much information as possible about specific characteristics of the data set. These descriptive statistics are most often selected to represent (1) the location of the center of the distribution of the data and (2) the degree of spread of the data set.

The most common location statistic is the mean. The *mean* ($\bar{y}$) of a sample is defined as the sum of the individual data points ($y_i$) divided by the number of points ($n$), or

$$\boxed{\bar{y} = \frac{\Sigma y_i}{n}} \qquad \text{[IV.1]}$$

where the summation is from $i = 1$ through $n$.

**TABLE IV.1** **Measurements of the coefficient of thermal expansion of a structural steel ($\times 10^{-6}$ in/in/°F)**

| | | | |
|---|---|---|---|
| 6.495 | 6.625 | 6.635 | 6.655 |
| 6.665 | 6.515 | 6.625 | 6.775 |
| 6.755 | 6.615 | 6.575 | 6.555 |
| 6.565 | 6.435 | 6.395 | 6.655 |
| 6.595 | 6.715 | 6.485 | 6.605 |
| 6.505 | 6.555 | 6.715 | 6.685 |

The most common measure of spread for a sample is the *standard deviation* ($s_y$) about the mean:

$$s_y = \sqrt{\frac{S_t}{n-1}} \qquad \text{[IV.2]}$$

where $S_t$ is the total sum of the squares of the residuals between the data points and the mean, or

$$S_t = \Sigma\,(y_i - \bar{y})^2 \qquad \text{[IV.3]}$$

Thus, if the individual measurements are spread out widely around the mean, $S_t$ (and, consequently, $s_y$) will be large. If they are grouped tightly, the standard deviation will be small. The spread can also be represented by the square of the standard deviation, which is called the *variance*:

$$s_y^2 = \frac{S_t}{n-1} \qquad \text{[IV.4]}$$

Note, that the denominator in both cases is $n - 1$. This takes into account the fact that one piece of information previously derived from the data (that is, the mean) was used to determine $S_t$. Formally, this is said to represent a loss of one *degree of freedom*. Another justification for dividing by $n - 1$ is the fact that there is no such thing as the spread of a single data point. Thus, for the case where $n = 1$, Eq. (IV.4) yields a meaningless result of infinity.

A final statistic that has utility in quantifying the spread of data is the *coefficient of variation* (*c.v.*). This statistic is the ratio of the standard deviation to the mean. As such it provides a normalized measure of the spread. It is often multiplied by 100 so that it can be expressed in the form of a percent:

$$\text{c.v.} = \frac{s_y}{\bar{y}}\,100\% \qquad \text{[IV.5]}$$

Notice that the coefficient of variation is similar in spirit to the percent relative error ($\epsilon_t$) discussed in Sec. 3.3. That is, it is the ratio of a measure of error ($s_y$) to an estimate of the true value ($\bar{y}$).

EXAMPLE IV.1
Simple Statistics of a Sample

Problem Statement: Compute the mean, variance, standard deviation, and coefficient of variation for the data in Table IV.1.

**TABLE IV.2 Computations for statistics and histogram for the readings of the coefficient of thermal expansion**

| $i$ | $y_i$ | $(y_i - \bar{y})^2$ | Frequency | INTERVAL Lower bound | INTERVAL Upper bound |
|---|---|---|---|---|---|
| 1 | 6.395 | 0.042025 | 1 | 6.36 | 6.40 |
| 2 | 6.435 | 0.027225 | 1 | 6.40 | 6.44 |
| 3 | 6.485 | 0.013225 | | | |
| 4 | 6.495 | 0.011025 | | | |
| 5 | 6.505 | 0.009025 | 4 | 6.48 | 6.52 |
| 6 | 6.515 | 0.007225 | | | |
| 7 | 6.555 | 0.002025 | 2 | 6.52 | 6.56 |
| 8 | 6.555 | 0.002025 | | | |
| 9 | 6.565 | 0.001225 | | | |
| 10 | 6.575 | 0.000625 | 3 | 6.56 | 6.60 |
| 11 | 6.595 | 0.000025 | | | |
| 12 | 6.605 | 0.000025 | | | |
| 13 | 6.615 | 0.000225 | | | |
| 14 | 6.625 | 0.000625 | 5 | 6.60 | 6.64 |
| 15 | 6.625 | 0.000625 | | | |
| 16 | 6.635 | 0.001225 | | | |
| 17 | 6.655 | 0.003025 | | | |
| 18 | 6.655 | 0.003025 | 3 | 6.64 | 6.68 |
| 19 | 6.665 | 0.004225 | | | |
| 20 | 6.685 | 0.007225 | | | |
| 21 | 6.715 | 0.013225 | 3 | 6.68 | 6.72 |
| 22 | 6.715 | 0.013225 | | | |
| 23 | 6.755 | 0.024025 | 1 | 6.72 | 6.76 |
| 24 | 6.775 | 0.030625 | 1 | 6.76 | 6.80 |
| $\Sigma$ | 158.400 | 0.217000 | | | |

Solution: The data is added (Table IV.2) and the results are used to compute [Eq. (IV.1)]:

$$\bar{y} = \frac{158.400}{24} = 6.6$$

As in Table IV.2, the sum of the squares of the residuals is 0.21700, which can be used to compute the standard deviation [Eq. (IV.2)]:

$$s_y = \sqrt{\frac{0.217000}{24 - 1}} = 0.097133$$

and the variance [Eq. (IV.4)]:

$$s_y^2 = 0.009435$$

and the coefficient of variation [Eq. (IV.5)]:

$$\text{c.v.} = \frac{0.097133}{6.6} 100\% = 1.47\%$$

## IV.2.2 The Normal Distribution

The final characteristic that bears on the present discussion is the *data distribution*—that is, the shape with which the data is spread around the mean. A *histogram* provides a simple visual representation of the distribution. As in Table IV.2, the histogram is constructed by sorting the measurements into intervals. The units of measurement are plotted on the abscissa and the frequency of occurrence of each interval is plotted on the ordinate. Thus, five of the measurements fall between 6.60 and 6.64. As in Fig. IV.2, the histogram suggests that most of the data is grouped close to the mean.

If we have a very large set of data, the histogram often will be transformed from a bar diagram into a smooth shape. The symmetric, bell-shaped curve superimposed on Fig. IV.2 is one such characteristic shape—the *normal distribution.* Given enough additional measurements, the histogram for this particular case could eventually approach the normal distribution.

The concepts of the mean, standard deviation, residual sum of the squares, and normal distribution all have great relevance to engineering practice. A very simple example is their use to quantify the confidence that can be ascribed to a particular measurement. If a quantity is normally distributed, the range defined by $\bar{y} - s_y$ to $\bar{y} + s_y$ will encompass approximately 68 percent of the total number of measurements. Similarly, the range defined by $\bar{y} - 2s_y$ to $\bar{y} + 2s_y$ will encompass approximately 95 percent.

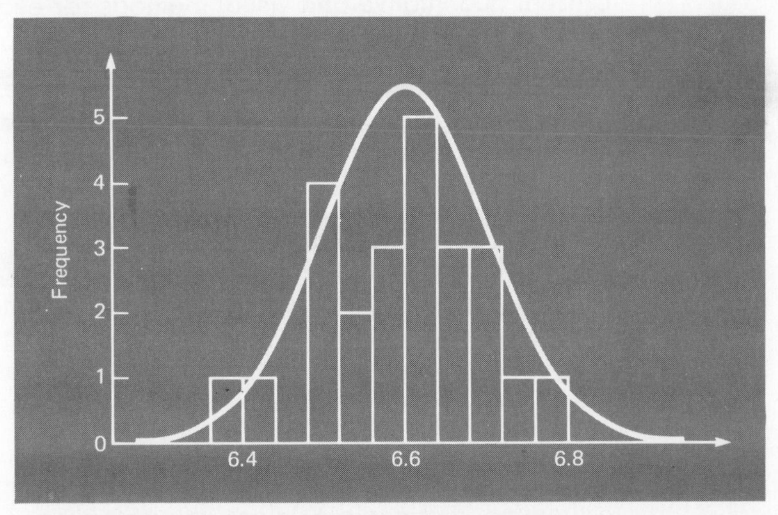


FIGURE IV.2 A histogram used to depict the distribution of data. As the number of data points increases, the histogram could approach the smooth, bell-shaped curve called the normal distribution.

For example, for the coefficients of thermal expansion in Table IV.1 ($\bar{y} = 6.6$ and $s_y = 0.097133$), we can make the statement that approximately 95 percent of the readings should fall between 6.405734 and 6.794266. If someone told us that they had measured a value of 7.35, we would suspect that the measurement could be erroneous.

The above is just one simple example of how statistics can be used to make judgments regarding uncertain data. These concepts will also have direct relevance to our discussion of regression models. You can consult any basic statistics book (for example, Ang and Tang, 1975, or Lapin, 1983) to obtain additional information on the subject.

# IV.3 ORIENTATION

Before proceeding to numerical methods for curve fitting, some orientation might be helpful. The following is intended as an overview of the material discussed in Part IV. In addition, we have formulated some objectives to help focus your efforts when studying the material.

## IV.3.1 Scope and Preview

Figure IV.3 provides a visual overview of the material to be covered in Part IV. *Chapter 10* is devoted to *least-squares regression.* We will first learn how to fit the "best" straight line through a set of uncertain data points. This technique is called *linear regression.* Besides discussing how to calculate the slope and intercept of this straight line, we also present quantitative and visual methods for evaluating the validity of the results.

In addition to fitting a straight line, we also present a general technique for fitting a "best" polynomial. Thus, you will learn to derive a parabolic, cubic, or higher-order polynomial that optimally fits uncertain data. Linear regression is a subset of this more general approach, which is called *polynomial regression.*

Finally, the last topic covered in Chap. 10 is *multiple linear regression.* It is designed for the case where the dependent variable $y$ is a linear function of two or more independent variables $x_1, x_2, \ldots, x_n$. This approach has special utility for evaluating experimental data where the variable of interest is dependent on a number of different factors.

In *Chapter 11,* the alternative curve-fitting technique called *interpolation* is described. As discussed previously, interpolation is used for estimating intermediate values between precise data points. In Chap. 11, polynomials are derived for this purpose. We introduce the basic concept of polynomial interpolation by using straight lines and parab-

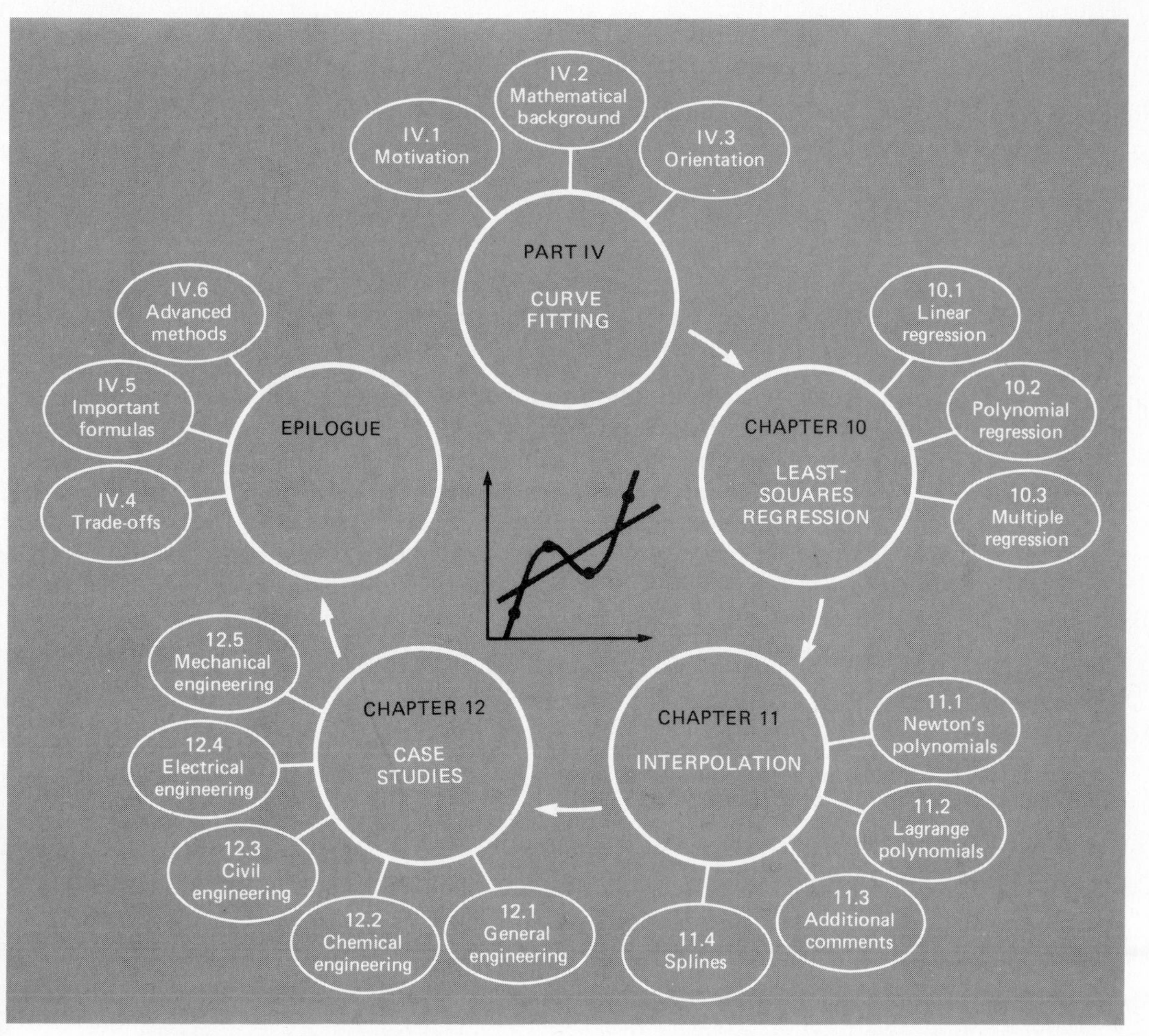


FIGURE IV.3 Schematic representation of the organization of material in Part IV: Curve fitting.

olas to connect points. Then, we develop a generalized procedure for fitting an $n$th-order polynomial. Two formats are presented for expressing these polynomials in equation form. The first, called *Newton's interpolating polynomial,* is preferable when the correct order of the polynomial is unknown. The second, called the *Lagrange interpolating polynomial* has advantages when the proper order is known beforehand.

The last section of Chap. 11 is devoted to an alternative technique for fitting precise data points. This technique, called *spline interpolation,* fits polynomials to data but in a piecewise fashion. As such, it is particularly well-suited for fitting data that is generally smooth but exhibits abrupt local changes.

*Chapter 12* is devoted to case studies that illustrate the utility of the numerical methods in engineering problem contexts. Examples are drawn from general engineering as well as from the four major specialty areas of chemical, civil, electrical, and mechanical engineering.

Finally, an epilogue is included at the end of Part IV. It includes a summary of the important formulas and concepts related to curve fitting as well as a discussion of trade-offs among the techniques and suggestions for future study.

Automatic computation capability is integrated into Part IV in a number of ways. First, the NUMERICOMP software package accompanying the text contains user-friendly programs for linear regression and Lagrange interpolation. Alternatively, computer codes for both methods, using both FORTRAN and BASIC, are included in the text. This provides you with the opportunity to copy and enhance the code for implementation on your own personal or mainframe computer. Flowcharts or algorithms are included for most of the other methods described in the text. This material can form the basis for a comprehensive software package that you can develop and apply to a number of engineering problems.

### IV.3.2 Goals and Objectives

*Study objectives.* After completing Part IV, you should have greatly enhanced your capability to fit curves to data. In general, you should have mastered the techniques, have learned to assess the reliability of the answers, and be capable of choosing the best method (or methods) for any particular problem. In addition to these general goals, the specific concepts in Table IV.3 should be assimilated and mastered.

*Computer objectives.* You have been provided with software, simple computer programs, algorithms, and flowcharts to implement the techniques discussed in Part IV. All have utility as learning tools.

The optional NUMERICOMP software includes linear regression and Lagrange interpolation programs. The graphics associated with this software will enable you to easily visualize your problem and the associated mathematical operations. The graphics are a critical part of your assessment of the validity of a regression fit. They also provide guidance regarding the proper order of polynomial interpolation and the potential dangers of extrapolation. The software is very easy to apply to solve practical problems and can be used to check the results of any computer programs you may develop yourself.

**TABLE IV.3 Specific study objectives for Part IV**

1. Understand the fundamental difference between regression and interpolation and realize why confusing the two could lead to serious problems
2. Understand the derivation of linear least-squares regression and be able to assess the reliability of the fit using graphical and quantitative assessments
3. Know how to linearize data by transformation
4. Understand situations where polynomial and multiple regression are appropriate
5. Understand that there is one and only one polynomial of degree $n$ or less that passes exactly through $n + 1$ points
6. Know how to derive the first-order Newton's interpolating polynomial
7. Understand the analogy between Newton's polynomial and the Taylor series expansion and how it relates to the truncation error
8. Recognize that the Newton and Lagrange equations are merely different formulations of the same interpolating polynomial and understand their respective advantages and disadvantages
9. Realize that more accurate results are obtained if points used for interpolation are centered around and close to the unknown
10. Realize that data points do not have to be equally spaced nor in any particular order for either the Newton or Lagrange polynomials
11. Know why equispaced interpolation formulas have utility
12. Recognize the liabilities and risks associated with extrapolation
13. Understand why spline functions have utility for data with local areas of abrupt change

In addition, computer codes, algorithms, or flowcharts are provided for most of the other methods in Part IV. This information will allow you to expand your software library to include techniques beyond linear regression and Lagrange interpolation. For example, you may find it useful from a professional viewpoint to have software to implement polynomial regression, Newton's interpolating polynomial, and cubic spline interpolation.

# CHAPTER TEN

# LEAST-SQUARES REGRESSION

Where substantial error is associated with data, polynomial interpolation is inappropriate and may yield unsatisfactory results when used to predict intermediate values. Experimental data is often of this type. For example, Fig. 10.1*a* shows seven experimentally-derived data points exhibiting significant variability. Visual inspection of the data suggests a positive relationship between $y$ and $x$. That is, the overall trend indicates that higher values of $y$ are associated with higher values of $x$. Now, if a sixth-order interpolating polynomial is fitted to this data (Fig. 10.1*b*), it will pass exactly through all of the points. However, because of the variability in the data, the curve oscillates widely in the interval between the points. In particular, the interpolated values at $x = 1.5$ and $x = 6.5$ appear to be well beyond the range suggested by the data.

A more appropriate strategy for such cases is to derive an approximating function that "adequately" fits the shape or general trend of the data without necessarily matching the individual points. Figure 10.1*c* illustrates how a straight line can be used to generally characterize the trend of the data without passing through any particular point.

One way to determine the line in Fig. 10.1*c* is to visually inspect the plotted data and then sketch a "best" line through the points. Although such "eyeball" approaches have commonsense appeal and are valid for "back-of-the-envelope" calculations, they are deficient because they are arbitrary. That is, unless the points define a perfect straight line (in which case, interpolation would be appropriate), different analysts would draw different lines.

In order to remove this subjectivity, some criterion must be devised to quantify the adequacy of the fit. One way to do this is to derive a curve that minimizes the discrepancy between the data points and the curve. A technique for accomplishing this objective, called *least-squares regression,* will be discussed in the present chapter.

## 10.1 LINEAR REGRESSION

The simplest example of a least-squares approximation is fitting a straight line to a set of paired observations: $(x_1, y_1)$, $(x_2, y_2)$, . . . , $(x_n, y_n)$. The mathe-

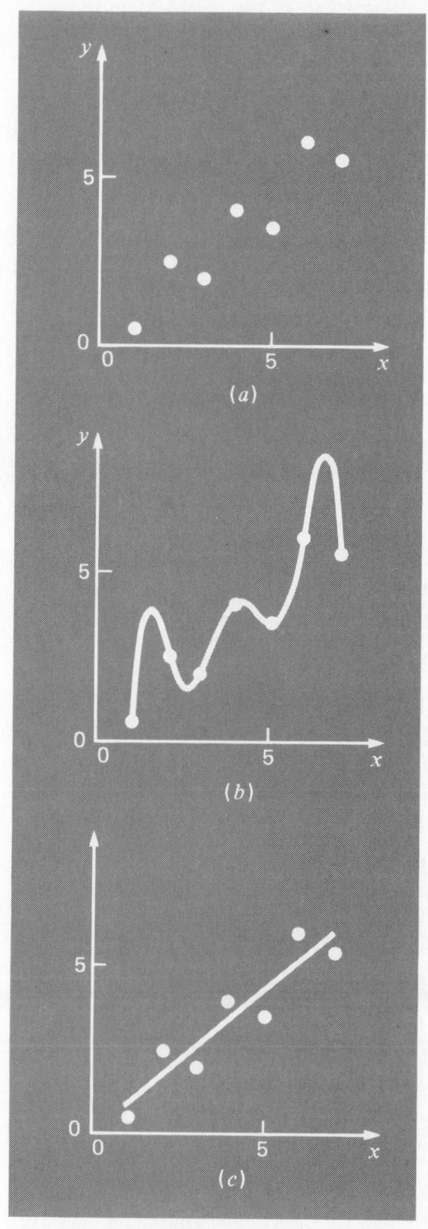


FIGURE 10.1 (a) Data exhibiting significant error. (b) Polynomial fit oscillating beyond the range of the data. (c) More satisfactory result using the least-squares fit.

matical expression for the straight line is

$$y = a_0 + a_1 x + E \tag{10.1}$$

where $a_0$ and $a_1$ are coefficients representing the intercept and the slope, respectively, and $E$ is the error, or residual between the model and the observations, which can be represented by rearranging Eq. (10.1) as

$$E = y - a_0 - a_1 x$$

Thus, the error, or residual, is the discrepancy between the true value of $y$ and the approximate value, $a_0 + a_1 x$, predicted by the linear equation.

### 10.1.1 Criteria for a "Best" Fit

One strategy for fitting a "best" line through the data would be to minimize the sum of the residual errors, as in

$$\sum_{i=1}^{n} E_i = \sum_{i=1}^{n} (y_i - a_0 - a_1 x_i) \qquad [10.2]$$

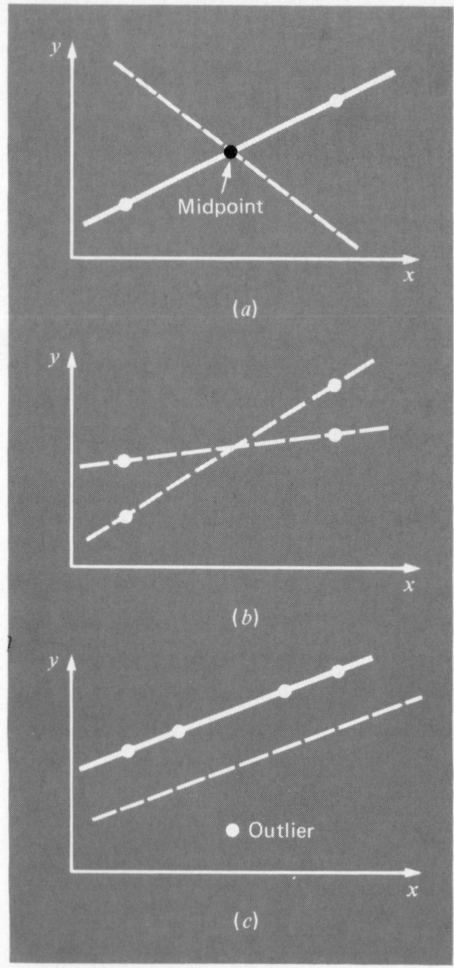


FIGURE 10.2 Examples of some criteria for "best fit" that are inadequate for regression: (*a*) minimizes the sum of the residuals; (*b*) minimizes the sum of the absolute values of the residuals; and (*c*) minimizes the maximum error of any individual point.

However, this is an inadequate criterion, as illustrated by Fig. 10.2*a*, which depicts the fit of a straight line to two points. Obviously, the best fit is the line connecting the points. However, any straight line passing through the midpoint of the connecting line (except a perfectly vertical line) results in a minimum value of Eq. (10.2) equal to zero because the errors cancel.

Another criterion would be to minimize the sum of the absolute values of the discrepancies, as in

$$\sum_{i=1}^{n} |E_i| = \sum_{i=1}^{n} |y_i - a_0 - a_1 x_i|$$

Figure 10.2*b* demonstrates why this criterion is also inadequate. For the four points shown, any straight line falling within the dashed lines will minimize the absolute value of the sum. Thus, this criterion also does not yield a unique best fit.

A third strategy for fitting a best line is the *minimax* criterion. In this technique, the line is chosen that minimizes the maximum distance that an individual point falls from the line. As depicted in Fig. 10.2*c*, this strategy is ill-suited for regression because it gives undue influence to an outlier, that is, a single point with a large error. It should be noted that the minimax principle is sometimes well-suited for fitting a simple function to a complicated function (Carnahan, Luther, and Wilkes, 1969).

A strategy that overcomes the shortcomings of the aforementioned approaches is to minimize the sum of the squares of the residuals, $S_r$, as in

$$S_r = \sum_{i=1}^{n} E_i^2 = \sum_{i=1}^{n} (y_i - a_0 - a_1 x_i)^2 \qquad [10.3]$$

This criterion has a number of advantages, including the fact that it yields a unique line for a given set of data. Before discussing these properties, we will present a technique for determining the values of $a_0$ and $a_1$ that minimize Eq. (10.3).

### 10.1.2 Least-Squares Fit of a Straight Line

In order to determine values for $a_0$ and $a_1$, Eq. (10.3) is differentiated with respect to each coefficient:

$$\frac{\partial S_r}{\partial a_0} = -2 \sum (y_i - a_0 - a_1 x_i)$$

$$\frac{\partial S_r}{\partial a_1} = -2 \sum [(y_i - a_0 - a_1 x_i) x_i]$$

Note that we have simplified the summation symbols; unless otherwise indicated, all summations are from $i = 1$ to $n$. Setting these derivatives equal to zero will result in a minimum $S_r$. If this is done, the equations can be expressed as

$$0 = \sum y_i - \sum a_0 - \sum a_1 x_i$$

$$0 = \sum y_i x_i - \sum a_0 x_i - \sum a_1 x_i^2$$

Now, realizing that $\Sigma a_0 = na_0$, the equations can be expressed as a set of two simultaneous linear equations with two unknowns ($a_0$ and $a_1$):

$$na_0 + \sum x_i a_1 = \sum y_i \qquad [10.4]$$

$$\sum x_i a_0 + \sum x_i^2 a_1 = \sum x_i y_i \qquad [10.5]$$

These are called the *normal equations*. They can be solved simultaneously for [recall Eq. (7.10)]

$$a_1 = \frac{n \Sigma x_i y_i - \Sigma x_i \Sigma y_i}{n \Sigma x_i^2 - (\Sigma x_i)^2} \qquad [10.6]$$

This result can then be used in conjunction with Eq. (10.4) to solve for

$$a_0 = \bar{y} - a_1 \bar{x} \qquad [10.7]$$

where $\bar{y}$ and $\bar{x}$ are the means of $y$ and $x$, respectively.

## EXAMPLE 10.1
### Linear Regression

Problem Statement: Fit a straight line to the $x$ and $y$ values in the first two columns of Table 10.1.

**TABLE 10.1 Computations for an error analysis of the linear fit**

| $x_i$ | $y_i$ | $(y_i - \bar{y})^2$ | $(y_i - a_0 - a_1 x_i)^2$ |
|---|---|---|---|
| 1 | 0.5 | 8.5765 | 0.1687 |
| 2 | 2.5 | 0.8622 | 0.5625 |
| 3 | 2.0 | 2.0408 | 0.3473 |
| 4 | 4.0 | 0.3265 | 0.3265 |
| 5 | 3.5 | 0.0051 | 0.5896 |
| 6 | 6.0 | 6.6122 | 0.7972 |
| 7 | 5.5 | 4.2908 | 0.1993 |
| Σ | 24 | 22.7143 | 2.9911 |

Solution: The following quantities can be computed:

$$n = 7 \qquad \sum x_i y_i = 119.5 \qquad \sum x_i^2 = 140$$

$$\sum x_i = 28 \qquad \bar{x} = \frac{28}{7} = 4$$

$$\sum y_i = 24 \qquad \bar{y} = \frac{24}{7} = 3.428571429$$

Using Eqs. (10.6) and (10.7),

$$a_1 = \frac{7(119.5) - 28(24)}{7(140) - (28)^2} = 0.839285714$$

$$a_0 = 3.428571429 - 0.839285714(4) = 0.07142857$$

Therefore, the least-squares fit is

$$y = 0.07142857 + 0.839285714x$$

The line, along with the data, is shown in Fig. 10.1*c*.

### 10.1.3 Quantification of Error of Linear Regression

Any line other than the one computed in Example 10.1 results in a larger sum of the squares of the residuals. Thus, the line is unique and in terms of our chosen criterion is a "best" line through the points. A number of additional properties of this fit can be elucidated by examining more closely the way in which residuals were computed. Recall that the sum of the squares is defined as [Eq. (10.3)]

$$S_r = \sum_{i=1}^{n} (y_i - a_0 - a_1 x_i)^2 \qquad [10.8]$$

Notice the similarity between Eqs. (IV.3) and (10.8). In the former case, the residuals represented the square of the discrepancy between the data and a single estimate of the measure of central tendency—the mean. In Eq. (10.8), the residuals represent the square of the vertical distance between the data and another measure of central tendency—the straight line (Fig. 10.3).

The analogy can be extended further for cases where (1) the spread of the points around the line is of similar magnitude along the entire range of the data and (2) the distribution of these points about the line is normal. It can be demonstrated that if these criteria are met, least-squares regression will provide the best (that is, the most likely) estimates of $a_0$ and $a_1$ (Draper and Smith, 1981). This is called the *maximum likelihood principle* in statistics. In addition, if these criteria are met, a "standard deviation" for the regression line can be determined as [compare with Eq. (IV.2)]

$$s_{y/x} = \sqrt{\frac{S_r}{n-2}} \qquad [10.9]$$

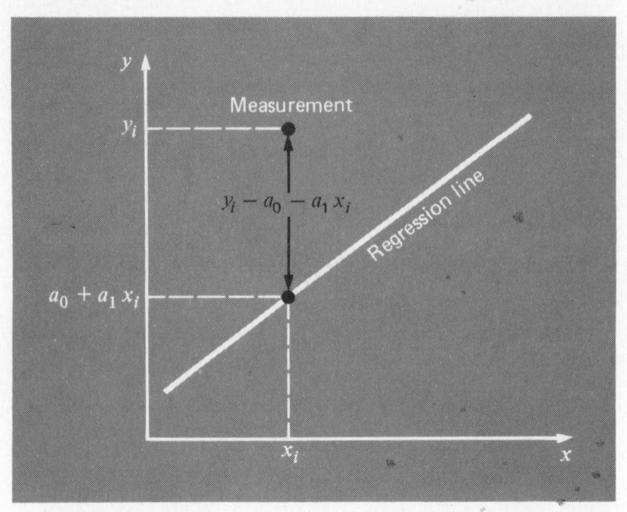


FIGURE 10.3 The residual in linear regression represents the square of the vertical distance between a data point and the straight line.

where $s_{y/x}$ is called the *standard error of the estimate.* The subscript notation "$y/x$" designates that the error is for a predicted value of $y$ corresponding to a particular value of $x$. Also, notice that we now divide by $n - 2$ because two data-derived estimates—$a_0$ and $a_1$—were used to compute $S_r$; thus, we have lost two degrees of freedom. As with our discussion of the standard deviation in Sec. IV.2.1, another justification for dividing by $n - 2$ is that there is no such thing as the "spread of data" around a straight line connecting two points. Thus, for the case where $n = 2$, Eq. (10.9) yields a meaningless result of infinity.

Just as with the standard deviation, the standard error of the estimate quantifies the spread of the data. However, $s_{y/x}$ quantifies the spread *around the regression line,* as shown in Fig. 10.4, as opposed to the original standard deviation, $s_y$, which quantifies the spread *around the mean.*

The above concepts can be used to quantify the "goodness" of our fit. This is particularly useful for comparison of several regressions (see Fig. 10.5). To do this, we return to the original data and determine the sum of the squares around the mean for the dependent variable (in our case, $y$). We can call this the total sum of the squares $S_t$. This is the amount of spread in the dependent variable that exists prior to regression. After performing the linear regression, we can compute $S_r$, which is the sum of the squares of the residuals around our regression line. This represents the spread that remains after regression. The difference between the two quantities, or $S_t - S_r$, quantifies the improvement or error reduction due to the straight-line model. This difference can be normalized to the total error to yield

$$r^2 = \frac{S_t - S_r}{S_t} \tag{10.10}$$

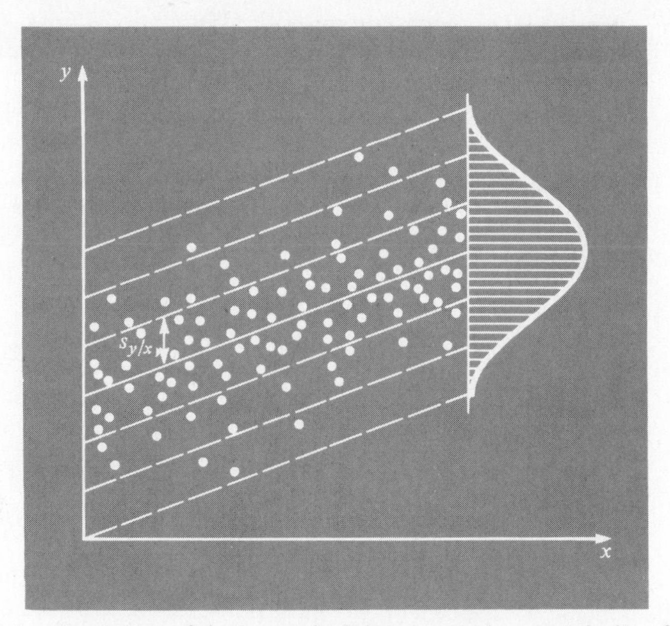


FIGURE 10.4 Depiction of the spread of data around a straight-line fit.

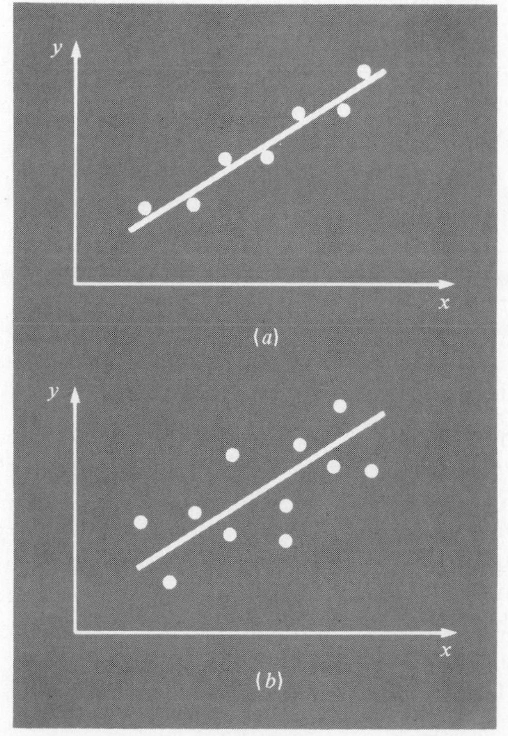


FIGURE 10.5 Examples of linear regression with (*a*) small and (*b*) large residual errors.

where $r$ is the *correlation coefficient* and $r^2$ is the *coefficient of determination.* For a perfect fit, $S_r = 0$ and $r^2 = 1$, signifying that the line explains 100 percent of the variability. For $r^2 = 0$, the fit represents no improvement.

### EXAMPLE 10.2
Estimation of Errors for the Linear Least-Squares Fit

Problem Statement: Compute the total standard deviation, the standard error of the estimate, and the correlation coefficient for the data in Example 10.1.

Solution: The summations are performed and presented in Table 10.1. The total standard deviation is [Eq. (IV.2)]

$$s_y = \sqrt{\frac{22.7143}{7 - 1}} = 1.9457$$

and the standard error of the estimate is [Eq. (10.9)]

$$s_{y/x} = \sqrt{\frac{2.9911}{7 - 2}} = 0.7735$$

Thus, because $s_{y/x} < s_y$, the linear regression model has merit. The extent of the improvement is quantified by [(Eq. (10.10)]

$$r^2 = \frac{22.7143 - 2.9911}{22.7143} = 0.868$$

or

$$r = \sqrt{0.868} = 0.932$$

These results indicate that 86.8 percent of the original uncertainty has been explained by the linear model.

Before proceeding to the computer program for linear regression, a word of caution is in order. Although the correlation coefficient provides a handy measure of goodness-of-fit, you should be careful not to ascribe more meaning to it than is warranted. Just because $r$ is "close" to 1 does not mean that the fit is necessarily "good." For example, it is possible to obtain a relatively high value of $r$ when the underlying relationship between $y$ and $x$ is not even linear. Draper and Smith (1981) provide guidance and additional material regarding assessment of results for linear regression. In addition, at the minimum, you should *always* inspect a plot of the data along with your regression line whenever you fit regression curves. As described in the next section, the NUMERICOMP software includes such a capability.

### 10.1.4 Computer Program for Linear Regression

It is a relatively trivial matter to develop a program for linear regression. FORTRAN and BASIC versions are contained in Fig. 10.6. Because the

ORTRAN

```
 DATA SX/0./,SY/0./,X2/0./,XY/0./
 READ(5,1)N
1 FORMAT(I5)
 DO 170 I=1,N
 READ(5,2)X,Y
2 FORMAT(2F10.0)
 SX=SX+X
 SY=SY+Y
 X2=X2+X*X
 XY=XY+X*Y
0 CONTINUE
 XM=SX/N
 YM=SY/N
 A1=(N*XY-SX*SY)/(N*X2-SX*SX)
 A0=YM-A1*XM
 WRITE(6,3)A0,A1
3 FORMAT(' ',2F10.3)
 STOP
 END
```

BASIC

```
100  INPUT N
110  FOR I = 1 TO N
120  INPUT X,Y
130 SX = SX + X
140 SY = SY + Y
150 X2 = X2 + X * X
160 XY = XY + X * Y
170  NEXT I
180 XM = SX / N
190 YM = SY / N
200 A1 = (N * XY - SX * SY) / (N *
     X2 - SX * SX)
210 A0 = YM - A1 * XM
220  PRINT A0,A1
230  END
```

N = number of data points
X = independent variable
Y = dependent variable
SX = sum of X's
SY = sum of Y's
X2 = sum of square of X's
XY = sum of product of X and Y
XM = mean of X's
YM = mean of Y's
A1 = slope
A0 = intercept

FIGURE 10.6 FORTRAN and BASIC programs for linear regression.

graphical capabilities of personal computers are so varied, we have not included plots in these programs. However, as mentioned above, such an option is critical to the effective use and interpretation of regression and is included in the supplementary NUMERICOMP software. If your computer system has plotting capabilities we recommend that you expand your program to include a plot of $y$ versus $x$ showing both the data and the regression line. The inclusion of the capability will greatly enhance the utility of the program in problem-solving contexts.

## EXAMPLE 10.3
### Linear Regression Using the Computer

Problem Statement: A user-friendly computer program to implement linear regression is contained in the NUMERICOMP software package associated with this text. We can use this software to solve a hypothesis-testing problem associated with the falling parachutist discussed in Chap. 1. A theoretical mathematical model for the velocity of the parachutist was given as the following [Eq. (1.9)]:

$$v(t) = \frac{gm}{c}[1 - e^{-(c/m)t}]$$

where $v$ is the velocity in centimeters per second, $g$ is the gravitational constant of 980 cm/s$^2$, $m$ is the mass of the parachutist equal to 68,100 g, and $c$ is the drag coefficient of 12,500 g/s. The model predicts the velocity of the parachutist as a function of time, as described in Example 1.1. A plot of the velocity variation was developed in Example 2.1.

An alternative empirical model for the velocity of the parachutist is given by the following

$$v(t) = \frac{gm}{c}\left[\frac{t}{3.75 + t}\right] \tag{E10.3.1}$$

Suppose that you would like to test and compare the adequacy of these two mathematical models. This might be accomplished by measuring the actual velocity of the parachutist at known values of time and comparing these results with the predicted velocities according to each model.

Such an experimental-data-collection program was implemented and the results are listed in column (*a*) of Table 10.2. Computed velocities for each model are listed in columns (*b*) and (*c*).

**TABLE 10.2 Measured and calculated velocities for the falling parachutist**

| Time, s | Measured *v*, cm/s (*a*) | Model-calculated *v*, cm/s [Eq. (1.9)] (*b*) | Model-calculated *v*, cm/s [Eq. (E10.3.1)] (*c*) |
|---|---|---|---|
| 1 | 1000 | 895.3 | 1124.0 |
| 2 | 1630 | 1640.5 | 1857.0 |
| 3 | 2300 | 2260.7 | 2372.9 |
| 4 | 2750 | 2776.9 | 2755.6 |
| 5 | 3100 | 3206.5 | 3050.9 |
| 6 | 3560 | 3564.1 | 3285.5 |
| 7 | 3900 | 3861.7 | 3476.6 |
| 8 | 4150 | 4109.5 | 3635.1 |
| 9 | 4290 | 4315.6 | 3768.7 |
| 10 | 4500 | 4487.2 | 3882.9 |
| 11 | 4600 | 4630.1 | 3981.6 |
| 12 | 4550 | 4749.0 | 4067.8 |
| 13 | 4600 | 4847.9 | 4143.7 |
| 14 | 4900 | 4930.3 | 4211.0 |
| 15 | 5000 | 4998.8 | 4271.2 |

Solution: The adequacy of the models can be tested by plotting the measured velocity versus the model-calculated velocity. Linear regression can be used to calculate a trend line for the plot. This trend line will have a slope of 1 and an intercept of 0 if the model matches the data perfectly. Any deviation from these values can be used as an indication of the inadequacy of the model.

Figure 10.7*a* and *b* are plots of the line and data for the regressions of column (*a*) versus columns (*b*) and (*c*), respectively. These plots indicate that the linear regression between the data and each of the models is highly significant. Both models match the data with a correlation coefficient of greater than 0.99.

However, the model described by Eq. (1.9) conforms to our hypothesis test criteria much better than that described by (E10.3.1) because the slope

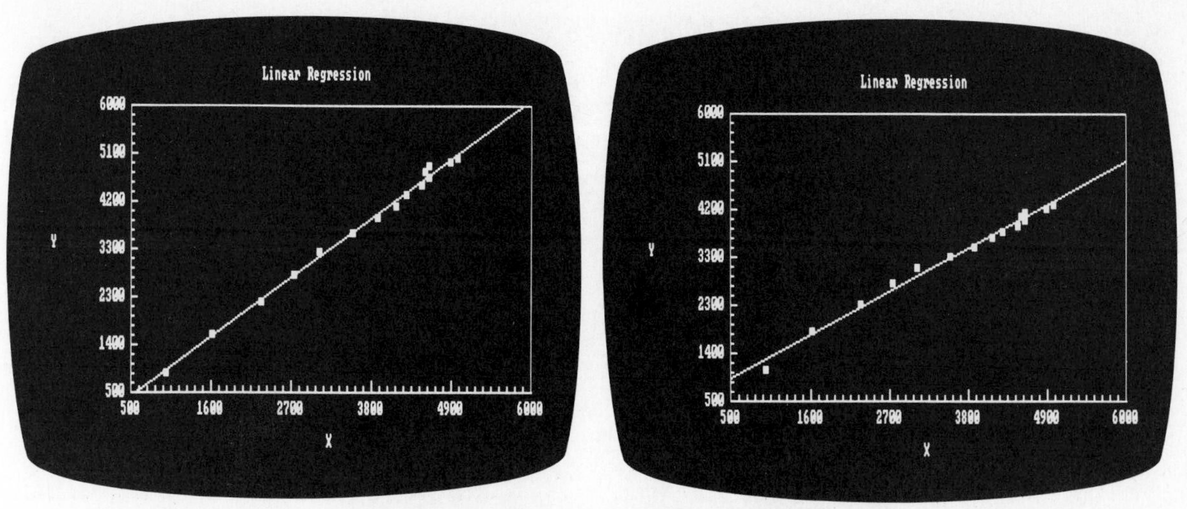


FIGURE 10.7 (*a*) Results using linear regression to compare measured values versus model predictions computed with the theoretical Eq. (1.9). (*b*) Results using linear regression to compare measured values versus model predictions computed with the empirical Eq. (E10.3.1).

and intercept are more nearly equal to 1 and 0. Thus, although each plot is well described by a straight line, Eq. (1.9) is a better model than Eq. (E10.3.1).

Model testing and selection are common and extremely important activities performed in all fields of engineering. The background material provided you in the present chapter together with the NUMERICOMP software and your own programs should allow you to address many practical problems of this type.

## 10.1.5 Applications of Linear Regression—Linearization of Nonlinear Relationships

Linear regression provides a powerful technique for fitting a "best" line to data. However, it is predicated on the fact that the relationship between the dependent and independent variables is linear. This is not always the case, and the first step in any regression analysis should be to plot and visually inspect the data to ascertain whether a linear model applies. For example, Fig. 10.8 shows some data that is obviously curvilinear. In some cases, techniques such as polynomial regression, which is described in Sec. 10.2, are appropriate. For others, transformations can be used to express the data in a form that is compatible with linear regression.

One example is the *exponential model*

$$y = a_1 e^{b_1 x} \qquad [10.11]$$

where $a_1$ and $b_1$ are constants. This model is used in many fields of engineering to characterize quantities that increase (positive $b_1$) or decrease

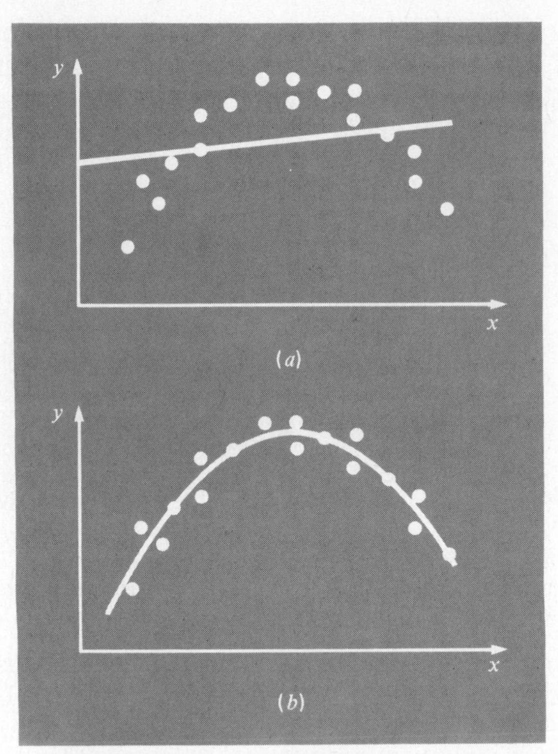


FIGURE 10.8 (*a*) Data that is ill-suited for linear least-squares regression. (*b*) Indication that a parabola is preferable.

(negative $b_1$) at a rate that is directly proportional to their own magnitude. For example, population growth or radioactive decay can exhibit such behavior. As depicted in Fig. 10.9*a*, the equation represents a nonlinear relationship (for $b_1 \neq 0$) between $y$ and $x$.

Another example of a nonlinear model is the simple *power equation*

$$y = a_2 x^{b_2} \tag{10.12}$$

where $a_2$ and $b_2$ are constant coefficients. This model has wide applicability in all fields of engineering. As depicted in Fig. 10.9*b*, the equation (for $b_2 \neq 0$ or 1) is nonlinear.

A third example of a nonlinear model is the *saturation-growth-rate equation*

$$y = a_3 \frac{x}{b_3 + x} \tag{10.13}$$

where $a_3$ and $b_3$ are constant coefficients. This model, which is particularly well-suited for characterizing population growth rate under limiting conditions, also represents a nonlinear relationship between $y$ and $x$ (Fig. 10.9*c*) that levels off, or "saturates," as $x$ increases.

Nonlinear regression techniques are available to fit these equations to

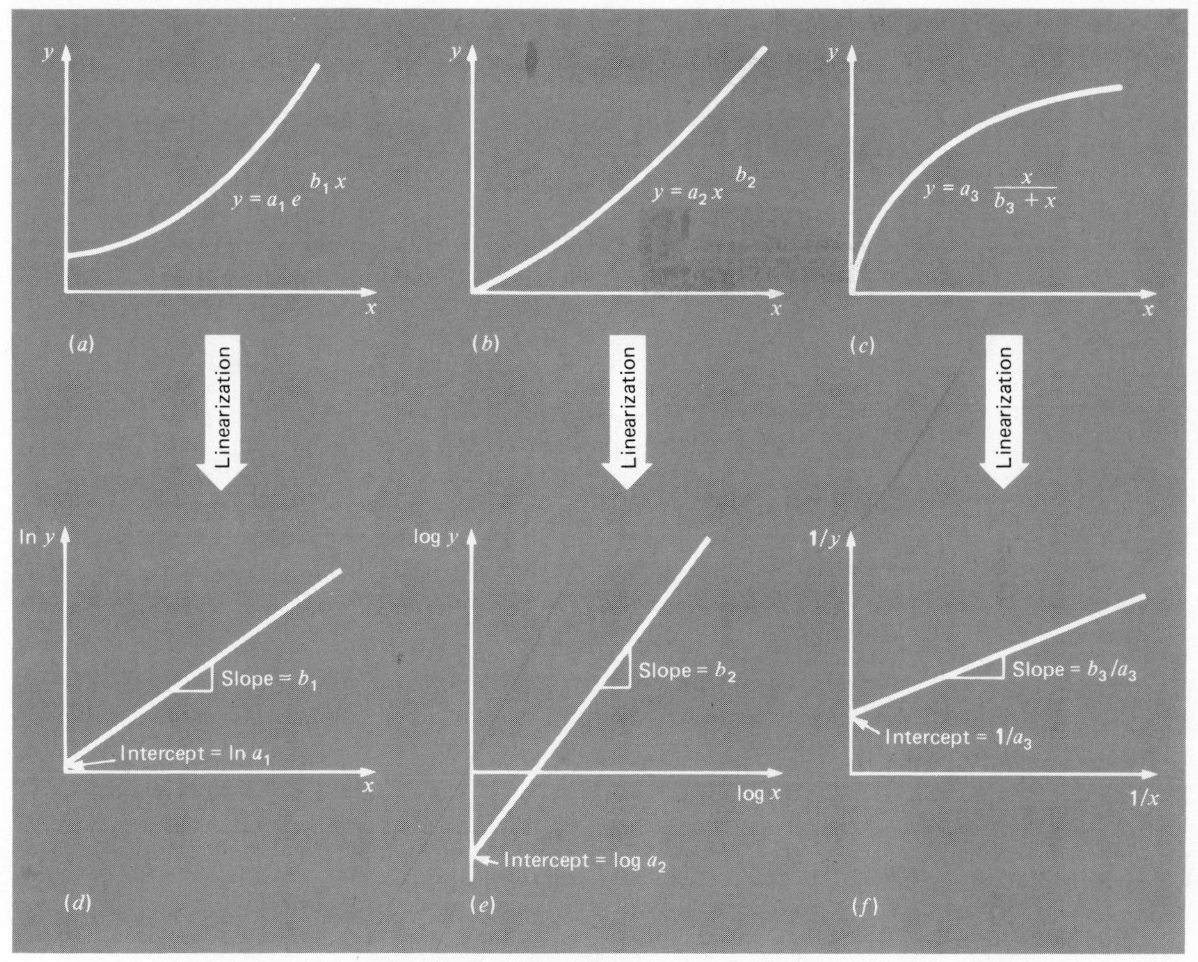


FIGURE 10.9 (*a*) The exponential equation, (*b*) the power equation, and (*c*) the saturation-growth-rate equation. Parts (*d*), (*e*), and (*f*) are linearized versions of these equations that result from simple transformations.

experimental data directly. However, a simpler alternative is to use mathematical manipulations to transform the equations into a linear form. Then simple linear regression can be employed to fit the equations to data.

For example, Eq. (10.11) can be linearized by taking its natural logarithm to yield

$$\ln y = \ln a_1 + b_1 x \ln e$$

But because $\ln e = 1$,

$$\ln y = \ln a_1 + b_1 x \qquad [10.14]$$

Thus a semilog plot of $\ln y$ versus $x$ will yield a straight line with a slope of $b_1$ and an intercept of $\ln a_1$ (Fig. 10.9*d*).

Equation (10.12) is linearized by taking its base 10 logarithm to give

$$\log y = b_2 \log x + \log a_2 \tag{10.15}$$

Thus, a log-log plot of log $y$ versus log $x$ will yield a straight line with a slope of $b_2$ and and intercept of log $a_2$ (Fig. 10.9*e*).

Equation (10.13) is linearized by inverting it to give

$$\frac{1}{y} = \frac{b_3}{a_3}\frac{1}{x} + \frac{1}{a_3} \tag{10.16}$$

Thus, a plot of $1/y$ versus $1/x$ will be linear, with a slope of $b_3/a_3$ and an intercept of $1/a_3$ (Fig. 10.9*f*).

In their transformed states, these models are fit using linear regression in order to evaluate the constant coefficients. They could then be transformed back to their original state and used for predictive purposes. Example 10.4 illustrates this procedure for Eq. (10.12). In addition, Case Studies 12.2 and 12.3 will provide engineering examples of the same sort of computation.

## EXAMPLE 10.4
### Linearization of a Power Equation

Problem Statement: Fit Eq. (10.12) to the data in Table 10.3 using a logarithmic transformation of the data.

**TABLE 10.3 Data to be fit to the power equation.**

| $x$ | $y$ | log $x$ | log $y$ |
|---|---|---|---|
| 1 | 0.5 | 0 | −0.301 |
| 2 | 1.7 | 0.301 | 0.226 |
| 3 | 3.4 | 0.477 | 0.534 |
| 4 | 5.7 | 0.602 | 0.753 |
| 5 | 8.4 | 0.699 | 0.922 |

Solution: Figure 10.10*a* is a plot of the original data in its untransformed state. Figure 10.10*b* shows the log-log plot of the transformed data. A linear regression of the log-transformed data yields the result

$$\log y = 1.75 \log x - 0.300$$

Thus, the intercept, log $a_2$, equals −0.300, and therefore, by taking the antilogarithm, $a_2 = 10^{-0.3} = 0.5$. The slope is $b_2 = 1.75$. Consequently, the power equation is

$$y = 0.5x^{1.75}$$

This curve, as plotted in Fig. 10.10*a*, indicates a good fit.

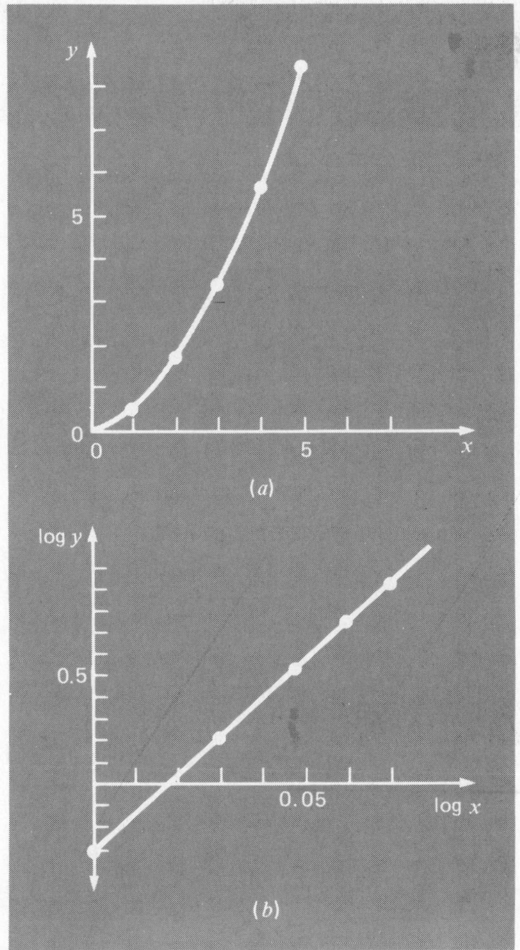


FIGURE 10.10 (*a*) Plot of untransformed data, along with the power equation that fits the data. (*b*) Plot of transformed data used to determine the coefficients of the power equation.

### 10.1.6 General Comments on Linear Regression

Before proceeding to curvilinear and multiple linear regression, we must emphasize the introductory nature of the foregoing material on linear regression. We have focused on the simple derivation and practical use of equations to fit data. You should be cognizant of the fact that there are theoretical aspects of regression that are of practical importance but are beyond the scope of this book. For example, some statistical assumptions that are inherent in the linear least-squares procedure are

1. $x$ has a fixed value; it is not random and is measured without error.

**2.** The $y$ values are independent random variables and all have the same variance.

**3.** The $y$ values for a given $x$ must be normally distributed.

Such assumptions are relevant to the proper derivation and use of regression. For example, the first assumption means that (1) the $x$'s must be error-free and (2) that the regression of $y$ versus $x$ is not the same as $x$ versus $y$ (try Prob. 10.4 at the end of the chapter).

You are urged to consult other references such as Draper and Smith (1981) in order to appreciate aspects and nuances of regression that are beyond the scope of this book.

## 10.2 POLYNOMIAL REGRESSION

In Sec. 10.1, a procedure was developed to derive the equation of a straight line using the least-squares criterion. Some engineering data, although exhibiting a marked pattern such as seen in Fig. 10.8, is poorly represented by a straight line. For these cases, a curve would be better suited to fit the data. As discussed in the previous section, one method to accomplish this objective is to use transformations. Another alternative is to fit polynomials to the data using *polynomial regression*.

The least-squares procedure can be readily extended to fit the data to an $m$th-degree polynomial:

$$y = a_0 + a_1 x + a_2 x^2 + \cdots + a_m x^m$$

For this case the sum of the squares of the residuals is [compare with Eq. (10.3)]

$$S_r = \sum_{i=1}^{n} (y_i - a_0 - a_1 x_i - a_2 x_i^2 - \cdots - a_m x_i^m)^2 \qquad [10.17]$$

Following the procedure of the previous section, we take the derivative of Eq. (10.17) with respect to each of the coefficients of the polynomial, as in

$$\frac{\partial S_r}{\partial a_0} = -2 \sum (y_i - a_0 - a_1 x_i - a_2 x_i^2 - \cdots - a_m x_i^m)$$

$$\frac{\partial S_r}{\partial a_1} = -2 \sum x_i (y_i - a_0 - a_1 x_i - a_2 x_i^2 - \cdots - a_m x_i^m)$$

$$\frac{\partial S_r}{\partial a_2} = -2 \sum x_i^2 (y_i - a_0 - a_1 x_i - a_2 x_i^2 - \cdots - a_m x_i^m)$$

$$\vdots \qquad\qquad \vdots$$

$$\frac{\partial S_r}{\partial a_m} = -2 \sum x_i^m (y_i - a_0 - a_1 x_i - a_2 x_i^2 - \cdots - a_m x_i^m)$$

These equations can be set equal to zero and rearranged to develop the following set of normal equations:

$$
\begin{aligned}
a_0 n + a_1 \sum x_i + a_2 \sum x_i^2 + \cdots + a_m \sum x_i^m &= \sum y_i \\
a_0 \sum x_i + a_1 \sum x_i^2 + a_2 \sum x_i^3 + \cdots + a_m \sum x_i^{m+1} &= \sum x_i y_i \\
a_0 \sum x_i^2 + a_1 \sum x_i^3 + a_2 \sum x_i^4 + \cdots + a_m \sum x_i^{m+2} &= \sum x_i^2 y_i \\
\vdots \qquad\qquad \vdots \qquad\qquad \vdots \qquad\qquad & \vdots \qquad \vdots \\
a_0 \sum x_i^m + a_1 \sum x_i^{m+1} + a_2 \sum x_i^{m+2} + \cdots + a_m \sum x_i^{2m} &= \sum x_i^m y_i
\end{aligned}
\qquad [10.18]
$$

where all summations are from $i = 1$ through $n$. Note, that the above $m + 1$ equations are linear and have $m + 1$ unknowns: $a_0, a_1, a_2, \ldots, a_m$. The coefficients of the unknowns can be calculated directly from the observed data. Thus, the problem of determining a least-squares polynomial of degree $m$ is equivalent to solving a system of $m + 1$ simultaneous linear equations. Techniques to solve such equations were discussed in Chaps. 7 and 8.

Just as for linear regression, the error of polynomial regression can be quantified by a standard error of the estimate:

$$
s_{y/x} = \sqrt{\frac{S_r}{n - (m + 1)}} \qquad [10.19]
$$

where $m$ is the order of the polynomial. This quantity is divided by $n - (m + 1)$ because $m + 1$ data-derived coefficients—$a_0, a_1, \ldots, a_m$—were used to compute $S_r$; thus, we have lost $m + 1$ degrees of freedom. In addition to the standard error, a correlation coefficient can also be computed for polynomial regression in the same manner as for the linear case:

$$
r^2 = \frac{S_t - S_r}{S_t}
$$

### EXAMPLE 10.5
Polynomial Regression

Problem Statement: Fit a second-order polynomial to the data in the first two columns of Table 10.4.

Solution: From the given data

$$
m = 2 \qquad \sum x_i = 15 \qquad \sum x_i^4 = 979
$$

$$
n = 6 \qquad \sum y_i = 152.6 \qquad \sum x_i y_i = 585.6
$$

$$\bar{x} = 2.5 \qquad \sum x_i^2 = 55 \qquad \sum x_i^2 y_i = 2488.8$$

$$\bar{y} = 25.433 \qquad \sum x_i^3 = 225$$

**TABLE 10.4 Computations for an error analysis of the quadratic least-squares fit.**

| $x_i$ | $y_i$ | $(y_i - \bar{y})^2$ | $(y_i - a_0 - a_1x_i - a_2x_i^2)^2$ |
|---|---|---|---|
| 0 | 2.1 | 544.44 | 0.14332 |
| 1 | 7.7 | 314.47 | 1.00286 |
| 2 | 13.6 | 140.03 | 1.08158 |
| 3 | 27.2 | 3.12 | 0.80491 |
| 4 | 40.9 | 239.22 | 0.61951 |
| 5 | 61.1 | 1272.11 | 0.09439 |
| Σ | 152.6 | 2513.39 | 3.74657 |

Therefore, the simultaneous linear equations are

$$6a_0 + 15a_1 + 55a_2 = 152.6$$

$$15a_0 + 55a_1 + 225a_2 = 585.6$$

$$55a_0 + 225a_1 + 979a_2 = 2488.8$$

Solving these equations through a technique such as Gauss elimination gives

$$a_0 = 2.47857$$

$$a_1 = 2.35929$$

$$a_2 = 1.86071$$

Therefore, the least-squares quadratic equation for this case is

$$y = 2.47857 + 2.35929x + 1.86071x^2$$

The standard error of the estimate based on the regression polynomial is [Eq. (10.19)]

$$s_{y/x} = \sqrt{\frac{3.74657}{6 - 3}} = 1.12$$

The coefficient of determination is

$$r^2 = \frac{2513.39 - 3.74657}{2513.39} = 0.99851$$

and the correlation coefficient is

$$r = 0.99925$$

These results indicate that 99.851 percent of the original uncertainty has

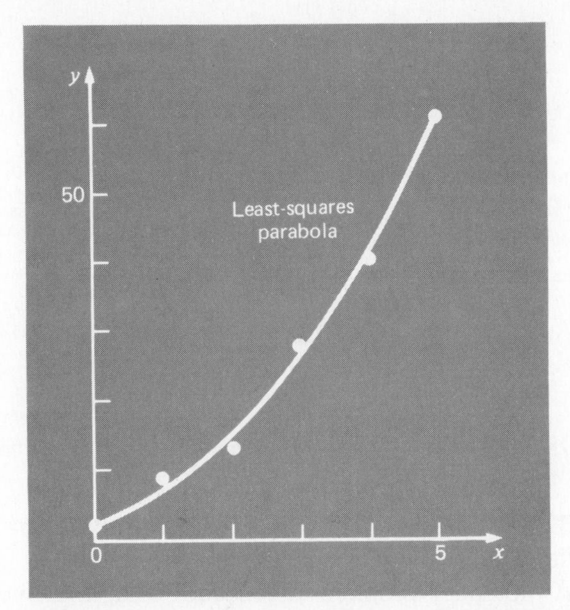


FIGURE 10.11 Fit of a second-order polynomial.

been explained by the model. This result supports the conclusion that the quadratic equation represents an excellent fit, as is also evident from Fig. 10.11.

### 10.2.1 Algorithm for Polynomial Regression

An algorithm for polynomial regression is delineated in Fig. 10.12. Note that the primary task is the generation of the coefficients of the normal equations [Eq. (10.18)]. (Subroutines for accomplishing this are presented in Fig.

**Step 1:** Input order of polynomial to be fit, $m$.

**Step 2:** Input number of data points, $n$.

**Step 3:** If $n \leq m$, print out an error message that polynomial regression is impossible and terminate the process. If $n > m$, continue.

**Step 4:** Compute sums of powers and products, as in Eq. (10.18).

**Step 5:** Set up these sums of powers and products in the form of an augmented matrix.

**Step 6:** Solve the augmented matrix for the coefficients $a_0, a_1, a_2, \ldots, a_m$, using an elimination method.

**Step 7:** Print out the coefficients.

FIGURE 10.12 Algorithm for implementation of polynomial regression.

FORTRAN

```
      SUBROUTINE POLREG(X,Y,A)
      DIMENSION X(15),Y(15),A(15,16)
      COMMON N,IO
      IP=IO+1
      DO 2100 I=1,IP
      DO 2060 J=1,IP
      K=I+J-2
      DO 2050 L=1,N
      A(I,J)=A(I,J)+X(L)**K
2050  CONTINUE
2060  CONTINUE
      DO 2090 L=1,N
      IR=IP+1
      A(I,IR)=A(I,IR)+Y(L)*X(L)**(I-1)
2090  CONTINUE
2100  CONTINUE
      RETURN
      END
```

BASIC

```
2000  FOR I = 1 TO IO + 1              IO = order of regression polynomial
2010  FOR J = 1 TO IO + 1
2020 K = I + J - 2
2030  FOR L = 1 TO N                   N = number of data points
2040 A(I,J) = A(I,J) + X(L) ^ K        (Determination of coefficients of normal equations and storage in matrix A)
2050  NEXT L
2060  NEXT J
2070  FOR L = 1 TO N
2080 A(I,IO + 2) = A(I,IO + 2) +       (Determination of right-hand-side constants for normal equations and storage in last column of matrix A)
     Y(L) * X(L) ^ (I - 1)
2090  NEXT L
2100  NEXT I
2110  RETURN
```

FIGURE 10.13 FORTRAN and BASIC subroutines to set up the normal equations for polynomial regression in matrix form.

10.13.) Then, techniques from Chaps. 7 or 8 can be applied to solve these simultaneous equations for the coefficients.

A potential problem associated with implementing polynomial regression on the computer is that the normal equations are sometimes ill-conditioned. This is particularly true for higher-order versions. For these cases, the computed coefficients may be highly susceptible to round-off error, and consequently the results can be inaccurate. Among other things, this problem is related to the fact that for higher-order polynomials the normal equations can have very large and very small coefficients. This is because the coefficients are summations of the data raised to powers.

Although some of the strategies for mitigating round-off error discussed in Chap. 7, such as pivoting and error equations, can help to partially remedy this problem, a simpler alternative is to use a computer with higher precision. This is one area where some personal computers may presently represent a limitation on the effective implementation of this particular numerical method. Fortunately, most practical problems are limited to lower-order polynomials for which round off is usually negligible. In situations where higher-order versions are required, other alternatives are available for certain types of data. However, these techniques (such as orthogonal polynomials) are beyond the scope of this book. The reader should consult texts on regression such as Draper and Smith (1981) for additional information regarding the problem and possible alternatives.

## 10.3 MULTIPLE LINEAR REGRESSION

A useful extension of linear regression is the case where $y$ is a linear function of two or more variables. For example, $y$ might be a linear function of $x_1$ and $x_2$, as in

$$y = a_0 + a_1x_1 + a_2x_2$$

Such an equation is particularly useful when fitting experimental data where the variable being studied is often a function of two other variables. For this two-dimensional case, the regression "line" becomes a "plane" (Fig. 10.14).

As with the previous cases, the "best" values of the coefficients are determined by setting up the sum of the squares of the residuals:

$$S_r = \sum_{i=1}^{n} (y_i - a_0 - a_1 x_{1,i} - a_2 x_{2,i})^2$$

and differentiating with respect to each of the coefficients:

$$\frac{\partial S_r}{\partial a_0} = -2 \sum (y_i - a_0 - a_1 x_{1,i} - a_2 x_{2,i})$$

$$\frac{\partial S_r}{\partial a_1} = -2 \sum x_{1,i}(y_i - a_0 - a_1 x_{1,i} - a_2 x_{2,i}) \qquad [10.20]$$

$$\frac{\partial S_r}{\partial a_2} = -2 \sum x_{2,i}(y_i - a_0 - a_1 x_{1,i} - a_2 x_{2,i})$$

The coefficients yielding the minimum sum of the squares of the residuals are obtained by setting the partial derivatives equal to zero and expressing Eq. (10.20) as a set of simultaneous linear equations:

$$na_0 + \sum x_{1,i} a_1 + \sum x_{2,i} a_2 = \sum y_i$$

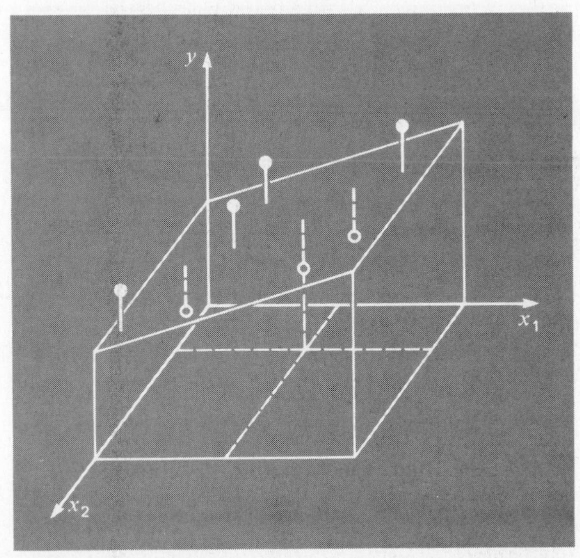


FIGURE 10.14 Graphical depiction of multiple linear regression where $y$ is a linear function of $x_1$ and $x_2$.

$$\sum x_{1,i}a_0 + \sum x_{1,i}^2 a_1 + \sum x_{1,i}x_{2,i}a_2 = \sum x_{1,i}y_i$$

$$\sum x_{2,i}a_0 + \sum x_{1,i}x_{2,i}a_1 + \sum x_{2,i}^2 a_2 = \sum x_{2,i}y_i$$

or as a matrix:

$$\begin{bmatrix} n & \Sigma x_{1,i} & \Sigma x_{2,i} \\ \Sigma x_{1,i} & \Sigma x_{1,i}^2 & \Sigma x_{1,i}x_{2,i} \\ \Sigma x_{2,i} & \Sigma x_{1,i}x_{2,i} & \Sigma x_{2,i}^2 \end{bmatrix} \begin{bmatrix} a_0 \\ a_1 \\ a_2 \end{bmatrix} = \begin{bmatrix} \Sigma y_i \\ \Sigma x_{1,i}y_i \\ \Sigma x_{2,i}y_i \end{bmatrix} \tag{10.21}$$

## EXAMPLE 10.6
### Multiple Linear Regression

Problem Statement: The following data was calculated from the equation $y = 5 + 4x_1 - 3x_2$.

| $x_1$ | $x_2$ | $y$ |
|---|---|---|
| 0 | 0 | 5 |
| 2 | 1 | 10 |
| 2.5 | 2 | 9 |
| 1 | 3 | 0 |
| 4 | 6 | 3 |
| 7 | 2 | 27 |

Use multiple linear regression to fit this data.

**TABLE 10.5 Computations required to develop the normal equations for Example 10.6**

| | $y$ | $x_1$ | $x_2$ | $x_1^2$ | $x_2^2$ | $x_1 x_2$ | $x_1 y$ | $x_2 y$ |
|---|---|---|---|---|---|---|---|---|
| | 5 | 0 | 0 | 0 | 0 | 0 | 0 | 0 |
| | 10 | 2 | 1 | 4 | 1 | 2 | 20 | 10 |
| | 9 | 2.5 | 2 | 6.25 | 4 | 5 | 22.5 | 18 |
| | 0 | 1 | 3 | 1 | 9 | 3 | 0 | 0 |
| | 3 | 4 | 6 | 16 | 36 | 24 | 12 | 18 |
| | 27 | 7 | 2 | 49 | 4 | 14 | 189 | 54 |
| Σ | 54 | 16.5 | 14 | 76.25 | 54 | 48 | 243.5 | 101 |

Solution: The summations required to develop Eq. (10.21) are computed in Table 10.5. They can be substituted into Eq. (10.21) to yield

$$\begin{bmatrix} 6 & 16.5 & 14 \\ 16.5 & 76.25 & 48 \\ 14 & 48 & 54 \end{bmatrix} \begin{bmatrix} a_0 \\ a_1 \\ a_2 \end{bmatrix} = \begin{bmatrix} 54 \\ 243.5 \\ 101 \end{bmatrix}$$

which can be solved using a method such as Gauss elimination for

$$a_0 = 5 \qquad a_1 = 4 \qquad a_2 = -3$$

which is consistent with the original equation from which the data was derived.

Multiple linear regression can be formulated for the more general case,

$$y = a_0 + a_1 x_1 + a_2 x_2 + \cdots + a_m x_m$$

where the coefficients that minimize the sum of the squares of the residuals are determined by solving

$$\begin{bmatrix} n & \sum x_{1,i} & \sum x_{2,i} & \cdots & \sum x_{m,i} \\ \sum x_{1,i} & \sum x_{1,i}^2 & \sum x_{2,i}x_{1,i} & \cdots & \sum x_{1,i}x_{m,i} \\ \sum x_{2,i} & \sum x_{2,i}x_{1,i} & \sum x_{2,i}^2 & \cdots & \sum x_{2,i}x_{m,i} \\ \vdots & \vdots & \vdots & & \vdots \\ \sum x_{m,i} & \sum x_{m,i}x_{1,i} & \sum x_{m,i}x_{2,i} & \cdots & \sum x_{m,i}^2 \end{bmatrix} \begin{bmatrix} a_0 \\ a_1 \\ a_2 \\ \vdots \\ a_m \end{bmatrix} = \begin{bmatrix} \sum y_i \\ \sum x_{1,i}y_i \\ \sum x_{2,i}y_i \\ \vdots \\ \sum x_{m,i}y_i \end{bmatrix} \tag{10.22}$$

The standard error of the estimate for multiple linear regression is formulated as

$$s_{y/x_1,x_2,\ldots,x_m} = \sqrt{\frac{S_r}{n-(m+1)}}$$

and the correlation coefficient computed as in Eq. (10.10).

Although there may be certain cases where a variable is linearly related to two or more other variables, multiple linear regression has additional utility in the derivation of power equations of the general form,

$$y = a_0 x_1^{a_1} x_2^{a_2} \cdots x_m^{a_m}$$

Such equations are extremely useful when fitting experimental data. In order to use multiple linear regression, the equation is transformed by taking its logarithm to yield

$$\log y = \log a_0 + a_1 \log x_1 + a_2 \log x_2 + \cdots + a_m \log x_m$$

This transformation is similar in spirit to the one used in Sec. 10.1.3 and Example 10.4 to fit a power equation when $y$ was a function of a single variable $x$. Case Study 12.5 provides an example of such an application.

# PROBLEMS

## Hand Calculations

**10.1** Given the data

| | | | | |
|---|---|---|---|---|
| 0.95 | 1.42 | 1.54 | 1.55 | 1.63 |
| 1.32 | 1.15 | 1.47 | 1.95 | 1.25 |
| 1.46 | 1.47 | 1.92 | 1.35 | 1.05 |

| | | | | |
|---|---|---|---|---|
| 1.85 | 1.74 | 1.65 | 1.78 | 1.71 |
| 2.39 | 1.82 | 2.06 | 2.14 | 2.27 |

determine (*a*) the mean, (*b*) the standard deviation, (*c*) the variance, and (*d*) the coefficient of variation.

**10.2** Construct a histogram for the data in Prob. 10.1. Use a range of 0.6 to 2.4 with intervals of 0.2.

**10.3** Given the data

| | | | | | | |
|---|---|---|---|---|---|---|
| 52 | 6 | 18 | 21 | 26 | 28 | 32 |
| 39 | 22 | 28 | 24 | 27 | 27 | 33 |
| 2 | 12 | 17 | 34 | 29 | 31 | 34 |
| 43 | 36 | 41 | 37 | 43 | 38 | 46 |

determine (*a*) the mean, (*b*) the standard deviation, (*c*) the variance, and (*d*) the coefficient of variation.
(*e*) Construct a histogram. Use a range from 0 to 55 with increments of 5.
(*f*) Assuming that the distribution is normal and that your estimate of the standard deviation is valid, compute the range (that is, the lower and the upper value) that encompasses 68 percent of the readings. Determine whether this is a valid estimate for the data in this problem.

**10.4** Use least-squares regression to fit a straight line to

| $x$ | 1 | 3 | 5 | 7 | 10 | 12 | 13 | 16 | 18 | 20 |
|---|---|---|---|---|---|---|---|---|---|---|
| $y$ | 3 | 2 | 6 | 5 | 8 | 7 | 10 | 9 | 12 | 10 |

Along with the slope and intercept, compute the standard error of the estimate and the correlation coefficient. Plot the data and the regression line. Then repeat the problem, but regress $x$ versus $y$—that is, switch the variables. Interpret your results.

**10.5** Use least-squares regression to fit a straight line to

| $x$ | 4 | 6 | 8 | 10 | 14 | 16 | 20 | 22 | 24 | 28 | 28 | 34 | 36 | 38 |
|---|---|---|---|---|---|---|---|---|---|---|---|---|---|---|
| $y$ | 30 | 18 | 22 | 28 | 14 | 22 | 16 | 8 | 20 | 8 | 14 | 14 | 0 | 8 |

Along with the slope and the intercept, compute the standard error of the estimate and the correlation coefficient. Plot the data and the regression line. If someone made an additional measurement of $x = 30$, $y = 30$, would you suspect, based on a visual assessment and the standard error, that the measurement was valid or faulty? Justify your conclusion.

**10.6** Use least-squares regression to fit a straight line to

| $x$ | 0 | 2 | 4 | 4 | 8 | 12 | 16 | 20 | 24 | 28 | 30 | 34 |
|---|---|---|---|---|---|---|---|---|---|---|---|---|
| $y$ | 10 | 12 | 18 | 22 | 20 | 30 | 26 | 30 | 26 | 28 | 22 | 20 |

(*a*) Along with the slope and intercept, compute the standard error of the estimate and the correlation coefficient. Plot the data and the straight line. Assess the fit.
(*b*) Recompute (*a*), but use polynomial regression to fit a parabola to the data. Compare the results with those of (*a*).

**10.7** Fit a saturation-growth-rate model to

| $x$ | 1 | 2 | 2.5 | 4 | 6 | 8 | 8.5 |
|---|---|---|---|---|---|---|---|
| $y$ | 0.4 | 0.7 | 0.8 | 1.0 | 1.2 | 1.3 | 1.4 |

Plot the data and the equation.

**10.8** Fit a power equation to the data from Prob. 10.7. Plot the data and the equation.

**10.9** Fit a parabola to the data from Prob. 10.7. Plot the data and the equation.

**10.10** Fit a power equation to

| $x$ | 2.5 | 3.5 | 5 | 6 | 7.5 | 10 | 12.5 | 15 | 17.5 | 20 |
|---|---|---|---|---|---|---|---|---|---|---|
| $y$ | 5 | 3.4 | 2 | 1.6 | 1.2 | 0.8 | 0.6 | 0.4 | 0.3 | 0.3 |

Plot $y$ versus $x$ along with the power equation.

**10.11** Fit an exponential model to

| $x$ | 0.05 | 0.4 | 0.8 | 1.2 | 1.6 | 2.0 | 2.4 |
|---|---|---|---|---|---|---|---|
| $y$ | 550 | 750 | 1000 | 1400 | 2000 | 2700 | 3750 |

Plot the data and the equation on both standard and semi-logarithmic graph paper. Discuss your results.

**10.12** Fit a power equation to the data in Prob. 10.11. Plot the data and the equation.

**10.13** Fit a parabola to the data in Prob. 10.11. Plot the data and the equation.

**10.14** Given the data

| $x$ | 5 | 10 | 15 | 20 | 25 | 30 | 35 | 40 | 45 | 50 |
|---|---|---|---|---|---|---|---|---|---|---|
| $y$ | 17 | 25 | 30 | 33 | 36 | 38 | 39 | 40 | 41 | 42 |

use least-squares regression to fit (*a*) a straight line, (*b*) a power equation, (*c*) a saturation-growth-rate equation, and (*d*) a parabola. Plot the data along with all the curves. Is any one of the curves superior? If so, justify.

**10.15** Fit a parabola to

| $x$ | 0 | 2 | 4 | 6 | 9 | 11 | 13 | 15 | 17 | 19 | 23 | 25 | 28 |
|---|---|---|---|---|---|---|---|---|---|---|---|---|---|
| $y$ | 1.2 | 0.6 | 0.4 | −0.2 | 0 | −0.6 | −0.4 | −0.2 | −0.4 | 0.2 | 0.4 | 1.2 | 1.8 |

Compute the coefficients, the standard error of the estimate, and the correlation coefficient. Plot the results and assess the fit.

**10.16** Use multiple linear regression to fit

| $x_1$ | 0 | 1 | 2 | 0 | 1 | 2 |
|---|---|---|---|---|---|---|
| $x_2$ | 2 | 2 | 4 | 4 | 6 | 6 |
| $y$ | 19 | 12 | 11 | 24 | 22 | 15 |

Compute the coefficients, the standard error of the estimate, and the correlation coefficient.

**10.17** Use multiple linear regression to fit

| $x_1$ | 1 | 1 | 2 | 2 | 3 | 3 | 4 | 4 |
|---|---|---|---|---|---|---|---|---|
| $x_2$ | 1 | 2 | 1 | 2 | 1 | 2 | 1 | 2 |
| $y$ | 18 | 12.8 | 25.7 | 20.6 | 35.0 | 29.8 | 45.5 | 40.3 |

Compute the coefficients, the standard error of the estimate, and the correlation coefficient.

## Computer-Related Problems

**10.18** Develop a user-friendly computer program for linear regression based on Fig. 10.6. Among other things:
(*a*) Add statements to document the code.
(*b*) Make the input and output more descriptive and user-oriented.
(*c*) Compute and print out the standard error of the estimate [Eq. (10.9)] and the correlation coefficient [the square root of Eq. (10.10)].
(*d*) (*Optional*) Include a computer plot of the data and the regression line.
(*e*) (*Optional*) Include an option that allows exponential, power, and saturation-growth-rate equations to be analyzed.

**10.19** Develop a user-friendly computer program for polynomial regression based on Figs. 10.12 and 10.13. Test the program by duplicating the computations from Example 10.5.

**10.20** Develop a user-friendly computer program for multiple regression based on Fig. 10.12, but with the matrix as specified by Eq. (10.22). Test the program by duplicating the computations from Example 10.6.

**10.21** Repeat Probs. 10.4 and 10.5 using the program from Prob. 10.18.

**10.22** Use the NUMERICOMP software to solve Probs. 10.4, 10.5, and 10.6a.

**10.23** Repeat Probs. 10.9, 10.13, and 10.15 using the program from Prob. 10.19.

**10.24** Repeat Probs. 10.16 and 10.17 using the program from Prob. 10.20.

# CHAPTER ELEVEN

# INTERPOLATION

You will frequently have occasion to estimate intermediate values between precise data points. The most common method used for this purpose is polynomial interpolation.

Recall that the general formula for an $n$th-order polynomial is

$$f(x) = a_0 + a_1x + a_2x^2 + \cdots + a_nx^n \qquad [11.1]$$

For $n + 1$ data points, there is one and only one polynomial of order $n$ or less that passes through all the points. For example, there is only one straight line (that is, a first-order polynomial) that connects two points (Fig. 11.1$a$). Similarly, only one parabola connects a set of three points (Fig. 11.1$b$). *Polynomial interpolation* consists of determining the unique $n$th-order polynomial that fits $n + 1$ data points. This polynomial then provides a formula to compute intermediate values.

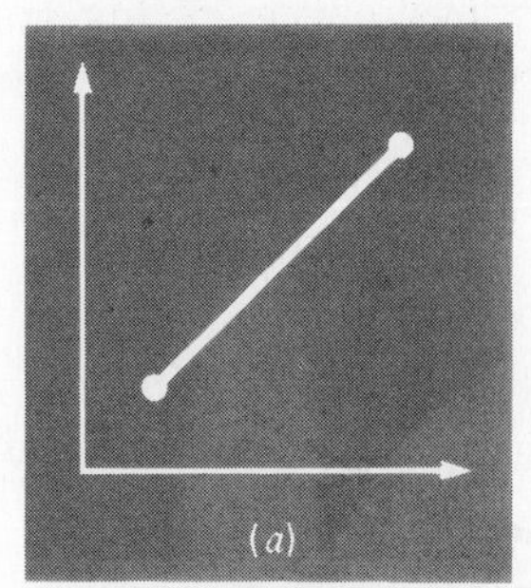

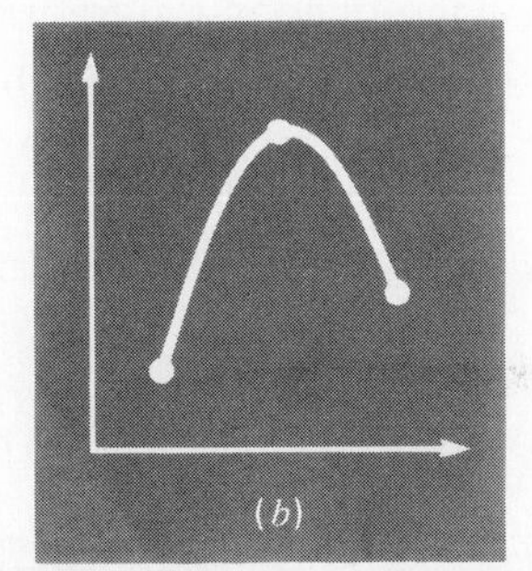

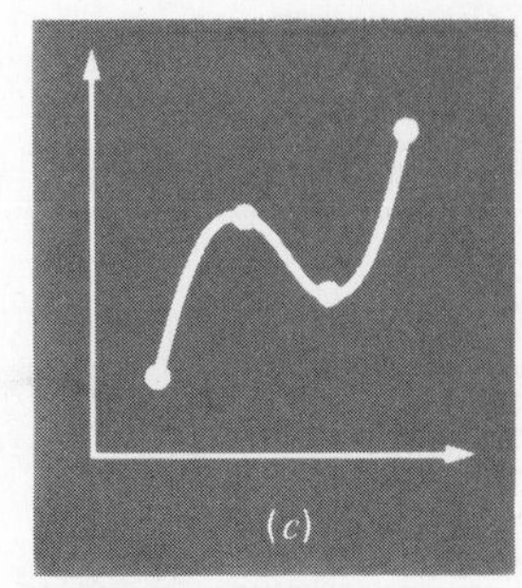


FIGURE 11.1 Examples of interpolating polynomials: ($a$) first-order (linear) connecting two points; ($b$) second-order (quadratic or parabolic) connecting three points; and ($c$) third-order (cubic) connecting four points.

Although there is one and only one $n$th-order polynomial that fits $n + 1$ points, there are a variety of mathematical formats in which this polynomial can be expressed. In the present chapter, we will describe two alternatives that are well-suited for implementation on personal computers. These are the Newton and the Lagrange polynomials.

## 11.1 NEWTON'S DIVIDED-DIFFERENCE INTERPOLATING POLYNOMIALS

As stated above, there are a variety of alternative forms for expressing an interpolating polynomial. *Newton's divided-difference interpolating polynomial* is among the most popular and useful forms. Before presenting the general equation, we will introduce the first- and second-order versions because of their simple visual interpretation.

### 11.1.1 Linear Interpolation

The simplest form of interpolation is to connect two data points with a straight line. This technique, called *linear interpolation*, is depicted graphically in Fig. 11.2. Using similar triangles,

$$\frac{f_1(x) - f(x_0)}{x - x_0} = \frac{f(x_1) - f(x_0)}{x_1 - x_0}$$

which can be rearranged to yield

$$f_1(x) = f(x_0) + \frac{f(x_1) - f(x_0)}{x_1 - x_0}(x - x_0) \qquad [11.2]$$

which is a *linear-interpolation formula*. The notation $f_1(x)$ designates that this is a first-order interpolating polynomial. Notice that besides representing the slope of the line connecting the points, the term $[f(x_1) - f(x_0)]/(x_1 - x_0)$ is a finite-divided-difference approximation of the first derivative [recall Eq. (3.24)]. In general, the smaller the interval between the data points, the better

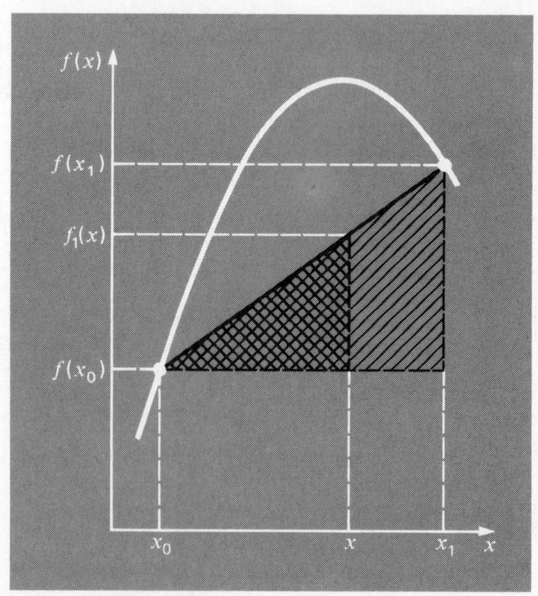


FIGURE 11.2 Graphical depiction of linear interpolation. The shaded areas indicate the similar triangles that are used to derive the linear-interpolation formula [Eq. (11.2)].

the approximation. This characteristic is demonstrated in the following example.

### EXAMPLE 11.1
Linear Interpolation

Problem Statement: Estimate the natural logarithm of 2 (ln 2) using linear interpolation. First, perform the computation by interpolating between ln 1 = 0 and ln 6 = 1.7917595. Then, repeat the procedure, but use a smaller interval from ln 1 to ln 4 (1.3862944). Note that the true value of ln 2 is 0.69314718.

Solution: Using Eq. (11.2), a linear interpolation from $x_0 = 1$ to $x_1 = 6$ gives

$$f_1(2) = 0 + \frac{1.7917595 - 0}{6 - 1}(2 - 1) = 0.35835190$$

which represents a percent error of $\epsilon_t = 48.3$ percent. Using the smaller interval from $x_0 = 1$ to $x_1 = 4$ yields

$$f_1(2) = 0 + \frac{1.3862944 - 0}{4 - 1}(2 - 1) = 0.46209813$$

Thus, using the shorter interval reduces the percent relative error to $\epsilon_t = 33.3$ percent. Both interpolations are shown in Fig. 11.3, along with the true function.

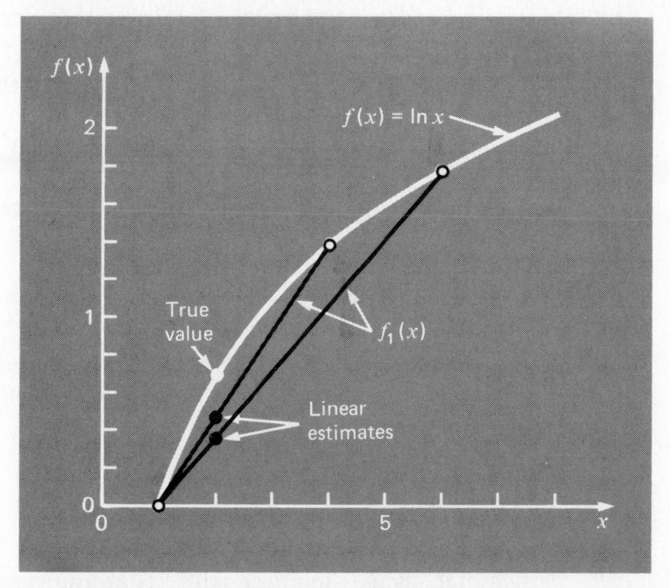


FIGURE 11.3 Two linear interpolations to estimate ln 2. Note how the smaller interval provides a better estimate.

### 11.1.2 Quadratic Interpolation

The error in Example 11.1 was due to the fact that we approximated a curve with a straight line. Consequently, a strategy for improving the estimate is to introduce some curvature into the line connecting the points. If three data points are available, this can be accomplished with a second-order polynomial (also called a quadratic polynomial or a parabola). A particularly convenient form for this purpose is

$$f_2(x) = b_0 + b_1(x - x_0) + b_2(x - x_0)(x - x_1) \qquad [11.3]$$

Note that although Eq. (11.3) might seem to differ from the general polynomial [Eq. (11.1)], the two equations are equivalent. This can be shown by multiplying the terms in Eq. (11.3) to yield

$$f_2(x) = b_0 + b_1x - b_1x_0 + b_2x^2 + b_2x_0x_1 - b_2xx_0 - b_2xx_1$$

or, collecting terms,

$$f_2(x) = a_0 + a_1x + a_2x^2$$

where

$$a_0 = b_0 - b_1x_0 + b_2x_0x_1$$
$$a_1 = b_1 - b_2x_0 - b_2x_1$$
$$a_2 = b_2$$

Thus, Eqs. (11.1) and (11.3) are alternative, equivalent formulations of the unique second-order polynomial joining the three points.

A simple procedure can be used to determine the values of the coefficients. For $b_0$, Eq. (11.3) with $x = x_0$ can be used to compute

$$b_0 = f(x_0) \qquad [11.4]$$

Equation (11.4) can be substituted into Eq. (11.3), which can be evaluated at $x = x_1$ for

$$b_1 = \frac{f(x_1) - f(x_0)}{x_1 - x_0} \qquad [11.5]$$

Finally, Eqs. (11.4) and (11.5) can be substituted into Eq. (11.3), which can be evaluated at $x = x_2$ and solved (after some algebraic manipulations) for

$$b_2 = \frac{\dfrac{f(x_2) - f(x_1)}{x_2 - x_1} - \dfrac{f(x_1) - f(x_0)}{x_1 - x_0}}{x_2 - x_0} \qquad [11.6]$$

Notice that, as was the case with linear interpolation, $b_1$ still represents the slope of the line connecting points $x_0$ and $x_1$. Thus, the first two terms of Eq. (11.3) are equivalent to linear interpolation from $x_0$ to $x_1$, as specified previously in Eq. (11.2). The last term, $b_2(x - x_0)(x - x_1)$, introduces the second-order curvature into the formula.

Before illustrating how to use Eq. (11.3), we should examine the form of the coefficient $b_2$. It is very similar to the finite-divided-difference approximation of the second derivative introduced previously in Eq. (3.31). Thus, Eq. (11.3) is beginning to manifest a structure that is very similar to the Taylor series expansion. This observation will be explored further when we relate Newton's interpolating polynomials to the Taylor series in Sec. 11.1.4. But first, we will show how Eq. (11.3) is used to interpolate between three points.

## EXAMPLE 11.2

### Quadratic Interpolation

Problem Statement: Fit a second-order polynomial to the three points used in Example 11.1:

$$x_0 = 1 \qquad f(x_0) = 0$$
$$x_1 = 4 \qquad f(x_1) = 1.3862944$$
$$x_2 = 6 \qquad f(x_2) = 1.7917595$$

Use the polynomial to evaluate ln 2.

Solution: Applying Eq. (11.4) yields

$$b_0 = 0$$

Equation (11.5) yields

$$b_1 = \frac{1.3862944 - 0}{4 - 1} = 0.46209813$$

and Eq. (11.6) gives

$$b_2 = \frac{\dfrac{1.7917595 - 1.3862944}{6 - 4} - 0.46209813}{6 - 1} = -0.051873116$$

Substituting these values into Eq. (11.3) yields the quadratic formula

$$f_2(x) = 0 + 0.46209813(x - 1) - 0.051873116(x - 1)(x - 4)$$

which can be evaluated at $x = 2$ for

$$f_2(2) = 0.56584436$$

which represents a percent relative error of $\epsilon_t = 18.4$ percent. Thus, the curvature introduced by the quadratic formula (Fig. 11.4) improves the interpolation compared with the result obtained using straight lines in Example 11.1 and Fig. 11.3.

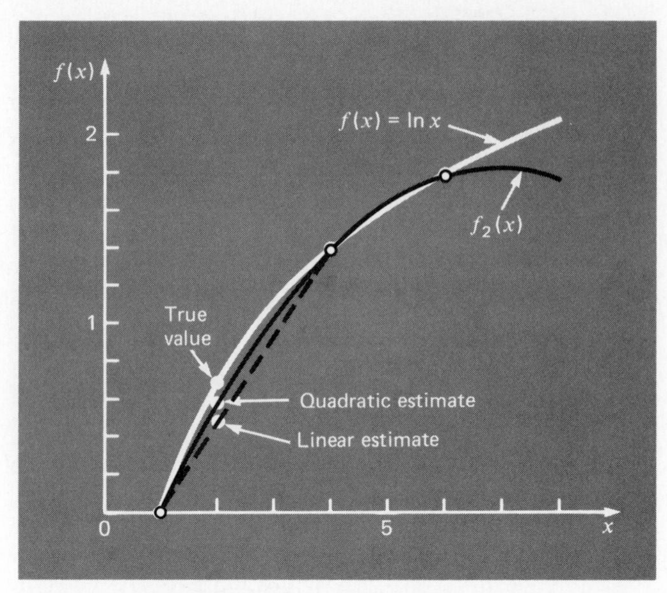


FIGURE 11.4 The use of quadratic interpolation to estimate ln 2. The linear interpolation from $x = 1$ to 4 is also included for comparison.

### 11.1.3 General Form of Newton's Interpolating Polynomials

The preceding analysis can be generalized to fit an $n$th-order polynomial to $n + 1$ data points. The $n$th-order polynomial is

$$f_n(x) = b_0 + b_1(x - x_0) + \cdots + b_n(x - x_0)(x - x_1) \ldots (x - x_{n-1}) \quad [11.7]$$

As was done previously with the linear and quadratic interpolations, data points can be used to evaluate the coefficients $b_0, b_1, \ldots, b_n$. For an $n$th-order polynomial, $n + 1$ data points are required: $x_0, x_1, x_2, \ldots, x_n$. Using these data points, the following equations are used to evaluate the coefficients:

$$b_0 = f(x_0) \quad [11.8]$$

$$b_1 = f[x_1, x_0] \quad [11.9]$$

$$b_2 = f[x_2, x_1, x_0] \quad [11.10]$$

$$\vdots$$

$$b_n = f[x_n, x_{n-1}, \ldots, x_1, x_0] \quad [11.11]$$

where the bracketed function evaluations are finite divided differences. For example, the *first finite divided difference* is represented generally as

$$f[x_i, x_j] = \frac{f(x_i) - f(x_j)}{x_i - x_j} \quad [11.12]$$

The *second finite divided difference,* which represents the difference of two first divided differences, is expressed generally as

$$f[x_i, x_j, x_k] = \frac{f[x_i, x_j] - f[x_j, x_k]}{x_i - x_k} \qquad [11.13]$$

Similarly, the *nth finite divided difference* is

$$f[x_n, x_{n-1}, \ldots, x_1, x_0] = \frac{f[x_n, x_{n-1}, \ldots, x_1] - f[x_{n-1}, x_{n-2}, \ldots, x_0]}{x_n - x_0} \qquad [11.14]$$

These differences can be used to evaluate the coefficients in Eqs. (11.8) through (11.11), which can then be substituted into Eq. (11.7) to yield the interpolating polynomial,

$$f_n(x) = f(x_0) + (x - x_0)f[x_1, x_0] + (x - x_0)(x - x_1)\, f[x_2, x_1, x_0] + \cdots + (x - x_0)(x - x_1) \ldots (x - x_{n-1})f[x_n, x_{n-1}, \ldots, x_0] \qquad [11.15]$$

which is called *Newton's divided-difference interpolating polynomial.* It should be noted that it is not necessary that the data points used in Eq. (11.15) be equally spaced or that the abscissa values necessarily be in ascending order, as illustrated in the following example. Also, notice how Eqs. (11.12) through (11.14) are recursive—that is, higher-order differences are composed of lower-order differences (Fig. 11.5). This property will be ex-

| *i* | $x_i$ | $f(x_i)$ | first | second | third |
|---|---|---|---|---|---|
| 0 | $x_0$ | $f(x_0)$ | $f[x_1, x_0]$ | $f[x_2, x_1, x_0]$ | $f[x_3, x_2, x_1, x_0]$ |
| 1 | $x_1$ | $f(x_1)$ | $f[x_2, x_1]$ | $f[x_3, x_2, x_1]$ | |
| 2 | $x_2$ | $f(x_2)$ | $f[x_3, x_2]$ | | |
| 3 | $x_3$ | $f(x_3)$ | | | |

FIGURE 11.5 Graphical depiction of the recursive nature of finite divided differences.

ploited when we develop an efficient computer program in Sec. 11.1.5 to implement the method.

## EXAMPLE 11.3
### Newton's Divided-Difference Interpolating Polynomials

Problem Statement: In Example 11.2, data points at $x_0 = 1$, $x_1 = 4$, and $x_2 = 6$ were used to estimate ln 2 with a parabola. Now, adding a fourth point [$x_3 = 5$; $f(x_3) = 1.6094379$], estimate ln 2 with a third-order Newton's divided-difference interpolating polynomial.

Solution: The third-order polynomial, Eq. (11.3) with $n = 3$, is

$$f_3(x) = b_0 + b_1(x - x_0) + b_2(x - x_0)(x - x_1) + b_3(x - x_0)(x - x_1)(x - x_2)$$

The first divided differences for the problem are [Eq. (11.12)]

$$f[x_1, x_0] = \frac{1.3862944 - 0}{4 - 1} = 0.46209813$$

$$f[x_2, x_1] = \frac{1.7917595 - 1.3862944}{6 - 4} = 0.20273255$$

$$f[x_3, x_2] = \frac{1.6094379 - 1.7917595}{5 - 6} = 0.18232160$$

The second divided differences are [Eq. (11.13)]

$$f[x_2, x_1, x_0] = \frac{0.20273255 - 0.46209813}{6 - 1} = -0.051873116$$

$$f[x_3, x_2, x_1] = \frac{0.18232160 - 0.20273255}{5 - 4} = -0.020410950$$

The third divided difference is [Eq. (11.14) with $n = 3$]

$$f[x_3, x_2, x_1, x_0] = \frac{-0.020410950 - (-0.051873116)}{5 - 1} = 0.0078655415$$

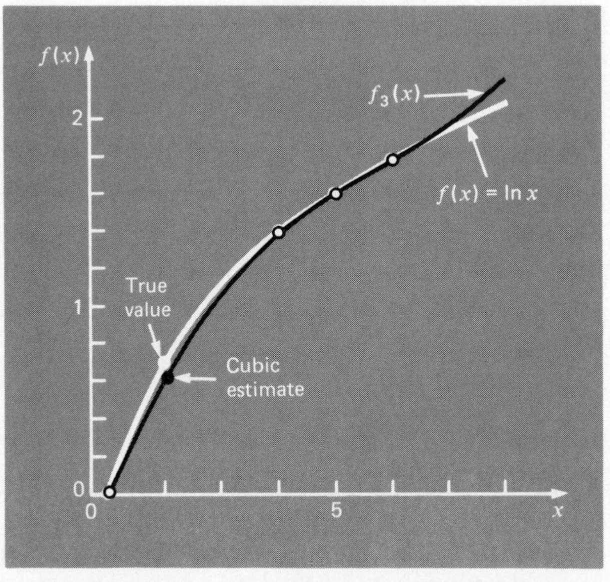


FIGURE 11.6 The use of cubic interpolation to estimate ln 2.

The results for $f[x_1, x_0]$, $f[x_2, x_1, x_0]$, and $f[x_3, x_2, x_1, x_0]$ represent the coefficients $b_1$, $b_2$, and $b_3$ of Eq. (11.7). Along with $b_0 = f(x_0) = 0.0$, Eq. (11.7) is

$$f_3(x) = 0 + 0.46209813\ (x - 1) - 0.051873116\ (x - 1)(x - 4) + 0.0078655415(x - 1)(x - 4)(x - 6)$$

which can be used to evaluate

$$f_3(2) = 0.62876869$$

which represents a percent relative error of $\epsilon_t = 9.3$ percent. The complete cubic polynomial is shown in Fig. 11.6.

### 11.1.4 Errors of Newton's Interpolating Polynomials

Notice that the structure of Eq. (11.15) is similar to the Taylor series expansion in the sense that terms are added sequentially in order to capture the higher-order behavior of the underlying function. These terms are finite divided differences and, thus, represent approximations of the higher-order derivatives. Consequently, as with the Taylor series, if the true underlying function is an $n$th-order polynomial, the $n$th-order interpolating polynomial based on $n + 1$ data points will yield exact results.

Also, as was the case with the Taylor series, a formulation for the truncation error can be obtained. Recall from Eq. (3.13) that the truncation error for the Taylor series could be expressed generally as

$$R_n = \frac{f^{(n+1)}(\xi)}{(n + 1)\,!}\,(x_{i+1} - x_i)^{n+1}$$

where $\xi$ is somewhere in the interval from $x_i$ to $x_{i+1}$. For an $n$th-order interpolating polynomial, an analogous relationship for the error is

$$R_n = \frac{f^{(n+1)}(\xi)}{(n + 1)\,!}\,(x - x_0)(x - x_1) \ldots (x - x_n) \tag{11.16}$$

where $\xi$ is somewhere in the interval containing the unknown and the data. For this formula to be of use, the function in question must be known and differentiable. This is not usually the case. Fortunately, an alternative formulation is available that does not require prior knowledge of the function. Rather, it uses a finite divided difference to approximate the $(n+1)$th derivative,

$$R_n = f[x, x_n, x_{n-1}, \ldots, x_0](x - x_0)(x - x_1) \ldots (x - x_n) \tag{11.17}$$

where $f[x, x_n, x_{n-1}, \ldots, x_0]$ is the $(n+1)$th finite divided difference. Because Eq. (11.17) contains the unknown $f(x)$, it cannot be solved for the error. However, if an additional data point $f(x_{n+1})$ is available, Eq. (11.17) can be used to estimate the error, as in

$$R_n \simeq f[x_{n+1}, x_n, x_{n-1}, \ldots, x_0](x - x_0)(x - x_1) \ldots (x - x_n) \qquad [11.18]$$

EXAMPLE 11.4
Error Estimation for Newton's Polynomial

Problem Statement: Use Eq. (11.18) to estimate the error for the second-order polynomial interpolation of Example 11.2. Use the additional data point $f(x_3) = f(5) = 1.6094379$ to obtain your results.

Solution: Recall that in Example 11.2, the second-order interpolating polynomial provided an estimate of $f(2) = 0.565844346$, which represents an error of $0.69314718 - 0.565844346 = 0.127302835$. If we had not known the true value, as is most usually the case, Eq. (11.18), along with the additional value at $x_3$, could have been used to estimate the error, as in

$$R_2 = f[x_3, x_2, x_1, x_0](x - x_0)(x - x_1)(x - x_2)$$

or

$$R_2 = 0.0078655415\ (x - 1)(x - 4)(x - 6)$$

where the value for the third-order finite divided difference is as computed previously in Example 11.3. This relationship can be evaluated at $x = 2$ for

$$R_2 = 0.0078655415\ (2 - 1)(2 - 4)(2 - 6) = 0.062924332$$

which is of the same order of magnitude as the true error.

## 11.1.5 Computer Program for Newton's Interpolating Polynomial

Three properties make Newton's interpolating polynomials extremely attractive for computer applications:

1. As in Eq. (11.7), higher-order versions can be developed sequentially by adding a single term to the next lower-order equation. This facilitates the evaluation of several different-order versions in the same program. Such a capability is especially valuable when the order of the polynomial is not known a priori. By adding new terms sequentially, we can determine when a point of diminishing returns is reached—that is, when addition of higher-order terms no longer significantly improves the estimate or in certain situations actually detracts from it. The error equations discussed below in (3) are useful in devising an objective criterion for identifying this point of diminishing terms.

2. The finite divided differences that constitute the coefficients of the polynomial [Eqs. (11.8) through (11.11)] can be computed using a recurrence relationship. That is, as in Eq. (11.14) and Fig. 11.5, lower-order differ-

FORTRAN

```
      DIMENSION FX(10,10),X(10)
      READ(5,1)N
    1 FORMAT(I5)
      DO 140 I=1,N
      READ(5,2)X(I),FX(I,1)
    2 FORMAT(2F10.0)
  140 CONTINUE
      M=N-1
      DO 200 J=1,M
      K=J+1
      NP=N-J
      DO 190 I=1,NP
      FX(I,K)=(FX(I+1,J)-FX(I,J))/(X(I+J)-X(I))
  190 CONTINUE
  200 CONTINUE
      DO 230 J=1,N
      WRITE(6,3)FX(1,J)
    3 FORMAT(' ',F10.3)
  230 CONTINUE
      READ(5,2)XI
      FA=1.
      Y=0.
      DO 340 J=1,N
      Y=Y+FX(1,J)*FA
      WRITE(6,3)Y
      FA=FA*(XI-X(J))
      IF(J.GE.N)GO TO 350
      JP=J+1
      ER=FA*FX(1,JP)
      WRITE(6,3)ER
  340 CONTINUE
  350 STOP
      END
```

BASIC

```
100  DIM FX(10,10),X(10)
110  INPUT N
120  FOR I = 1 TO N
130  INPUT X(I),FX(I,1)
140  NEXT I
150  FOR J = 1 TO N - 1
160 K = J + 1
170  FOR I = 1 TO N - J
180 FX(I,K) = (FX(I + 1,J) - FX(I
   ,J)) / (X(I + J) - X(I))
190  NEXT I
200  NEXT J
210  FOR J = 1 TO N
220  PRINT FX(1,J)
230  NEXT J
240  INPUT XI
250 FA = 1
260 Y = 0
270  FOR J = 1 TO N
280 Y = Y + FX(1,J) * FA
290  PRINT Y
300 FA = FA * (XI - X(J))
310  IF J > = N THEN 350
320 ER = FA * FX(1,J + 1)
330  PRINT ER
340  NEXT J
350  END
```



FIGURE 11.7 Computer program for Newton's interpolating polynomial.

ences are used to compute higher-order differences. By utilizing this previously determined information, the coefficients can be computed efficiently. The program in Fig. 11.7 contains such a scheme.

3. The error equation [Eq. (11.18)] is expressed in terms of the finite divided differences that were already computed to determine the coefficients of the polynomial. Therefore, if this information is retained, the error estimate can be calculated without recomputing these quantities.

All the above characteristics can be exploited and incorporated into a general computer program for implementing Newton's polynomial (Fig. 11.7). As with the other programs throughout the book, this version is not documented. In addition, it does not include the error estimates mentioned in (3) above. One of your tasks in making this program more user-friendly (see Prob. 11.11) will be to incorporate the error equation. The utility of that equation is demonstrated in the following example.

## EXAMPLE 11.5
### Using Error Estimates to Determine the Appropriate Order of Interpolation

Problem Statement: After incorporating the error [Eq. (11.18)], utilize the computer program given in Fig. 11.7 and the following information to evaluate $f(x) = \ln x$ at $x = 2$:

| $x$ | $f(x) = \ln x$ |
|---|---|
| 1 | 0 |
| 4 | 1.3862944 |
| 6 | 1.7917595 |
| 5 | 1.6094379 |
| 3 | 1.0986123 |
| 1.5 | 0.40546511 |
| 2.5 | 0.91629073 |
| 3.5 | 1.2527630 |

Solution: The results of employing the program in Fig. 11.7 to obtain a solution are shown in Fig. 11.8. The error estimates, along with the true error

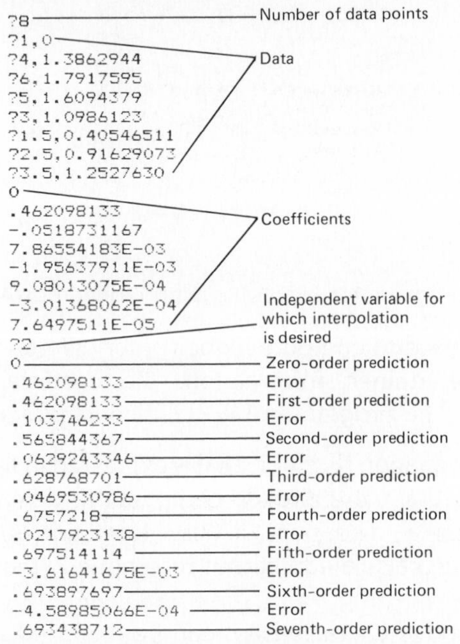


FIGURE 11.8 The output of the BASIC program to evaluate ln 2.

(based on the fact that $\ln 2 = 0.69314718$), are depicted in Fig. 11.9. Note that the estimated and the true error are similar and that their agreement improves as the order increases. From this plot, it can be concluded that the fifth-order version yields a good estimate and that higher-order terms do not significantly enhance the prediction.

This exercise also illustrates the importance of the positioning and ordering of the points. For example, up through the third-order estimate, the rate of improvement is slow because the points that are added (at $x = 4$, 6, and 5) are distant and on one side of the point in question at $x = 2$. The fourth-order estimate shows a somewhat greater improvement because the new point at $x = 3$ is closer to the unknown. However, the most dramatic decrease in the error is associated with the inclusion of the fifth-order term using

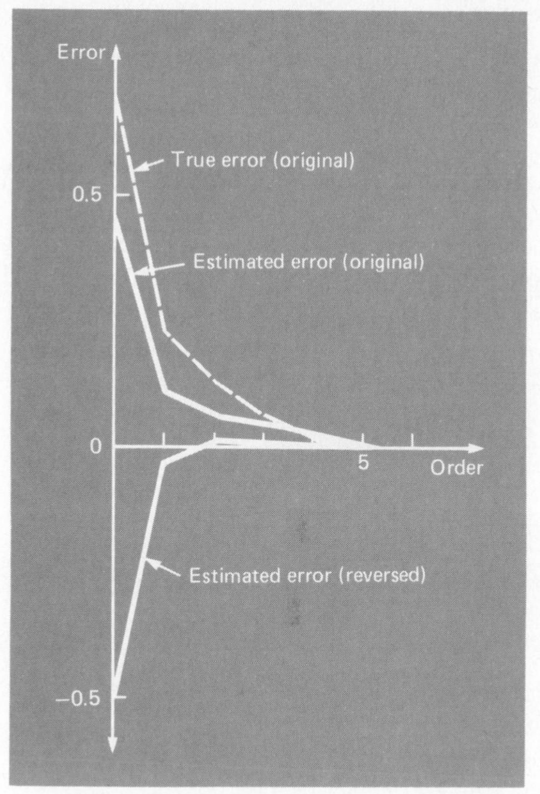


FIGURE 11.9 Percent relative errors for the prediction of ln 2 as a function of the order of the interpolating polynomial.

the data point at $x = 1.5$. Not only is this point close to the unknown but it is also positioned on the opposite side from most of the other points. As a consequence, the error is reduced almost an order of magnitude.

The significance of the position and sequence of the data can also be illustrated by using the same data to obtain an estimate for ln 2, but considering the points in a different sequence. Figure 11.9 shows results for the case of reversing the order of the original data, that is, $x_0 = 3.5$, $x_1 = 2.5$, $x_3 = 1.5$, and so forth. Because the initial points for this case are closer to and spaced on either side of ln 2, the error decreases much more rapidly than for the original situation. By the second-order term, the error has been reduced to a percent-relative level of less than $\epsilon_t = 2$ percent. Other combinations could be employed to obtain different rates of convergence.

The foregoing example illustrates the importance of the choice of base points. As should be intuitively obvious, the points should be centered around and as close as possible to the unknown. This observation is also supported by direct examination of the error equation [Eq. (11.17)]. Assuming that the finite divided difference does not vary markedly along the range of the data, the error is proportional to the product: $(x - x_0)(x - x_1) \cdots (x - x_n)$. Obviously, the closer the base points are to $x$, the smaller the magnitude of this product.

## 11.2 LAGRANGE INTERPOLATING POLYNOMIALS

The *Lagrange interpolating polynomial* is simply a reformulation of the Newton polynomial that avoids the computation of divided differences. It can be represented concisely as

$$f_n(x) = \sum_{i=0}^{n} L_i(x) f(x_i) \tag{11.19}$$

where

$$L_i(x) = \prod_{\substack{j=0 \\ j \neq i}}^{n} \frac{x - x_j}{x_i - x_j} \tag{11.20}$$

where $\Pi$ designates the "product of." For example, the linear version ($n = 1$) is

$$f_1(x) = \frac{x - x_1}{x_0 - x_1} f(x_0) + \frac{x - x_0}{x_1 - x_0} f(x_1) \tag{11.21}$$

and the second-order version is

$$f_2(x) = \frac{(x - x_1)(x - x_2)}{(x_0 - x_1)(x_0 - x_2)} f(x_0) + \frac{(x - x_0)(x - x_2)}{(x_1 - x_0)(x_1 - x_2)} f(x_1) + \frac{(x - x_0)(x - x_1)}{(x_2 - x_0)(x_2 - x_1)} f(x_2) \tag{11.22}$$

As with Newton's method, the Lagrange version has an estimated error of [Eq. (11.17)]

$$R_n = f[x, x_n, x_{n-1}, \ldots, x_0] \prod_{i=0}^{n} (x - x_i)$$

Equation (11.19) can be derived directly from Newton's polynomial (Box 11.1). However, the rationale underlying the Lagrange formulation can be grasped directly by realizing that each term $L_i(x)$ will be 1 at $x = x_i$ and 0 at all other sample points. Thus, each product $L_i(x) f(x_i)$ takes on the value of $f(x_i)$ at the sample point $x_i$. Consequently, the summation of all the products designated by Eq. (11.19) is the unique $n$th-order polynomial that passes exactly through all $n + 1$ data points.

### BOX 11.1 Derivation of the Lagrange Form Directly from Newton's Interpolating Polynomial

The Lagrange interpolating polynomial can be derived directly from Newton's formulation. We will do this for the first-order case,

$$f_1(x) = f(x_0) + (x - x_0)f[x_1, x_0] \qquad \text{[B11.1.1]}$$

In order to derive the Lagrange form, we reformulate the divided differences. For example, the first divided difference,

$$f[x_1, x_0] = \frac{f(x_1) - f(x_0)}{x_1 - x_0}$$

can be reformulated as

$$f[x_1, x_0] = \frac{f(x_1)}{x_1 - x_0} + \frac{f(x_0)}{x_0 - x_1} \qquad \text{[B11.1.2]}$$

which is referred to as the *symmetric form*. Substituting Eq. (B11.1.2) into Eq. (B11.1.1) yields

$$f_1(x) = f(x_0) + \frac{(x - x_0)}{(x_1 - x_0)} f(x_1) + \frac{(x - x_0)}{(x_0 - x_1)} f(x_0)$$

Finally, grouping similar terms and simplifying yields the Lagrange form,

$$f_1(x) = \frac{x - x_1}{x_0 - x_1} f(x_0) + \frac{x - x_0}{x_1 - x_0} f(x_1)$$

### EXAMPLE 11.6
Lagrange Interpolating Polynomials

Problem Statement: Use a Lagrange interpolating polynomial of the first and second order to evaluate ln 2 on the basis of the data given in Example 11.2:

$$x_0 = 1 \qquad f(x_0) = 0$$
$$x_1 = 4 \qquad f(x_1) = 1.3862944$$
$$x_2 = 6 \qquad f(x_2) = 1.7917595$$

Solution: The first-order polynomial [Eq. (11.21)] is

$$f_1(x) = \frac{x - x_1}{x_0 - x_1} f(x_0) + \frac{x - x_0}{x_1 - x_0} f(x_1)$$

and, therefore, the estimate at $x = 2$ is

$$f_1(2) = \frac{2 - 4}{1 - 4} 0 + \frac{2 - 1}{4 - 1} 1.3862944 = 0.4620981$$

In a similar fashion, the second-order polynomial is developed as [Eq. (11.22)]

$$f_2(2) = \frac{(2 - 4)(2 - 6)}{(1 - 4)(1 - 6)} 0 + \frac{(2 - 1)(2 - 6)}{(4 - 1)(4 - 6)} 1.3862944 + \frac{(2 - 1)(2 - 4)}{(6 - 1)(6 - 4)} 1.7917595 = 0.56584437$$

As expected, both these results agree closely with those obtained previously using Newton's interpolating polynomial.

In summary, for cases where the order of the polynomial is unknown, the Newton method has advantages because of the insight it provides into the behavior of the different-order formulas. In addition, the error estimate represented by Eq. (11.18) can usually be integrated easily into the Newton computation because the estimate employs a finite difference (Example 11.5). Thus, for single exploratory computations, Newton's method is often preferable.

When only one interpolation is to be performed, the Lagrange and Newton formulations require comparable computational effort. However, the Lagrange version is somewhat easier to program. Also, there are cases where the Newton form is more susceptible to round-off error (Ruckdeschel, 1981). Because of this and because it does not require computation and storage of divided differences, the Lagrange form is often used when the order of the polynomial is known a priori.

EXAMPLE 11.7
Lagrange Interpolation Using the Computer

Problem Statement: A user-friendly computer program to implement Lagrange interpolation is contained in the NUMERICOMP software package associated with this text. We can use this software to study a trend analysis problem associated with our now-familiar falling parachutist. Assume that we have developed instrumentation to measure the velocity of the parachutist. The measured data obtained for a particular test case is

| **Time, s** | **Measured velocity *v*, cm/s** |
|---|---|
| 1 | 800 |
| 3 | 2310 |
| 5 | 3090 |
| 7 | 3940 |
| 13 | 4755 |

Our problem is to estimate the velocity of the parachutist at $t = 10$ s to fill in the large gap in the measurements between $t = 7$ and $t = 13$ s. We are aware that the behavior of interpolating polynomials can be unexpected. Therefore, we will construct polynomials of orders 4, 3, 2, and 1 and compare the results.

Solution: The NUMERICOMP program can be used to construct fourth-, third-, second-, and first-order interpolating polynomials. These results are

| Polynomial order | COEFFICIENT | | | | | Estimated value of $v$ at $t$ = 10 s |
|---|---|---|---|---|---|---|
| | 4th order | 3d order | 2d order | 1st order | 0 order | |
| 4 | −1.76302 | 44.87501 | −392.87 | 1813.625 | −663.867 | 5430.195 |
| 3 | | −4.498586 | 76.09375 | 1.239258 | 1742.656 | 4874.838 |
| 2 | | | −36.14584 | 858.75 | −300.1035 | 4672.812 |
| 1 | | | | 135.8333 | 2989.167 | 4347.5 |

The fourth-order polynomial and the input data can be plotted as shown in Fig. 11.10*a*. It is evident from this plot that the estimated value of $y$ at $x = 10$ is higher than the overall trend of the data.

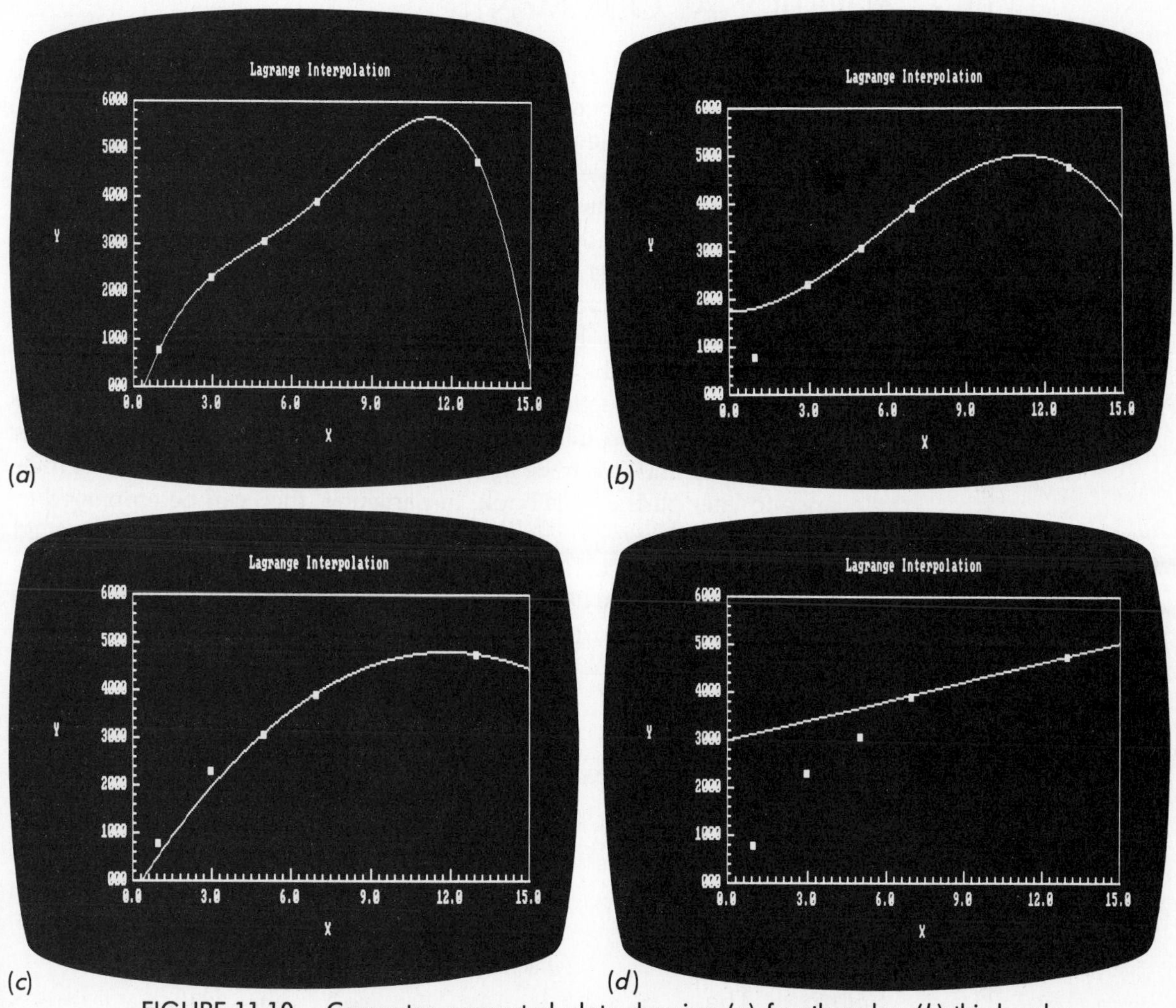


FIGURE 11.10 Computer-generated plots showing (*a*) fourth-order, (*b*) third-order, (*c*) second-order, and (*d*) first-order interpolations.

Figure 11.10*b* through *d* shows plots of the results of the computations for third-, second-, and first-order interpolating polynomials. It is noted that the lower the order of the interpolating polynomial the lower the estimated value of the velocity at $t = 10$ s. The plots of the interpolating polynomials indicate that the higher-order polynomials tend to overshoot the trend of the data. This suggests that the first- or second-order polynomials are most appropriate for this particular trend analysis. It should be remembered, however, that because we are dealing with uncertain data, regression would actually be more appropriate.

## 11.3 ADDITIONAL COMMENTS

Before proceeding to the next section, we must mention two additional topics: interpolation with equally spaced data and extrapolation.

Because both the Newton and Lagrange polynomials are compatible with arbitrarily spaced data, you might wonder why we address the special case of equally spaced data (Box 11.2). Prior to the advent of digital computers, these techniques had great utility for interpolation from tables with equally spaced arguments. In fact, a computational framework known as a divided-difference table was developed to facilitate the implementation of these techniques. (Figure 11.5 is an example of such tables.)

However, because the formulas are subsets of the computer-compatible Newton and Lagrange schemes and because many tabular functions are available as library subroutines, the need for the equispaced versions has waned. In spite of this, we have included them at this point because of their relevance to later parts of this book. In particular, they can be employed to derive numerical integration formulas which typically employ equispaced data (Chap. 13). Because the numerical integration formulas have relevance to the solution of ordinary differential equations, the material in Box 11.2 also has significance to Chap. 17.

### BOX 11.2 Interpolation with Equally Spaced Data

If data is equally spaced and in ascending order, then the independent variable assumes values of

$$x_1 = x_0 + h$$

$$x_2 = x_0 + 2h$$

$$\vdots$$

$$x_n = x_0 + nh$$

Where $h$ is the interval, or step size, between the data. On this basis, the finite divided differences can be expressed in concise form. For example, the second forward divided difference is

$$f[x_2, x_1, x_0] = \frac{\dfrac{f(x_0) - f(x_1)}{x_0 - x_1} - \dfrac{f(x_1) - f(x_2)}{x_1 - x_2}}{x_0 - x_2}$$

which can be expressed as

$$f[x_2, x_1, x_0] = \frac{f(x_0) - 2f(x_1) + f(x_2)}{2h^2} \qquad \text{[B11.2.1]}$$

because $x_0 - x_1 = x_1 - x_2 = (x_0 - x_2)/2 = h$. Now recall that the second forward difference $\Delta^2 f(x_0)$ is equal to [numerator of Eq. (3.31)]

$$\Delta^2 f(x_0) = f(x_0) - 2f(x_1) + f(x_2)$$

Therefore, Eq. (B11.2.1) can be represented as

$$f[x_0, x_1, x_2] = \frac{\Delta f^2(x_0)}{2!h^2}$$

or, in general,

$$f[x_0, x_1, \ldots, x_n] = \frac{\Delta f^n(x_0)}{n!h^n} \quad \text{[B11.2.2]}$$

Using Eq. (B11.2.2), Newton's interpolating polynomial [Eq. (11.15)] can be expressed for the case of equispaced data as

$$f_n(x) = f(x_0) + \frac{\Delta f(x_0)}{h}(x - x_0)$$
$$+ \frac{\Delta^2 f(x_0)}{2!h^2}(x - x_0)(x - x_0 - h) + \cdots$$
$$+ \frac{\Delta^n f(x_0)}{n!h^n}(x - x_0)(x - x_0 - h) \cdots$$
$$\{x - x_0 - (n - 1)h\}$$
$$+ R_n \quad \text{[B11.2.3]}$$

where the remainder is the same as Eq. (11.16). This equation is known as *Newton's formula* or the *Newton-Gregory forward formula.* It can be simplified further by defining a new quantity, $\alpha$:

$$\alpha = \frac{x - x_0}{h}$$

This definition can be used to develop the following simplified expressions for the terms in Eq. (B11.2.3):

$$x - x_0 = \alpha h$$
$$x - x_0 - h = \alpha h - h = h(\alpha - 1)$$
$$\vdots$$
$$x - x_0 - (n - 1) = \alpha h - (n - 1)h$$
$$= h(\alpha - n + 1)$$

which can be substituted into Eq. (B11.2.3) to give

$$f_n(x) = f(x_0) + \Delta f(x_0)\alpha + \frac{\Delta^2 f(x_0)}{2!}\alpha(\alpha - 1)$$
$$+ \cdots + \frac{\Delta^n f(x_0)}{n!}\alpha(\alpha - 1) \ldots$$
$$(\alpha - n + 1) + R_n \quad \text{[B11.2.4]}$$

where

$$R_n = \frac{f^{(n+1)}(\xi)}{(n + 1)!} h^{n+1} \alpha(\alpha - 1)(\alpha - 2) \ldots$$
$$(\alpha - n)$$

This concise notation will have utility in our derivation and error analyses of the integration formulas in Chap. 13.

In addition to the forward formula, backward and central Newton-Gregory formulas are also available. Carnahan, Luther, and Wilkes (1969) can be consulted for further information regarding interpolation for equally spaced data.

---

*Extrapolation* is the process of estimating a value of $f(x)$ that lies outside the range of the known base points, $x_0, x_1, \ldots, x_n$ (Fig. 11.11). In a previous section, we mentioned that the most accurate interpolation is usually obtained when the unknown lies near the center of the base points. Obviously, this is violated when the unknown lies outside the range, and consequently, the error in extrapolation can be very large. As depicted in Fig. 11.11, the open-ended nature of extrapolation represents a step into the unknown because the process extends the curve beyond the known region. As such, the true curve could easily diverge from the prediction. Extreme care should, therefore, be exercised whenever a case arises where one must extrapolate. Case Study 12.1 in the next chapter presents an example of the liabilities involved in projecting beyond the limits of the data.

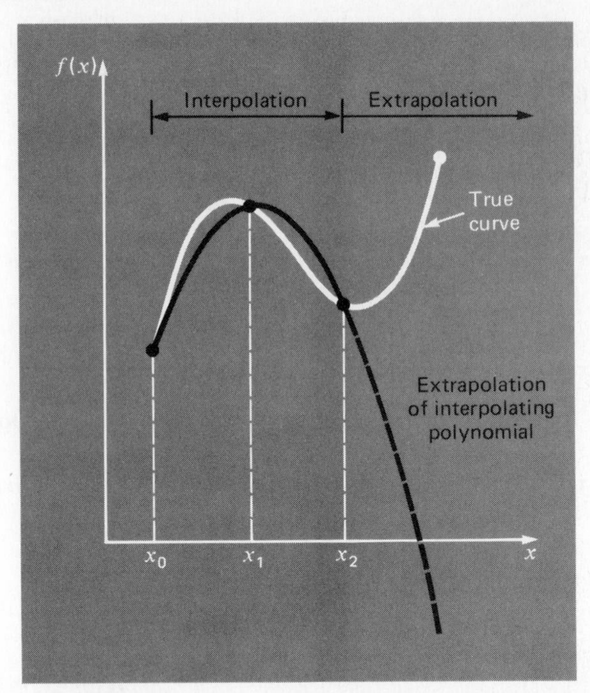


FIGURE 11.11 Illustration of the possible divergence of an extrapolated prediction. The extrapolation is based on fitting a parabola through the first three points.

## 11.4 SPLINE INTERPOLATION

In the previous sections, $n$th-order polynomials were used to interpolate between $n + 1$ data points. For example, for eight points, we can derive a perfect seventh-order polynomial. This curve would capture all the meanderings (at least up to and including seventh derivatives) suggested by the points. However, there are cases where these functions can lead to erroneous results. An alternative approach is to apply lower-order polynomials to subsets of data points. Such connecting polynomials are called *spline functions.*

For example, third-order curves employed to connect each pair of data points are called *cubic splines.* These functions have the additional property that the connection between adjacent cubic equations are visually smooth. On the surface, it would seem that the third-order approximation of the splines would be inferior to the seventh-order expression. You might wonder why a spline would ever be preferable.

Figure 11.12 illustrates a situation where a spline performs better than a higher-order polynomial. This is the case where a function is generally smooth but undergoes an abrupt change somewhere along the region of interest. The step increase depicted in Fig. 11.12 is an extreme example of such a change and serves to illustrate the point.

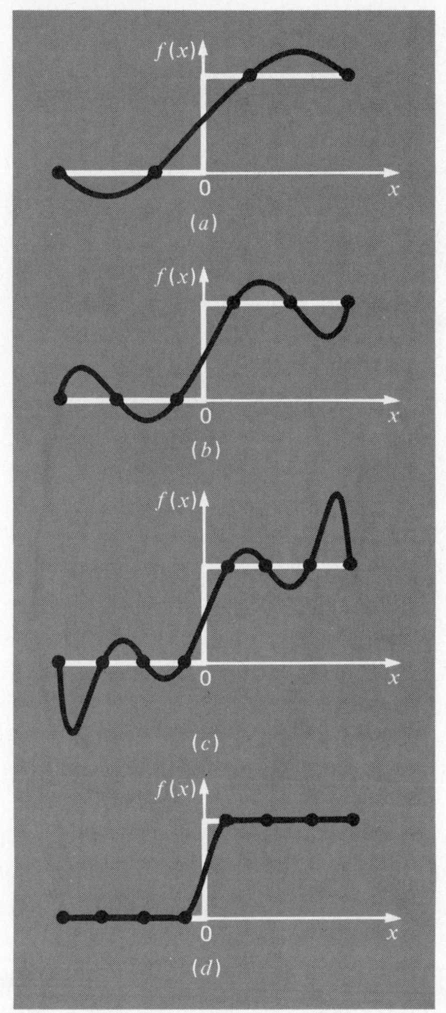


FIGURE 11.12 A visual representation of a situation where splines are superior to higher-order interpolating polynomials. The function to be fit undergoes an abrupt increase at $x = 0$. Parts (*a*) through (*c*) indicate that the abrupt change induces oscillations in interpolating polynomials. In contrast, because it is limited to third-order curves with smooth transitions, the cubic spline (*d*) provides a much more acceptable approximation.

Figure 11.12*a* through *c* illustrates how higher-order polynomials tend to swing through wild oscillations in the vicinity of an abrupt change. In contrast, the spline also connects the points, but because it is limited to third-order changes, the oscillations are kept to a minimum. As such the spline usually provides a superior approximation of the behavior of functions that have local, abrupt changes.

The concept of the spline originated from the drafting technique of using a thin, flexible strip (called a *spline*) to draw smooth curves through a set of

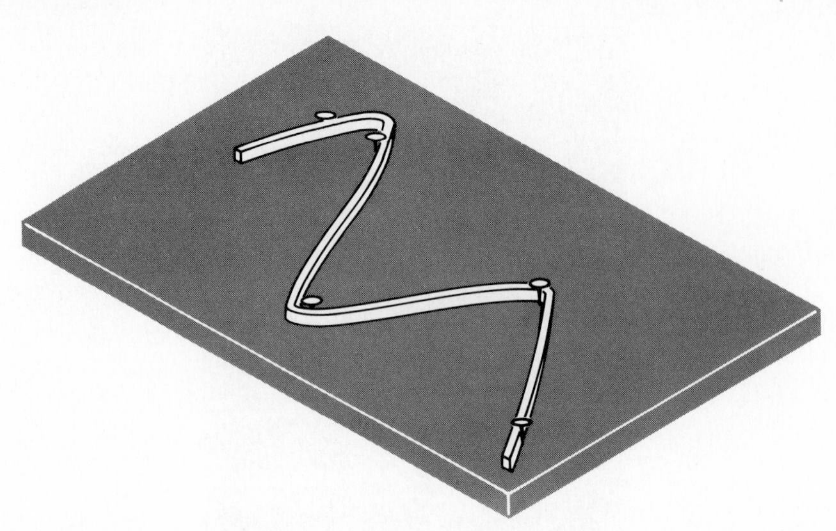

FIGURE 11.13 The drafting technique of using a spline to draw smooth curves through a series of points. Notice how, at the end points, the spline straightens out. This is called a "natural" spline.

points. The process is depicted in Fig. 11.13 for a series of five pins (data points). In this technique, the drafter places paper over a wooden board and hammers nails or pins into the paper (and board) at the location of the data points. A smooth cubic curve results from interweaving the strip between the pins. Hence, the name "cubic spline" has been adopted for polynomials of this type.

In the present section, simple linear functions will first be used to introduce some basic concepts and problems associated with spline interpolation. Then we derive an algorithm for fitting quadratic splines to data. Finally, we present material on the cubic spline, which is the most common and useful version in engineering practice.

### 11.4.1 Linear Splines

The simplest connection between two points is a straight line. The first-order splines for a group of ordered data points can be defined as a set of linear functions joining the points:

$$
\begin{aligned}
f(x) &= f(x_0) + m_0(x - x_0) && x_0 \le x \le x_1 \\
f(x) &= f(x_1) + m_1(x - x_1) && x_1 \le x \le x_2 \\
&\vdots && \vdots \\
f(x) &= f(x_{n-1}) + m_{n-1}(x - x_{n-1}) && x_{n-1} \le x \le x_n
\end{aligned}
$$

where $m_i$ is the slope of the straight line connecting the points:

$$
m_i = \frac{f(x_{i+1}) - f(x_i)}{x_{i+1} - x_i} \qquad [11.23]
$$

These equations can be used to evaluate the function at any point between $x_0$ and $x_n$ by first locating the interval within which the point lies. Then the appropriate equation is used to determine the function value within the interval. The method is obviously identical to linear interpolation.

EXAMPLE 11.8
First-Order Splines

Problem Statement: Fit the data in Table 11.1 with first-order splines. Evaluate the function at $x = 5$.

**TABLE 11.1 Data to be fit with spline functions**

| $x$ | $f(x)$ |
|---|---|
| 3.0 | 2.5 |
| 4.5 | 1.0 |
| 7.0 | 2.5 |
| 9.0 | 0.5 |

Solution: The data can be used to determine the slopes between points. For example, for the interval from $x = 4.5$ to $x = 7$ the slope can be computed using Eq. (11.23):

$$m = \frac{2.5 - 1.0}{7.0 - 4.5} = 0.60$$

The slopes for the other intervals can be computed, and the resulting first-order splines are plotted in Fig. 11.14*a*. The value at $x = 5$ is 1.3.

Visual inspection of Fig. 11.14*a* indicates that the primary disadvantage of first-order splines is that they are not smooth. In essence, at the data points where two splines meet (called a *knot*), the slope changes abruptly. In formal terms, the first derivative of the function is discontinuous at these points. This deficiency is overcome by using higher-order polynomial splines that ensure smoothness at the knots by equating derivatives at these points, as discussed in the next section.

### 11.4.2 Quadratic Splines

In order to ensure that the $m$th derivatives are continuous at the knots, a spline of at least $m + 1$ order must be used. Third-order polynomials or cubic splines that ensure continuous first and second derivatives are most frequently used in practice. Although third and higher derivatives could be discontinuous when using cubic splines, they usually cannot be detected visually and consequently are ignored.

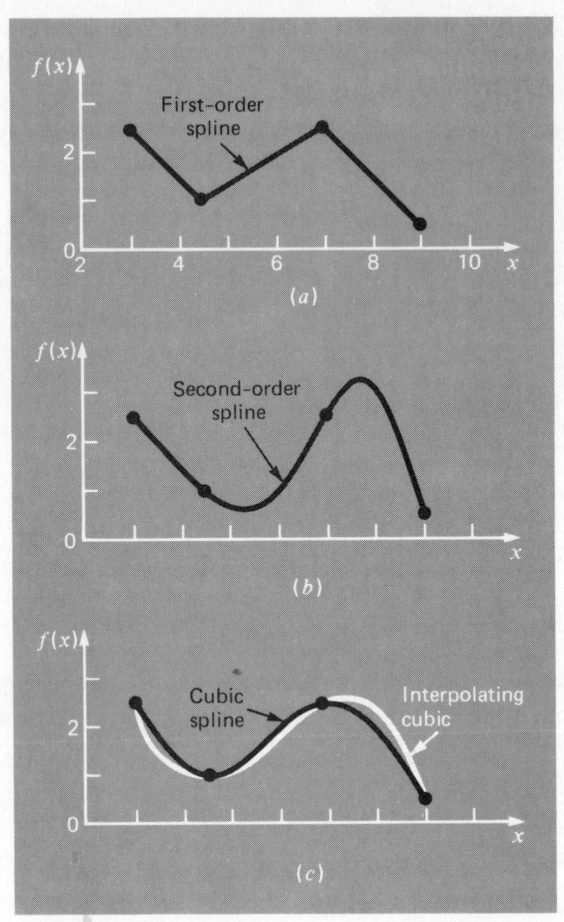


FIGURE 11.14 Spline fits of a set of four points. (*a*) Linear spline, (*b*) quadratic spline, and (*c*) cubic spline, with a cubic interpolating polynomial also plotted.

Because the derivation of cubic splines is somewhat involved, we have chosen to include them in a subsequent section. We have decided to first illustrate the concept of spline interpolation using second-order polynomials. These "quadratic splines" have continuous first derivatives at the knots. Although quadratic splines do not ensure equal second derivatives at the knots, they serve nicely to demonstrate the general procedure for developing higher-order splines.

The objective in quadratic splines is to derive a second-order polynomial for each interval between data points. The polynomial for each interval can be represented generally as

$$f_i(x) = a_i x^2 + b_i x + c_i \tag{11.24}$$

Figure 11.15 has been included to help clarify the notation. For $n + 1$ data points ($i = 0, 1, 2, \ldots, n$), there are $n$ intervals and, consequently, $3n$

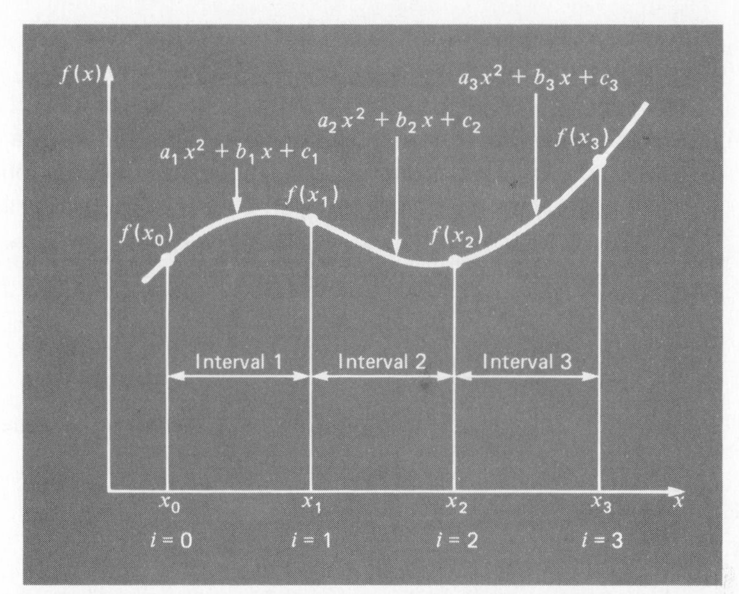


FIGURE 11.15 Notation used to derive quadratic splines. Notice that there are $n$ intervals and $n + 1$ data points. The example shown is for $n = 3$.

unknown constants (the $a$'s, $b$'s and $c$'s) to evaluate. Therefore, $3n$ equations or conditions are required to evaluate the unknowns. These are:

1. *The function values must be equal at the interior knots.* This condition can be represented as

$$a_{i-1}x_{i-1}^2 + b_{i-1}x_{i-1} + c_{i-1} = f(x_{i-1}) \quad [11.25]$$

$$a_i x_{i-1}^2 + b_i x_{i-1} + c_i = f(x_{i-1}) \quad [11.26]$$

for $i = 2$ to $n$. Because only interior knots are used, Eqs. (11.25) and (11.26) each provide $n - 1$ for a total of $2n - 2$ conditions.

2. *The first and last functions must pass through the end points.* This adds two additional equations:

$$a_1 x_0^2 + b_1 x_0 + c_1 = f(x_0) \quad [11.27]$$

$$a_n x_n^2 + b_n x_n + c_n = f(x_n) \quad [11.28]$$

for a total of $2n - 2 + 2 = 2n$ conditions.

3. *The first derivatives at the interior knots must be equal.* The first derivative of Eq. (11.22) is

$$f'(x) = 2ax + b$$

Therefore, the condition can be represented generally as

$$2a_{i-1}x_i + b_{i-1} = 2a_i x_i + b_i \quad [11.29]$$

for $i = 2$ to $n$. This provides another $n - 1$ conditions for a total of $2n + n - 1 = 3n - 1$. Because we have $3n$ unknowns, we are one condition short. Unless we have some additional information regarding the functions or their derivatives, we must make an arbitrary choice in order to successfully compute the constants. Although there are a number of different choices that can be made, we select the following:

**4.** *Assume that the second derivative is zero at the first point.* Because the second derivative of Eq. (11.24) is $2a_i$, this condition can be expressed mathematically as

$$a_1 = 0 \tag{11.30}$$

The visual interpretation of this condition is that the first two points will be connected by a straight line.

## EXAMPLE 11.9
### Quadratic Splines

Problem Statement: Fit quadratic splines to the same data used in Example 11.8 (Table 11.1). Use the results to estimate the value at $x = 5$.

Solution: For the present problem, we have four data points and $n = 3$ intervals. Therefore, $3(3) = 9$ unknowns must be determined. Equations (11.25) and (11.26) yield $2(3) - 2 = 4$ conditions:

$$20.25a_1 + 4.5b_1 + c_1 = 1.0$$

$$20.25a_2 + 4.5b_2 + c_2 = 1.0$$

$$49a_2 + 7b_2 + c_2 = 2.5$$

$$49a_3 + 7b_3 + c_3 = 2.5$$

Passing the first and last functions through the initial and final values adds 2 more: [Eq. (11.27)]

$$9a_1 + 3b_1 + c_1 = 2.5$$

and [Eq. (11.28)]

$$81a_3 + 9b_3 + c_3 = 0.5$$

Continuity of derivatives creates an additional $3 - 1 = 2$ [Eq. (11.29)]:

$$9a_1 + b_1 = 9a_2 + b_2$$

$$14a_2 + b_2 = 14a_3 + b_3$$

Finally, Eq. (11.30) specifies that $a_1 = 0$. Because this equation specifies $a_1$ exactly, the problem reduces to solving eight simultaneous equations. These conditions can be expressed in matrix form as

$$\begin{bmatrix} 4.5 & 1.0 & 0.0 & 0.0 & 0.0 & 0.0 & 0.0 & 0.0 \\ 0.0 & 0.0 & 20.25 & 4.5 & 1.0 & 0.0 & 0.0 & 0.0 \\ 0.0 & 0.0 & 49.00 & 7.0 & 1.0 & 0.0 & 0.0 & 0.0 \\ 0.0 & 0.0 & 0.0 & 0.0 & 0.0 & 49.00 & 7.00 & 1.00 \\ 3.0 & 1.0 & 0.0 & 0.0 & 0.0 & 0.0 & 0.0 & 0.0 \\ 0.0 & 0.0 & 0.0 & 0.0 & 0.0 & 81.00 & 9.00 & 1.00 \\ 1.0 & 0.0 & -9.00 & -1.0 & 0.0 & 0.0 & 0.0 & 0.0 \\ 0.0 & 0.0 & 14.00 & 1.0 & 0.0 & -14.00 & -1.00 & 0.0 \end{bmatrix} \begin{bmatrix} b_1 \\ c_1 \\ a_2 \\ b_2 \\ c_2 \\ a_3 \\ b_3 \\ c_3 \end{bmatrix} = \begin{bmatrix} 1.0 \\ 1.0 \\ 2.5 \\ 2.5 \\ 2.5 \\ 0.5 \\ 0.0 \\ 0.0 \end{bmatrix}$$

These equations can be solved using techniques from Part III with the results:

$$a_1 = 0 \qquad b_1 = -1 \qquad c_1 = 5.5$$

$$a_2 = 0.64 \qquad b_2 = -6.76 \qquad c_2 = 18.46$$

$$a_3 = -1.6 \qquad b_3 = 24.6 \qquad c_3 = -91.3$$

which can be substituted into the original quadratic equations to develop the following relationships for each interval:

$$f_1(x) = -x + 5.5 \qquad 3.0 \le x \le 4.5$$

$$f_2(x) = 0.64x^2 - 6.76x + 18.46 \qquad 4.5 \le x \le 7.0$$

$$f_3(x) = -1.6x^2 + 24.6x - 91.3 \qquad 7.0 \le x \le 9.0$$

The prediction for $x = 5$ is, therefore,

$$f_2(5) = 0.64(5)^2 - 6.76(5) + 18.46 = 0.66$$

The total spline fit is depicted in Fig. 11.14*b*. Notice that there are two shortcomings that detract from the fit: (1) the straight line connecting the first two points and (2) the spline for the last interval seems to swing too high. The cubic splines in the next section do not exhibit these shortcomings and as a consequence are usually better methods for spline interpolation.

### 11.4.3 Cubic Splines

The objective in cubic splines is to derive a third-order polynomial for each interval between knots, as in

$$f_i(x) = a_i x^3 + b_i x^2 + c_i x + d_i \tag{11.31}$$

Thus, for $n + 1$ data points ($i = 0, 1, 2, \ldots, n$), there are $n$ intervals and, consequently, $4n$ unknown constants to evaluate. Just as for quadratic splines, $4n$ conditions are required to evaluate the unknowns. These are:

1. *The function values must be equal at the interior knots ($2n - 2$ conditions).*

2. *The first and last functions must pass through the end points (2 conditions).*

3. *The first derivatives at the interior knots must be equal ($n - 1$ conditions).*

4. *The second derivatives at the interior knots must be equal ($n - 1$ conditions).*

5. *The second derivatives at the end knots are zero (2 conditions).*

The visual interpretation of condition 5 is that the function becomes a straight line at the end knots. Specification of such an end condition leads to what is termed a "natural" spline. It is given this name because the drafting spline naturally behaves in this fashion (Fig. 11.13). If the value of the second derivative at the end knots is nonzero (that is, there is some curvature), this information can be used alternatively to supply the two necessary conditions.

The above five types of conditions provide the total of $4n$ equations required to solve for the $4n$ coefficients. Whereas it is certainly possible to develop cubic splines in this fashion, we will present an alternative technique that only requires the solution of $n - 1$ equations. Although the derivation of this method (Box 11.3) is somewhat less straightforward than that for quadratic splines, the gain in efficiency is well worth the effort.

---

BOX 11.3 Derivation of Cubic Splines

The first step in the derivation (Cheney and Kincaid, 1980) is based on the observation that because each pair of knots is connected by a cubic, the second derivative within each interval is a straight line. Equation (11.31) can be differentiated twice to verify this observation. On this basis, the second derivatives can be represented by a first-order Lagrange interpolating polynomial [Eq. (11.21)]:

$$f_i''(x) = f''(x_{i-1}) \frac{x - x_i}{x_{i-1} - x_i} + f''(x_i) \frac{x - x_{i-1}}{x_i - x_{i-1}} \qquad \text{[B11.3.1]}$$

where $f_i''(x)$ is the value of the second derivative at any point $x$ within the $i$th interval. Thus, this equation is a straight line connecting the second derivative at the first knot $f''(x_{i-1})$ with the second derivative at the second knot $f''(x_i)$.

Next, Eq. (B11.3.1) can be integrated twice to yield an expression for $f_i(x)$. However, this expression will contain two unknown constants of integration. These constants can be evaluated by invoking the function-equality conditions—$f(x)$ must equal $f(x_{i-1})$ at $x_{i-1}$ and $f(x)$ must equal $f(x_i)$ at $x_i$. By performing these evaluations, the following cubic equation results:

$$\begin{aligned} f_i(x) &= \frac{f''(x_{i-1})}{6(x_i - x_{i-1})}(x_i - x)^3 \\ &+ \frac{f''(x_i)}{6(x_i - x_{i-1})}(x - x_{i-1})^3 \\ &+ \left[\frac{f(x_{i-1})}{x_i - x_{i-1}} - \frac{f''(x_{i-1})(x_i - x_{i-1})}{6}\right](x_i - x) \\ &+ \left[\frac{f(x_i)}{x_i - x_{i-1}} - \frac{f''(x_i)(x_i - x_{i-1})}{6}\right](x - x_{i-1}) \end{aligned} \qquad \text{[B11.3.2]}$$

Now, admittedly, this relationship is a much more complicated expression for the cubic spline for the $i$th interval than, say, Eq. (11.31). However, notice that it contains only two unknown "coefficients," the second derivatives at the beginning and the end of the interval—$f''(x_{i-1})$ and $f''(x_i)$. Thus, if we can determine the proper second derivative at each knot, Eq.

(B11.3.2) is a third-order polynomial that can be used to interpolate within the interval.

The second derivatives can be evaluated by invoking the condition that the first derivatives at the knots must be continuous:

$$f'_{i-1}(x_i) = f'_i(x_i) \quad \text{[B11.3.3]}$$

Equation (B11.3.2) can be differentiated to give an expression for the first derivative. If this is done for both the $(i-1)$th and the $i$th intervals and the two results are set equal according to Eq. (B11.3.3), the following relationship results:

$$(x_i - x_{i-1}) f''(x_{i-1}) + 2(x_{i+1} - x_{i-1}) f''(x_i) + (x_{i+1} - x_i) f''(x_{i+1})$$

$$= \frac{6}{(x_{i+1} - x_i)} [f(x_{i+1}) - f(x_i)] + \frac{6}{(x_i - x_{i-1})} [f(x_{i-1}) - f(x_i)] \quad \text{[B11.3.4]}$$

If Eq. (B11.3.4) is written for all interior knots, $n - 1$ simultaneous equations result with $n + 1$ unknown second derivatives. However, because this is a natural cubic spline, the second derivatives at the end knots are zero and the problem reduces to $n - 1$ equations with $n - 1$ unknowns. In addition, notice that the system of equations will be tridiagonal. Thus, not only have we reduced the number of equations but we have also cast them in a form that is extremely easy to solve (recall Box 7.2).

The derivation from Box 11.3 results in the following cubic equation for each interval:

$$f_i(x) = \frac{f''(x_{i-1})}{6(x_i - x_{i-1})}(x_i - x)^3 + \frac{f''(x_i)}{6(x_i - x_{i-1})}(x - x_{i-1})^3 + \left[\frac{f(x_{i-1})}{x_i - x_{i-1}} - \frac{f''(x_{i-1})(x_i - x_{i-1})}{6}\right](x_i - x) + \left[\frac{f(x_i)}{x_i - x_{i-1}} - \frac{f''(x_i)(x_i - x_{i-1})}{6}\right](x - x_{i-1}) \quad \text{[11.32]}$$

This equation contains only two unknowns—the second derivatives at the end of each interval. These unknowns can be evaluated using the following equation:

$$(x_i - x_{i-1}) f''(x_{i-1}) + 2(x_{i+1} - x_{i-1}) f''(x_i) + (x_{i+1} - x_i) f''(x_{i+1}) = \frac{6}{(x_{i+1} - x_i)} [f(x_{i+1}) - f(x_i)] + \frac{6}{(x_i - x_{i-1})} [f(x_{i-1}) - f(x_i)] \quad \text{[11.33]}$$

If this equation is written for all the interior knots, $n - 1$ simultaneous equations result with $n - 1$ unknowns. (Remember, the second derivatives at the end knots are zero.) The application of these equations is illustrated in the following example.

### EXAMPLE 11.10
### Cubic Splines

Problem Statement: Fit cubic splines to the same data used in Examples 11.8 and 11.9 (Table 11.1). Utilize the results to estimate the value at $x = 5$.

Solution: The first step is to employ Eq. (11.33) to generate the set of simultaneous equations that will be utilized to determine the second derivatives at the knots. For example, for the first interior knot, the following data is used:

$$x_0 = 3 \qquad f(x_0) = 2.5$$

$$x_1 = 4.5 \qquad f(x_1) = 1$$

$$x_2 = 7 \qquad f(x_2) = 2.5$$

These values can be substituted into Eq. (11.33) to yield

$$(4.5 - 3)f''(3) + 2(7 - 3)f''(4.5) + (7 - 4.5)f''(7)$$
$$= \frac{6}{7 - 4.5}(2.5 - 1) + \frac{6}{4.5 - 3}(2.5 - 1)$$

Because of the natural spline condition, $f''(3) = 0$, and the equation reduces to

$$8f''(4.5) + 2.5f''(7) = 9.6$$

In a similar fashion, Eq. (11.33) can be applied to the second interior point to give

$$3.5f''(4.5) + 9f''(7) = -9.6$$

These two equations can be solved simultaneously for

$$f''(4.5) = 1.74545$$

$$f''(7) = -1.74545$$

These values can then be substituted into Eq. (11.32), along with values for the $x$'s and the $f(x)$'s, to yield

$$f_1(x) = \frac{1.74545}{6(4.5 - 3)}(x - 3)^3 + \frac{2.5}{4.5 - 3}(4.5 - x)$$
$$+ \left[\frac{1}{4.5 - 3} - \frac{1.74545\,(4.5 - 3)}{6}\right](x - 3)$$

or

$$f_1(x) = 0.193939(x - 3)^3 + 1.666667(4.5 - x) + 0.230303(x - 3)$$

This equation is the cubic spline for the first interval. Similar substitutions can be made to develop the equations for the second and third intervals:

$$f_2(x) = 0.116364(7 - x)^3 - 0.116364(x - 4.5)^3$$
$$- 0.327273(7 - x) + 1.727273(x - 4.5)$$

and

$$f_3(x) = -0.145455(9 - x)^3 + 1.831818(9 - x) + 0.25(x - 7)$$

The three equations can then be employed to compute values within each interval. For example, the value at $x = 5$, which falls within the second interval, is calculated as

$$f_2(5) = 0.116364(7 - 5)^3 - 0.116364(5 - 4.5)^3 - 0.327273(7 - 5) + 1.727273(5 - 4.5) = 1.1255$$

Other values are computed and the results are plotted in Fig. 11.14c.

The results of Examples 11.8 through 11.10 are summarized in Fig. 11.14. Notice the progressive improvement of the fit as we move from linear to quadratic to cubic splines. We have also superimposed a cubic interpolating polynomial on Fig. 11.14c. Although the cubic spline consists of a series of third-order curves, the resulting fit differs from that obtained using the third-order polynomial. This is due to the fact that the natural spline requires zero second derivatives at the end knots, whereas the cubic polynomial has no such constraint.

### 11.4.4 Computer Algorithm for Cubic Splines

The method for calculating cubic splines outlined in the previous section is ideal for implementation on personal computers. Recall that by some clever manipulations, the method reduced to solving $n - 1$ simultaneous equations. An added benefit of the derivation is that, as specified by Eq. (11.33), the system of equations is *tridiagonal*. As described in Box 7.2, algorithms are available to solve such systems in an extremely efficient manner. Figure 11.16 outlines a computational framework that incorporates these features.

**Step 1:** Input data

**Step 2:** Apply Eq. (11.33) in order to generate a tridiagonal system of equations.

**Step 3:** Solve the system of equations for the unknown second derivatives using an algorithm of the sort described in Box 7.2.

**Step 4:** Use Eq. (11.32) to interpolate at a given value of $x$.

**Step 5:** If you require another interpolation, return to step 4. If not, end the computation.

FIGURE 11.16 Algorithm for cubic spline interpolation.

## PROBLEMS

### Hand Calculations

**11.1** Estimate the logarithm of 4 to the base 10 (log 4) using linear interpolation.
(*a*) Interpolate between log 3 = 0.4771213 and log 5 = 0.6989700.

(*b*) Interpolate between log 3 and log 4.5 = 0.6532125.
For each of the interpolations, compute the percent relative error based on the true value of log 4 = 0.6020600.

**11.2** Fit a second-order Newton's interpolating polynomial to estimate log 4 using the data from Prob. 11.1. Compute the percent relative error.

**11.3** Fit a third-order Newton's interpolating polynomial to estimate log 4 using the data from Prob. 11.1 along with the additional point, log 3.5 = 0.5440680. Compute the percent relative error.

**11.4** Given the data

| $x$ | 0 | 0.5 | 1.0 | 1.5 | 2.0 | 2.5 |
|---|---|---|---|---|---|---|
| $f(x)$ | 1 | 2.119 | 2.910 | 3.945 | 5.720 | 8.695 |

(*a*) Calculate $f(1.6)$ using Newton's interpolating polynomials of order 1 through 3. Choose the sequence of the points for your estimates to attain good accuracy.
(*b*) Utilize Eq. (11.18) to estimate the error for each prediction.

**11.5** Given the data

| $x$ | 1 | 2 | 3 | 5 | 6 |
|---|---|---|---|---|---|
| $f(x)$ | 4.75 | 4 | 5.25 | 19.75 | 36 |

Calculate $f(3.5)$ using Newton's interpolating polynomials of order 1 through 4. Choose your base points to attain good accuracy. What do your results indicate regarding the order of the polynomial used to generate the data in the table?

**11.6** Repeat Probs 11.1 through 11.3 using the Lagrange polynomial.

**11.7** Repeat Prob. 11.4*a* using the Lagrange polynomial.

**11.8** Repeat Prob. 11.5 using the Lagrange polynomial of order 1 through 3.

**11.9** Develop quadratic splines for the data in Prob. 11.5 and predict $f(3.5)$.

**11.10** Develop cubic splines for the data in Prob. 11.5 and predict $f(3.5)$.

## Computer-Related Problems

**11.11** Reprogram Fig. 11.7 so that it is user-friendly. Among other things:
(*a*) Place documentation statements throughout the program to identify what each section is intended to accomplish.
(*b*) Label the input and output.

**11.12** Test the program you developed in Prob. 11.11 by duplicating the computation from Example 11.5.

**11.13** Use the program you developed in Prob. 11.11 to solve Probs. 11.1 through 11.3.

**11.13** Use the program you developed in Prob. 11.11 to solve Probs. 11.1 through 11.3.

**11.14** Use the program you developed in Prob. 11.11 to solve Probs. 11.4 and 11.5. In Prob. 11.4, utilize all the data to develop first- through fifth-order polynomials. For both problems, plot the estimated error versus order.

**11.15** Repeat Probs. 11.12 and 11.13, except use the NUMERICOMP software associated with the text.

**11.16** Use the NUMERICOMP software to duplicate Examples 11.6 and 11.7.

**11.17** Develop a user-friendly program for Lagrange interpolation. Test it by duplicating Example 11.7.

**11.18** Develop a user-friendly program for cubic spline interpolation based on Fig. 11.16 and Sec. 11.4.4. Test the program by duplicating Example 11.10.

**11.19** Use the software developed in Prob. 11.18 to fit cubic splines through the data in Probs. 11.4 and 11.5. For both cases, predict $f(2.25)$.

# CHAPTER TWELVE

# CASE STUDIES: CURVE FITTING

The purpose of this chapter is to use the numerical methods for curve fitting to solve some engineering problems. As with the other case study chapters, the first example is taken from the general area of engineering economics and management. This is followed by case studies from the four major fields of engineering: chemical, civil, electrical, and mechanical.

*Case Study 12.1* deals with a trend analysis of computer sales data. The example illustrates two important points regarding curve fitting: (1) interpolating polynomials are ill-suited for fitting imprecise data and (2) extrapolation is a particularly tenuous proposition when the cause-effect relationships underlying a trend are unknown.

*Case Study 12.2,* which is taken from chemical engineering, demonstrates how a nonlinear model can be linearized and fit to data using linear regression. *Case Study 12.3* uses a similar approach but also employs polynomial interpolation in order to determine the stress-strain relationship for a civil engineering structures problem.

*Case Study 12.4* illustrates how a simple interpolating polynomial can be used to approximate a more complicated function in electrical engineering. Finally, *Case Study 12.5* demonstrates how multiple linear regression is used to analyze experimental data for a fluids problem taken from mechanical engineering.

## CASE STUDY 12.1 ENGINEERING PRODUCT SALES MODEL (GENERAL ENGINEERING)

Background: Engineers concerned with the design and manufacture of products such as automobiles, TV sets, and computers may become involved with many other aspects of the business. These involvements include sales, marketing, and distribution of the product.

You are an engineer working for the Ultimate Computer Company (see Case Study 6.1). Planning and resource allocation considerations (Case Study 9.1) require that you be able to predict how long your computers will stay on the market as a function of time. In this case study, you are provided with measured data that describes the number of computers on the market for various times up to 60 days (Fig. 12.1). You are asked to examine this

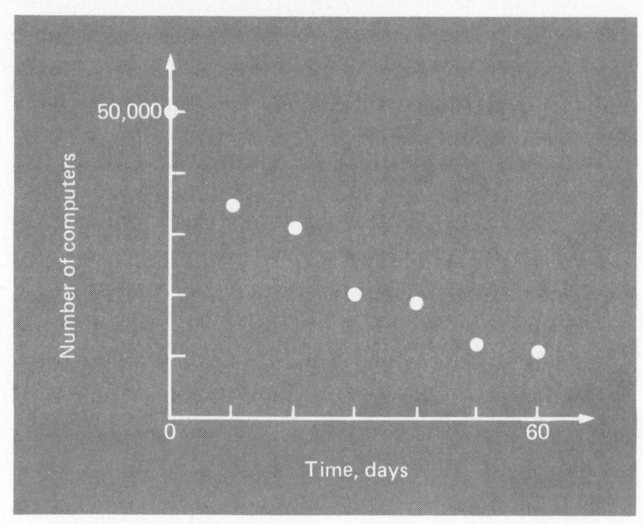


FIGURE 12.1 Number of computers on the market versus time.

data and, using extrapolation techniques, estimate how many computers will be available at day 90. The data is listed in Table 12.1.

**TABLE 12.1 Number of computers on the market as a function of time**

| Time, days | Number of computers on the market |
|---|---|
| 0 | 50,000 |
| 10 | 35,000 |
| 20 | 31,000 |
| 30 | 20,000 |
| 40 | 19,000 |
| 50 | 12,050 |
| 60 | 11,000 |

This trend analysis and extrapolation problem will be addressed using first- through sixth-order interpolating polynomials as well as with first- through sixth-order polynomial regression. The resulting curves will be used to make predictions at days 55, 65, and 90 in order to contrast interpolation and extrapolation.

Solution: Inspection of Fig. 12.1 indicates that the data is not smooth. Although the number of computers decreases with time, the rate of decrease varies from interval to interval in a seemingly random fashion. Thus, even before the analysis begins, we might suspect that extrapolation of this data could pose difficulties.

The results in Table 12.2 bear this suspicion out. Notice that there is a great deal of discrepancy between the predictions using the various tech-

**TABLE 12.2** **Results of fitting various interpolating and least-squares polynomials to the data in Table 12.1. An interpolation at $t$ = 55 and extrapolations at $t$ = 65 and 90 are shown. Notice that because the normal equations are ill-conditioned, the sixth-order least-squares polynomial differs from the more precise interpolating polynomial (recall Sec. 10.2.1).**

| | INTERPOLATION | EXTRAPOLATION | |
|---|---|---|---|
| | $t$ = 55 | $t$ = 65 | $t$ = 90 |
| Interpolating polynomials | | | |
| First-order | 11,525 | 10,475 | 7,850 |
| Second-order | 10,788 | 12,688 | 43,230 |
| Third-order | 10,047 | 16,391 | 161,750 |
| Fourth-order | 8,961 | 23,992 | 578,750 |
| Fifth-order | 7,300 | 38,942 | 1,854,500 |
| Sixth-order | 4,660 | 67,975 | 5,458,100 |
| Mean | 8,880 | 28,411 | 1,350,700 |
| Standard deviation | 2,542 | 21,951 | 2,128,226 |
| Coefficient of variation | 29% | 77% | 157% |
| Least-squares polynomials | | | |
| First-order | 9,820 | 3,573 | −12,045 |
| Second-order | 11,829 | 10,939 | 16,529 |
| Third-order | 12,040 | 8,872 | −3,046 |
| Fourth-order | 11,101 | 12,733 | 83,104 |
| Fifth-order | 11,768 | 9,366 | −78,906 |
| Sixth-order | 4,261 | 71,266 | 5,768,460 |
| Mean | 10,203 | 18,910 | 910,623 |
| Standard deviation | 2,834 | 24,233 | 2,228,408 |
| Coefficient of variation | 28% | 128% | 245% |

niques. In order to quantify this discrepancy, we have computed a mean, a standard deviation, and a coefficient of variation for the predictions. The coefficient of variation, which is the mean divided by the standard deviation (multiplied by 100 percent), provides a relative measure of the variability for each set of predictions [Eq. (IV.5)]. Notice how the coefficient of variation is lowest for the interpolated value at day 55. Also, notice how the highest discrepancy results for day 90, which represents the farthest extrapolation.

In addition, the results for the interpolating polynomials deteriorate as the order is increased, to the point that the sixth-order case yields the ridiculous prediction that 5,458,100 computers would be available on day 90. The reason for this poor performance is illustrated by Fig. 12.2, which shows the sixth-order polynomial. Because the trend suggested by the data is not smooth, the higher-order polynomial loops up and down in order to intersect each point. These "oscillations" can lead to both spurious interpolations and extrapolations of the type manifested in Fig. 12.2.

Because it is not constrained to pass through every point, regression is sometimes useful in remedying this situation. Figure 12.3, which shows the quadratic and cubic regression results, suggests that this is true for lower-order regression. Within the range of the data ($t$ = 0 to 60 days), the results

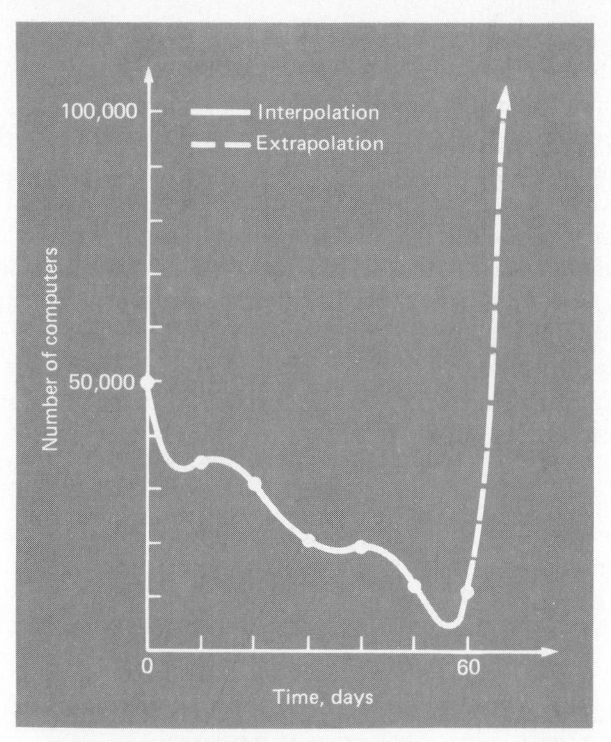


FIGURE 12.2 Plot of the sixth-order interpolating polynomial used to interpolate and extrapolate computer sales data.

of the two regressions yield fairly consistent results. However, when they are extrapolated beyond this range, the predictions diverge. At $t = 90$, the second-order regression yields the absurd result that the number of computers has increased, whereas the third-order version yields the equally ridiculous projection that there will be a negative number of computers.

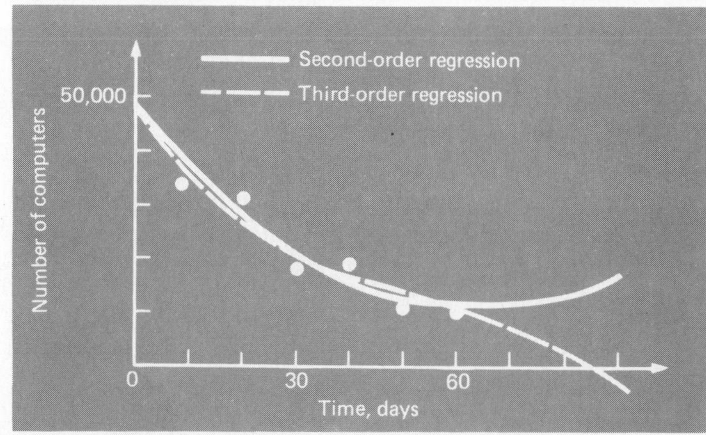


FIGURE 12.3 Plot of the second- and third-order regression curves used to interpolate and extrapolate computer sales data.

The primary reason that both interpolation and regression are ill-suited for the present case is that neither is based on a model of physical reality. In both instances, the behavior of the predictions is purely an artifact of the behavior of the numbers. For example, neither model takes account of the simple fact that beyond $t = 60$, the number of computers must lie somewhere between 0 and 11,000. Thus, if you were interested in a quick estimate of the number of computers on the market at a future time, an "eyeball" fit and extrapolation of the data would actually provide more realistic results. This is due to the fact that you are cognizant of the physical constraints of the problem and would, therefore, incorporate these constraints into your simple graphical solution. In Case Study 18.1, we will use a differential equation to develop a model that has a theoretical basis and, consequently, yields more satisfactory projections.

On the positive side, it should be noted that this example illustrates how regression has some utility for interpolation between noisy data points. However, the primary conclusion of the present case study is that extrapolation should always be performed with care and caution.

## CASE STUDY 12.2 LINEAR REGRESSION AND POPULATION MODELS (CHEMICAL ENGINEERING)

Background: Population growth models are important in many fields of engineering. Fundamental to many of the models is the assumption that the rate of change of the population ($dp/dt$) is proportional to the actual population ($p$) at any time ($t$), or, in equation form,

$$\frac{dp}{dt} = kp \qquad [12.1]$$

where $k$ is a proportionality factor called the specific growth rate and has units of time$^{-1}$. If $k$ is a constant, then the solution of Eq. (12.1) can be obtained from the theory of differential equations:

$$p(t) = p_0 e^{kt} \qquad [12.2]$$

where $p_0$ is the population when $t = 0$. It is observed that $p(t)$ in Eq. (12.2) approaches infinity as $t$ becomes large. This behavior is clearly impossible for real systems. Therefore, the model must be modified to make it more realistic.

Solution: First, it must be recognized that the specific growth rate $k$ cannot be constant as the population becomes large. This is the case because, as $p$ approaches infinity, the organism being modeled will become limited by factors such as food shortages and toxic waste production. One way to express this mathematically is to use a saturation-growth-rate model such that

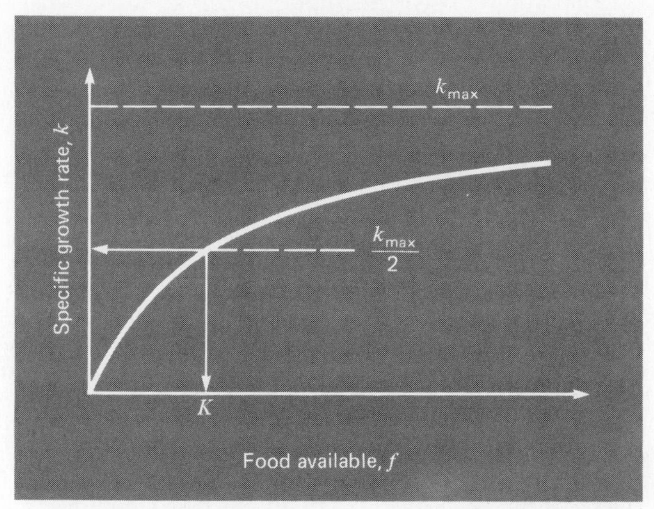


FIGURE 12.4 Plot of specific growth rate versus available food for the saturation-growth-rate model used to characterize microbial kinetics. The value $K$ is called a half-saturation constant because it conforms to the concentration where the specific growth rate is half its maximum value.

$$k = k_{max}\frac{f}{K + f} \tag{12.3}$$

where $k_{max}$ is the maximum attainable growth rate for large values of food ($f$) and $K$ is the half-saturation constant. The plot of Eq. (12.3) in Fig. 12.4 shows that when $f = K$, $k = k_{max}/2$. Therefore, $K$ is that amount of available food which supports a population growth rate equal to one-half the maximum rate.

The constants $K$ and $k_{max}$ are empirical values based on experimental measurements of $k$ for various values of $f$. As an example, suppose the population $p$ represents a yeast employed in the commercial production of beer and $f$ is the concentration of the carbon source to be fermented. Measurements of $k$ versus $f$ for the yeast are shown in Table 12.3. It is required

**TABLE 12.3 Data used to evaluate the constants for a saturation-growth-rate model to characterize microbial kinetics**

| $f$, mg/L | $k$, day$^{-1}$ | $1/f$, L/mg | $1/k$, day |
|---|---|---|---|
| 7 | 0.29 | 0.14286 | 3.448 |
| 9 | 0.37 | 0.11111 | 2.703 |
| 15 | 0.48 | 0.06666 | 2.083 |
| 25 | 0.65 | 0.04000 | 1.538 |
| 40 | 0.80 | 0.02500 | 1.250 |
| 75 | 0.97 | 0.01333 | 1.031 |
| 100 | 0.99 | 0.01000 | 1.010 |
| 150 | 1.07 | 0.00666 | 0.935 |

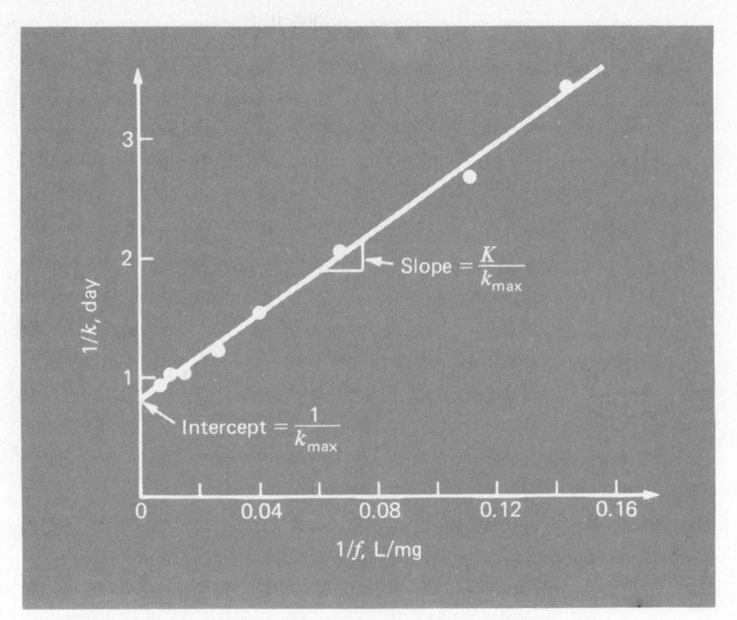


FIGURE 12.5 Linearized version of the saturation-growth-rate model. The line is a least-squares fit that is used to evaluate the model coefficients $k_{max} = 1.23$ days$^{-1}$ and $k = 22.18$ mg/L for a yeast that is used to produce beer.

to calculate $k_{max}$ and $K$ from this empirical data. This is accomplished by inverting Eq. (12.3) in a manner similar to Eq. (10.16) to yield

$$\frac{1}{k} = \frac{K + f}{k_{max} f} = \frac{K}{k_{max}} \frac{1}{f} + \frac{1}{k_{max}} \qquad [12.4]$$

By this manipulation, we have transformed Eq. (12.3) into a linear form; that is, $1/k$ is a linear function of $1/f$, with slope $K/k_{max}$ and intercept $1/k_{max}$. These values are plotted in Fig. 12.5.

Because of this transformation, the linear least-squares procedures described in Chap. 10 can be used to determine $k_{max} = 1.23$ days$^{-1}$ and $K = 22.18$ mg/L. These results combined with Eq. (12.3) are compared to the untransformed data in Fig. 12.6, and when substituted into the model in Eq. (12.1) give

$$\frac{dp}{dt} = 1.23 \frac{f}{22.18 + f} p \qquad [12.5]$$

This equation can be solved using the theory of differential equations or using numerical methods discussed in Chaps. 16 and 17 when $f(t)$ is known. If $f$ approaches zero as $p$ becomes large, then $dp/dt$ approaches zero and the population stabilizes.

The linearization of Eq. (12.3) is one way to evaluate the constants $k_{max}$ and $K$. An alternative approach, which fits the relationship in its original form, is called *nonlinear regression* (Draper and Smith, 1981). In either case, you

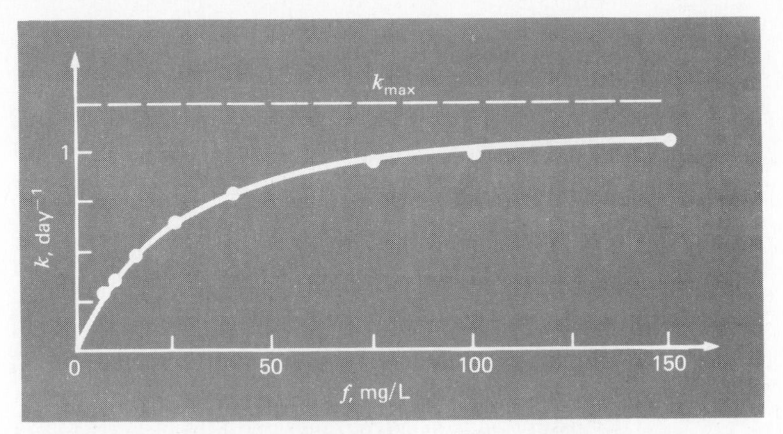


FIGURE 12.6 Fit of the saturation-growth-rate model to a yeast employed in the commercial production of beer.

can use regression analysis to calculate model coefficients using measured data. This is an example of the use of regression to test a hypothesis, as discussed in Sec. IV.1.2.

## CASE STUDY 12.3 CURVE FITTING TO DESIGN A SAILBOAT MAST (CIVIL ENGINEERING)

Background: The mast of a racing sailboat (see Case Study 15.3 for more details) has a cross-sectional area of 0.876 in$^2$ and is constructed of an experimental aluminum alloy. Tests were performed to define the relationship between *stress* (force per area) applied to the material and *strain* (deflection per unit length). These test results are shown in Fig. 12.7 and summarized in Table 12.4. It is necessary to estimate the change in length of the mast due to stress caused by wind force. The stress caused by wind can be computed using

$$\text{Stress} = \frac{\text{force in mast}}{\text{cross-sectional area of mast}}$$

For this case, a wind force of 6440.6 lb is given (note that in Case Study 15.3 we will use numerical methods to determine this value directly from wind data), and stress is calculated as

$$\text{Stress} = \frac{6440.6}{0.876} = 7350 \text{ lb/in}^2$$

This stress can then be used to determine a strain from Fig. 12.7, which, in turn, can be substituted into Hooke's law to compute the change in length of the mast:

$$\Delta L = (\text{strain})(\text{length}) \qquad [12.6]$$

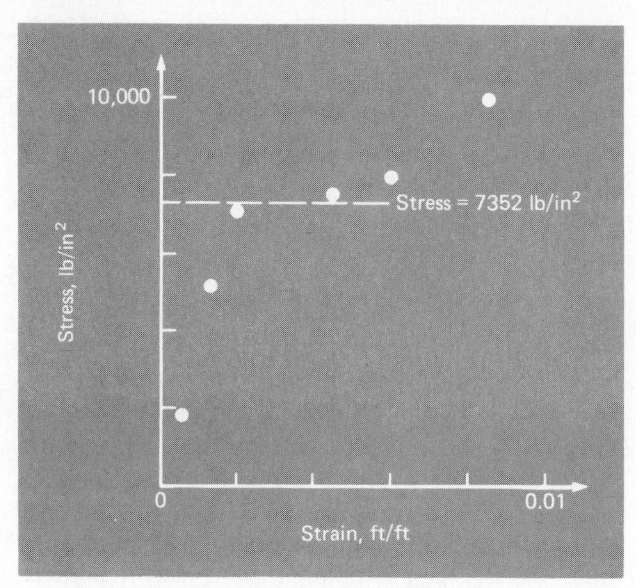


FIGURE 12.7 Stress-strain curve for an experimental aluminum alloy. For Case Study 12.3, we must obtain a strain estimate from this data conforming to a stress of 7350 lb/in$^2$.

**TABLE 12.4 Stress-strain data ordered so that the points used for interpolation are always closest to the stress of 7350 lb/in$^2$**

| Point number | Stress, lb/in$^2$ | Strain, ft/ft |
|---|---|---|
| 1 | 7,200 | 0.0020 |
| 2 | 7,500 | 0.0045 |
| 3 | 8,000 | 0.0060 |
| 4 | 5,200 | 0.0013 |
| 5 | 10,000 | 0.0085 |
| 6 | 1,800 | 0.0005 |

where length refers to the height of the mast. Thus, the problem reduces to determining the value of strain from the data in Fig. 12.7. Because no data point is available at the given stress value of 7350, the problem calls for some sort of curve fit. In the present case study, we will use two approaches: polynomial interpolation and least-squares regression.

Solution: The first approach will be to use interpolating polynomials of order 0 through 5 to estimate the strain at a stress of 7350 lb/in$^2$. To do this, the data points are ordered so that the interpolation will always use information that is closest to the unknown point (Table 12.4). Newton's interpolating polynomial can then be applied, with the results summarized in Table 12.5.

All but the zero-order polynomial yield results in fairly close agreement.

**TABLE 12.5 Results of using Newton's interpolating polynomial to predict a strain conforming to a stress of 7350 lb/in$^2$ on the basis of the information from Table 12.4**

| Order of polynomial ($n$) | $n$th order coefficient | Strain (at stress = 7350) |
|---|---|---|
| 0 | $2 \times 10^{-3}$ | $2 \times 10^{-3}$ |
| 1 | $8.33 \times 10^{-6}$ | $3.27 \times 10^{-3}$ |
| 2 | $-6.67 \times 10^{-9}$ | $3.42 \times 10^{-3}$ |
| 3 | $-3.62 \times 10^{-12}$ | $3.36 \times 10^{-3}$ |
| 4 | $1.198 \times 10^{-15}$ | $3.401 \times 10^{-3}$ |
| 5 | $2.292 \times 10^{-19}$ | $3.38 \times 10^{-3}$ |

On the basis of the analysis, we would conclude that a strain of approximately $3.4 \times 10^{-3}$ ft/ft is a reasonable estimate.

However, a cautionary note is in order. It is actually fortuitous that the estimates of strain are in such close agreement. This can be seen by examining Fig. 12.8, where the fifth-order polynomial is shown along with the

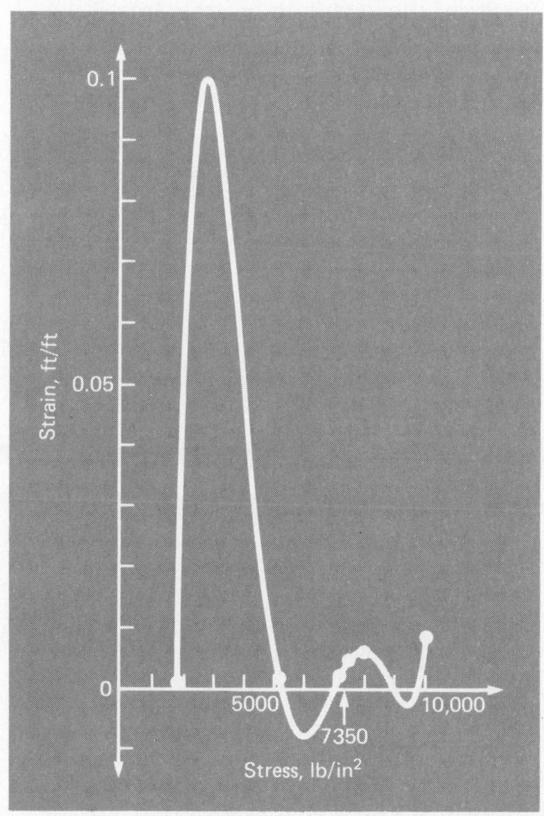


FIGURE 12.8 Plot of the fifth-order interpolating polynomial that perfectly fits the data from Table 12.4. Note that although the curve passes nicely through the three points in the vicinity of the stress of 7350, the curve oscillates wildly in other parts of the data range.

data. Notice that because three points are in close proximity to the unknown value of 7350, the interpolation would not be expected to vary significantly at this point. However, if estimates were required at other stresses, oscillations of the polynomial could lead to wild results.

The foregoing results illustrate that higher-order interpolation is ill-suited for uncertain or "noisy" data of the sort in this problem. Regression provides an alternative that is usually more appropriate for such situations.

For example, a linear regression can be used to fit a straight line through the data. The line of best fit is

$$\text{Strain} = -0.002527 + 9.562 \times 10^{-7} \text{ (stress)} \tag{12.7}$$

The line along with the data is shown in Fig. 12.9. Substituting stress = 7350 $\text{lb/in}^2$ into Eq. (12.7) yields a prediction of $4.5 \times 10^{-3}$ in/in.

One problem with a linear regression is that it yields the unrealistic result that strain is negative at a stress of zero. An alternative form of regression that avoids this unrealistic result is to fit a straight line to the logarithm (base 10) of strain versus the logarithm of stress (recall Sec. 10.1.5). The result for this case is

$$\log \text{(strain)} = -8.565 + 1.586 \log \text{(stress)}$$

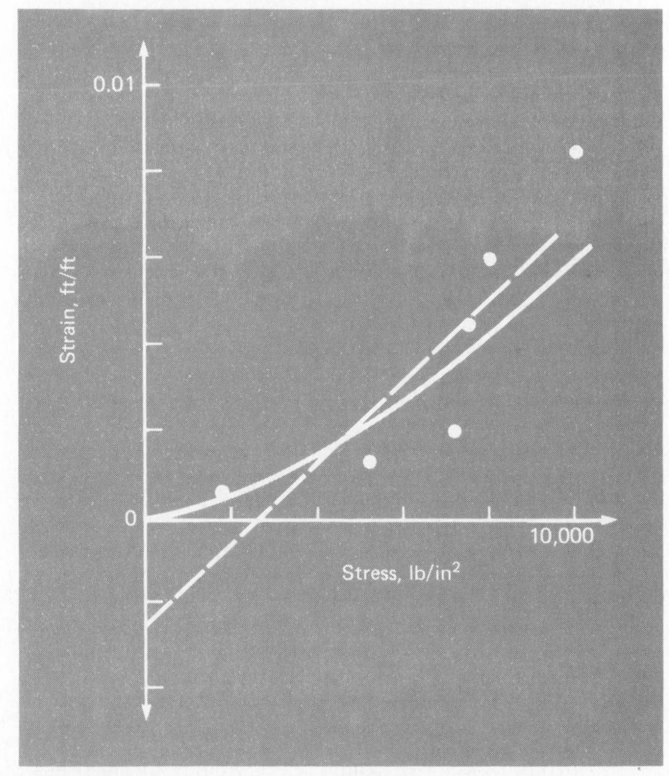


FIGURE 12.9 Plots of a linear regression line and a log-linear regression line for the strain-stress data for the sailboat mast.

This equation can be transformed back to a form that predicts strain by taking its antilogarithm to yield

$$\text{strain} = 2.723 \times 10^{-9}\,(\text{stress})^{1.586} \tag{12.8}$$

This curve is also superimposed on Fig. 12.9 where it can be seen that this version has the more physically realistic result that strain is equal to zero for a stress of zero. The curve is also somewhat more realistic because it captures some of the curvature suggested by the data. Substituting stress $= 7350$ into Eq. (12.8) yields a prediction of strain $= 3.7 \times 10^{-3}$ in/in.

Thus, polynomial interpolation and the two types of regression all yield different predictions for strain. Because of its physical realism and its more satisfactory behavior across the entire range of data, we will opt for Eq. (12.8) as providing the best prediction. Using a value of length $= 30$ ft and Eq. (12.6) yields the following result for the change in length of the mast:

$$\Delta L = (3.7 \times 10^{-3}\ \text{ft/ft})(30\ \text{ft}) = 0.11\ \text{ft}$$

## CASE STUDY 12.4 CURVE FITTING TO ESTIMATE THE RMS CURRENT (ELECTRICAL ENGINEERING)

Background: The average value of an oscillating electric current over one period may be zero. For example, suppose that the current is described by a simple sinusoid: $i(t) = \sin(2\pi t/T)$ where $T$ is the period. The average value of this function can be determined by the following equation:

$$\begin{aligned} i &= \frac{\int_0^T \sin\left(\frac{2\pi t}{T}\right) dt}{T - 0} \\ &= \frac{-\cos 2\pi + \cos 0}{T} \\ &= 0 \end{aligned}$$

The same estimate is depicted graphically in Fig. 12.10*a*. As can be seen, a net current of zero results because the positive and negative areas under the curves cancel out.

Despite the fact that the net result is zero, such current is capable of performing work and generating heat. Therefore, electrical engineers often characterize such current by

$$I_{\text{RMS}} = \sqrt{\frac{\int_0^T i^2(t)\, dt}{T}} \tag{12.9}$$

where $I_{\text{RMS}}$ is called the RMS (root-mean-square) current. By squaring the

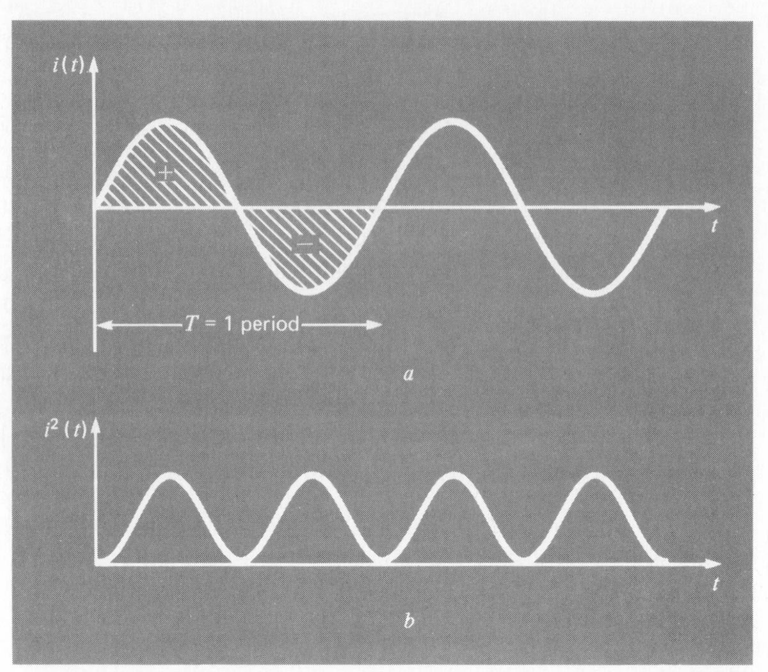


FIGURE 12.10 (*a*) Plot of an oscillating electric current. Over one period $T$ (that is, a complete cycle), the integral of such a function is zero because the positive and negative areas are equal and, hence, cancel out. In order to circumvent this result, the current is squared as in (*b*). The square root of the average of the square, which is called the root-mean-square current, provides a measure of the magnitude of the current.

current before taking the average, the problem of positive and negative signs canceling is circumvented.

For the present case study, suppose that the current for a circuit is

$$i(t) = 10e^{-t/T}\left(\sin \frac{2\pi t}{T}\right) \qquad \text{for } 0 \le t \le T/2$$
$$i(t) = 0 \qquad \text{for } T/2 < t \le T \tag{12.10}$$

Determine the RMS current by fitting a second-order polynomial that matches $i^2(t)$ exactly at $t = 0$, $T/4$, and $T/2$. Next, integrate this polynomial analytically and calculate the RMS current over the interval from 0 to $T$ using Eq. (12.9). Assume that $T = 1$ s. This result will be compared to Case Study 15.4 where other techniques will be employed to calculate the RMS current.

Solution: Using Eq. (12.10), the following points can be generated.

| $t$ | $i(t)$ | $i^2(t)$ |
|---|---|---|
| 0 | 0.000000000 | 0.00000000 |
| 1/4 | 7.788007831 | 60.65306598 |
| 1/2 | 0.000000000 | 0.00000000 |

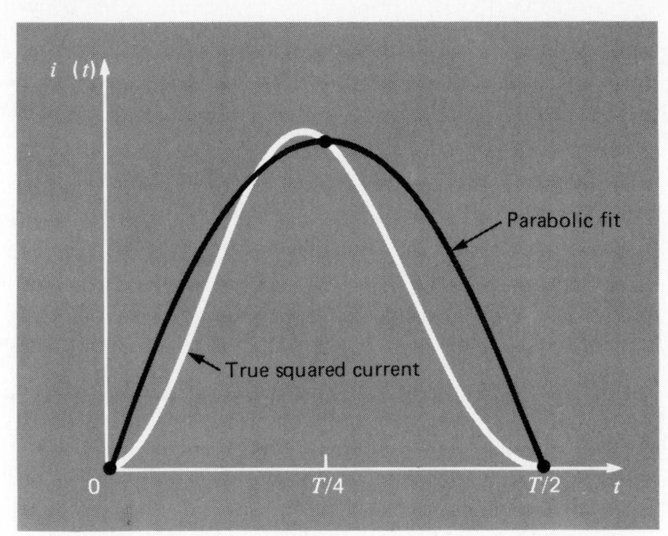


FIGURE 12.11 Plot of the actual current [Eq. (12.10)], along with the parabola that is used as an approximation.

A second-order Newton's interpolating polynomial can be fit to this data (Fig. 12.11). The polynomial is

$$i^2(t) = 242.612264t - 970.449056t(t - 1/4)$$

which can be integrated from $t = 0$ to $T/2$ ($T = 1$ s):

$$\int_0^{1/2} i^2(t)\, dt = 121.306132t^2 - 323.4830187t^3 + 121.306132t^2 \Bigg|_0^{1/2}$$

which can be evaluated as 20.21768866, which, in turn can be substituted into Eq. (12.9) to obtain an $I_{RMS} = 4.496408418$. In Case Study 15.4, we will use several numerical integration schemes to perform the same computation.

## CASE STUDY 12.5 MULTIPLE LINEAR REGRESSION FOR ANALYSIS OF EXPERIMENTAL DATA (MECHANICAL ENGINEERING)

Background: Engineering design variables are often dependent on several independent variables. Often this functional dependence is best characterized by multivariable power equations. As discussed in Sec. 10.3, a multiple linear regression of log-transformed data provides a means to evaluate such relationships.

For example, a mechanical engineering study indicates that fluid flow through a pipe is related to pipe diameter and slope (Table 12.6). Use multiple linear regression to analyze this data. Then use the resulting model to predict the flow for a pipe with a diameter of 2.5 ft and a slope of 0.025 ft/ft.

**TABLE 12.6 Experimental data for diameter, slope, and flow of concrete circular pipes**

| Experiment | Diameter, ft | Slope, ft/ft | Flow, $ft^3/s$ |
|---|---|---|---|
| 1 | 1 | 0.001 | 1.4 |
| 2 | 2 | 0.001 | 8.3 |
| 3 | 3 | 0.001 | 24.2 |
| 4 | 1 | 0.01 | 4.7 |
| 5 | 2 | 0.01 | 28.9 |
| 6 | 3 | 0.01 | 84.0 |
| 7 | 1 | 0.05 | 11.1 |
| 8 | 2 | 0.05 | 69.0 |
| 9 | 3 | 0.05 | 200.0 |

Solution: The power equation to be evaluated is

$$Q = a_0 D^{a_1} S^{a_2} \tag{12.11}$$

where $Q$ is flow (in cubic feet per second), $S$ is slope (in feet per foot), $D$ is pipe diameter (in feet), and $a_0$, $a_1$, and $a_2$ are coefficients. Taking the logarithm of this equation yields

$$\log Q = \log a_0 + a_1 \log D + a_2 \log S$$

In this form, the equation is suited for multiple linear regression because $\log Q$ is a linear function of $\log S$ and $\log D$. Using the logarithm (base 10) of the data in Table 12.6, we can generate the following normal equations expressed in matrix form [recall Eq. (10.21)]:

$$\begin{bmatrix} 9 & 2.334 & -18.903 \\ 2.334 & 0.954 & -4.903 \\ -18.903 & -4.903 & 44.079 \end{bmatrix} \begin{bmatrix} \log a_0 \\ a_1 \\ a_2 \end{bmatrix} = \begin{bmatrix} 11.691 \\ 3.945 \\ -22.207 \end{bmatrix}$$

This system can be solved using Gauss elimination for

$$\log a_0 = 1.7475$$

$$a_1 = 2.62$$

$$a_2 = 0.54$$

If $\log a_0 = 1.7475$, $a_0 = 55.9$ and Eq. (12.11) is

$$Q = 55.9 D^{2.62} S^{0.54} \tag{12.12}$$

Equation (12.12) can be used to predict flow for the case of $D = 2.5$ ft and $S = 0.025$ ft/ft, as in

$$Q = 55.9(2.5)^{2.62}(0.025)^{0.54} = 84.1 \text{ ft}^3/\text{s}$$

It should be noted that Eq. (12.12) can be used for other purposes besides computing flow. For example, the slope is related to head loss $h_L$ and pipe length $L$ by $S = h_L/L$. If this relationship is substituted into Eq. (12.12) and the resulting formula solved for $h_L$, the following equation can be developed:

$$h_L = \frac{L}{1721} Q^{1.85} D^{4.85}$$

This relationship is called the *Hazen-Williams equation.*

# PROBLEMS

## General Engineering

**12.1** Reproduce the computations performed in Case Study 12.1 using your own software.

**12.2** Perform the same computation as in Case Study 12.1, but with the number of computers on the market on days 50 and 60 modified slightly to 12,000 and 11,050.

**12.3** If a sum of money is deposited at a certain interest rate, economic tables can be used to determine the accumulated sum at a later time. For example, the following information is contained in an economics table for the future worth of a deposit after 20 years:

| Interest rate, % | $F/P$ ($n$ = 20 yr) |
|---|---|
| 15 | 16.366 |
| 20 | 38.337 |
| 25 | 86.736 |
| 30 | 190.05 |

where $F/P$ is the ratio of the future worth to the present value. Thus, if $P$ = \$10,000 was deposited, after 20 years at 20 percent interest it would be worth:

$$F = (F/P)P = 38.337(10{,}000) = \$383{,}370$$

Use linear, quadratic, and cubic interpolation to determine the future value of \$25,000 deposited at 23.6 percent interest. Interpret your results from the perspective of the lending institution.

**12.4** Use the information given in Prob. 12.3, but suppose that you have invested \$40,000 and are told that 20 years later the lender will pay you back \$2,800,000. Use linear, quadratic, and cubic interpolation to determine what interest rate you are being given.

**12.5** Suppose you win a sweepstakes and are given the choice of accepting \$2 million now or \$700,000 per year for the next 5 years. The relationship between a present value $P$ and an annual series of payments $A$ is given by the following information from an economics table:

| **Interest rate, %** | **$A/P$ ($n = 5$ yr)** |
|---|---|
| 15 | 0.29832 |
| 20 | 0.33438 |
| 25 | 0.37185 |
| 30 | 0.41058 |

where $A/P$ is the ratio of the annual payment to the present worth. Thus, if the interest rate is 15 percent, the five annual payments $A$ that are equivalent to a single present payment ($P$ = \$2 million) is calculated as

$$A = (A/P)P = 0.29832(2{,}000{,}000) = \$596{,}640$$

Use interpolation to determine the interest rate at which taking the \$2 million becomes the better decision.

**12.6** You are performing a study to determine the relationship between the upward drag force and velocity for the falling parachutist. A number of experiments yields the following information for velocity ($v$ in centimeters per second) and upward drag force ($F_u$ in $10^6$ dynes):

| $v$ | 1000 | 2000 | 3000 | 4000 | 5000 |
|---|---|---|---|---|---|
| $F_u$ | 5 | 15.3 | 29.3 | 46.4 | 66.3 |

Plot $F_u$ versus $v$ and then use regression to determine the relationship between upward drag and velocity.

## Chemical Engineering

**12.7** Reproduce the computations performed in Case Study 12.2 using your own software.

**12.8** Perform the same computation as in Case Study 12.2, but use polynomial regression to fit a parabola to the data. Discuss your results.

**12.9** Perform the same computation as in Case Study 12.2, but use linear regression and transformations to fit the data with a power equation. Ignore the first point when fitting the equation.

**12.10** You perform experiments and determine the following values of heat capacity ($c$) at various temperatures ($T$) for a metal:

| $T$ | −50 | −20 | 10 | 70 | 100 | 120 |
|---|---|---|---|---|---|---|
| $c$ | 0.125 | 0.128 | 0.134 | 0.144 | 0.150 | 0.155 |

Use regression to determine a model to predict $c$ as a function of $T$.

**TABLE P12.11 Dependency of dissolved oxygen concentration on temperature and chloride concentration**

| | DISSOLVED OXYGEN (mg/L) FOR STATED CONCENTRATIONS OF CHLORIDE | | |
|---|---|---|---|
| Temperature, °C | Chloride = 0 mg/L | Chloride = 10,000 mg/L | Chloride = 20,000 mg/L |
| 5 | 12.8 | 11.6 | 10.5 |
| 10 | 11.3 | 10.3 | 9.2 |
| 15 | 10.0 | 9.1 | 8.2 |
| 20 | 9.0 | 8.2 | 7.4 |
| 25 | 8.2 | 7.4 | 6.7 |
| 30 | 7.4 | 6.8 | 6.1 |

**12.11** The saturation concentration of dissolved oxygen in water as a function of temperature and chloride concentration is listed in Table P12.11. Use interpolation to estimate the dissolved oxygen level for $T = 22.4$°C with chloride = 10,000 mg/L.

**12.12** For the data in Table P12.11, use polynomial interpolation to derive a predictive equation for dissolved oxygen concentration as a function of temperature for the case where chloride concentration is equal to 20,000 mg/L.

**12.13** Use polynomial regression to perform the same task as in Prob. 12.12.

**12.14** Use multiple linear regression and logarithmic transformations to derive an equation to predict dissolved oxygen concentration as a function of temperature and chloride concentration. Evaluate your results.

## Civil Engineering

**12.15** Reproduce the computations performed in Case Study 12.3 using your own software.

**12.16** Perform the same computations as in Case Study 12.3, but use second-order polynomial regression to relate strain to stress.

**12.17** Perform the same computations as in Case Study 12.3, but use an exponential formulation to relate strain to stress.

**12.18** Perform the same computations as in Case Study 12.3, but use polynomial interpolation to evaluate the $\Delta L$ if the stress is 7700 lb/in$^2$.

## Electrical Engineering

**12.19** Reproduce the computations performed in Case Study 12.4 using your own software.

**12.20** Perform the same computations as in Case Study 12.4, but fit and integrate a third-order polynomial that matches $i^2(t)$ exactly at $t = 0$, $T/6$, $T/3$, and $T/2$.

**12.21** You measure the voltage drop $v$ across a resistor for a number of different values of current $i$. The results are

| $i$ | 0.25 | 0.75 | 1.25 | 1.5 | 2.0 |
|---|---|---|---|---|---|
| $v$ | −0.23 | −0.33 | 0.70 | 1.88 | 6.00 |

Use polynomial interpolation to estimate the voltage drop for $i = 0.9$. Interpret your results.

**12.22** Duplicate the computation for Prob. 12.21, but use polynomial regression to derive a cubic equation to fit the data. Plot and evaluate your results.

## Mechanical Engineering

**12.23** Reproduce the computations performed in Case Study 12.5 using your own software.

**12.24** Based on Table 12.6, use linear and quadratic interpolation to compute $Q$ for $D = 1.23$ ft and $S = 0.01$ ft/ft. Compare your results with the same value computed with the formula derived in Case Study 12.5.

**12.25** Reproduce Case Study 12.5, but develop an equation to predict diameter as a function of slope and flow. Compare your results with the formula from Case Study 12.5 and discuss your results.

**12.26** Kinematic viscosity of water, $\nu$, is related to temperature in the following manner:

| $T$(°F) | 40 | 50 | 60 | 70 | 80 |
|---|---|---|---|---|---|
| $\nu$ ($10^{-5}$ ft$^2$/s) | 1.66 | 1.41 | 1.22 | 1.06 | 0.93 |

Plot this data. Use interpolation to predict $\nu$ at $T = 62$°F.

**12.27** Repeat Prob. 12.26, but use regression.

## Miscellaneous

**12.28** Read all the case studies in Chap. 12. On the basis of your reading and experience, make up your own case study for any one of the fields of engineering. This may involve modifying or reexpressing one of our case studies. However, it can also be totally original. As with our examples, it must be drawn from an engineering problem context and must demonstrate the use of numerical methods for curve fitting. Write up your results using our case studies as models.

# EPILOGUE: PART IV

## IV.4 TRADE-OFFS

Table IV.4 provides a summary of the trade-offs involved in curve fitting. The techniques are divided into two broad categories depending on the uncertainty of the data. For imprecise measurements, regression is used to develop a "best" curve that fits the overall trend of the data without necessarily passing through any of the individual points. For precise measurements, interpolation is used to develop a curve that passes directly through each of the points.

All the regression methods are designed to fit functions that minimize the sum of the squares of the residuals between the data and the function. Such methods are termed least-squares regression. Linear least-squares regression is used for cases where a dependent and an independent variable are related to each other in a linear fashion. For situations where a dependent and an independent variable exhibit a curvilinear relationship, several options are available. In some cases, transformations can be used to linearize the relationship. In these instances, linear regression can be applied to the transformed variables in order to determine the best straight line. Alternatively, polynomial regression can be employed to fit a curve directly to the data.

Multiple linear regression is utilized when a dependent variable is a linear function of two or more independent variables. Logarithmic transformations can also be applied to this type of regression for some cases where the multiple dependency is curvilinear.

Polynomial interpolation is designed to fit a unique $n$th-order polynomial that passes exactly through $n + 1$ precise data points. This polynomial is presented in two alternative formats. Newton's divided-difference interpolating polynomial is ideally suited for those cases where the proper order of the polynomial is unknown. Newton's polynomial is appropriate for such situations because it is easily programmed in a format to compare results with different orders. In

**TABLE IV.4 Comparison of the characteristics of alternative methods for curve fitting**

| Method | Error associated with data | Match of individual data points | Number of points matched exactly | Programming effort | Comments |
|---|---|---|---|---|---|
| Regression | | | | | |
| Linear regression | Large | Approximate | 0 | Easy | |
| Polynomial regression | Large | Approximate | 0 | Moderate | Round-off error becomes pronounced for higher-order versions |
| Multiple linear regression | Large | Approximate | 0 | Moderate | |
| Interpolation | | | | | |
| Newton's divided-difference polynomials | Small | Exact | $n + 1$ | Easy | Usually preferred for exploratory analyses |
| Lagrange polynomials | Small | Exact | $n + 1$ | Easy | Usually preferred when order is known |
| Cubic splines | Small | Exact | Piecewise fit of data points | Moderate | First and second derivatives equal at knots |

addition, an error estimate can be simply incorporated into the technique. Thus, you can compare and choose from results using several different-order polynomials.

The Lagrange interpolating polynomial is an alternative formulation that is appropriate when the order is known a priori. For these situations, the Lagrange version is somewhat simpler to program and does not require the computation and storage of finite divided differences.

The final method of curve fitting is spline interpolation. This technique fits a low-order polynomial to each interval between data points. The fit is made smooth by setting the derivatives of adjacent polynomials to the same value at their connecting points. The cubic spline is the most common version. Splines are of great utility when fitting data that is generally smooth but exhibits local areas of roughness. Such data tends to induce wild oscillations in higher-order interpolating polynomials. Cubic splines are less prone to these oscillations because they are limited to third-order variations.

## IV.5 IMPORTANT RELATIONSHIPS AND FORMULAS

Table IV.5 summarizes important information that was presented in Part IV. This table can be consulted to quickly access important relationships and formulas.

## IV.6 ADVANCED METHODS AND ADDITIONAL REFERENCES

Although we have reviewed a number of curve-fitting techniques, there are other methods that have utility in engineering practice. For example, *orthogonal polynomials* can be employed to develop an alternative method for polynomial regression. This technique has utility because it is not susceptible to ill conditioning when higher-order polynomials must be fit. Information on *orthogonal polynomials* can be found in Shampine and Allen (1973) and Guest (1961).

There are a number of methods to directly develop a least-squares fit for a nonlinear equation. These nonlinear regression techniques include the *Gauss-Newton, Marquardt's,* and *steepest descent methods.* Information regarding these techniques and regression in general can be found in Draper and Smith (1981).

All the methods in Part IV have been couched in terms of fitting a curve to data points. In addition, you may also desire to fit a curve to another

**TABLE IV.5 Summary of important information presented in Part IV**

| Method | Formulation | Graphical interpretation | Errors |
|---|---|---|---|
| Linear regression | $y = a_0 + a_1 x$<br>where $a_1 = \dfrac{n\Sigma x_i y_i - \Sigma x_i \Sigma y_i}{n\Sigma x_i^2 - (\Sigma x_i)^2}$<br>$a_0 = \bar{y} - a_1 \bar{x}$ | | $s_{y/x} = \sqrt{\dfrac{S_r}{n-2}}$<br>$r^2 = \dfrac{S_t - S_r}{S_t}$ |
| Polynomial regression | $y = a_0 + a_1 x + \cdots + a_m x^m$<br>(Evaluation of $a$'s equivalent to solution of $m + 1$ linear simultaneous equations) | | $s_{y/x} = \sqrt{\dfrac{S_r}{n-(m+1)}}$<br>$r^2 = \dfrac{S_t - S_r}{S_t}$ |
| Multiple linear regression | $y = a_0 + a_1 x + \cdots + a_m x_m$<br>(Evaluation of $a$'s equivalent to solution of $m + 1$ linear simultaneous equations) | | $s_{y/x_1,\ldots,x_m} = \sqrt{\dfrac{S_r}{n-(m+1)}}$<br>$r^2 = \dfrac{S_t - S_r}{S_t}$ |
| Newton's divided-difference interpolating polynomial* | $f_2(x) = b_0 + b_1(x - x_0) + b_2(x - x_0)(x - x_1)$<br>where $b_0 = f(x_0)$<br>$b_1 = f[x_1, x_0]$<br>$b_2 = f[x_2, x_1, x_0]$ | | $R_2 = (x - x_0)(x - x_1)(x - x_2)\dfrac{f^{(3)}(\xi)}{6}$<br>or<br>$R_2 = (x - x_0)(x - x_1)(x - x_2)f[x_3, x_2, x_1, x_0]$ |
| Lagrange interpolating polynomial* | $f_2(x) = f(x_0)\left(\dfrac{x - x_1}{x_0 - x_1}\right)\left(\dfrac{x - x_2}{x_0 - x_2}\right)$<br>$+ f(x_1)\left(\dfrac{x - x_0}{x_1 - x_0}\right)\left(\dfrac{x - x_2}{x_1 - x_2}\right)$<br>$+ f(x_2)\left(\dfrac{x - x_0}{x_2 - x_0}\right)\left(\dfrac{x - x_1}{x_2 - x_1}\right)$ | | $R_2 = (x - x_0)(x - x_1)(x - x_2)\dfrac{f^{(3)}(\xi)}{6}$<br>or<br>$R_2 = (x - x_0)(x - x_1)(x - x_2)f[x_3, x_2, x_1, x_0]$ |
| Cubic Splines | A cubic:<br>$a_i x^3 + b_i x^2 + c_i x + d_i$<br>is fit to each interval between knots. First and second derivatives are equal at each knot | $a_1 x^3 + b_1 x^2 + c_1 x + d_1$<br>knot<br>$a_2 x^3 + b_2 x^2 + c_2 x + d_2$ | |

* Note, for simplicity, second-order versions are shown.

curve. The primary motivation for such *functional approximation* is to represent a complicated function by a simpler version that is easier to manipulate. One way to do this is to use the complicated function to generate a table of values. Then the techniques discussed in this part of the book can be used to fit polynomials to these discrete values.

Beyond this approach, there are a variety of alternative, and usually preferable, methods for functional approximation. For example, if the function is continuous and differentiable, it can be fit with a truncated Taylor series. However, this strategy is flawed because the truncation error increases as we move away from the base point of the expansion. Thus, we will have a very good prediction at one point in the interval and poorer predictions elsewhere.

An alternative approach is based on the *minimax principle* (recall Fig. 10.2*c*). This principle specifies that the coefficients of the approximating polynomial be chosen so that the maximum discrepancy is as small as possible. Thus, although the approximation may not be as good as that given by the Taylor series at the base point, it is generally better across the entire range of the fit. *Chebyshev economization* is an example of an approach for functional approximation based on such a strategy (Ralston and Rabinowitz, 1978; Gerald and Wheatley, 1984; and Carnahan, Luther, and Wilkes, 1969).

A final method for functional approximation is to use trigonometric functions. *Fast Fourier transforms* are an example of such an approach that has broad applicability in engineering practice (Brigham, 1974; Davis and Rabinowitz, 1975; and Gerald and Wheatley, 1984).

In summary, the foregoing is intended to provide you with avenues for deeper exploration of the subject. Additionally, all the above references provide descriptions of the basic technques covered in Part IV. We urge you to consult these alternative sources to broaden your understanding of numerical methods for curve fitting.*

*Books are referenced only by author here; a complete bibliography will be found at the back of this text.

# PART FIVE

# INTEGRATION

## V.1 MOTIVATION

According to the dictionary definition, *to integrate* means "to bring together, as parts, into a whole; to unite; to indicate the total amount . . . ." Mathematically, integration is represented by

$$I = \int_a^b f(x)\, dx \qquad \text{[V.1]}$$

which stands for the integral of the function $f(x)$ with respect to the variable $x$, evaluated between the limits $x = a$ to $x = b$.

As suggested by the dictionary definition, the "meaning" of Eq. (V.1) is the *total value* or *summation* of $f(x)\, dx$ over the range from $x = a$ to $b$. In fact, the symbol $\int$ is actually a stylized capital S that is intended to signify the close connection between integration and summation (Thomas and Finney, 1979).

Figure V.1 represents a graphical manifestation of the concept. For functions lying above the $x$ axis, the integral expressed by Eq. (V.1) corresponds to the area under the curve of $f(x)$ between $x = a$ and $b$. We will have numerous occasions to refer back to this graphical conception as we develop mathematical formulas for numerical integration. In fact, most numerical methods for integration, including the simple approaches discussed in the next section, can be interpreted from a graphical perspective.

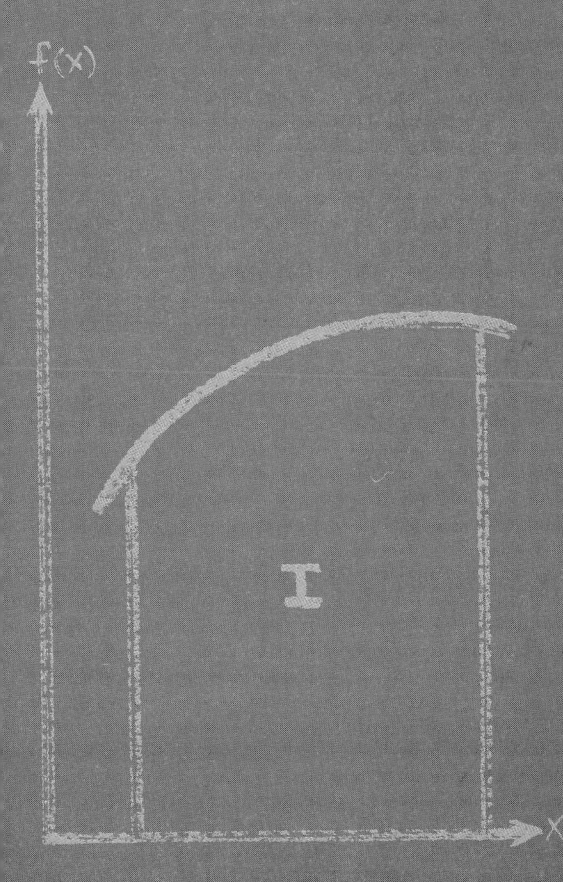


### V.1.1 Precomputer Methods for Integration

The function to be integrated will typically be in one of three forms. It will be

1. A simple continuous function such as a polynomial, an exponential, or a trigonometric function
2. A complicated continuous function that is difficult or impossible to integrate directly, or

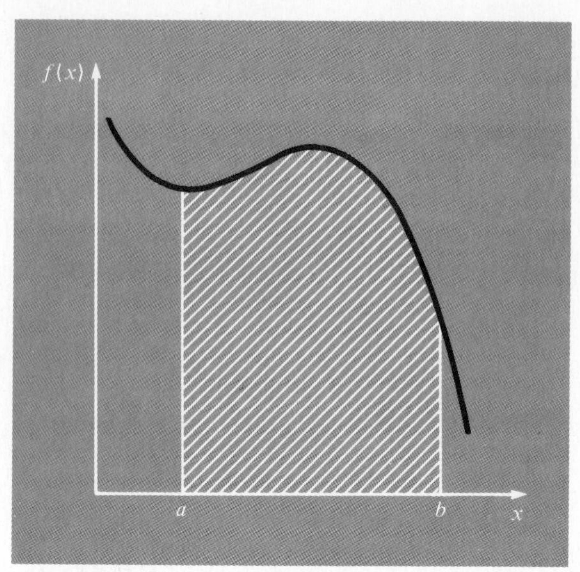


FIGURE V.1 Graphical representation of the integral of $f(x)$ between the limits $x = a$ to $b$. The integral is equivalent to the area under the curve.

**3.** A tabulated function where values of $x$ and $f(x)$ are given at a number of discrete points, as is often the case with experimental data

In the first case, the integral of a simple function may be evaluated exactly using analytical techniques you learned in calculus. For the latter two cases, however, approximate methods must be employed.

A simple, intuitive approach is to plot the function on a grid (Fig. V.2) and count the number of boxes that approximate the area. This number multiplied by the area of each box provides a rough estimate of the total area under the curve. This estimate can be refined, at the expense of additional effort, by using a finer grid.

Another commonsense approach is to divide the area into vertical segments, or strips, with a height equal to the function value at the midpoint of each strip (Fig. V.3). The area of the rectangles can then be calculated and summed to estimate the total area. In this approach, it is assumed that the value at the midpoint provides a valid approximation of the average height of the function for each strip. As with the grid method, refined estimates are possible by using more (and thinner) strips to approximate the integral.

Although such simple approaches have utility for quick estimates, alternative techniques called *numerical integration,* or *quadrature, methods* are available for the same purpose. These methods, which

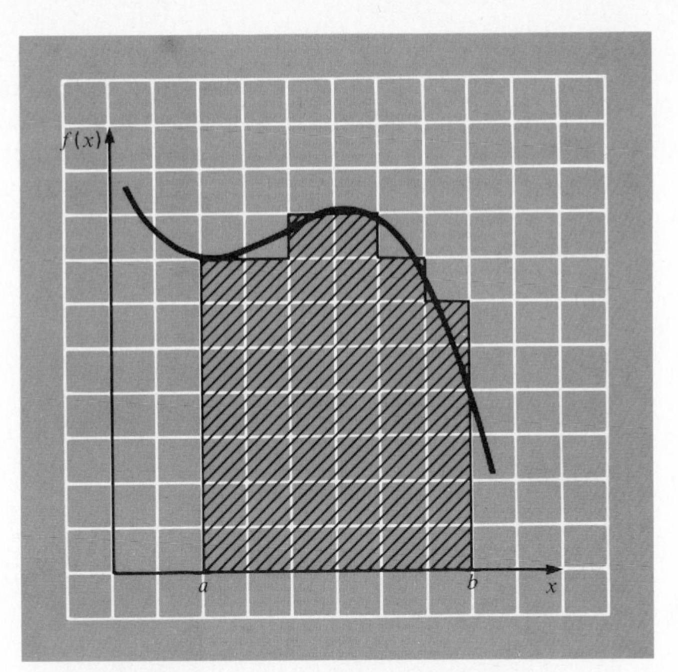


FIGURE V.2 The use of a grid to approximate an integral.

are actually easier to implement than the grid approach, are similar in spirit to the strip method. That is, function heights are multiplied by strip widths and summed in order to estimate the integral. However, through clever choices of weighting factors, the resulting estimate can be made more accurate than that from the simple "strip method."

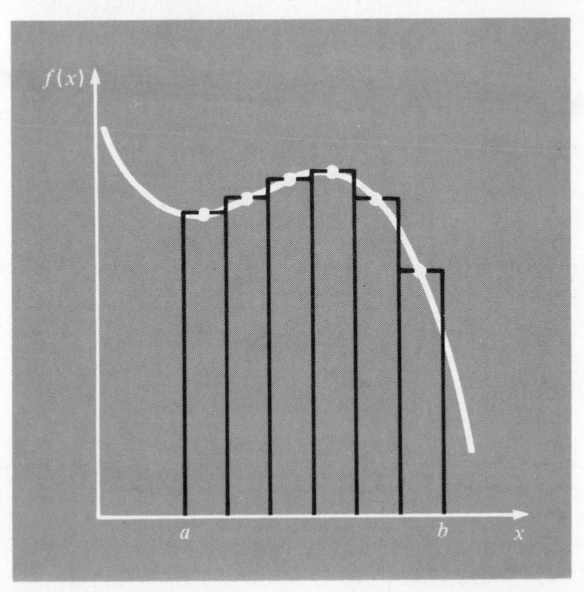


FIGURE V.3 The use of rectangles, or strips, to approximate the integral.

As in the simple strip method, numerical integration techniques utilize data at discrete points. Because tabulated information is already in such a form, it is naturally compatible with many numerical integration methods. Although continuous functions are not originally in discrete form, it is usually a simple proposition to use the given equation to generate a table of values. As depicted in Fig. V.4, this table can then be employed to perform a numerical integration.

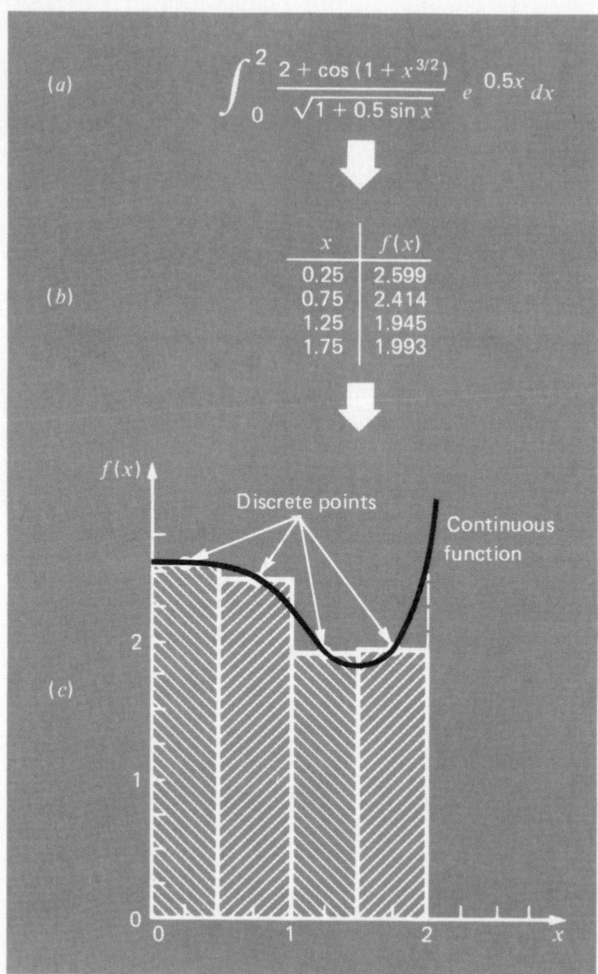


FIGURE V.4 Application of a numerical integration method: (*a*) A complicated, continuous function. (*b*) Table of discrete values of $f(x)$ generated from the function. (*c*) Use of a numerical method (the strip method here) to estimate the integral on the basis of the discrete points. For a tabulated function, the data is already in tabular form (*b*); therefore, step (*a*) is unnecessary.

## V.1.2 Numerical Integration and Engineering Practice

The integration of a function has so many engineering applications that you were probably required to take integral calculus in your first year at college. Many specific examples of such applications could be given in all fields of engineering.

One of these is the use of integration to determine the mean of continuous functions. In Part IV, you were introduced to the concept of the mean of *n* *discrete* data points [recall Eq. (IV.1)]:

$$\text{Mean} = \frac{\sum_{i=1}^{n} y_i}{n} \qquad [\text{V.2}]$$

where $y_i$ are individual measurements. The determination of the mean of discrete points is depicted in Fig. V.5*a*.

In contrast, suppose that *y* is a *continuous function* of an independent variable *x*, as depicted in Fig. V.5*b*. For this case, there are an infinite number of values between *a* and *b*. Just as Eq. (V.2) can be applied to determine the mean of the discrete readings, you might also be inter-

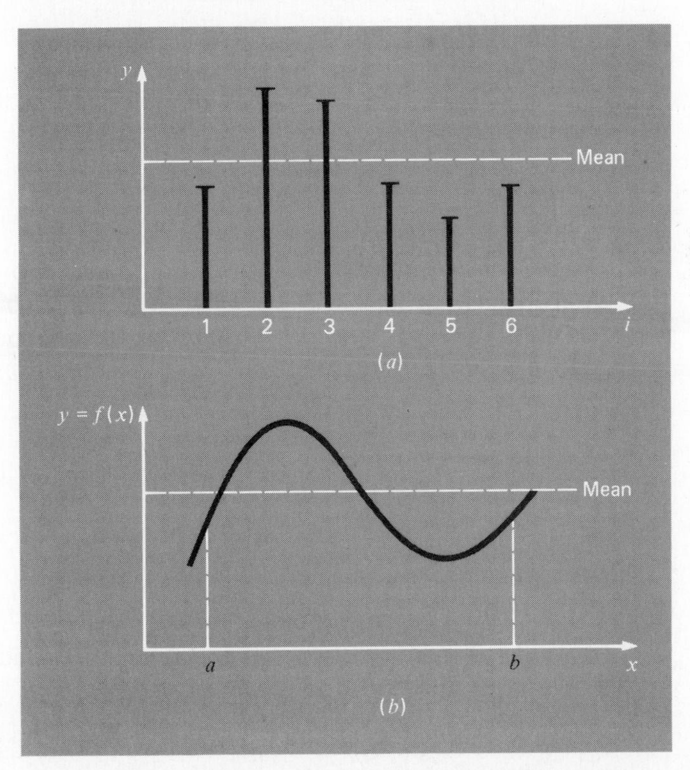


FIGURE V.5 An illustration of the mean of (*a*) discrete and (*b*) continuous data.

ested in computing the mean or average of the continuous function $y = f(x)$ for the interval from $a$ to $b$. Integration is used for this purpose, as specified by the formula

$$\text{Mean} = \frac{\int_a^b f(x)\, dx}{b - a} \qquad [\text{V.3}]$$

This formula has hundreds of engineering applications. For example, it is used to calculate the center of gravity of irregular objects in mechanical and civil engineering and to determine the root-mean-square current in electrical engineering.

Integrals are also employed by engineers to evaluate the total amount or quantity of a given physical variable. The integral may be evaluated over a line, an area, or a volume. For example, the total mass of chemical contained in a reactor is given as the product of the concentration of chemical and the reactor volume, or

$$\text{Mass} = \text{concentration} \times \text{volume}$$

where concentration has units of mass per volume. However, suppose that concentration varies from location to location within the reactor. In this case, it is necessary to sum the products of local concentrations $c_i$ and corresponding elemental volumes ($\Delta V_i$):

$$\text{Mass} = \sum_{i=1}^{n} c_i \, \Delta V_i$$

where $n$ is the number of discrete volumes. For the continuous case, where $c(x, y, z)$ is a known function and $x$, $y$, and $z$ are independent variables designating position in cartesian coordinates, integration can be used for the same purpose:

$$\text{Mass} = \iiint c(x, y, z)\, dx\, dy\, dz$$

or,

$$\text{Mass} = \iiint_V c(V)\, dV$$

which is referred to as a *volume integral.* Notice the strong analogy between summation and integration.

Similar examples could be given in other fields of engineering. For example, the total rate of energy transfer across a plane where the flux (in calories per square centimeter per second) is a function of position is given by

$$\text{Heat transfer} = \iint_A \text{flux}\, dA$$

which is referred to as an *areal integral* where $A$ = area.

Similarly, for the one-dimensional case, the total weight of a variable density rod is given by

$$w = A \int_0^L \rho(x)\, dx$$

where $w$ is the total weight (in pounds), $L$ is the length of the rod (in feet), $\rho(x)$ is the known density (in pounds per cubic foot) as a function of length $x$ (in feet), and $A$ is the cross-sectional area of rod (in square feet).

Finally, integrals are used to evaluate rate equations. Suppose the velocity of a particle is a known continuous function of time $v(t)$. The total distance $d$ traveled by this particle over a time $t$ is given by

$$d = \int_0^t v(t)\, dt \qquad \text{[V.4]}$$

These are just a few of the applications of integrals that you might face regularly in the pursuit of your profession. When the functions to be integrated are simple, you will normally choose to integrate them analytically. For example, in the falling parachutist problem, we determined the solution for velocity as a function of time [Eq. (1.8)]. This relationship could be substituted into Eq. (V.4), which could then be integrated easily to determine how far the parachutist fell over a time period $t$. For this case, the integral is simple to evaluate. However, it is difficult or impossible when the function is complicated, as is typically the case in more realistic examples. In addition, the underlying function is often unknown and defined only by measurement at discrete points. For both these cases, you must have the ability to obtain approximate values for integrals using numerical techniques. Several such techniques will be discussed in this part of the book.

## V.2 MATHEMATICAL BACKGROUND

In high school or during your first years of college, you were introduced to *integral calculus.* There you learned techniques to obtain analytical or exact solutions of both indefinite and definite integrals. *Indefinite integration,* which primarily involves determining a function whose derivative is given, will be discussed in Part VI.

The present part of the book is devoted to *definite integration,* which deals with determining an integral between specified limits, as in

$$I = \int_a^b f(x)\,dx \tag{V.5}$$

According to the *fundamental theorem* of integral calculus, Eq. (V.5) is evaluated as

$$\int_a^b f(x)\,dx = F(x)\Big|_a^b$$

where $F(x)$ is the integral of $f(x)$—that is, any function such that $F'(x) = f(x)$. The nomenclature on the right-hand side stands for

$$F(x)\Big|_a^b = F(b) - F(a) \tag{V.6}$$

An example of a definite integral is

$$I = \int_0^{0.8} (0.2 + 25x - 200x^2 + 675x^3 - 900x^4 + 400x^5)\,dx \tag{V.7}$$

For this case, the function is a simple polynomial that can be integrated analytically by evaluating each term according to the rule

$$\int_a^b x^n\,dx = \frac{x^{n+1}}{n+1}\Big|_a^b \tag{V.8}$$

where $n$ cannot equal $-1$. Applying this rule to each term in Eq. (V.7) yields

$$I = 0.2x + 12.5x^2 - \frac{200}{3}x^3 + 168.75x^4 - 180x^5 + \frac{400}{6}x^6\Big|_0^{0.8}$$

which can be evaluated according to Eq. (V.6) as $I = 1.64053334$. This value is equal to the area under the original polynomial [Eq. (V.7)] between $x = 0$ and 0.8.

The foregoing integration depends on knowledge of the rule expressed by Eq. (V.8). Other functions follow different rules. These "rules" are all merely instances of antidifferentiation, that is, finding $F(x)$ so that $F'(x) = f(x)$. Consequently, analytical integration depends on prior knowledge of the answer. Such knowledge is acquired by training and experience. Many of the rules are summarized in handbooks and in tables of integrals. We list some commonly encountered integrals in Table V.1. However, many functions of practical importance are too complicated to be contained in such tables. One reason why the tech-

**TABLE V.1 Some simple integrals that are used in Part V. The *a* and *b* in this table are constants and should not be confused with the limits of integration discussed in the text**

$$\int u\,dv = uv - \int v\,du$$

$$\int u^n\,du = \frac{u^{n+1}}{n+1} + C \qquad n \neq -1$$

$$\int a^{bx}\,dx = \frac{a^{bx}}{b \ln a} + C \qquad a > 0,\ a \neq 1$$

$$\int \frac{dx}{x} = \ln|x| + C$$

$$\int \sin(ax + b) = -\frac{1}{a}\cos(ax + b) + C$$

$$\int \cos(ax + b)\,dx = \frac{1}{a}\sin(ax + b) + C$$

$$\int \ln|x|\,dx = x\ln|x| - x + C$$

$$\int e^{ax}\,dx = \frac{e^{ax}}{a} + C$$

$$\int xe^{ax}\,dx = \frac{e^{ax}}{a^2}(ax - 1) + C$$

$$\int \frac{dx}{a + bx^2} = \frac{1}{\sqrt{ab}}\tan^{-1}\frac{\sqrt{ab}}{a}x + C$$

niques in the present part of the book are so valuable is that they provide a means to evaluate relationships such as Eq. (V.7) without knowledge of the rules.

# V.3 ORIENTATION

Before proceeding to the numerical methods for integration, some further orientation might be helpful. The following is intended as an overview of the material discussed in Part V. In addition, we have formulated some objectives to help focus your efforts when studying the material.

## V.3.1 Scope and Preview

Figure V.6 provides an overview of Part V. *Chapter 13* is devoted to the most common approaches for numerical integration—the *Newton-Cotes formulas.* These relationships are based on replacing a

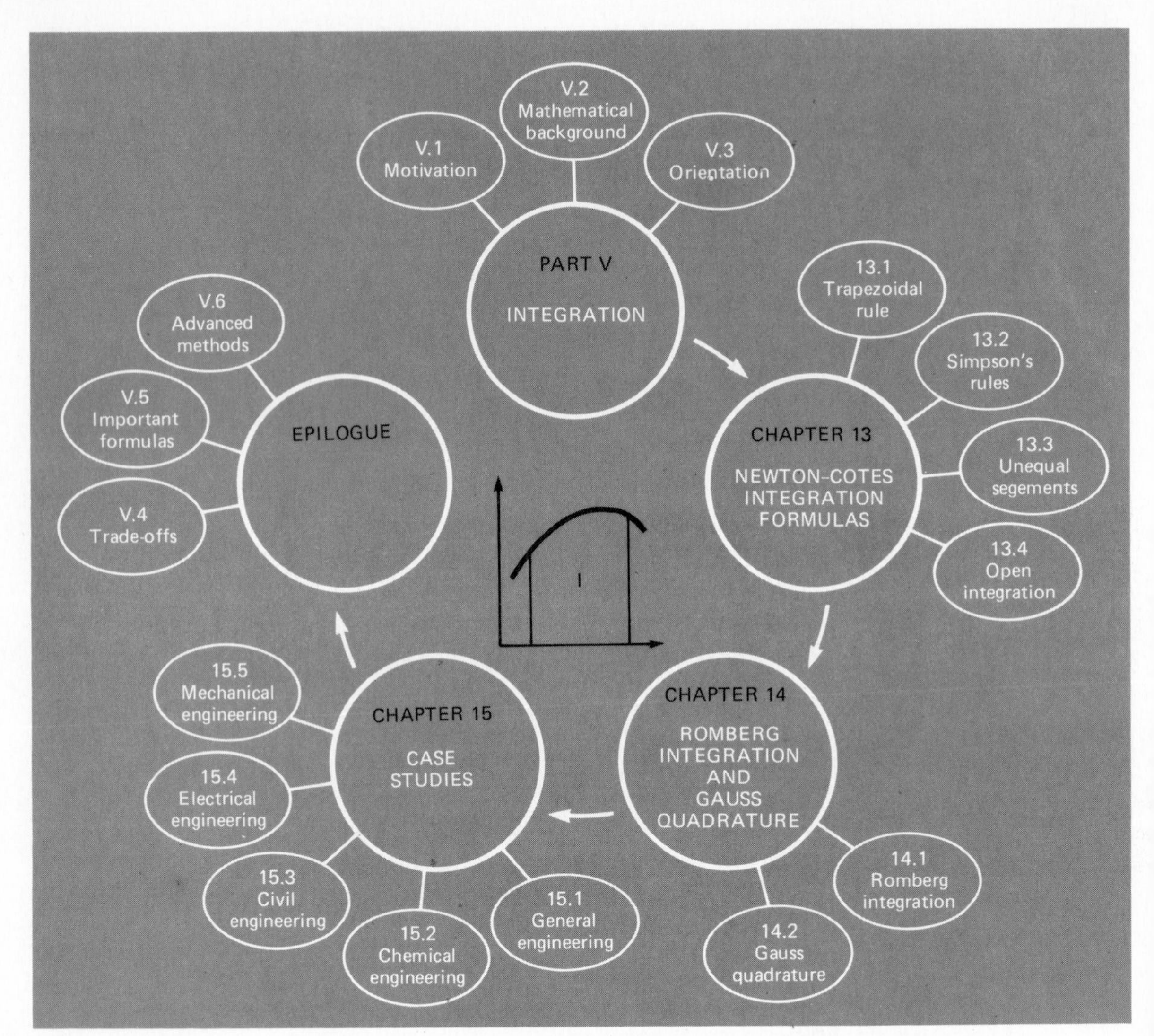


FIGURE V.6 Schematic of the organization of material in Part V: Numerical integration.

complicated function or tabulated data with a simple polynomial that is easy to integrate. Three of the most widely used Newton-Cotes formulas are discussed in detail: the *trapezoidal rule, Simpson's 1/3 rule,* and *Simpson's 3/8 rule.* All of these formulas are designed for cases where the data to be integrated is evenly spaced. In addition, we also include a discussion of numerical integration of unequally spaced data. This is a very important topic because many real-world applications deal with data that is in this form.

All the above material relates to closed integration, where the function values at the ends of the limits of integration are known. At the end of

Chap. 13, we present *open integration formulas,* where the integration limits extend beyond the range of the known data. Although they are not commonly used for definite integration, open integration formulas are presented here because they are utilized extensively in the solution of ordinary differential equations in Part VI.

The formulations covered in Chap. 13 can be employed to analyze both tabulated and continuous functions. *Chapter 14* deals with two techniques that are expressly designed to integrate continuous functions: *Romberg integration* and *Gauss quadrature.* Computer algorithms are provided for both of these methods.

*Chapter 15* demonstrates how the methods can be applied for problem solving. As with other parts of the book, case studies are drawn from all fields of engineering.

A review section, or *epilogue,* is included at the end of Part V. This review includes a discussion of trade-offs that are relevant to implementation in engineering practice. In addition, the important formulas and concepts related to numerical integration are summarized. Finally, we present a short review of advanced methods and alternative references that will facilitate your further studies of numerical integration.

Automatic computation capability is provided for in a number of ways. First, the NUMERICOMP software for the trapezoidal rule is available on an optional basis for the APPLE II and IBM-PC computers. Alternatively, computer codes for the trapezoidal rule, using both the FORTRAN and BASIC languages, are given directly in the text. This gives you the opportunity to copy this code for implementation on a personal or mainframe computer. Flowcharts are provided for most of the other methods described in the text. These flowcharts, combined with your own well-written computer code in any language, should provide programs that can be applied to a number of engineering problems.

### V.3.2 Goals and Objectives

*Study objectives.* After completing Part V, you should be able to solve many numerical integration problems and appreciate their application for engineering problem solving. You should strive to master several techniques and assess their reliability. You should understand the trade-offs involved in selecting the "best" method (or methods) for any particular problem. In addition to these general objectives, the specific concepts listed in Table V.2 should be assimilated and mastered.

*Computer objectives.* You have been provided with software, simple computer programs, algorithms, and flowcharts to implement the techniques discussed in Part V. All have utility as learning tools.

**TABLE V.2 Specific study objectives for Part V**

1. Understand the derivation of the Newton-Cotes formulas; know how to derive the trapezoidal rule and how to set up the derivation of both of Simpson's rules; recognize that the trapezoidal and Simpson's 1/3 and 3/8 rules represent the areas under first-, second-, and third-order polynomials, respectively
2. Know the formulas and error equations for
   (*a*) The trapezoidal rule
   (*b*) The multiple-segment trapezoidal rule
   (*c*) Simpson's 1/3 rule
   (*d*) Simpson's 3/8 rule
   (*e*) The multiple-segment Simpson's rule
   Be able to choose the "best" among these formulas for any particular problem context
3. Recognize that Simpson's 1/3 rule is fourth-order accurate even though it is based on only three points; realize that all the even-segment–odd-point Newton-Cotes formulas have similar enhanced accuracy
4. Know how to evaluate the integral of unequally spaced data
5. Recognize the difference between open and closed integration formulas
6. Understand the theoretical basis of Richardson extrapolation and how it is applied in the Romberg integration algorithm
7. Understand the fundamental difference between Newton-Cotes and Gauss quadrature formulas
8. Recognize why both Romberg integration and Gauss quadrature have utility for integrating continuous (as opposed to tabular) functions

The NUMERICOMP personal computer software is user-friendly. It employs the trapezoidal rule to evaluate the integral of either continuous or tabular functions. The graphics associated with this software will enable you to easily visualize your problem and the associated mathematical operations as the area between the curve and the *x*-axis. The software is very easy to apply to solve many practical problems and can be used to check the results of any computer programs you may develop yourself.

Alternatively, FORTRAN and BASIC programs for the trapezoidal rule are supplied directly in the text. In addition, general algorithms or flowcharts are provided for most of the methods in Part V. This information will allow you to expand your software library to include techniques beyond the trapezoidal rule. For example, you may find it useful from a professional viewpoint to develop software that can handle unequally spaced data. You may also want to develop your own software for Simpson's rules, Romberg integration, and Gauss quadrature, which are usually more efficient and accurate than the trapezoidal rule.

# CHAPTER THIRTEEN

# NEWTON-COTES INTEGRATION FORMULAS

The *Newton-Cotes formulas* are the most common numerical integration schemes. They are based on the strategy of replacing a complicated function or tabulated data with some approximating function that is easy to integrate:

$$I = \int_a^b f(x)\,dx \simeq \int_a^b f_n(x)\,dx \qquad [13.1]$$

where $f_n(x)$ is a polynomial of the form

$$f_n(x) = a_0 + a_1 + \cdots + a_{n-1}\,x^{n-1} + a_n\,x^n$$

where $n$ is the order of the polynomial. For example, in Fig. 13.1*a*, a first-order polynomial (a straight line) is used as an approximation. In Fig. 13.1*b*, a parabola is employed for the same purpose.

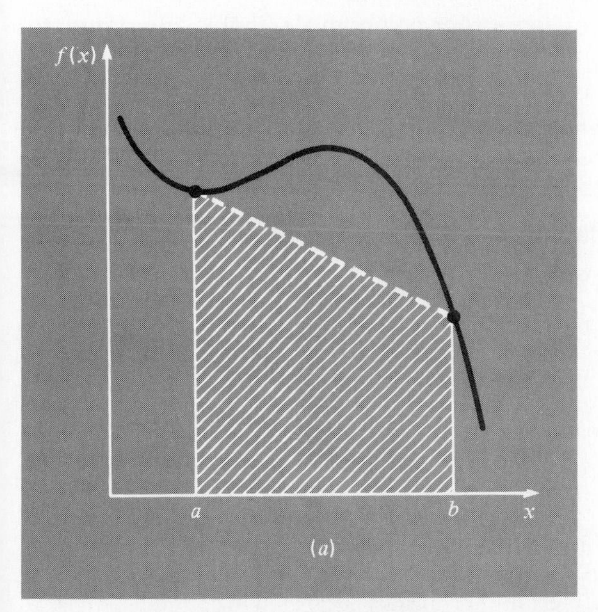


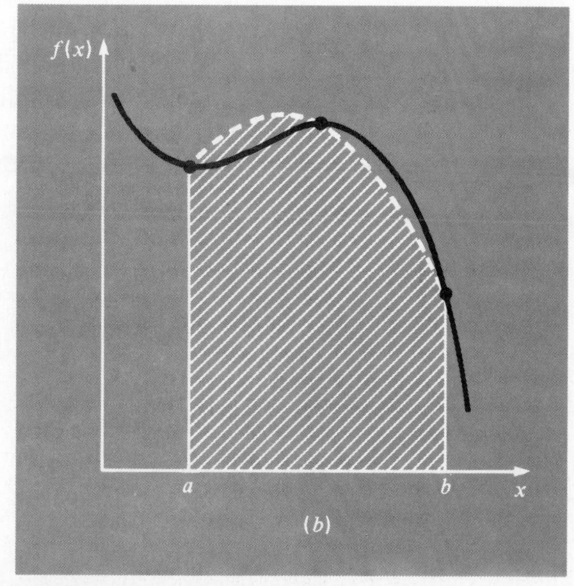


FIGURE 13.1 The approximation of an integral by the area under (*a*) a single straight line and (*b*) a single parabola.

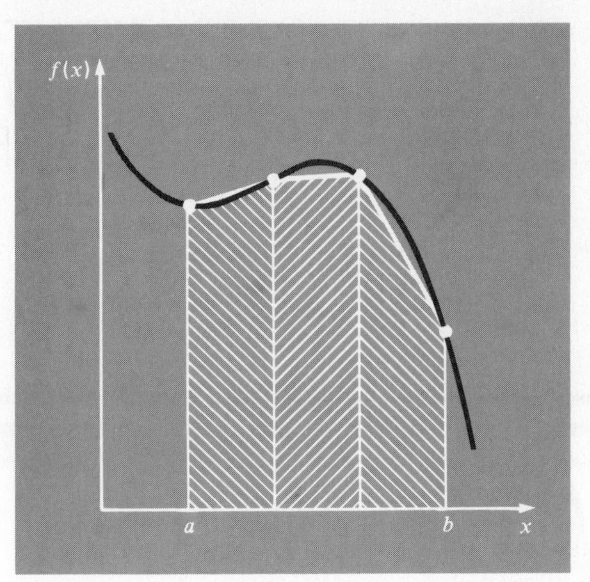


FIGURE 13.2 The approximation of an integral by the area under three straight-line segments.

The integral can also be approximated using a series of polynomials applied piecewise to the function or data over segments of constant length. For example, in Fig. 13.2, three straight-line segments are used to approximate the integral. Higher-order polynomials can be utilized for the same purpose. With this background, we now recognize that the "strip method" in Fig. V.3 employed a series of zero-order polynomials (that is, constants) to approximate the integral.

Closed and open forms of the Newton-Cotes formulas are available. The *closed forms* are those where the data points at the beginning and end of the limits of integration are known (Fig. 13.3*a*). The *open forms* have

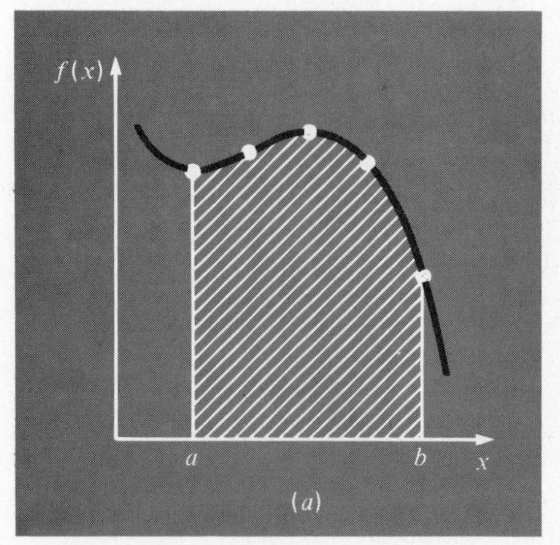


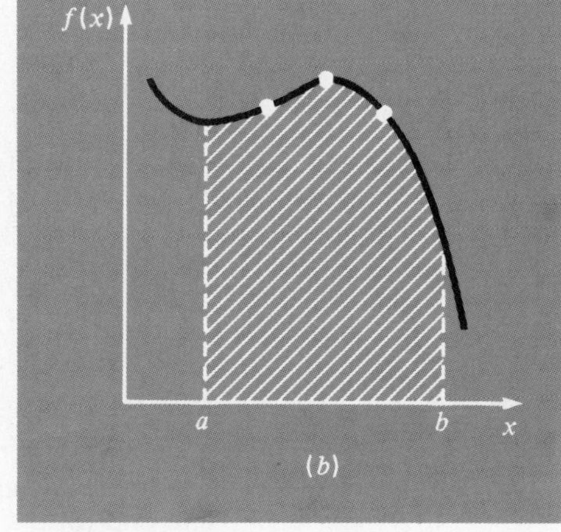


FIGURE 13.3 The difference between (*a*) closed and (*b*) open integration formulas.

integration limits that extend beyond the range of the data (Fig. 13.3*b*). In this sense, they are akin to extrapolation as discussed at the end of Chap. 11. Open Newton-Cotes formulas are not generally used for definite integration. However, they are utilized extensively for the solution of ordinary differential equations. The present chapter emphasizes the closed forms. However, material on open Newton-Cotes formulas is briefly introduced at the end of this chapter.

## 13.1 THE TRAPEZOIDAL RULE

The *trapezoidal rule* is the first of the Newton-Cotes closed integration formulas. It corresponds to the case where the polynomial in Eq. (13.1) is first order.

$$I = \int_a^b f(x)\,dx \simeq \int_a^b f_1(x)\,dx$$

Recall from Chap. 11 that a straight line can be represented as [Eq. (11.2)]

$$f_1(x) = f(a) + \frac{f(b) - f(a)}{b - a}(x - a) \tag{13.2}$$

The area under this straight line is an estimate of the integral of $f(x)$ between the limits $a$ and $b$:

$$I \simeq \int_a^b \left[ f(a) + \frac{f(b) - f(a)}{b - a}(x - a) \right] dx$$

The result of the integration (see Box 13.1 for details) is

$$I \simeq (b - a)\frac{f(a) + f(b)}{2} \tag{13.3}$$

which is called the *trapezoidal rule*.

Geometrically, the trapezoidal rule is equivalent to approximating the area of the trapezoid under the straight line connecting $f(a)$ and $f(b)$ in Fig. 13.4. Recall from geometry that the formula for computing the area of a trapezoid is the height times the average of the bases (Fig. 13.5*a*). In our case, the concept is the same but the trapezoid is on its side (Fig. 13.5*b*). Therefore, the integral estimate can be represented as

$$I \simeq \text{width} \times \text{average height} \tag{13.4}$$

or

$$I \simeq (b - a) \times \text{average height} \tag{13.5}$$

### BOX 13.1 Derivation of Trapezoidal Rule

Before integration, Eq. (13.2) can be expressed as

$$f(x) = \frac{f(b) - f(a)}{b - a} x + f(a) - \frac{af(b) - af(a)}{b - a}$$

Grouping the last two terms gives

$$f(x) = \frac{f(b) - f(a)}{b - a} x + \frac{bf(a) - af(a) - af(b) + af(a)}{b - a}$$

or

$$f(x) = \frac{f(b) - f(a)}{b - a} x + \frac{bf(a) - af(b)}{b - a}$$

which can be integrated between $x = a$ and $x = b$ to yield

$$I \simeq \left. \frac{f(b) - f(a)}{b - a} \frac{x^2}{2} + \frac{bf(a) - af(b)}{b - a} x \right|_a^b$$

This result can be evaluated to give

$$I \simeq \frac{f(b) - f(a)}{b - a} \frac{(b^2 - a^2)}{2} + \frac{bf(a) - af(b)}{b - a} (b - a)$$

Now, realizing that $b^2 - a^2 = (b - a)(b + a)$

$$I \simeq [f(b) - f(a)] \frac{(b + a)}{2} + bf(a) - af(b)$$

Multiplying and collecting terms yields

$$I \simeq (b - a) \frac{f(a) + f(b)}{2}$$

which is the formula for the trapezoidal rule.

---

where for the trapezoidal rule, the average height is the average of the function values at the end points, or $[f(a) + f(b)]/2$.

All the Newton-Cotes closed formulas can be expressed in the general format of Eq. (13.5). In fact, they only differ with respect to the formulation of the average height.

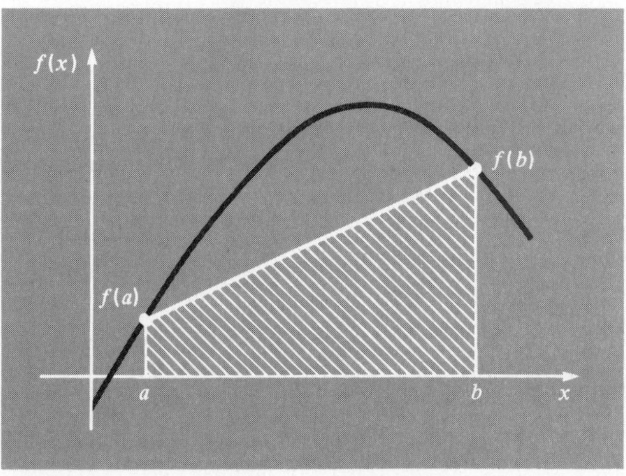


FIGURE 13.4 Graphical depiction of the trapezoidal rule.

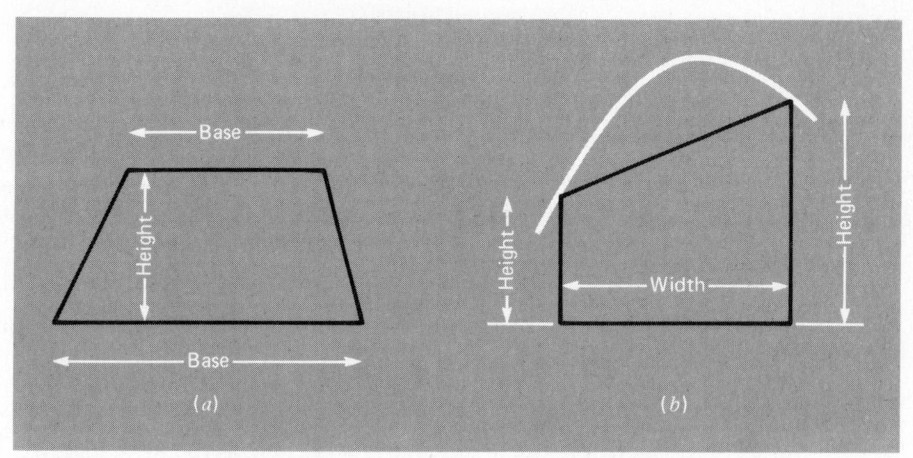


FIGURE 13.5 (*a*) The formula for computing the area of a trapezoid: height times the average of the bases. (*b*) For the trapezoidal rule, the concept is the same but the trapezoid is on its side.

### 13.1.1 Error of the Trapezoidal Rule

When we employ the integral under a straight-line segment to approximate the integral under a curve, we obviously can incur an error that may be substantial (Fig. 13.6). An estimate for the local truncation error of a single application of the trapezoidal rule is (Box 13.2)

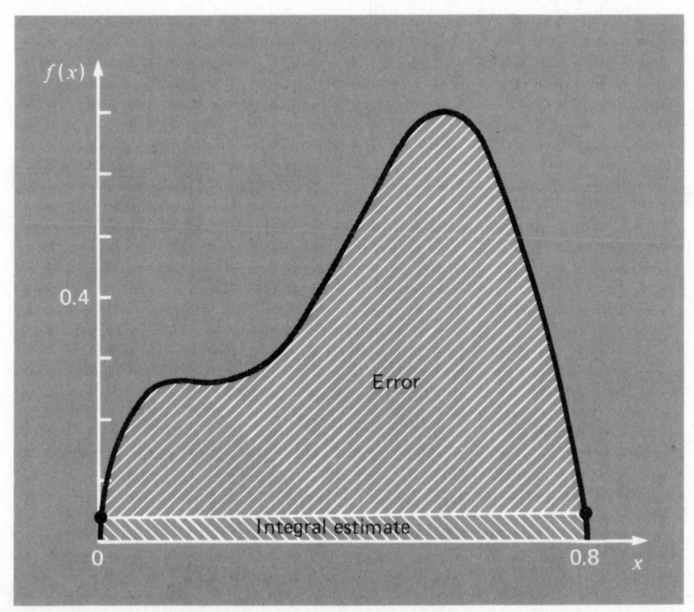


FIGURE 13.6 Graphical depiction of the use of a single application of the trapezoidal rule to approximate the integral of $f(x) = 0.2 + 25x - 200x^2 + 675x^3 - 900x^4 + 400x^5$ from $x = 0$ to 0.8.

$$E_t = -\frac{1}{12} f''(\xi)\,(b - a)^3 \tag{13.6}$$

where $\xi$ lies somewhere in the interval from $a$ to $b$. Equation (13.6) indicates that if the function being integrated is linear, the trapezoidal rule will be exact. Otherwise, for functions with second- and higher-order derivatives (that is, with curvature), some error can occur.

---

BOX 13.2 Derivation and Error Estimate of the Trapezoidal Rule Based on Integrating the Forward Newton-Gregory Interpolating Polynomial

An alternative derivation of the trapezoidal rule is possible by integrating the forward Newton-Gregory interpolating polynomial. Recall that for the first-order version with error term, the integral would be (Box 11.2)

$$I = \int_a^b \left[ f(a) + \Delta f(a)\alpha + \frac{f''(\xi)}{2}\,\alpha(\alpha - 1)h^2 \right] dx \tag{B13.2.1}$$

In order to simplify the analysis, realize that because $\alpha = (x - a)/h$,

$$dx = h\, d\alpha$$

In as much as $h = b - a$ (for the one-segment trapezoidal rule), the limits of integration, $a$ and $b$, correspond to 0 and 1, respectively. Therefore, Eq. (B13.2.1) can be expressed as

$$I = h \int_0^1 \left[ f(a) + \Delta f(a)\,\alpha + \frac{f''(\xi)}{2}\,\alpha(\alpha - 1)\,h^2 \right] d\alpha$$

If it is assumed that, for small $h$, the term $f''(\xi)$ is approximately constant, this equation can be integrated:

$$I = h\left[ \alpha f(a) + \frac{\alpha^2}{2}\Delta f(a) + \left(\frac{\alpha^3}{6} - \frac{\alpha^2}{4}\right) f''(\xi)h^2 \right]_0^1$$

and evaluated as

$$I = h\left[ f(a) + \frac{\Delta f(a)}{2} \right] - \frac{1}{12} f''(\xi)h^3$$

Because $\Delta f(a) = f(b) - f(a)$, the result can be written as

$$I = \underbrace{h\,\frac{f(a) + f(b)}{2}}_{\text{Trapezoidal rule}} - \underbrace{\frac{1}{12} f''(\xi)h^3}_{\text{Truncation error}}$$

Thus, the first term is the trapezoidal rule and the second is an approximation for the error.

---

EXAMPLE 13.1
Single Application of the Trapezoidal Rule

Problem Statement: Use Eq. (13.3) to numerically integrate

$$f(x) = 0.2 + 25x - 200x^2 + 675x^3 - 900x^4 + 400x^5$$

from $a = 0$ to $b = 0.8$. Recall from Sec. V.2 that the exact value of the integral can be determined analytically to be 1.64053334.

Solution: The function values

$$f(0) = 0.2$$

$$f(0.8) = 0.232$$

can be substituted into Eq. (13.3) to yield

$$I \simeq 0.8\frac{0.2 + 0.232}{2} = 0.1728$$

which represents an error of

$$E_t = 1.64053334 - 0.1728 = 1.46773334$$

which corresponds to a percent relative error of $\epsilon_t = 89.5$ percent. The reason for this large error is evident from the graphical depiction in Fig. 13.6. Notice that the area under the straight line neglects a significant portion of the integral lying above the line.

In actual situations, we would have no foreknowledge of the true value. Therefore, an approximate error estimate is required. To obtain this estimate, the function's second derivative over the interval can be computed by differentiating the original function twice to give

$$f''(x) = -400 + 4050x - 10{,}800x^2 + 8000x^3$$

The average value of the second derivative can be computed using Eq. (V.3):

$$\bar{f}'' = \frac{\int_0^{0.8} (-400 + 4050x - 10{,}800x^2 + 8000x^3)dx}{0.8 - 0} = -60$$

which can be substituted into Eq. (13.6) to yield

$$E_a = -\frac{1}{12}(-60)(0.8)^3 = 2.56$$

which is of the same order of magnitude and sign as the true error. A discrepancy does exist, however, because of the fact that for an interval of this size, the average second derivative is not necessarily an accurate approximation of $f''(\xi)$. Thus, we denote that the error is approximate by using the notation $E_a$, rather than exact by using $E_t$.

### 13.1.2 The Multiple-Segment Trapezoidal Rule

One way to improve the accuracy of the trapezoidal rule is to divide the integration interval from $a$ to $b$ into a number of segments and apply the method to each segment (Fig. 13.7). The areas of individual segments can then be added to yield the integral for the entire interval. The resulting equations are called *multiple-segment*, or *composite, integration formulas*.

Figure 13.8 shows the general format and nomenclature we will use to characterize multiple-segment integrals. There are $n + 1$ equally spaced base points $(x_0, x_1, x_2, \ldots, x_n)$. Consequently, there are $n$ segments of equal width:

$$h = \frac{b - a}{n} \qquad [13.7]$$

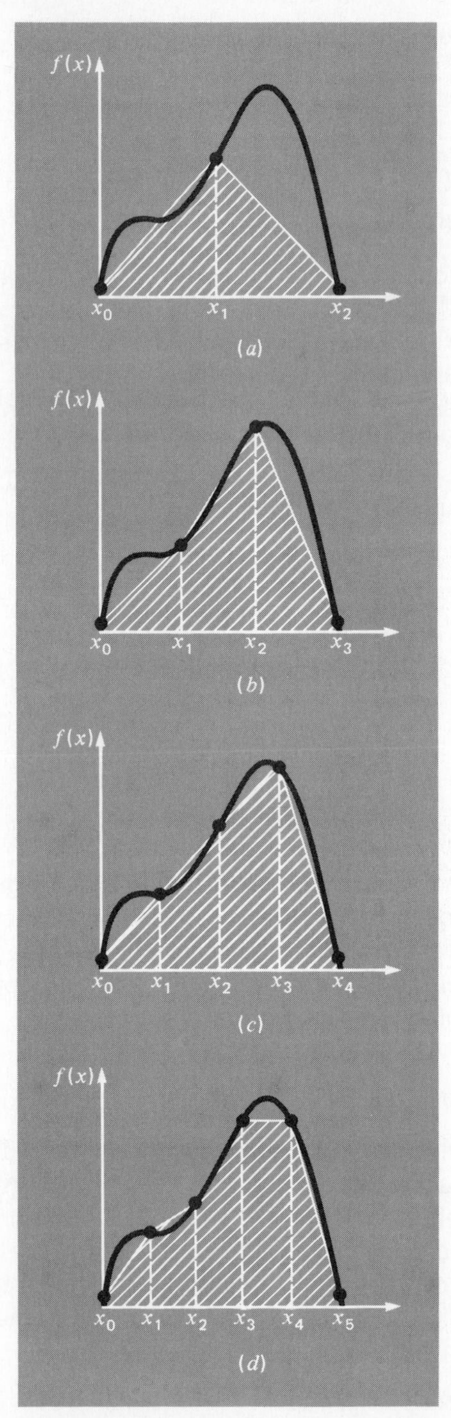


FIGURE 13.7 Illustration of the multiple-segment trapezoidal rule. (*a*) Two segments; (*b*) three segments; (*c*) four segments; and (*d*) five segments.

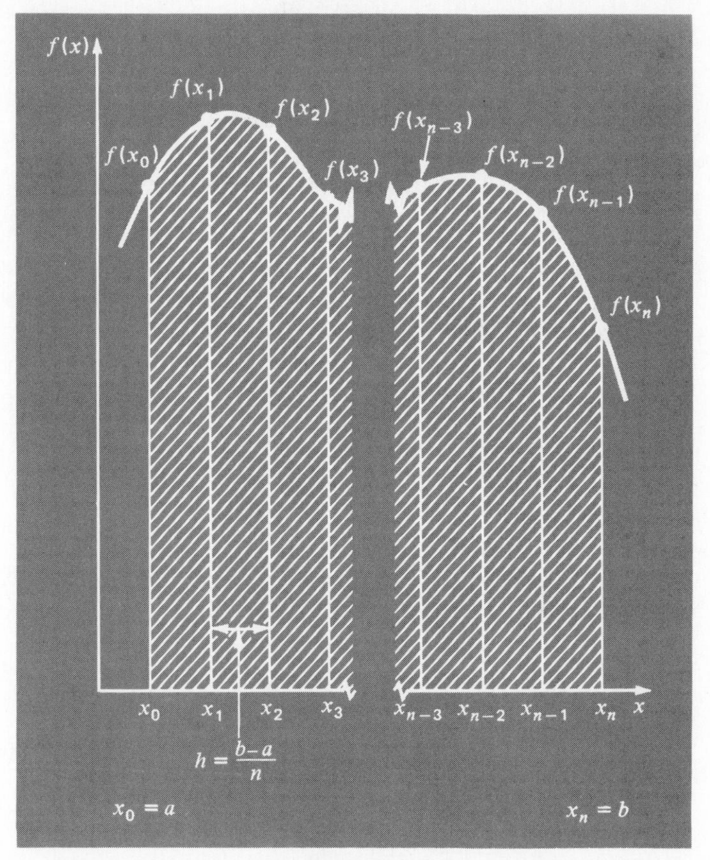


FIGURE 13.8 The general format and nomenclature for multiple-segment integrals.

If $a$ and $b$ are designated as $x_0$ and $x_n$, respectively, the total integral can be represented as

$$I = \int_{x_0}^{x_1} f(x)dx + \int_{x_1}^{x_2} f(x)dx + \cdots + \int_{x_{n-1}}^{x_n} f(x)dx$$

Substituting the trapezoidal rule for each integral yields

$$I \simeq h\frac{f(x_1) + f(x_0)}{2} + h\frac{f(x_2) + f(x_1)}{2} + \cdots + h\frac{f(x_n) + f(x_{n-1})}{2} \tag{13.8}$$

or, grouping terms,

$$I \simeq \frac{h}{2}\left[f(x_0) + 2\sum_{i=1}^{n-1} f(x_i) + f(x_n)\right] \tag{13.9}$$

or, using Eq. (13.7) to express Eq. (13.9) in the general form of Eq. (13.5),

$$I \simeq \underbrace{(b-a)}_{\text{Width}} \underbrace{\frac{f(x_0) + 2\sum_{i=1}^{n-1} f(x_i) + f(x_n)}{2n}}_{\text{Average height}} \tag{13.10}$$

Because the summation of the coefficients of $f(x)$ in the numerator divided by $2n$ is equal to 1, the average height represents a *weighted average* of the function values. According to Eq. (13.10), the interior points are given twice the weight of the two end points $f(x_0)$ and $f(x_n)$.

An error for the multiple-segment trapezoidal rule can be obtained by summing the individual errors for each segment to give

$$E_t = -\frac{(b-a)^3}{12n^3} \sum_{i=1}^{n} f''(\xi_i) \tag{13.11}$$

where $f''(\xi_i)$ is the second derivative at a point $\xi_i$ located in segment $i$. This result can be simplified by estimating the mean or average value of the second derivative for the entire interval as [Eq. (V.2)]

$$\bar{f}'' \simeq \frac{\sum_{i=1}^{n} f''(\xi_i)}{n} \tag{13.12}$$

Therefore, $\Sigma f''(\xi_i) \simeq n\bar{f}''$ and Eq. (13.11) can be rewritten as

$$E_a = -\frac{(b-a)^3}{12n^2} \bar{f}'' \tag{13.13}$$

Thus, if the number of segments is doubled, the truncation error will be quartered. Note that Eq. (13.13) is an approximate error because of the approximate nature of Eq. (13.12).

### EXAMPLE 13.2
Multiple-Segment Trapezoidal Rule

Problem Statement: Use the two-segment trapezoidal rule to estimate the integral of

$$f(x) = 0.2 + 25x - 200x^2 + 675x^3 - 900x^4 + 400x^5$$

from $a = 0$ to $b = 0.8$. Employ Eq. (13.13) to estimate the error. Recall from Sec. V.2 that the correct value for the integral is 1.64053334.

Solution: $n = 2$ $(h = 0.4)$:

$$f(0) = 0.2$$

$$f(0.4) = 2.456$$

$$f(0.8) = 0.232$$

$$I \simeq 0.8\frac{0.2 + 2(2.456) + 0.232}{4} = 1.0688$$

$$E_t = 1.64053334 - 1.0688 = 0.57173 \qquad \epsilon_t = 34.9\%$$

$$E_a = -\frac{0.8^3}{12(2)^2}(-60) = 0.64$$

where $-60$ is the average second derivative determined previously in Example 13.1.

The results of the previous example, along with three- through ten-segment applications of the trapezoidal rule, are summarized in Table 13.1. Notice how the error decreases as the number of segments increases. However, also notice that the rate of decrease is gradual. This is because the error is inversely related to the square of $n$ [Eq. (13.13)]. Therefore, doubling the number of segments quarters the error. In subsequent sections we develop higher-order formulas that are more accurate and that converge more quickly on the true integral as the segments are increased. However, before investigating these formulas, we will first discuss a computer program to implement the trapezoidal rule.

### 13.1.3 Computer Program for the Multiple-Segment Trapezoidal Rule

A simple computer program to implement the trapezoidal rule is listed in Fig. 13.9. This program has a number of shortcomings. First, it is limited to data that is in tabulated form. A general program should have the capability to evaluate known functions as well. Additionally, the program is not user-friendly; it is designed strictly to come up with the answer. In Prob. 13.21, you

**TABLE 13.1** **Results for multiple-segment trapezoidal rule to estimate the integral of $f(x) = 0.2 + 25x - 200x^2 + 675x^3 - 900x^4 + 400x^5$ from $x = 0$ to 0.8. The exact value is 1.64053334**

| $n$ | $h$ | $I$ | $\epsilon_t$ % |
|---|---|---|---|
| 2 | 0.4 | 1.0688 | 34.9 |
| 3 | 0.2667 | 1.3695 | 16.5 |
| 4 | 0.2 | 1.4848 | 9.5 |
| 5 | 0.16 | 1.5399 | 6.1 |
| 6 | 0.1333 | 1.5703 | 4.3 |
| 7 | 0.1143 | 1.5887 | 3.2 |
| 8 | 0.1 | 1.6008 | 2.4 |
| 9 | 0.0889 | 1.6091 | 1.9 |
| 10 | 0.08 | 1.6150 | 1.6 |

FORTRAN

```
      DIMENSION F(20),Y(20)
      REAL IN
      COMMON N,A,B
      READ(5,1)N
    1 FORMAT(I5)
      NI=N-1
      READ(5,2)A,B
    2 FORMAT(2F10.0)
      H=(B-A)/NI
      DO 170 I=1,N
      READ(5,3)Y(I)
    3 FORMAT(F10.0)
  170 CONTINUE
      CALL TRAP(Y,IN)
      WRITE(6,4)IN
    4 FORMAT(' ',F10.3)
      STOP
      END

      SUBROUTINE TRAP(Y,IN)
      DIMENSION Y(20)
      REAL IN
      COMMON N,A,B
      NI=N-1
      SU=Y(1)
      DO 1030 I=2,NI
      SU=SU+2*Y(I)
 1030 CONTINUE
      HT=(SU+Y(N))/(2*NI)
      IN=(B-A)*HT
      RETURN
      END
```

BASIC

```
100  DIM F(20),Y(20)
110  INPUT N                    N = number of points
120 NI = N - 1                  NI = number of segments
130  INPUT A,B                  A, B = limits of integration
140 H = (B - A) / NI            H = segment width
150  FOR I = 1 TO N
160  INPUT Y(I)                 Y = value of dependent variable
170  NEXT I
180  GOSUB 1000
190  PRINT IN
200  END

1000 SU = Y(1)                  (Subroutine to compute trapezoidal rule)
1010  FOR I = 2 TO NI
1020 SU = SU + 2 * Y(I)
1030  NEXT I
1040 HT = (SU + Y(N)) / (2 * NI)
1050 IN = (B - A) * HT
1060  RETURN
```

FIGURE 13.9 Computer program for the multiple-segment trapezoidal rule for tabulated data.

will have the task of making this skeletal computer code easier to use and understand. Also, you will have the optional task of modifying the program so that it is capable of evaluating the integral of known functions.

An example of a user-friendly program for the trapezoidal rule is included in the supplementary NUMERICOMP software associated with this text. This software can evaluate the integrals of either tabulated data or user-defined functions. The following example demonstrates its utility for evaluating integrals. It also provides a good reference for assessing and testing your own software.

## EXAMPLE 13.3
### Evaluating Integrals with the Computer

Problem Statement: A user-friendly computer program to implement the multiple-segment trapezoidal rule is contained in the NUMERICOMP software associated with the text. We can use this software to solve a problem associated with our friend the falling parachutist. As you recall from Example 1.1, the velocity of the parachutist is given as the following function of time:

$$v(t) = \frac{gm}{c}\left[1 - e^{-(c/m)t}\right] \qquad \text{[E13.3.1]}$$

where $v$ is the velocity in centimeters per second, $g$ is the gravitational constant of 980 cm/s$^2$, $m$ is the mass of the parachutist equal to 68,100 g, and $c$ is the drag coefficient of 12,500 g/s. The model predicts the velocity of the parachutist as a function of time as described in Example 1.1. A plot of the velocity variation was developed in Example 2.1.

Suppose we would like to know how far the parachutist has fallen after a certain time $T$. This distance is given by [Eq. (V.4)]

$$d = \int_0^T v(t)\, dt$$

where $d$ is the distance in centimeters. Substituting Eq. (E.13.3.1) and setting $T = 10$ s,

$$d = \frac{gm}{c} \int_0^{10} [1 - e^{-(c/m)t}]\, dt$$

Performing the integration and substituting known values results in

$$d = 28{,}943.5147 \text{ cm}$$

This exact result can be used to study the efficiency of the multiple-segment trapezoidal rule. Figure 13.10*a* shows the computer screen that requests the upper and lower limits of integration and the step size. After the computations are complete, the integral is printed as 28,874.91. The integral is equivalent to the area under the $v(t)$ versus $t$ curve, as shown in Fig. 13.10*b*. Visual inspection confirms that the integral is the width of the interval (10 s) times the average height (about 2900 cm/s).

Other numbers of segments can be easily tested by repeating the com-

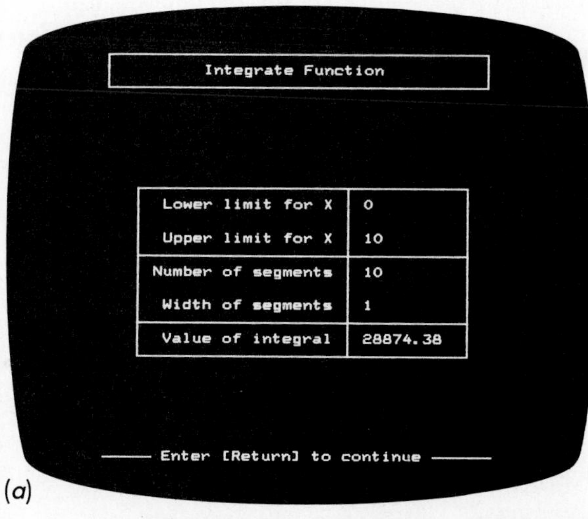


(*a*)

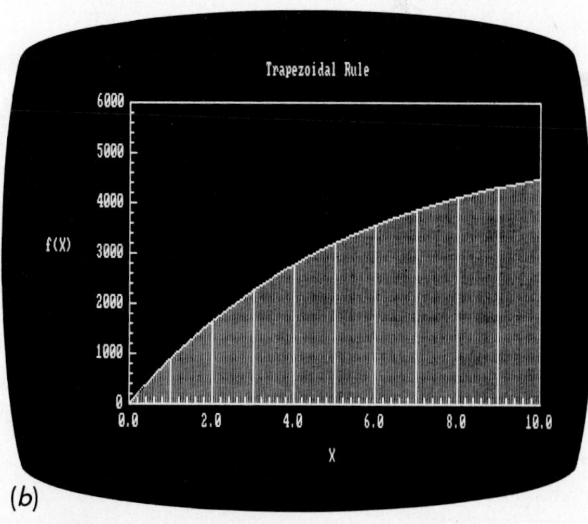


(*b*)

FIGURE 13.10 Computer screens showing (*a*) the entry of integration parameters and integration results; and (*b*) a graph of the integral as the area between the function and the $x$ axis.

putations. The results indicate how the estimated distance the parachutist falls approaches the exact result as the segment size decreases:

| Segments | Segment size | Estimated $d$, cm | $\epsilon_t$, % |
|---|---|---|---|
| 10 | 1.0 | 28,874.9146 | 0.237 |
| 20 | 0.5 | 28,926.3574 | 0.0593 |
| 50 | 0.2 | 28,940.7692 | $9.49 \times 10^{-3}$ |
| 100 | 0.1 | 28,942.8282 | $2.37 \times 10^{-3}$ |
| 200 | 0.05 | 28,943.3431 | $5.93 \times 10^{-4}$ |
| 500 | 0.02 | 28,943.4871 | $9.52 \times 10^{-5}$ |
| 1,000 | 0.01 | 28,943.5076 | $2.44 \times 10^{-5}$ |
| 2,000 | 0.005 | 28,943.5133 | $4.66 \times 10^{-6}$ |
| 5,000 | 0.002 | 28,943.5157 | $-3.63 \times 10^{-6}$ |
| 10,000 | 0.001 | 28,943.5159 | $-4.32 \times 10^{-6}$ |

Thus, the multiple-segment trapezoidal rule attains excellent accuracy. However, notice how the error changes sign and begins to increase in absolute value beyond the 5000-segment case. This is due to the intrusion of round-off error because of the great number of computations for this many segments. Thus, the level of precision is limited, and we would never reach the exact result of 28,943.5147 obtained analytically. This limitation will be discussed in further detail in Chap. 14.

## 13.2 SIMPSON'S RULES

Aside from applying the trapezoidal rule with finer segmentation, another way to obtain a more accurate estimate of an integral is to use higher-order polynomials to connect the points. For example, if there is an extra point midway between $f(a)$ and $f(b)$, the three points can be connected with a parabola (Fig. 13.11*a*). If there are two points equally spaced between $f(a)$ and $f(b)$, the four points can be connected with a third-order polynomial (Fig. 13.11*b*). The formulas that result from taking the integrals under these polynomials are called *Simpson's rules.*

### 13.2.1 Simpson's 1/3 Rule

Simpson's 1/3 rule results when a second-order interpolating polynomial is substituted into Eq. (13.1):

$$I = \int_a^b f(x)\,dx \simeq \int_a^b f_2(x)\,dx$$

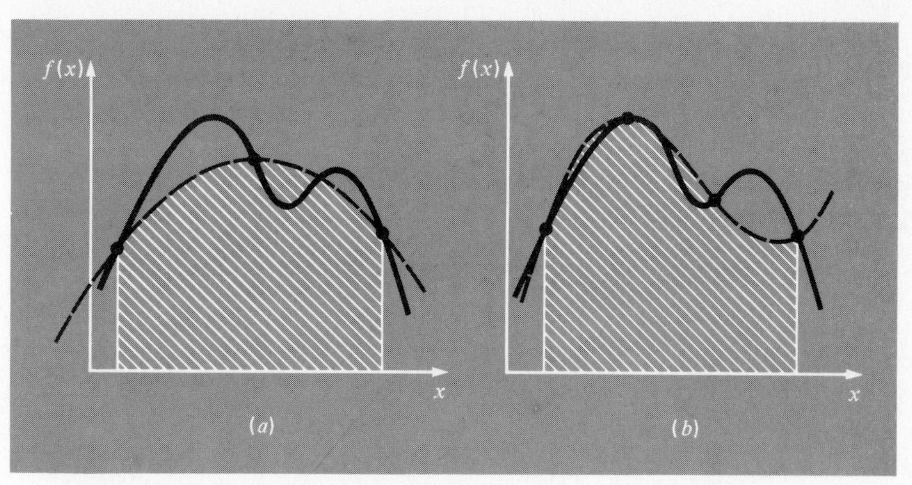


FIGURE 13.11 (*a*) Graphical depiction of Simpson's 1/3 rule: it consists of taking the area under a parabola connecting three points. (*b*) Graphical depiction of Simpson's 3/8 rule: it consists of taking the area under a cubic equation connecting four points.

If $a$ and $b$ are designated as $x_0$ and $x_2$ and $f_2(x)$ is represented by a second-order Lagrange polynomial [Eq. (11.22)], the integral becomes

$$I \simeq \int_{x_0}^{x_2} \left[ \frac{(x - x_1)(x - x_2)}{(x_0 - x_1)(x_0 - x_2)} f(x_0) + \frac{(x - x_0)(x - x_2)}{(x_1 - x_0)(x_1 - x_2)} f(x_1) + \frac{(x - x_0)(x - x_1)}{(x_2 - x_0)(x_2 - x_1)} f(x_2) \right] dx$$

After integration and algebraic manipulation, the following formula results:

$$I \simeq \frac{h}{3} [f(x_0) + 4f(x_1) + f(x_2)] \qquad [13.14]$$

where, for this case, $h = (b - a)/2$. This equation is known as *Simpson's 1/3 rule.* It is the second Newton-Cotes closed integration formula. The label "1/3" stems from the fact that $h$ is divided by 3 in Eq. (13.14). An alternative derivation is shown in Box 13.3 where the Gregory-Newton polynomial is integrated to obtain the same formula.

Simpson's 1/3 rule can also be expressed using the format of Eq. (13.5):

$$I \simeq \underbrace{(b - a)}_{\text{Width}} \underbrace{\frac{f(x_0) + 4f(x_1) + f(x_2)}{6}}_{\text{Average height}} \qquad [13.15]$$

where $a = x_0$, $b = x_2$, and $x_1$ is the point midway between $a$ and $b$ which is given by $(b + a)/2$. Notice that according to Eq. (13.15), the middle point is weighted by two-thirds and the two end points by one-sixth.

### BOX 13.3 Derivation and Error Estimate of Simpson's Rule Based on Integrating the Forward Newton-Gregory Interpolating Polynomial

As was done in Box 13.2 for the trapezoidal rule, Simpson's 1/3 rule can be derived by integrating the forward Newton-Gregory interpolating polynomial:

$$I = \int_{x_0}^{x_2} \left[ f(x_0) + \Delta f(x_0)\,\alpha + \frac{\Delta^2 f(x_0)}{2}\,\alpha\,(\alpha-1) + \frac{\Delta^3 f(x_0)}{6}\,\alpha\,(\alpha-1)(\alpha-2) + \frac{f^{(4)}(\xi)}{24}\,\alpha(\alpha-1)(\alpha-2)(\alpha-3)h^4 \right] dx$$

Notice that we have written the polynomial up to the fourth-order term rather than to the third-order term as would be expected. The reason for this will be apparent shortly. Also notice that the limits of integration are from $x_0$ to $x_2$. Therefore, when the simplifying substitutions are made (recall Box 13.2), the integral is from $\alpha = 0$ to 2:

$$I = h \int_0^2 \left[ f(x_0) + \Delta f(x_0)\,\alpha + \frac{\Delta^2 f(x_0)}{2}\,\alpha\,(\alpha-1) + \frac{\Delta^3 f(x_0)}{6}\,\alpha\,(\alpha-1)(\alpha-2) + \frac{f^{(4)}(\xi)}{24}\,\alpha\,(\alpha-1)(\alpha-2)(\alpha-3)\,h^4 \right] d\alpha$$

which can be integrated to yield

$$I = h\left[ \alpha f(x_0) + \frac{\alpha^2}{2}\,\Delta f(x_0) + \left(\frac{\alpha^3}{6} - \frac{\alpha^2}{4}\right)\Delta^2 f(x_0) + \left(\frac{\alpha^4}{24} - \frac{\alpha^3}{6} + \frac{\alpha^2}{6}\right)\Delta^3 f(x_0) + \left(\frac{\alpha^5}{120} - \frac{\alpha^4}{16} + \frac{11\alpha^3}{72} - \frac{\alpha^2}{8}\right) f^{(4)}(\xi)\,h^4 \right]_0^2$$

and evaluated for the limits to give

$$I = h\left[ 2f(x_0) + 2\Delta f(x_0) + \frac{\Delta^2 f(x_0)}{3} + (0)\Delta^3 f(x_0) - \frac{1}{90} f^{(4)}(\xi)\,h^4 \right] \quad \text{[B13.3.1]}$$

Notice the significant result that the coefficient of the third divided difference is zero. Because $\Delta f(x_0) = f(x_1) - f(x_0)$ and $\Delta^2 f(x_0) = f(x_2) - 2f(x_1) + f(x_0)$, Eq. (B.13.3.1) can be rewritten as

$$I = \underbrace{\frac{h}{3}\left[ f(x_0) + 4f(x_1) + f(x_2) \right]}_{\text{Simpson's 1/3 rule}} \underbrace{- \frac{1}{90} f^{(4)}(\xi)\,h^5}_{\text{Truncation error}}$$

Thus, the first term is Simpson's 1/3 rule and the second is the truncation error. Because the third divided difference dropped out, we obtain the significant result that the formula is third-order accurate.

---

It can be shown that a single-segment application of Simpson's 1/3 rule has a truncation error of (Box 13.3)

$$E_t = -\frac{1}{90}\,h^5 f^{(4)}(\xi)$$

or, because $h = (b-a)/2$,

$$E_t = -\frac{(b-a)^5}{2880}\,f^{(4)}(\xi) \quad \text{[13.16]}$$

where $\xi$ lies somewhere in the interval from $a$ to $b$. Thus, Simpson's 1/3 rule is more accurate than the trapezoidal rule. However, comparison with Eq. (13.6) indicates that it is *much* more accurate than expected. Rather than being proportional to the third derivative, the error is proportional to the fourth derivative. This is because, as shown in Box 13.3, the coefficient of the third-order term goes to zero during the integration of the interpolating polynomial. Consequently, Simpson's 1/3 rule is third-order accurate even though it is based on only three points.

EXAMPLE 13.4
Single Application of Simpson's 1/3 Rule

Problem Statement: Use Eq. (13.15) to integrate

$$f(x) = 0.2 + 25x - 200x^2 + 675x^3 - 900x^4 + 400x^5$$

from $a = 0$ to $b = 0.8$. Recall that the exact integral is 1.64053334.

Solution:

$$f(0) = 0.2 \qquad f(0.4) = 2.456 \qquad f(0.8) = 0.232$$

Therefore, Eq. (13.15) can be used to compute

$$I \simeq 0.8\frac{0.2 + 4(2.456) + 0.232}{6} = 1.36746667$$

which represents an exact error of

$$E_t = 1.64053334 - 1.36746667 = 0.27306666 \qquad \epsilon_t = 16.6\%$$

which is approximately five times more accurate than for a single application of the trapezoidal rule (Example 13.1).

The estimated error is [Eq. (13.16)]

$$E_a = -\frac{(0.8)^5}{2880}(-2400) = 0.27306667$$

where $-2400$ is the average fourth derivative for the interval as obtained using Eq. (V.3). As was the case in Example 13.1, the error is approximate ($E_a$) because the average fourth derivative is not an exact estimate of $f^{(4)}(\xi)$. However, because this case deals with a fifth-order polynomial, the discrepancy is not large and the estimated and exact errors are almost identical.

### 13.2.2 The Multiple-Segment Simpson's 1/3 Rule

Just as with the trapezoidal rule, Simpson's rule can be improved by dividing the integration interval into a number of segments of equal width (Fig. 13.12):

$$h = \frac{b - a}{n} \qquad [13.17]$$

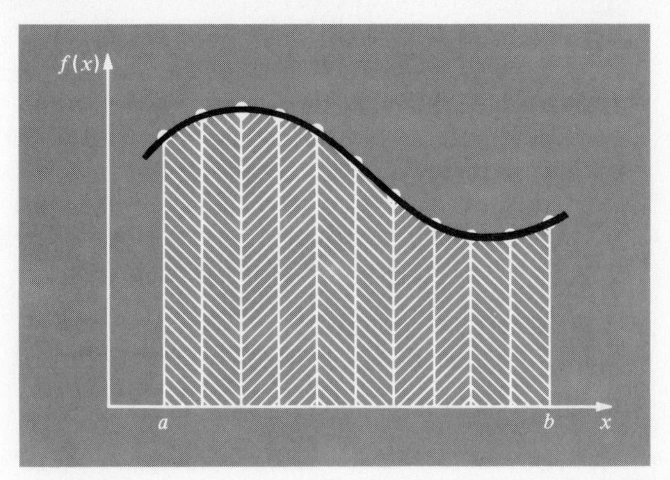


FIGURE 13.12 Graphical representation of the use of multiple segments for Simpson's 1/3 rule. Note that the method can only be employed if the number of segments is even.

The total integral can be represented as

$$I = \int_{x_0}^{x_2} f(x)dx + \int_{x_2}^{x_4} f(x)dx + \cdots + \int_{x_{n-2}}^{x_n} f(x)dx$$

Substituting Simpson's rule for the individual integral yields

$$I \simeq 2h\frac{f(x_0) + 4f(x_1) + f(x_2)}{6}$$

$$+ 2h\frac{f(x_2) + 4f(x_3) + f(x_4)}{6}$$

$$+ \cdots + 2h\frac{f(x_{n-2}) + 4f(x_{n-1}) + f(x_n)}{6}$$

or, combining terms and using Eq. (13.17),

$$I \simeq \underbrace{(b - a)}_{\text{Width}} \underbrace{\frac{f(x_0) + 4\sum_{i=1,3,5}^{n-1} f(x_i) + 2\sum_{j=2,4,6}^{n-2} f(x_j) + f(x_n)}{3n}}_{\text{Average Height}} \qquad [13.18]$$

Notice that, as illustrated in Fig. 13.12, an even number of segments must be utilized to implement the method.

An error estimate for the multiple-segment Simpson's rule is obtained in the same fashion as for the trapezoidal rule by summing the individual errors for the segments and averaging the derivative to yield

$$E_a = -\frac{(b-a)^5}{180n^4}\bar{f}^{(4)} \qquad [13.19]$$

where $\bar{f}^{(4)}$ is the average fourth derivative for the interval.

### EXAMPLE 13.5
### Multiple-Segment Application of Simpson's 1/3 Rule

Problem Statement: Use Eq. (13.18) with $n = 4$ to estimate the integral of

$$f(x) = 0.2 + 25x - 200x^2 + 675x^3 - 900x^4 + 400x^5$$

from $a = 0$ to $b = 0.8$. Recall that the exact integral is 1.64053334.

Solution: $n = 4$ $(h = 0.2)$:

$$f(0) = 0.2 \qquad f(0.2) = 1.288$$

$$f(0.4) = 2.456 \qquad f(0.6) = 3.464$$

$$f(0.8) = 0.232$$

from Eq. (13.18)

$$I = 0.8\frac{0.2 + 4(1.288 + 3.464) + 2(2.456) + 0.232}{12}$$

$$= 1.62346667$$

$$E_t = 1.64053334 - 1.62346667 = 0.01706667 \qquad \epsilon_t = 1.04\%$$

The estimated error [Eq. (13.19)] is

$$E_a = -\frac{(0.8)^5}{180(4)^4}(-2400) = 0.01706667$$

The previous example illustrates that the multiple-segment version of Simpson's 1/3 rule yields very accurate results. For this reason, it is considered superior to the trapezoidal rule for most applications. However, as mentioned previously, it is limited to cases with an even number of segments and an odd number of points. Consequently, as discussed in the next section, an odd-segment–even-point formula known as Simpson's 3/8 rule is used in conjunction with the 1/3 rule to permit evaluation of both even and odd numbers of segments.

### 13.2.3 Simpson's 3/8 Rule

In a similar manner to the derivation of the trapezoidal and Simpson's 1/3 rule, a third-order Lagrange polynomial can be fit to four points and integrated:

$$I = \int_a^b f(x)\,dx \cong \int_a^b f_3(x)\,dx$$

to yield

$$I \cong \frac{3h}{8}\left[f(x_0) + 3f(x_1) + 3f(x_2) + f(x_3)\right]$$

where $h = (b - a)/3$. This equation is called *Simpson's 3/8 rule* because $h$ is multiplied by 3/8. It is the third Newton-Cotes closed integration formula. The 3/8 rule can also be expressed in the form of Eq. (13.5):

$$I \cong \underbrace{(b - a)}_{\text{Width}} \underbrace{\frac{f(x_0) + 3f(x_1) + 3f(x_2) + f(x_3)}{8}}_{\text{Average Height}} \qquad [13.20]$$

Thus, the two interior points are given weights of three-eighths, whereas the end points are weighted with one-eighth. Simpson's 3/8 rule has an error of

$$E_t = -\frac{3}{80} h^5 f^{(4)}(\xi)$$

or, because $h = (b - a)/3$,

$$E_t = -\frac{(b - a)^5}{6480} f^{(4)}(\xi) \qquad [13.21]$$

Thus, the 3/8 rule is somewhat more accurate than the 1/3 rule [Eq. (13.16)].

Simpson's 1/3 rule is usually the method of preference because it attains third-order accuracy with three points rather than the four points required for the 3/8 version. However, the 3/8 rule has utility in multiple-segment applications when the number of segments is odd. For instance, in Example 13.5 we used Simpson's rule to integrate the function for four segments. Suppose that you desired an estimate for five segments. One option would be to use a multiple-segment application of the trapezoidal rule as was done in Example 13.3. This may not be advisable, however, because of the large truncation error associated with this method. An alternative would be to apply Simpson's 1/3 rule to the first two segments and Simpson's 3/8 rule to the last three (Fig. 13.13). In this way, we could obtain an estimate with third-order accuracy across the entire interval.

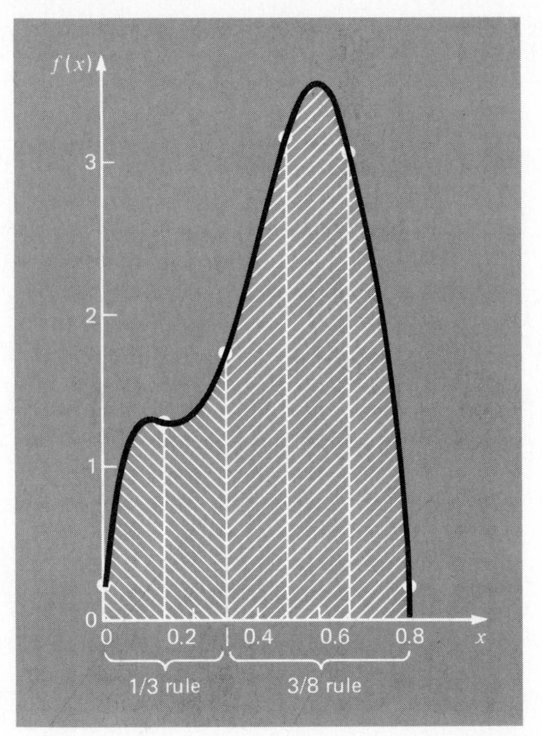


FIGURE 13.13 Illustration of how Simpson's 1/3 and 3/8 rules can be applied in tandem to handle multiple-segment applications with odd numbers of intervals.

## EXAMPLE 13.6
### Simpson's 3/8 Rule

Problem Statement:
(*a*) Use Simpson's 3/8 rule to integrate

$$f(x) = 0.2 + 25x - 200x^2 + 675x^3 - 900x^4 + 400x^5$$

from $a = 0$ to $b = 0.8$.
(*b*) Use it in conjunction with Simpson's 1/3 rule to integrate the same function for five segments.

Solution:
(*a*) A single application of Simpson's 3/8 rule requires four equally spaced points:

$$f(0) = 0.2 \qquad f(0.2667) = 1.43272428$$

$$f(0.5333) = 3.48717696 \qquad f(0.8) = 0.232$$

Using Eq. (13.20),

$$I \simeq 0.8\,\frac{0.2 + 3(1.43272428 + 3.48717696) + 0.232}{8}$$

$$= 1.51917037$$

$$E_t = 1.64053334 - 1.51917037 = 0.12136297 \qquad \epsilon_t = 7.4\%$$

$$E_a = -\frac{(0.8)^5}{6480}(-2400) = 0.12136296$$

(*b*) The data needed for a five-segment application ($h = 0.16$) is

$f(0) = 0.2$ $\qquad f(0.16) = 1.29691904$

$f(0.32) = 1.74339328$ $\qquad f(0.48) = 3.18601472$

$f(0.64) = 3.18192896$ $\qquad f(0.80) = 0.232$

The integral for the first two segments is obtained using Simpson's 1/3 rule:

$$I \simeq 0.32\,\frac{0.2 + 4(1.29691904) + 1.74339328}{6} = 0.38032370$$

For the last three segments, the 3/8 rule can be used to obtain

$$I \simeq 0.48\,\frac{1.74339328 + 3(3.18601472 + 3.18192896) + 0.232}{8}$$

$$= 1.26475346$$

The total integral is computed by summing the two results:

$$I = 0.38032370 + 1.26475346 = 1.64507716$$

$$E_t = 1.64053334 - 1.64507716 = -0.00454383 \qquad \epsilon_t = -0.28\%$$

### 13.2.4 Computer Algorithm for Simpson's Rule

A flowchart for Simpson's rule is outlined in Fig. 13.14. Notice that the program is set up so that either an even or odd number of segments may be used. For the former case, Simpson's 1/3 rule is applied to each pair of segments, and the results are summed to compute the total integral. For the latter, Simpson's 3/8 rule is applied to the last three segments and the 1/3 rule is applied to all the previous segments.

### 13.2.5 Higher-Order Newton-Cotes Closed Formulas

As noted previously, the trapezoidal rule and both of Simpson's rules are members of a family of integrating equations known as the *Newton-Cotes closed integration formulas*. Some of the formulas are summarized in Table 13.2, along with their truncation-error estimates.

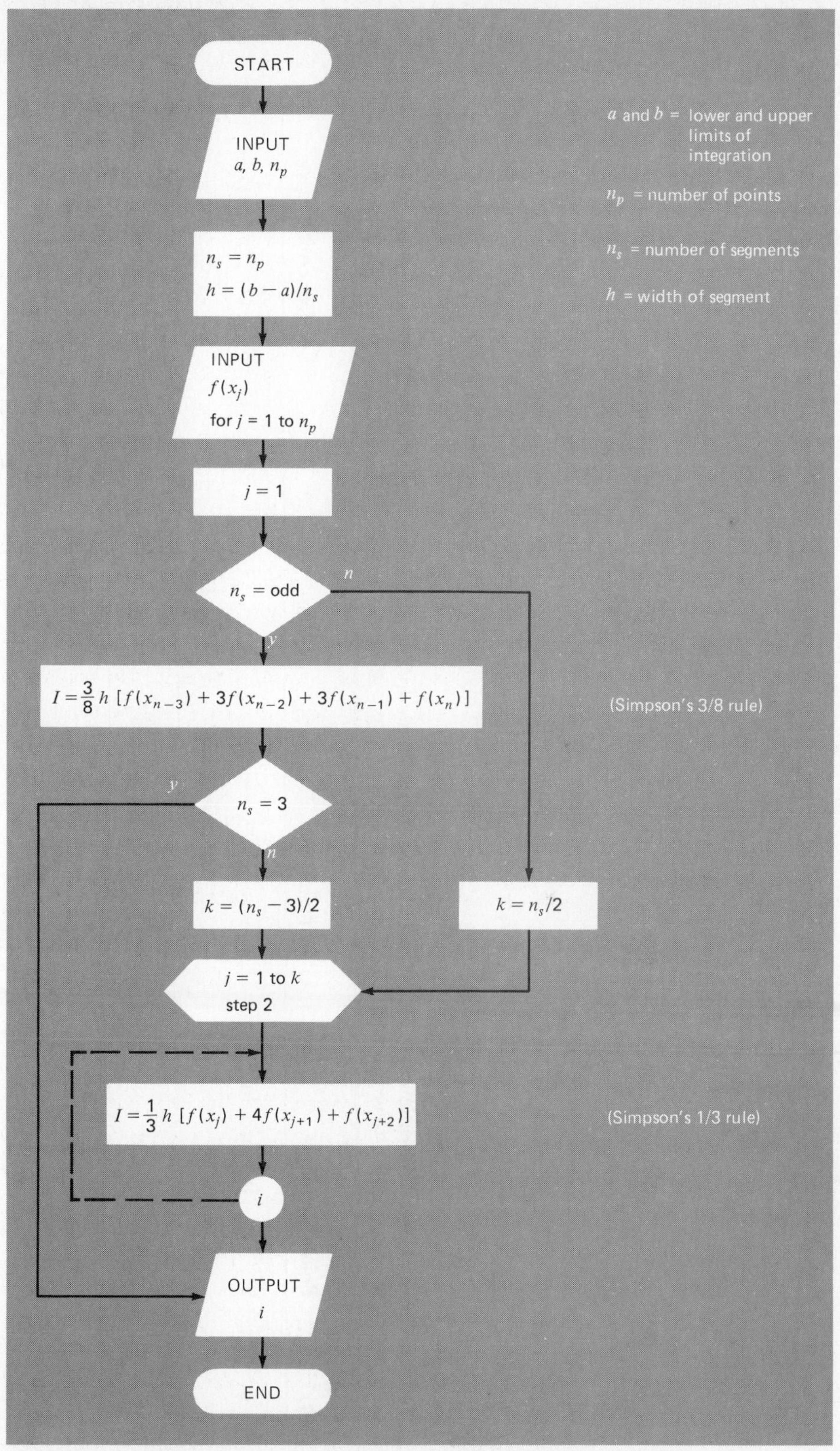


FIGURE 13.14 Computer flowchart for a multisegment version of Simpson's rule.

**TABLE 13.2**
**Newton-Cotes closed integration formulas. The formulas are presented in the format of Eq. (13.6) so that the weighting of the data points to estimate the average height is apparent. The step size is given by $h = (b - a)/n$.**

| Segments ($n$) | Points | Name | Formula | Truncation error |
|---|---|---|---|---|
| 1 | 2 | Trapezoidal rule | $(b-a)\dfrac{f(x_0)+f(x_1)}{2}$ | $-(1/12)h^3 f''(\xi)$ |
| 2 | 3 | Simpson's 1/3 rule | $(b-a)\dfrac{f(x_0)+4f(x_1)+f(x_2)}{6}$ | $-(1/90)h^5 f^{(4)}(\xi)$ |
| 3 | 4 | Simpson's 3/8 rule | $(b-a)\dfrac{f(x_0)+3f(x_1)+3f(x_2)+f(x_3)}{8}$ | $-(3/80)h^5 f^{(4)}(\xi)$ |
| 4 | 5 | Boole's rule | $(b-a)\dfrac{7f(x_0)+32f(x_1)+12f(x_2)+32f(x_3)+7f(x_4)}{90}$ | $-(8/945)h^7 f^{(6)}(\xi)$ |
| 5 | 6 | | $(b-a)\dfrac{19f(x_0)+75f(x_1)+50f(x_2)+50f(x_3)+75f(x_4)+19f(x_5)}{288}$ | $-(275/12{,}096)h^7 f^{(6)}(\xi)$ |

Notice that, as was the case with Simpson's 1/3 and 3/8 rules, the five- and six-point formulas have the same order error. This general characteristic holds for the higher-point formulas and leads to the result that the even-segment–odd-point formulas (for example, 1/3 rule and Boole's rule) are usually the methods of preference.

However, it must also be stressed that in engineering practice, the higher-order (that is, greater than four-point) formulas are rarely used. Simpson's rules are sufficient for most applications. Accuracy can be improved by using the multi-segment version rather than opting for the higher-point formulas. Furthermore, when the function is known and high accuracy is required, methods such as Romberg integration or Gauss quadrature, described in Chap. 14, offer viable and attractive alternatives.

## 13.3 INTEGRATION WITH UNEQUAL SEGMENTS

To this point, all formulas for numerical integration have been based on equally spaced data points. In practice, there are many situations where this assumption does not hold and we must deal with unequal-sized segments. For example, experimentally derived data is often of this type. For these cases, one method is to apply the trapezoidal rule to each segment and sum the results:

$$I = h_1 \frac{f(x_1) + f(x_0)}{2} + h_2 \frac{f(x_2) + f(x_1)}{2} + \cdots + h_n \frac{f(x_n) + f(x_{n-1})}{2} \qquad [13.22]$$

where $h_i$ is the width of segment $i$. Note that this was the same approach used for the multiple-segment trapezoidal rule. The only difference between Eqs. (13.8) and (13.22) is that the $h$'s in the former are constant. Consequently, Eq. (13.8) could be simplified by grouping terms to yield Eq. (13.9). Although this simplification cannot be applied to Eq. (13.22), a computer program can be easily developed to accommodate unequal-sized segments. Before describing such a program, we will illustrate in the following example how Eq. (13.22) is applied to evaluate an integral.

### EXAMPLE 13.7
Trapezoidal Rule with Unequal Segments

Problem Statement: The information in Table 13.3 was generated using the same polynomial employed in Example 13.1. Use Eq. (13.22) to determine the integral for this data. Recall that the correct answer is 1.64053334.

Solution: Applying Eq. (13.22) to the data in Table 13.3 yields

$$I = 0.12 \frac{1.30972928 + 0.2}{2} + 0.10 \frac{1.30524128 + 1.30972928}{2}$$

**TABLE 13.3** **Data for $f(x) = 0.2 + 25x - 200x^2 + 675x^3 - 900x^4 + 400x^5$, with unequally spaced values of $x$.**

| $x$ | $f(x)$ | $x$ | $f(x)$ |
|---|---|---|---|
| 0.0 | 0.20000000 | 0.44 | 2.84298496 |
| 0.12 | 1.30972928 | 0.54 | 3.50729696 |
| 0.22 | 1.30524128 | 0.64 | 3.18192896 |
| 0.32 | 1.74339328 | 0.70 | 2.36300000 |
| 0.36 | 2.07490304 | 0.80 | 0.23200000 |
| 0.40 | 2.45600000 | | |

$$+ \cdots + 0.1\frac{0.232 + 2.363}{2}$$

$$= 0.09058376 + 0.13074853 + \cdots + 0.12975$$

$$= 1.56480098$$

which represents an absolute percent relative error of $\epsilon_t = 4.6$ percent.

The data from Example 13.7 is depicted in Fig. 13.15. Notice that some adjacent segments are of equal width and, consequently, could have been evaluated using Simpson's rules. This usually leads to more accurate results, as illustrated by the following example.

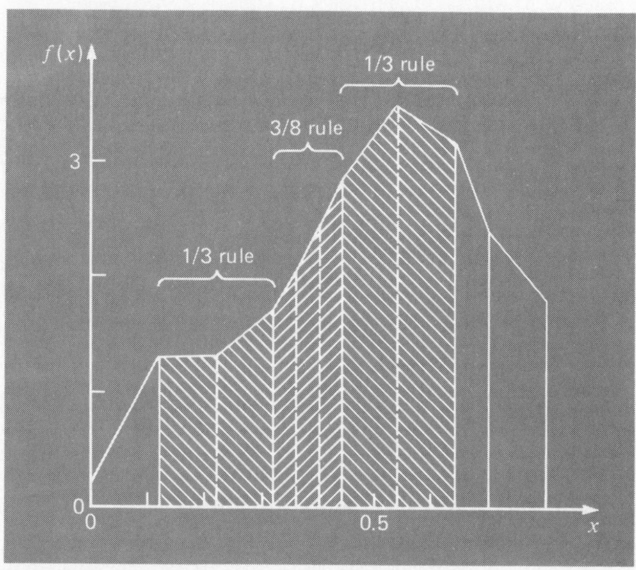


FIGURE 13.15 Use of the trapezoidal rule to determine the integral of unevenly spaced data. Notice how the shaded segments could be evaluated with Simpson's rules to attain higher accuracy.

### EXAMPLE 13.8
### Inclusion of Simpson's Rules in the Evaluation of Uneven Data

Problem Statement: Recompute the integral for the data in Table 13.3, but use Simpson's rules for those segments where they are appropriate.

Solution: The first segment is evaluated with the trapezoidal rule:

$$I = 0.12\,\frac{1.30972928 + 0.2}{2} = 0.09058376$$

Because the next two segments from $x = 0.22$ to $0.36$ are of equal length, their integral can be computed with Simpson's 1/3 rule:

$$I = 0.2\,\frac{1.74339328 + 4(1.30524128) + 1.30972928}{6}$$

$$= 0.27580292$$

The next three segments are also equal and, as such, may be evaluated with the 3/8 rule to give $I = 0.27268631$. Similarly, the 1/3 rule can be applied to the two segments from $x = 0.44$ to $0.64$ to yield $I = 0.66847006$. Finally, the last two segments, which are of unequal length, can be evaluated with the trapezoidal rule to give values of 0.16634787 and 0.12975000, respectively. The area of these individual segments can be summed to yield a total integral of 1.60364092. This represents an error of $\epsilon_t = 2.2$ percent, which is superior to the result using the trapezoidal rule in Example 13.7.

Computer Program for Unequally Spaced Data. It is a fairly simple proposition to program Eq. (13.22). However, as demonstrated in Example 13.8, the approach is enhanced if it implements Simpson's rules wherever possible. For this reason, we have developed a computer algorithm that incorporates this capability.

As depicted in Fig. 13.16, the flowchart checks the length of adjacent segments. If two consecutive segments are of equal length, Simpson's 1/3 rule is applied. If three are equal, the 3/8 rule is used. When adjacent segments are of unequal length, the trapezoidal rule is implemented.

We suggest that you develop your own computer program from this flowchart. Not only does it allow evaluation of unequal segment data, but if equal spaced information is used, it reduces to Simpson's rules. As such, it represents a basic, all-purpose algorithm for the determination of the integral of tabulated data.

## 13.4 OPEN INTEGRATION FORMULAS

Recall from Fig. 13.3*b* that open integration formulas have limits that extend beyond the range of the data. Table 13.4 summarizes the *Newton-Cotes*

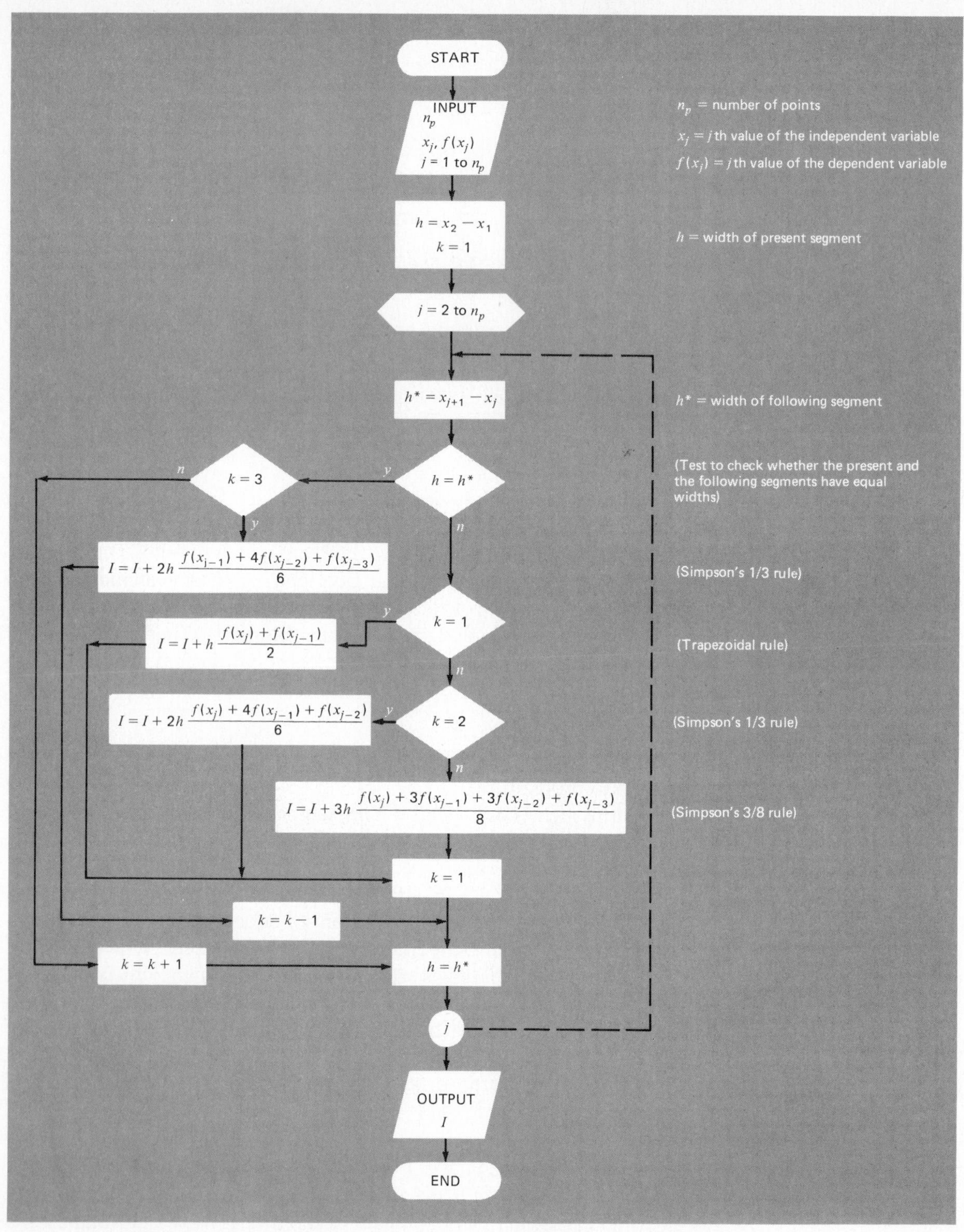


FIGURE 13.16 Flowchart for integrating unequally spaced data.

**TABLE 13.4**
**Newton-Cotes open integration formulas. The formulas are presented in the format of Eq. (13.6) so that the weighting of the data points to estimate the average height is apparent. The step size is given by $h = (b - a)/n$.**

| Segments ($n$) | Points | Name | Formula | Truncation error |
|---|---|---|---|---|
| 2 | 1 | Midpoint method | $(b - a)f(x_1)$ | $(1/3)h^3f''(\xi)$ |
| 3 | 2 | | $(b - a)\dfrac{f(x_1) + f(x_2)}{2}$ | $(3/4)h^3f''(\xi)$ |
| 4 | 3 | | $(b - a)\dfrac{2f(x_1) - f(x_2) + 2f(x_3)}{3}$ | $(14/45)h^5f^{(4)}(\xi)$ |
| 5 | 4 | | $(b - a)\dfrac{11f(x_1) + f(x_2) + f(x_3) + 11f(x_4)}{24}$ | $(95/144)h^5f^{(4)}(\xi)$ |
| 6 | 5 | | $(b - a)\dfrac{11f(x_1) - 14f(x_2) + 26f(x_3) - 14f(x_4) + 11f(x_5)}{20}$ | $(41/140)h^7f^{(6)}(\xi)$ |

*open integration formulas.* The formulas are expressed in the form of Eq. (13.5) so that the weighting factors are evident. As with the closed versions, successive pairs of the formulas have the same order error. The even-segment–odd-point formulas are usually the methods of preference because they require fewer points to attain the same accuracy as the odd-segment–even-point formulas. Notice that the strip method depicted in Fig. V.3 is actually a multisegment version of the midpoint method in Table 13.4.

As mentioned previously, the open formulas are rarely used for integration. However, they will have direct application to the multistep methods for solving ordinary differential equations discussed in Chap. 17.

# PROBLEMS

## Hand Calculations

**13.1** Use analytical means to evaluate

(*a*) $\int_0^{10} (10 + 2x - 6x^2 + 5x^4)\, dx$

(*b*) $\int_{-3}^{5} (1 - x - 4x^3 + 3x^5)\, dx$

(*c*) $\int_0^{\pi} (8 + 5 \sin x)\, dx$

**13.2** Use a single application of the trapezoidal rule to evaluate the integrals from Prob. 13.1.

**13.3** Evaluate the integrals from Prob. 13.1 with a multiple-segment trapezoidal rule, with $n = 2$, 4, and 6.

**13.4** Evaluate the integrals from Prob. 13.1 with a single application of Simpson's 1/3 rule.

**13.5** Evaluate the integrals from Prob. 13.1 with a multiple-segment Simpson's 1/3 rule, with $n = 4$ and 6.

**13.6** Evaluate the integrals from Prob. 13.1 with a single application of Simpson's 3/8 rule.

**13.7** Evaluate the integrals from Prob. 13.1, but use a multiple-segment Simpson's rule, with $n = 5$.

**13.8** Integrate the following function both analytically and using the trapezoidal rule, with $n = 1$, 2, 3, and 4:

$$\int_0^{3\pi/20} [\sin (5x + 1)]\, dx$$

Compute percent relative errors to evaluate the accuracy of the trapezoidal approximations.

**13.9** Integrate the following function both analytically and using Simpson's rules, with $n = 4$ and 5:

$$\int_{-4}^{6} [(4x + 8)^3]\, dx$$

Discuss the results.

**13.10** Integrate the following function both analytically and numerically. Use both the trapezoidal and Simpson's 1/3 rules to numerically integrate the function. For both cases, use the multisegment version, with $n = 4$.

$$\int_{0}^{4} xe^{2x}\, dx$$

Compute percent relative errors for the numerical results.

**13.11** Integrate the following function both analytically and numerically. Use a single application of the trapezoidal rule, Simpson's 1/3 and 3/8 rules, and Boole's rule (see Table 13.2).

$$\int_{0}^{1} 15.3^{2.5x}\, dx$$

Compute percent relative errors for the numerical results.

**13.12** Evaluate the integral

$$\int_{0}^{\pi} (4 + 2 \sin x)\, dx$$

(*a*) Analytically.
(*b*) By single application of trapezoidal rule.
(*c*) By multiple application of trapezoidal rule ($n = 5$).
(*d*) By single application of Simpson's 1/3 rule.
(*e*) By single application of Simpson's 3/8 rule.
(*f*) By multiple application of Simpson's rules ($n = 5$).
For (*b*) through (*f*), compute the percent relative error ($\epsilon_t$) based on (*a*).

**13.13** Evaluate the integral of the following tabular data with the trapezoidal rule:

| $x$ | 0 | 0.1 | 0.2 | 0.3 | 0.4 | 0.5 |
|---|---|---|---|---|---|---|
| $f(x)$ | 1 | 7 | 4 | 3 | 5 | 9 |

**13.14** Perform the same evaluation as in Prob. 13.13, but use Simpson's rules.

**13.15** Evaluate the integral of the following tabular data using the trapezoidal rule:

| $x$ | −3 | −1 | 1 | 3 | 5 | 7 | 9 | 11 |
|---|---|---|---|---|---|---|---|---|
| $f(x)$ | 1 | −4 | −5 | 2 | 4 | 8 | 6 | −3 |

**13.16** Perform the same evaluation as in Prob. 13.15, but use Simpson's rules.

**13.17** Determine the mean value of the function

$$f(x) = -46 + 45.4x - 13.8x^2 + 1.71x^3 - 0.0729x^4$$

between $x = 2$ and 10 by
(*a*) Graphing the function and visually estimating the mean value.
(*b*) Using Eq. (V.3) and the analytical evaluation of the integral.
(*c*) Using Eq. (V.3) and a four-segment version of the trapezoidal rule to estimate the integral.
(*d*) Using Eq. (V.3) and a four-segment version of Simpson's 1/3 rule.

**13.18** The function

$$f(x) = 10 - 38.6x + 74.07x^2 - 40.1x^3$$

can be used to generate the following table of unequally spaced data:

| $x$ | 0 | 0.1 | 0.3 | 0.5 | 0.7 | 0.95 | 1.2 |
|---|---|---|---|---|---|---|---|
| $f(x)$ | 10 | 6.84 | 4 | 4.20 | 5.51 | 5.77 | 1 |

Evaluate the integral from $a = 0$ to $b = 1.2$ using
(*a*) Analytical means.
(*b*) The trapezoidal rule.
(*c*) A combination of the trapezoidal and Simpson's rules; employ Simpson's rules wherever possible to obtain the highest accuracy.
For (*b*) and (*c*), compute the percent relative error ($\epsilon_t$).

**13.19** Evaluate the following double integral:

$$\int_{-2}^{2}\int_{0}^{4} (x^3 - 3y^2 + xy^3)\, dx\, dy$$

(*a*) Analytically.
(*b*) Using a multiple-segment trapezoidal rule ($n = 2$).
(*c*) Using single applications of Simpson's 1/3 rule.
For (*b*) and (*c*) compute the percent relative error ($\epsilon_t$).

**13.20** Evaluate the triple integral

$$\int_{-4}^{4}\int_{0}^{6}\int_{-1}^{3} (x^4 - 2yz)\, dx\, dy\, dz$$

(*a*) Analytically.
(*b*) Using single applications of Simpson's 1/3 rule.
For (*b*), compute the percent relative error ($\epsilon_t$).

## Computer-Related Problems

**13.21** Develop a user-friendly computer program for the multiple-segment trapezoidal rule based on Fig. 13.9. Among other things,
(*a*) Add documentation statements to the code.
(*b*) Make the input and output more descriptive and user-oriented.
(*c*) (*Optional*) Modify the program so that it is capable of evaluating given functions in addition to tabular data.
Test your program by duplicating the computation from Example 13.2.

**13.22** Develop a user-friendly computer program for the multiple-segment version of Simpson's rule based on Fig. 13.14. Test it by duplicating the computations from Examples 13.5 and 13.6.

**13.23** Develop a user-friendly computer program for integrating unequally spaced data based on Fig. 13.16. Test it by duplicating the computation from Example 13.7.

**13.24** Use the TRAPEZOIDAL RULE program on the NUMERICOMP disk (or your own program from Prob. 13.21) to repeat (*a*) Prob. 13.2, (*b*) Prob. 13.3, (*c*) Prob. 13.8, (*d*) Prob. 13.10, and (*e*) Prob. 13.13. Use the graphical option to help you visualize the concept that $I = \int_a^b f(x)\, dx$ is the area between the $f(x)$ curve and the axis. Try several different step sizes for each problem.

**13.25** Make up five of your own functions. Use the NUMERICOMP software (or your own program) to calculate the integral of each function over some limits based on your inputs. Try step sizes of $h = (b - a)/n$ for $n = 1, \ldots, n = 10$. Plot $I$ as a function of $n$.

**13.26** Use the NUMERICOMP software (or your own program) to calculate the integral of tabular data. Make up your own data for both $x$ and $f(x)$. Use negative values and zero values for both $x$ and $f(x)$. Look at your function using the plot option and convince yourself that the NUMERICOMP software is working properly.

CHAPTER FOURTEEN

# ROMBERG INTEGRATION AND GAUSS QUADRATURE

In the introduction to Part V, we noted that functions to be integrated numerically will typically be of two forms: a table of values or an equation. The form of the data has an important influence on the approaches that can be used to evaluate the integral. For tabulated information, you are limited by the number of points that are given. In contrast, if the function is available in equation form, you can generate as many values of $f(x)$ as are required to attain acceptable accuracy (recall Fig. V.4).

The present chapter is devoted to two techniques that are expressly designed to analyze cases where the function is given. Both capitalize on the ability to generate function values in order to develop efficient schemes for numerical integration. The first is based on *Richardson's extrapolation,* which is a method for combining two numerical integral estimates in order to obtain a third, more accurate value. The computational algorithm for implementing Richardson's extrapolation in a highly efficient manner is called *Romberg integration.* This technique is recursive and can be used to generate an integral estimate within a prespecified error tolerance.

The second method is called *Gauss quadrature.* Recall that in the last chapter, $f(x)$ values for the Newton-Cotes formulas were determined at specified values of $x$. For example, if we used the trapezoidal rule to determine an integral we were constrained to take the weighted average of $f(x)$ at the ends of the interval. Gauss quadrature formulas employ $x$ values that are positioned between $a$ and $b$ in such a manner that a much more accurate integral estimate results.

## 14.1 ROMBERG INTEGRATION

In Chap. 13, we presented multiple-segment versions of the trapezoidal rule and Simpson's rules. For an analytical (as opposed to a tabular) function, the error equations [Eqs. (13.13) and (13.19)] indicate that increasing the number of segments $n$ will result in more accurate integral estimates. This observation is borne out by Fig. 14.1, which is a plot of true error versus $n$ for the integral of $f(x) = 0.2 + 25x - 200x^2 + 675x^3 - 900x^4 + 400x^5$. Notice how the error drops as $n$ increases. However, also notice that at large values of $n$, the error starts to increase as round-off errors begin to dominate. Also

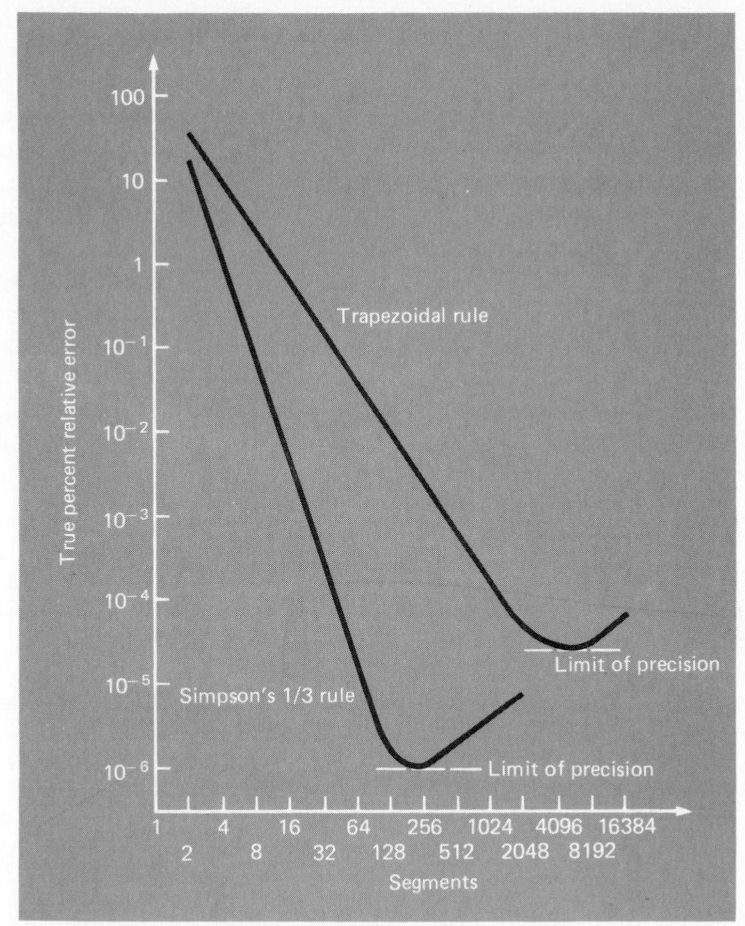


FIGURE 14.1 Absolute value of the true percent relative error versus number of segments for the determination of the integral of $f(x) = 0.2 + 25x - 200x^2 + 675x^3 - 900x^4 + 400x^5$, evaluated from $a = 0$ to $b = 0.8$ using the multiple-segment trapezoidal rule and the multiple-segment Simpson's 1/3 rule. Note that both results indicate that for a large number of segments, round-off errors limit precision.

observe that a very large number of segments (and, hence, computational effort) is required to attain high levels of accuracy. As a consequence of these shortcomings, the multiple-segment trapezoidal rule and Simpson's rules are sometimes inadequate for problem contexts where high efficiency and low errors are needed.

*Romberg integration* is one technique that is designed to obviate these shortcomings. It is quite similar to the techniques discussed in Chap. 13, in the sense that it is based on successive application of the trapezoidal rule. However, through mathematical manipulations, superior results are attained for less effort.

### 14.1.1 Richardson's Extrapolation

Recall that in Sec. 7.4.4, we used error equations to improve the solution of a set of simultaneous linear equations. In the same spirit, error-correction techniques are available to improve the results of numerical integration on the basis of the integral estimates themselves. Generally called *Richardson's extrapolation,* these methods use two estimates of an integral to compute a third, more accurate approximation.

The estimate and error associated with a multiple-segment trapezoidal rule can be represented generally as

$$I = I(h) + E(h)$$

where $I$ is the exact value of the integral, $I(h)$ is the approximation from an $n$-segment application of the trapezoidal rule with step size $h = (b - a)/n$, and $E(h)$ is the truncation error. If we make two separate estimates using step sizes of $h_1$ and $h_2$ and have exact values for the error,

$$I(h_1) + E(h_1) = I(h_2) + E(h_2) \tag{14.1}$$

Now recall that the error of the multiple-segment trapezoidal rule can be represented approximately by Eq. (13.13) [with $n = (b - a)/h$]:

$$E \simeq -\frac{b - a}{12} h^2 \bar{f}'' \tag{14.2}$$

If it is assumed that $\bar{f}''$ is constant regardless of step size, Eq. (14.2) can be used to determine that the ratio of the two errors will be

$$\frac{E(h_1)}{E(h_2)} \simeq \frac{h_1^2}{h_2^2} \tag{14.3}$$

This calculation has the important effect of removing the term $\bar{f}''$ from the computation. In so doing, we have made it possible to utilize the information embodied by Eq. (14.2) without prior knowledge of the function's second derivative. To do this, we rearrange Eq. (14.3) to give

$$E(h_1) \simeq E(h_2)\left(\frac{h_1}{h_2}\right)^2$$

which can be substituted into Eq. (14.1):

$$I(h_1) + E(h_2)\left(\frac{h_1}{h_2}\right)^2 \simeq I(h_2) + E(h_2)$$

which can be solved for

$$E(h_2) \simeq \frac{I(h_1) - I(h_2)}{1 - (h_1/h_2)^2}$$

Thus, we have developed an estimate of the truncation error in terms of the integral estimates and their step sizes. This estimate can then be substituted into

$$I = I(h_2) + E(h_2)$$

to yield an improved estimate of the integral:

$$I \simeq I(h_2) + \left[\frac{1}{(h_1/h_2)^2 - 1}\right][I(h_2) - I(h_1)] \tag{14.4}$$

It can be shown (Ralston and Rabinowitz, 1978) that the error of this estimate is $O(h^4)$. Thus, we have combined two trapezoidal rule estimates of $O(h^2)$ to yield a new estimate of $O(h^4)$. For the special case where the interval is halved ($h_2 = h_1/2$), this equation becomes

$$I \simeq I(h_2) + \frac{1}{2^2 - 1}[I(h_2) - I(h_1)]$$

or, collecting terms,

$$I \simeq \frac{4}{3}I(h_2) - \frac{1}{3}I(h_1) \tag{14.5}$$

EXAMPLE 14.1

Error Corrections of the Trapezoidal Rule

Problem Statement: In the previous chapter (Example 13.1 and Table 13.1), single and multiple-segment applications of the trapezoidal rule yielded the following results:

| Segments | $h$ | Integral | $\epsilon_t$, % |
|---|---|---|---|
| 1 | 0.8 | 0.1728 | 89.5 |
| 2 | 0.4 | 1.0688 | 34.9 |
| 4 | 0.2 | 1.4848 | 9.5 |

Use this information along with Eq. (14.5) to compute improved estimates of the integral.

Solution: The estimates for one and two segments can be combined to yield

$$I \simeq \frac{4}{3}(1.0688) - \frac{1}{3}(0.1728) = 1.36746667$$

The error of the improved integral is

$$E_t = 1.64053334 - 1.36746667 = 0.27306667 \qquad \epsilon_t = 16.6\%$$

which is superior to the estimates upon which it was based.

In the same manner, the estimates for two and four segments can be combined to give

$$I \simeq \frac{4}{3}(1.4848) - \frac{1}{3}(1.0688) = 1.62346667$$

which represents an error of

$$E_t = 1.64053334 - 1.62346667 = 0.01706667 \qquad \epsilon_t = 1.0\%$$

Equation (14.4) provides a way to combine two applications of the trapezoidal rule with error $0(h^2)$ in order to compute a third estimate with error $0(h^4)$. This approach is a subset of a more general method for combining integrals to obtain improved estimates. For instance, in Example 14.1, we computed two improved integrals of $0(h^4)$ on the basis of three trapezoidal rule estimates. These two improved estimates can, in turn, be combined to yield an even better value with $0(h^6)$. For the special case where the original trapezoidal estimates are based on successive halving of the step size, the equation used for $0(h^6)$ accuracy is

$$I = \frac{16}{15}I_m - \frac{1}{15}I_l \qquad [14.6]$$

where $I_m$ and $I_l$ are the more and less accurate estimates, respectively. Similarly, two $0(h^6)$ results can be combined to compute an integral that is $0(h^8)$ using

$$I = \frac{64}{63}I_m - \frac{1}{63}I_l \qquad [14.7]$$

## EXAMPLE 14.2
### Higher-Order Error Correction of Integral Estimates

Problem Statement: In Example 14.1, we used Richardson's extrapolation to compute two integral estimates of $0(h^4)$. Utilize Eq. (14.6) to combine these estimates to compute an integral with $0(h^6)$.

Solution: The two integral estimates of $0(h^4)$ obtained in Example 14.1 were 1.36746667 and 1.62346667. These values can be substituted into Eq. (14.6) to yield

$$I = \frac{16}{15}(1.62346667) - \frac{1}{15}(1.36746667) = 1.64053334$$

which is the correct answer to the nine significant figures that are carried in this example.

### 14.1.2 The Romberg Integration Algorithm

Notice that the coefficients in each of the extrapolation equations [Eqs. (14.5), (14.6), and (14.7)] add up to 1. Thus, they represent weighting factors that, as accuracy increases, place progressively greater weight on the superior integral estimate. These formulations can be expressed in a general form that is well-suited for computer implementation:

$$I_{j,k} \simeq \frac{4^{k-1} I_{j+1,k-1} - I_{j,k-1}}{4^{k-1} - 1} \tag{14.8}$$

where $I_{j+1,k-1}$ and $I_{j,k-1}$ are the more and less accurate integrals, respectively, and $I_{j,k}$ is the improved integral. The index $k$ signifies the level of the integration where $k = 1$ corresponds to the original trapezoidal rule estimates, $k = 2$ corresponds to $O(h^4)$, $k = 3$ to $O(h^6)$, and so forth. The index $j$ is used to distinguish between the more $(j + 1)$ and the less $(j)$ significant estimates. For example, for $k = 2$ and $j = 1$, Eq. (14.8) becomes

$$I_{1,2} \simeq \frac{4I_{2,1} - I_{1,1}}{3}$$

which is equivalent to Eq. (14.5).

The general form represented by Eq. (14.8) is attributed to Romberg, and its systematic application to evaluate integrals is known as *Romberg integration.* Figure 14.2 is a graphical depiction of the sequence of integral estimates generated using this approach. Each matrix corresponds to a single iteration. The first column contains the trapezoidal rule evaluations that are designated $I_{j,1}$, where $j = 1$ is for a single-segment application (step size is

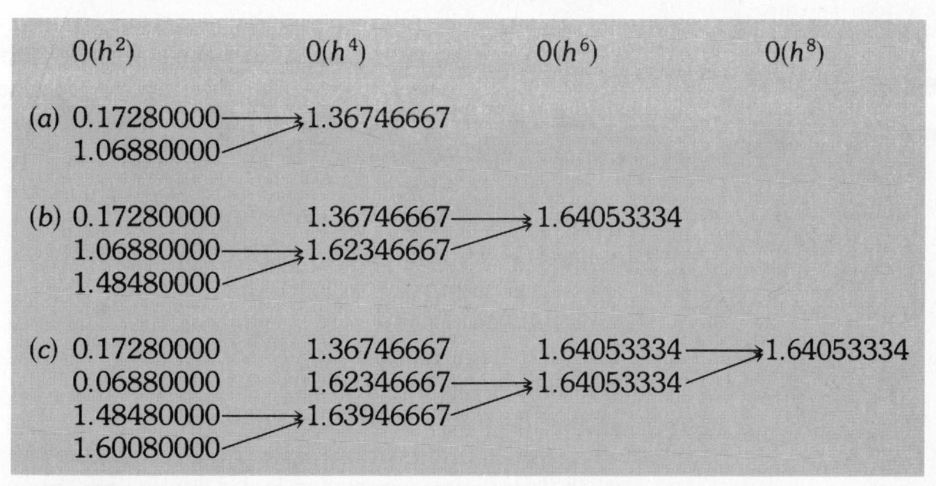


FIGURE 14.2 Graphical depiction of the sequence of integral estimates generated using Romberg integration.

$b - a$); $j = 2$ is for a two-segment application [step size is $(b - a)/2$]; $j = 3$ is for a four-segment application [step size is $(b - a)/4$]; and so forth. The other columns of the matrix are generated by systematically applying Eq. (14.8) to obtain successively better estimates of the integral.

For example, the first iteration (Fig. 14.2*a*) involves computing the one- and two-segment trapezoidal rule estimates ($I_{1,1}$ and $I_{2,1}$). Equation (14.8) is then used to compute the element $I_{1,2} = 1.36746667$, which has an error of $O(h^4)$.

Now, we must check to determine whether this result is adequate for our needs. As is done in other approximate methods in this book, a termination, or stopping, criterion is required to assess the accuracy of the results. One method that can be employed for the present purposes is [Eq. (3.5)]

$$\epsilon_a = \left| \frac{I_{j,k} - I_{j,k-1}}{I_{j,k}} \right| 100\% \qquad [14.9]$$

where $\epsilon_a$ is an estimate of the percent relative error. Thus, as was done previously in other iterative processes, we compare the new estimate with a previous value. When the change between the old and new values as represented by $\epsilon_a$ is below a prespecified error criterion $\epsilon_s$, the computation is terminated. For Fig. 14.2*a*, this evaluation indicates an 87.4 percent change over the course of the first iteration.

The object of the second iteration (Fig. 14.2*b*) is to obtain the $O(h^6)$ estimate—$I_{1,3}$. To do this, an additional trapezoidal rule estimate, $I_{3,1} = 1.4848$, is determined. Then it is combined with $I_{2,1}$ using Eq. (14.8) to generate $I_{2,2} = 1.62346667$. This result is, in turn, combined with $I_{1,2}$ to yield $I_{1,3} = 1.64053334$. Equation (14.9) can be applied to determine that this result represents a change of 16.6 percent when compared with the previous result $I_{1,2}$.

The third iteration (Fig. 14.2*c*) continues the process in the same fashion. In this case, a trapezoidal estimate is added to the first column and then Eq. (14.8) is applied to compute successively more accurate integrals along the lower diagonal. After only three iterations, the result, $I_{1,5} = 1.64053334$, is known to be accurate to at least nine significant figures.

Romberg integration is more efficient than the trapezoidal rule and Simpson's rules discussed in Chap. 13. For example, for determination of the integral as shown in Fig. 14.1, Simpson's 1/3 rule would require a 256-segment application to yield an estimate of 1.64053332. Finer approximations would not be possible because of round-off error. In contrast, Romberg integration yields an exact result (to nine significant figures) based on combining two-, four-, and eight-segment trapezoidal rules.

Figure 14.3 presents a flowchart for Romberg integration. By using loops, this algorithm implements the method in an efficient manner. Remember that Romberg integration is designed for cases where the function to be integrated is known. This is because knowledge of the function permits the evaluations required for the initial implementations of the trapezoidal rule.

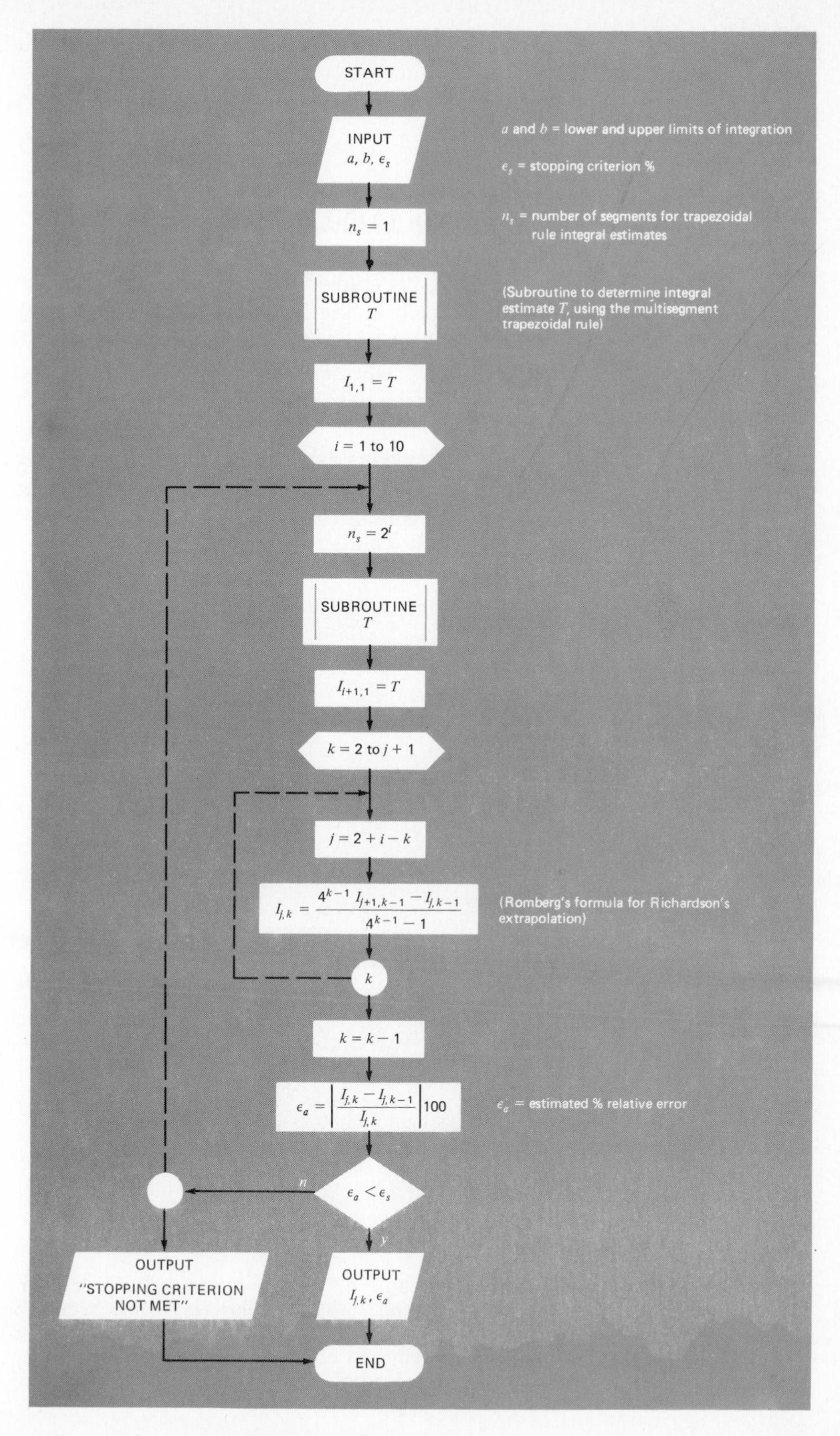


FIGURE 14.3
Flowchart for Romberg integration.

Tabulated data is rarely in the form needed to make the necessary successive evaluations.

## 14.2 GAUSS QUADRATURE

In Chap. 13, we studied the group of numerical integration or quadrature formulas known as the Newton-Cotes equations. A characteristic of these formulas (with the exception of the special case of Sec. 13.3) was that the integral estimate was based on evenly spaced function values. Consequently, the location of the base points used in these equations was predetermined or fixed.

For example, as depicted in Fig. 14.4*a*, the trapezoidal rule is based on taking the area under the straight line connecting the function values at the ends of the integration interval. The formula that is used to compute this area is

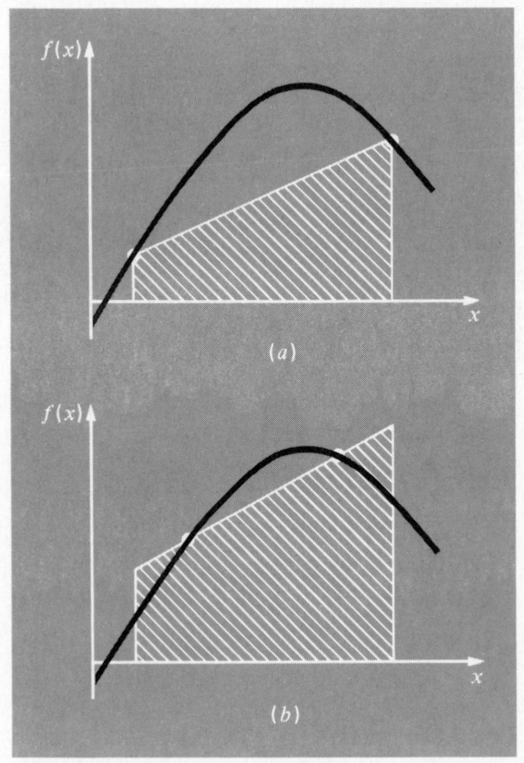


FIGURE 14.4 (*a*) Graphical depiction of the trapezoidal rule as the area under the straight line joining fixed end points. (*b*) An improved integral estimate obtained by taking the area under the straight line passing through two intermediate points. By positioning these points wisely, the positive and negative errors are balanced, and an improved integral estimate results.

$$I \simeq (b - a)\frac{f(a) + f(b)}{2} \qquad [14.10]$$

where $a$ and $b$ are the limits of integration and $b - a$ is the width of the integration interval. Because the trapezoidal rule must pass through the end points, there are cases such as Fig. 14.4*a* where the formula results in a large error.

Now, suppose that the constraint of fixed base points was removed and we were free to evaluate the area under a straight line joining *any* two points on the curve. By positioning these points wisely, we could define a straight line that would balance the positive and negative errors. Hence, as in Fig. 14.4*b*, we would arrive at an improved estimate of the integral.

*Gauss quadrature* is the name for one class of techniques to implement such a strategy. The particular Gauss quadrature formulas described in the present section are called *Gauss-Legendre formulas*. Before describing the approach, we will show how numerical integration formulas such as the trapezoidal rule can be derived using the method of undetermined coefficients. This method will then be employed to develop the Gauss-Legendre formulas.

### 14.2.1 Method of Undetermined Coefficients

In Chap. 13, we derived the trapezoidal rule by integrating a linear interpolating polynomial and by geometrical reasoning. The *method of undetermined coefficients* offers a third approach that also has utility in deriving other integration techniques such as Gauss quadrature.

In order to illustrate the approach, Eq. (14.10) is expressed as

$$I \simeq c_1 f(a) + c_2 f(b) \qquad [14.11]$$

where the $c$'s are constants. Now realize that the trapezoidal rule should yield exact results when the function being integrated is a constant or a straight line. Two simple equations that represent these cases are $y = 1$ and $y = x$. Both are illustrated in Fig. 14.5. Thus, the following equalities should hold

$$c_1 f(a) + c_2 f(b) = \int_{-(b-a)/2}^{(b-a)/2} 1\, dx$$

and

$$c_1 f(a) + c_2 f(b) = \int_{-(b-a)/2}^{(b-a)/2} x\, dx$$

or, evaluating the integrals,

$$c_1 f(a) + c_2 f(b) = b - a$$

and

$$c_1 f(a) + c_2 f(b) = 0$$

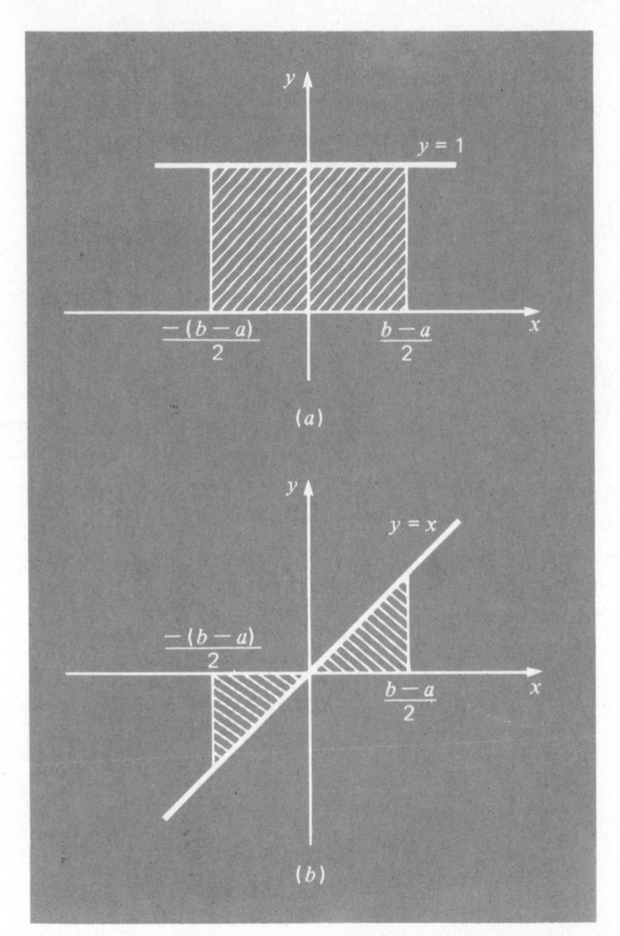


FIGURE 14.5 Two integrals that should be evaluated exactly by the trapezoidal rule: (*a*) a constant and (*b*) a straight line.

These are two equations with two unknowns that can be solved for

$$c_1 = c_2 = \frac{b - a}{2}$$

which, when substituted back into Eq. (14.11), gives

$$I \simeq \frac{b - a}{2} f(a) + \frac{b - a}{2} f(b)$$

which is equivalent to the trapezoidal rule.

### 14.2.2 Derivation of the Two-Point Gauss-Legendre Formula

Just as was the case for the above derivation of the trapezoidal rule, the object of Gauss quadrature is to determine the coefficients of an equation of the form

$$I \simeq c_1 f(x_1) + c_2 f(x_2) \qquad [14.12]$$

where the $c$'s are the unknown coefficients. However, in contrast to the trapezoidal rule that used fixed end points $a$ and $b$, the function arguments $x_1$ and $x_2$ are not fixed at the end points, but are unknowns (Fig. 14.6). Thus, we now have a total of four unknowns that must be evaluated, and consequently, we require four conditions to determine them exactly.

Just as for the trapezoidal rule, we can obtain two of these conditions by assuming that Eq. (14.12) fits the integral of a constant and a linear function exactly. Then, to arrive at the other two conditions, we merely extend this reasoning by assuming that it also fits the integral of a parabolic ($y = x^2$) and a cubic ($y = x^3$) function. By doing this, we determine all four unknowns and in the bargain derive a linear two-point integration formula that is exact for cubics. The four equations to be solved are

$$c_1 f(x_1) + c_2 f(x_2) = \int_{-1}^{1} 1\, dx = 2 \qquad [14.13]$$

$$c_1 f(x_1) + c_2 f(x_2) = \int_{-1}^{1} x\, dx = 0 \qquad [14.14]$$

$$c_1 f(x_1) + c_2 f(x_2) = \int_{-1}^{1} x^2\, dx = \frac{2}{3} \qquad [14.15]$$

$$c_1 f(x_1) + c_2 f(x_2) = \int_{-1}^{1} x^3\, dx = 0 \qquad [14.16]$$

Equations (14.13) through (14.16) can be solved simultaneously for

$$c_1 = c_2 = 1$$

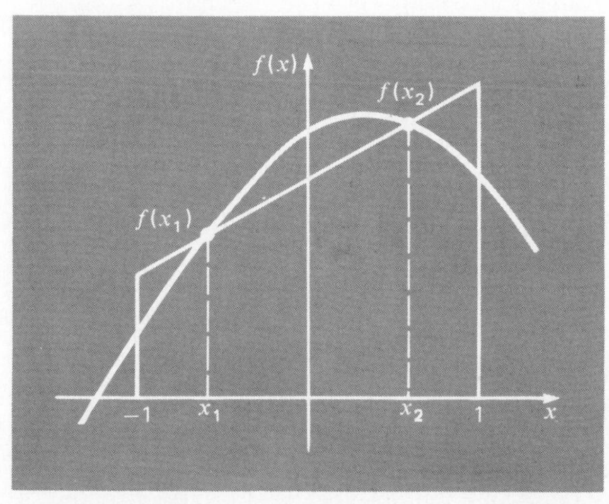


FIGURE 14.6 Graphical depiction of the unknown variables—$x_1$ and $x_2$—for integration using Gauss quadrature.

$$x_1 = \frac{-1}{\sqrt{3}} = -0.577350629\ldots$$

$$x_2 = \frac{1}{\sqrt{3}} = 0.577350269\ldots$$

which can be substituted into Eq. (14.12) to yield the two-point Gauss-Legendre formula

$$I \simeq f\left(\frac{-1}{\sqrt{3}}\right) + f\left(\frac{1}{\sqrt{3}}\right) \quad [14.17]$$

Thus, we arrive at the interesting result that the simple addition of the function values at $x = 1/\sqrt{3}$ and $-1/\sqrt{3}$ yields an integral estimate that is third-order accurate.

Notice that the integration limits in Eqs. (14.13) through (14.16) are from $-1$ to 1. This was done to simplify the arithmetic and to make the formulation as general as possible. A simple change of variable can be used to translate other limits of integration into this form. This is accomplished by assuming that a new variable $x_d$ is related to the original variable $x$ in a linear fashion, as in

$$x = a_0 + a_1 x_d \quad [14.18]$$

If the lower limit, $x = a$, corresponds to $x_d = -1$, these values can be substituted into Eq. (14.18) to yield

$$a = a_0 + a_1(-1) \quad [14.19]$$

Similarly, the upper limit, $x = b$, corresponds to $x_d = 1$, to give

$$b = a_0 + a_1(1) \quad [14.20]$$

Equations (14.19) and (14.20) can be solved simultaneously for

$$a_0 = \frac{b + a}{2} \quad [14.21]$$

and

$$a_1 = \frac{b - a}{2} \quad [14.22]$$

which can be substituted into Eq. (14.18) to yield

$$x = \frac{(b + a) + (b - a)x_d}{2} \quad [14.23]$$

This equation can be differentiated to give

$$dx = \frac{b - a}{2}\, dx_d \quad [14.24]$$

Equations (14.23) and (14.24) can be substituted for $x$ and $dx$, respectively, in the equation to be integrated. These substitutions effectively transform the integration interval without changing the value of the integral. The following example illustrates how this is done in practice.

### EXAMPLE 14.3
Two-Point Gauss-Legendre Formula

Problem Statement: Use Eq. (14.14) to evaluate the integral of

$$f(x) = 0.2 + 25x - 200x^2 + 675x^3 - 900x^4 + 400x^5$$

between the limits $x = 0$ to 0.8. Recall that this was the same problem that we solved in Chap. 13 using a variety of Newton-Cotes formulations. The exact value of the integral is 1.64053334.

Solution: Before integrating the function, we must perform a change of variable so that the limits are from $-1$ to $+1$. To do this, we substitute $a = 0$ and $b = 0.8$ into Eq. (14.23) to yield

$$x = 0.4 + 0.4x_d$$

The derivative of this relationship is [Eq. (14.24)]

$$dx = 0.4\,dx_d$$

Both of these can be substituted into the original equation to yield

$$\int_0^{0.8} (0.2 + 25x - 200x^2 + 675x^3 - 900x^4 + 400x^5)\,dx$$

$$= \int_{-1}^{1} \{[0.2 + 25(0.4 + 0.4x_d) - 200(0.4 + 0.4x_d)^2$$

$$+ 675(0.4 + 0.4x_d)^3 - 900(0.4 + 0.4x_d)^4$$

$$+ 400(0.4 + 0.4x_d)^5]0.4\}\,dx_d$$

Therefore, the right-hand side is in the form that is suitable for evaluation using Gauss quadrature. The transformed function can be evaluated at $-1/\sqrt{3}$ to be equal to 0.51674055 and at $1/\sqrt{3}$ to be equal to 1.30583723. Therefore, the integral according to Eq. (14.17) is

$$I \simeq 0.51674055 + 1.30583723 = 1.82257778$$

which represents a percent relative error of $-11.1$ percent. This result is comparable in magnitude to a four-segment application of the trapezoidal rule (Table 13.1) or a single application of Simpson's 1/3 and 3/8 rules (Examples 13.4 and 13.6). This latter result is to be expected because Simpson's rules are also third-order accurate. However, because of the clever choice of base points, Gauss quadrature attains this accuracy on the basis of only two function evaluations.

**TABLE 14.1 Weighting factors *c* and function arguments *x* used in Gauss-Legendre formulas**

| Points | Weighting factors | Function arguments | Truncation error |
|---|---|---|---|
| 2 | $c_1 = 1.000000000$ | $x_1 = -0.577350269$ | $\simeq f^{(4)}(\xi)$ |
| | $c_2 = 1.000000000$ | $x_2 = 0.577350269$ | |
| 3 | $c_1 = 0.555555556$ | $x_1 = -0.774596669$ | $\simeq f^{(6)}(\xi)$ |
| | $c_2 = 0.888888889$ | $x_2 = 0.0$ | |
| | $c_3 = 0.555555556$ | $x_3 = 0.774596669$ | |
| 4 | $c_1 = 0.347854845$ | $x_1 = -0.861136312$ | $\simeq f^{(8)}(\xi)$ |
| | $c_2 = 0.652145155$ | $x_2 = -0.339981044$ | |
| | $c_3 = 0.652145155$ | $x_3 = 0.339981044$ | |
| | $c_4 = 0.347854845$ | $x_4 = 0.861136312$ | |
| 5 | $c_1 = 0.236926885$ | $x_1 = -0.906179846$ | $\simeq f^{(10)}(\xi)$ |
| | $c_2 = 0.478628670$ | $x_2 = -0.538469310$ | |
| | $c_3 = 0.568888889$ | $x_3 = 0.0$ | |
| | $c_4 = 0.478628670$ | $x_4 = 0.538469310$ | |
| | $c_5 = 0.236926885$ | $x_5 = 0.906179846$ | |
| 6 | $c_1 = 0.171324492$ | $x_1 = -0.932469514$ | $\simeq f^{(12)}(\xi)$ |
| | $c_2 = 0.360761573$ | $x_2 = -0.661209386$ | |
| | $c_3 = 0.467913935$ | $x_3 = -0.238619186$ | |
| | $c_4 = 0.467913935$ | $x_4 = 0.238619186$ | |
| | $c_5 = 0.360761573$ | $x_5 = 0.661209386$ | |
| | $c_6 = 0.171324492$ | $x_6 = 0.932469514$ | |

### 14.2.3 Higher-Point Formulas

Beyond the two-point formula described in the previous section, higher-point versions can be developed in the general form

$$I \simeq c_1 f(x_1) + c_2 f(x_2) + \cdots + c_n f(x_n) \qquad [14.25]$$

Values for $c$'s and $x$'s for up to and including the six-point formula are summarized in Table 14.1.

#### EXAMPLE 14.4
Three-Point Gauss-Legendre Formula

Problem Statement: Use the three-point formula from Table 14.1 to estimate the integral for the same function as in Example 14.3.

Solution: According to Table 14.1, the three-point formula is

$$I = 0.555555556\, f(-0.774596669) + 0.888888889\, f(0) + 0.555555556\, f(0.774596669)$$

which is equal to

$$I = 0.281301290 + 0.873244444 + 0.485987599 = 1.64053334$$

which is exact.

Because Gauss quadrature requires function evaluations at nonuniformly spaced points within the integration interval, it is not appropriate for cases where the function is unknown. Thus, it is not suited for the many engineering problems that deal with tabulated data. However, where the function is known, its efficiency can be a decided advantage. This is particularly true when numerous integral evaluations must be performed.

## FORTRAN

```
      DIMENSION C(11),X(11),JO(5),J1(5)
      FC(XD)=AO+A1*XD
      F(X)=0.2+25*X-200*X**2+675*X**3
     C-900*X**4+400*X**5
      DATA C/1.,.888888,.555555,.652145,
     C.347855,.568889,.478629,.236927,
     C.467914,.360762,.171324/
      DATA X/.577350,0.,.774597,.339981,
     C.861136,0.,.538469,.906180,.238619,
     C.661209,.932470/
      DATA JO/1,3,4,7,9/
      DATA J1/1,3,5,8,11/
      WRITE(6,1)
    1 FORMAT('0',5X,'GAUSS QUADRATURE')
      READ(5,4)A,B
    4 FORMAT(2F10.0)
      AO=(B+A)/2
      A1=(B-A)/2
      DO 410 I=1,5
      SM=0.
      JA=JO(I)
      JB=J1(I)
      FI=(I/2)-I/2.
      IF(FI.NE.0.)GO TO 350
      K=(I-1)*2
      SM=SM+C(K)*F(FC(X(K))
  350 DO 380 J=JA,JB
      SM=SM+C(J)*F(-FC(X(J))
      SM=SM+C(J)*F(FC(X(J))
  380 CONTINUE
      SM=SM*A1
      M=I+1
      WRITE(6,5)M,SM
    5 FORMAT('0',I5,'POINT ESTIMATE = ',F10.3)
  410 CONTINUE
      STOP
      END
```

## BASIC

```
100  DIM X(11),C(11),JO(5),J1(5)
110  DEF  FN C(XD) = AO + A1 * XD—— (Function to implement
                                      change of variable)
120  DEF  FN F(X) = 0.2 + 25 * X - — (Function specifying
     200 * X ^ 2 + 675 * X ^ 3 -      the equation to
     900 * X ^ 4 + 400 * X ^ 5        be integrated)
130  PRINT : PRINT "     GAUSS-QU
     ADRATURE": PRINT
140  FOR I = 1 TO 11
150  READ C(I)——————————————————— C(I) = vector containing
160  NEXT I                                weighting factors
170  FOR I = 1 TO 11                       (Table 14.1)
180  READ X(I)——————————————————— X(I) = vector containing
190  NEXT I                                function arguments
200  FOR I = 1 TO 5                        (Table 14.1)
210  READ JO(I)
220  NEXT I
230  FOR I = 1 TO 5
240  READ J1(I)
250  NEXT I
260  INPUT "INTEGRATION LIMITS(A,
     B)=";A,B
270 AO = (B + A) / 2
280 A1 = (B - A) / 2
290  PRINT
300  FOR I = 1 TO 5
310 SM = 0
320  IF  INT (I / 2) - I / 2 < >
     0 THEN 350
330 K = (I - 1) * 2
340 SM = SM + C(K) * FN F(FN C(X(K))
350  FOR J = JO(I) TO J1(I)
360 SM = SM + C(J) * FN F( -
     FNC(X(J))
370 SM = SM + C(J) * FN F(FNC(X(J))
380  NEXT J
390 SM = SM * A1
400  PRINT I + 1;" POINT ESTIMATE
     = ";SM
410  NEXT I
420  DATA 1,.888888889,.555555556
     ,.652145155,.347854845,.5688
     88889,.478628671,.236926885,
     .467913935,.360761573,.17132
     4492
430  DATA .577350269,0,.774596669
     ,.339981044,.861136312,0,.53
     8469310,.906179846,.23861918
     6,.661209386,.932469514
440  DATA 1,3,4,7,9
450  DATA 1,3,5,8,11
460  END
```

FIGURE 14.7 Annotated computer programs in FORTRAN and BASIC to implement Gauss quadrature using Gauss-Legendre formulas.

### 14.2.4 Computer Program for Gauss Quadrature

FORTRAN and BASIC programs to implement the Gauss quadrature method are contained in Fig. 14.7. Notice that the programs are designed to capitalize on the symmetry of the weighting factors and function arguments in Table 14.1.

The programs shown in Fig. 14.7 are set up to solve the same equation studied in Examples 14.3 and 14.4. Estimates are computed up to and including the six-point formula. In order to apply these programs to another case, the function specifying the equation to be integrated must be changed. When this is done, the programs can then be employed to analyze a wide variety of engineering problems.

#### EXAMPLE 14.5
#### Applying Gauss Quadrature to the Falling Parachutist Problem

Problem Statement: In Example 13.3, we used the multiple-segment trapezoidal rule to evaluate

$$d = \frac{gm}{c} \int_0^{10} [1 - e^{-(c/m)t}]\, dt$$

where $g = 980$, $c = 12{,}500$, and $m = 68{,}100$. The exact value of the integral was determined by calculus to be 28,943.5147. Recall that the best estimate obtained using a 5000-segment trapezoidal rule was 28,943.5157 with an $|\epsilon_t| \simeq 4 \times 10^{-6}$ percent. Repeat this computation using the Gauss quadrature computer program from Fig. 14.7.

Solution: After modifying the function, the following results are obtained:

Two-point estimate = 29,001.4478

Three-point estimate = 28,943.9297

Four-point estimate = 28,943.5162

Five-point estimate = 28,943.5147

Six-point estimate = 28,943.5147

Thus, the five- and six-point estimates yield results that are exact to nine significant figures.

### 14.2.5 Error Analysis for Gauss Quadrature

The error for the Gauss-Legendre formulas is specified generally by (Carnahan et al., 1969)

$$E_t = \frac{2^{2n+3}[(n+1)!]^4}{(2n+3)[(2n+2)!]^3} f^{(2n+2)}(\xi) \qquad [14.26]$$

where $n$ is the number of points minus one and $f^{(2n+2)}(\xi)$ is the $(2n+2)$th derivative of the function after the change of variable and $\xi$ is located somewhere on the interval from $-1$ to 1. Comparison of Eq. (14.26) with Table 13.2 indicates the superiority of Gauss quadrature to Newton-Cotes formulas, provided the higher-order derivatives do not increase substantially with increasing $n$. Problem 14.8 at the end of this chapter illustrates a case where the Gauss-Legendre formulas perform poorly. In these situations, the multiple-segment Simpson's rule or Romberg integration would be preferable. However, for many functions confronted in engineering practice, Gauss quadrature provides an efficient means for evaluating integrals.

# PROBLEMS

## Hand Calculations

**14.1** Use Romberg integration to evaluate

$$\int_0^{3\pi/20} [\sin(5x + 1)]\, dx$$

to an accuracy of $\epsilon_s = 0.5$ percent. Your results should be presented in the form of Fig. 14.1. Compute the analytical solution and use it to determine the true error $\epsilon_t$ of the result obtained with Romberg integration. Check that $\epsilon_t$ is less than the stopping criterion $\epsilon_s$.

**14.2** Perform the same computations as in Prob. 14.1, but for the integral

$$\int_0^4 xe^{2x}\, dx$$

**14.3** Use Romberg integration to evaluate

$$\int_0^3 \frac{e^x \sin x}{1 + x^2}\, dx$$

to an accuracy of 0.1 percent. Your results should be presented in the form of Fig. 14.2.

**14.4** Obtain an estimate of the integral from Prob. 14.1, but using two- three-, and four-point Gauss-Legendre formulas. Compute $\epsilon_t$ for each case on the basis of the analytical solution.

**14.5** Obtain an estimate of the integral from Prob. 14.2, but using two-, three-, and four-point Gauss-Legendre formulas. Compute $\epsilon_t$ for each case on the basis of the analytical solution.

**14.6** Obtain an estimate of the integral from Prob. 14.3, but using two- through five-point Gauss-Legendre formulas.

**14.7** Perform the computation in Examples 13.3 and 14.5 for the falling parachutist, but use Romberg integration ($\epsilon_s = 0.01$ percent).

**14.8** Employ analytical methods (recall Table V.1) and two- through six-point Gauss-Legendre formulas to solve

$$\int_{-3}^{3} \frac{2}{1 + 2x^2}\, dx$$

Interpret your results in light of Eq. (14.26).

## Computer-Related Problems

**14.9** Develop a user-friendly computer program for Romberg integration based on Fig. 14.3. Test it by duplicating the computation depicted in Fig. 14.2.

**14.10** Develop a user-friendly computer program for Gauss quadrature based on Fig. 14.7. Test it by duplicating the results of Examples 14.3 and 14.4.

**14.11** Use the program developed in Prob. 14.9 to solve Probs. 14.1, 14.2, and 14.3.

**14.12** Use the program developed in Prob. 14.10 to solve Probs. 14.4, 14.5, and 14.6.

# CHAPTER FIFTEEN

# CASE STUDIES: INTEGRATION

The purpose of this chapter is to apply the methods of numerical integration discussed in Part V to practical engineering problems. Two situations are most frequently encountered. In the first case, the function under study can be expressed in analytic form but is too complicated to be readily integrated using the methods of calculus. Numerical integration is applied to situations of this type by using the analytic expression to generate a table of argument and function values. In the second case, the function to be integrated is inherently tabular in nature. This type of function usually represents a series of measurements, observations, or some other empirical information. Data for either case is directly compatible with several numerical integration schemes discussed in Chaps. 13 and 14.

*Case Study 15.1,* which analyzes cash flow for a computer company, is an example of the integration of tabular data. The trapezoidal rule and Simpson's 1/3 rule are used to determine cash flow. *Case Study 15.2,* which deals with heat calculations from chemical engineering, involves analytic data. In this case study, an analytic function is integrated numerically to determine the heat required to raise the temperature of a material.

Case Studies 15.3 and 15.4 involve functions that are available in analytic form. *Case Study 15.3,* which is taken from civil engineering, uses numerical integration to determine the total wind force acting on the mast of a racing sailboat. *Case Study 15.4* determines the root-mean-square current for an electric circuit. This example is used to demonstrate the utility of Romberg integration and Gauss quadrature.

Finally, *Case Study 15.5* returns to the analysis of tabular information to determine the work required to move a block. Although this example has a direct connection with mechanical engineering, it is germane to all other areas of engineering. Among other things, we use this case study to illustrate the integration of unequally spaced data.

## CASE STUDY 15.1 CASH-FLOW ANALYSIS (GENERAL ENGINEERING)

Background: An important part of any engineering or business project is cash-flow analysis. Available cash may affect many aspects of the problem, for instance, resource allocation (see Case Study 9.1). Your engineering

**TABLE 15.1 Computer sales data and cash-flow analysis. Column (*c*) is calculated using numerical differentiation of the information in column (*b*). The first and last values in column (*c*) are determined using forward and backward differences of order $h^2$, the middle values by centered differences of order $h^2$.**

| Numbers of computers available on the market (*a*) | Number of computers sold (*b*) | Computer sales rate per day (*c*) | Cost per computer, $ [Based on column (*a*) and Eq. (15.1)] (*d*) | Cash generated per day, $ [(*c*) × (*d*)] (*e*) | Time days (*f*) |
|---|---|---|---|---|---|
| 50,000 | 0 | 2050.0 | 1542 | 3,161,100 | 0 |
| 35,000 | 15,000 | 950.0 | 1639 | 1,557,050 | 10 |
| 31,000 | 19,000 | 1500.0 | 1677 | 2,515,500 | 20 |
| 20,000 | 30,000 | 600.0 | 1833 | 1,099,800 | 30 |
| 19,000 | 31,000 | 397.5 | 1853 | 736,568 | 40 |
| 12,050 | 37,950 | 400.0 | 2040 | 816,000 | 50 |
| 11,000 | 39,000 | −190.0 | 2083 | −395,770 | 60 |

position at Ultimate Computer Company requires that you calculate the total cash generated from computer sales for the first 60 days following the introduction of the new computer on the market. (See Table 15.1 for computer sales data.)

Your problem is complicated by the fact that the cost of the computer is very sensitive to the supply or availability. Your sales and marketing research teams have given you the information that the base sales price at infinite supply is $1250 per computer. As the supply diminishes, the price increases to a maximum of $3000 per computer. Furthermore, the continuous variation of cost with supply $N$ is defined by the empirically derived equation

$$\text{Cost per computer (\$)} = 3000 - 1750 \frac{N}{10{,}000 + N} \qquad [15.1]$$

which is plotted in Fig. 15.1.

Solution: The total cash generated is given by

$$\text{Total cash} = \int_0^{60} (\text{cash generated per day})\, dt$$

or

$$\text{Total cash} = \int_0^{60} (\text{sales rate} \times \text{unit cost})\, dt$$

In this case, the rate of sales on days 0 through 60 is given by column (*c*) in Table 15.1. The rate is determined using finite divided differences (recall Sec. 3.5.4) to estimate the first derivative of column (*b*). Notice how, because of the noise in the data, the derivative estimates in column (*c*) vary greatly. In

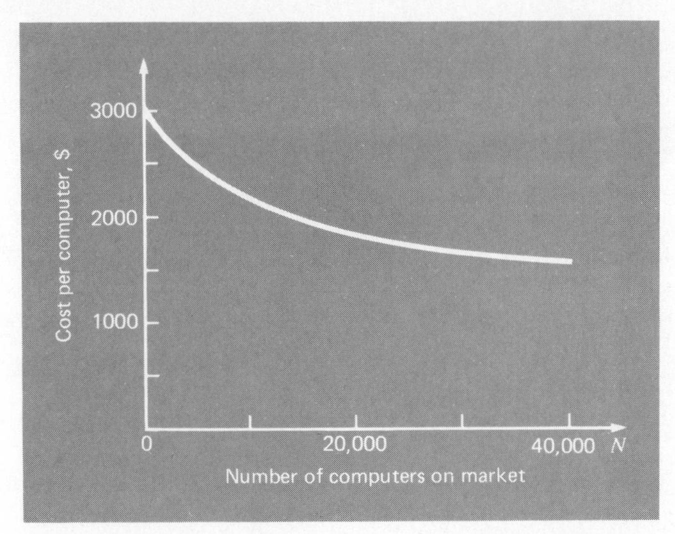


FIGURE 15.1 Cost of a computer versus the number of computers on the market. The curve is based on Eq. (15.1).

fact, even though the number of computers sold always increases, the noise in the data results in a negative sales rate at day 60. This outcome is due to the fact that numerical estimates of derivatives are highly sensitive to data noise.

The cost per computer on each day is computed on the basis of Eq. (15.1) and the number of computers available as listed in column (*a*) in Table 15.1. The cost per computer for each day 0 through 60 is given in column (*d*). Column (*e*) gives the cash generated per day. This data can be used in conjunction with the numerical integration procedures discussed in Chap. 13.

Table 15.2 gives the results of applying the trapezoidal rule and Simpson's 1/3 rule to this problem. Notice how the results vary widely depending on how many segments are employed for the analysis. In particular, the estimate from the three-segment version of the trapezoidal rule is much higher than other estimates because of the selective inclusion of the high cash-flow estimate on day 20.

**TABLE 15.2 Results of applying the trapezoidal rule and Simpson's 1/3 rule to calculate the cash generated from computer sales**

| Technique | Segments | Cash generated, $ |
|---|---|---|
| Trapezoidal rule | 1 | 82,959,900 |
| | 2 | 74,473,950 |
| | 3 | 96,294,660 |
| | 6 | 81,075,830 |
| Simpson's 1/3 rule | 2 | 71,645,300 |
| | 6 | 77,202,887 |

On the basis of this analysis, we can conclude that cash flow is approximately $77 million. However, the results indicate that care should be exercised when applying numerical integration formulas and that estimates from tabular data can usually be improved only by obtaining additional information. This conclusion is reiterated in Case Study 15.5 where we demonstrate how the number of data points can have a significant effect on the final outcome of an integral estimate.

## CASE STUDY 15.2 THE USE OF INTEGRALS TO DETERMINE THE TOTAL QUANTITY OF HEAT OF MATERIALS (CHEMICAL ENGINEERING)

Background: Heat calculations are employed routinely in chemical engineering as well as in many other fields of engineering. The present case study provides a simple but useful example of such computations.

One problem that is oftén encountered is the determination of the quantity of heat required to raise the temperature of a material. The characteristic that is needed to carry out this computation is the heat capacity $c$. This parameter represents the quantity of heat required to raise a unit mass by a unit temperature. If $c$ is constant over the range of temperatures being examined, the required heat $\Delta H$ (in calories) can be calculated by

$$\Delta H = mc\,\Delta T \qquad [15.2]$$

where $c$ has units of calories per gram per degree Celsius, $m$ is mass (in grams), and $\Delta T$ is change in temperature (in degrees Celsius). For example, the amount of heat required to raise 20 g of water from 5 to 10°C is equal to

$$\Delta H = (20)1(10 - 5) = 100 \text{ cal}$$

where the heat capacity of water is approximately 1 cal/g/°C. Such a computation is adequate when the $\Delta T$ is small. However, for large ranges of temperature, the heat capacity is not constant and, in fact, varies as a function of temperature. For example, the heat capacity of a material could increase with temperature according to a relationship such as

$$c(T) = 0.132 + 1.56 \times 10^{-4}T + 2.64 \times 10^{-7}T^2 \qquad [15.3]$$

In this instance you are asked to compute the heat required to raise 1000 g of this material from $-100$ to 200°C.

Solution: Equation (V.3) provides a way to calculate the average value of $c(T)$:

$$\bar{c}(T) = \int_{T_1}^{T_2} \frac{c(T)}{T_2 - T_1}\,dT$$

which can be substituted into Eq. (15.2) to yield

$$\Delta H = m \int_{T_1}^{T_2} c(T)\, dT \tag{15.4}$$

where $\Delta T = T_2 - T_1$. Now because, for the present case, $c(T)$ is a simple quadratic, $\Delta H$ can be determined analytically. Equation (15.3) is substituted into Eq. (15.4) and the result integrated to yield an exact value of $\Delta H = 42{,}732$ cal. It is useful and instructive to compare this result with the numerical methods developed in Chap. 13. To accomplish this, it is necessary to generate a table of values of $c$ for various values of $T$:

| *T*, °C | *c*, cal/g/°C |
|---|---|
| −100 | 0.11904 |
| −50 | 0.12486 |
| 0 | 0.13200 |
| 50 | 0.14046 |
| 100 | 0.15024 |
| 150 | 0.16134 |
| 200 | 0.17376 |

These points can be used in conjunction with a six-segment Simpson's 1/3 rule to compute an integral estimate of 42.732. This result can be substituted into Eq. (15.4) to yield a value of $\Delta H = 42{,}732$ cal, which agrees exactly with the analytical solution. This exact agreement would occur no matter how many segments are used. This is to be expected because $c$ is a quadratic function and Simpson's rule is exact for polynomials of the third order or less (see Sec. 13.2).

The results using the trapezoidal rule are listed in Table 15.3. It is seen that the trapezoidal rule is also capable of estimating the total heat very accurately. However, a small step (< 10°C) is required for five-place accuracy. This example is a good illustration of why Simpson's rule is very popular. It is easy to perform with either a hand calculator or, better yet, a

**TABLE 15.3 Results using the trapezoidal rule with various step sizes**

| Step size, °C | $\Delta H$ | $\epsilon_t$, % |
|---|---|---|
| 300 | 96,048 | 125 |
| 150 | 43,029 | 0.7 |
| 100 | 42,864 | 0.3 |
| 50 | 42,765 | 0.07 |
| 25 | 42,740 | 0.018 |
| 10 | 42,733.3 | < 0.01 |
| 5 | 42,732.3 | < 0.01 |
| 1 | 42,732.01 | < 0.01 |
| 0.05 | 42,732.0003 | < 0.01 |

personal computer. In addition, it is usually sufficiently accurate for relatively large step sizes and is exact for polynomials of the third order or less.

## CASE STUDY 15.3 EFFECTIVE FORCE ON THE MAST OF A RACING SAILBOAT (CIVIL ENGINEERING)

Background: A cross section of a racing sailboat is shown in Fig. 15.2*a*. Wind forces ($f$) exerted per foot of mast from the sails vary as a function of distance above the deck of the boat ($z$) as in Fig. 15.2*b*. Calculate the tensile force $T$ in the left mast support cable, assuming that the right support cable is completely slack and the mast joins the deck in a manner that transmits horizontal or vertical forces but no moments. Assume that the mast remains vertical.

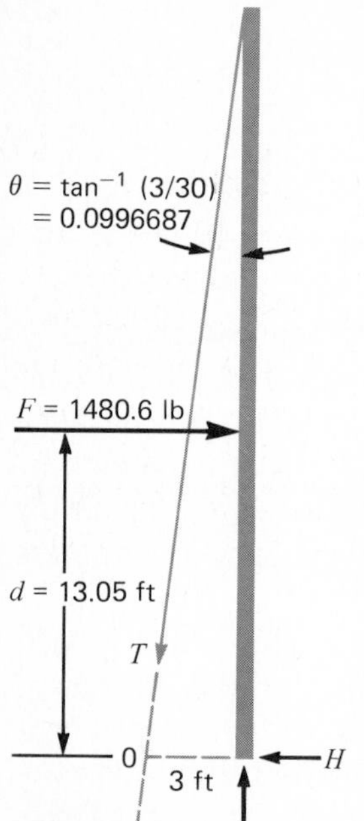


FIGURE 15.3
Free-body diagram of the forces exerted on the mast of a sailboat.

Solution: In order to proceed with this problem, it is required that the distributed force $f$ be converted to an equivalent total force $F$ and that its effective location above the deck $d$ be calculated (Fig. 15.3). This computation is complicated by the fact that the force exerted per foot of mast varies with the distance above the deck. The total force exerted on the mast can be expressed as the integral of the following continuous function:

$$F = \int_0^{30} 200 \left(\frac{z}{5 + z}\right) e^{-2z/30}\, dz$$

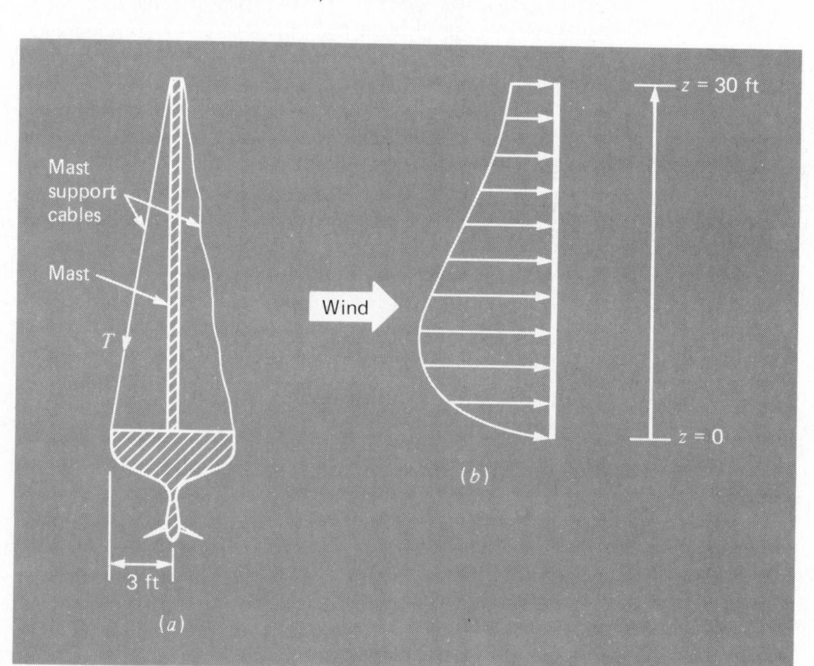


FIGURE 15.2 (*a*) Cross section of a racing sailboat. (*b*) Wind forces $f$ exerted per foot of mast as a function of distance $z$ above the deck of the boat.

**TABLE 15.4 Values of $f(z)$ for a step size of 3 ft that provide data for the trapezoidal rule and Simpson's 1/3 rule**

| $z$, ft | $f(z)$, lb/ft |
|---|---|
| 0 | 0 |
| 3 | 61.40 |
| 6 | 73.13 |
| 9 | 70.56 |
| 12 | 63.43 |
| 15 | 55.18 |
| 18 | 47.14 |
| 21 | 39.83 |
| 24 | 33.42 |
| 27 | 27.89 |
| 30 | 23.20 |

This nonlinear integral is difficult to evaluate analytically. Therefore, it is convenient to employ numerical approaches such as Simpson's rule and the trapezoidal rule for this problem. This is accomplished by calculating $f(z)$ for various values of $z$ and then using Eqs. (13.10) or (13.18). For example, Table 15.4 has values of $f(z)$ for a step size of 3 ft that provide data for Simpson's 1/3 rule or the trapezoidal rule. Results for several step sizes are given in Table 15.5. It is observed that both methods give a value of $F = 1480.6$ lb as the step size becomes small. In this case, a step size of 0.05 ft for the trapezoidal rule and 0.5 for Simpson's rule provides good results.

**TABLE 15.5 Values of $F$ computed on the basis of various versions of the trapezoidal rule and Simpson's 1/3 rule**

| Technique | Step size, ft | Segments | $F$, lb |
|---|---|---|---|
| Trapezoidal rule | 15 | 2 | 1001.7 |
| | 10 | 3 | 1222.3 |
| | 6 | 5 | 1372.3 |
| | 3 | 10 | 1450.8 |
| | 1 | 30 | 1477.1 |
| | 0.5 | 60 | 1479.7 |
| | 0.25 | 120 | 1480.3 |
| | 0.1 | 300 | 1480.5 |
| | 0.5 | 600 | 1480.6 |
| Simpson's 1/3 rule | 15 | 2 | 1219.6 |
| | 5 | 6 | 1462.9 |
| | 3 | 10 | 1476.9 |
| | 1 | 30 | 1480.5 |
| | 0.5 | 60 | 1480.6 |

The effective line of action $F$ (Fig. 15.3) can be calculated by evaluation of the integral

$$d = \frac{\int_0^{30} z\,f(z)\,dz}{\int_0^{30} f(z)\,dz}$$

or

$$d = \frac{\int_0^{30} 200z[z/(5+z)]e^{-2z/30}\,dz}{1480.6}$$

This integral can be evaluated using methods similar to the above. For example, Simpson's 1/3 rule with a step size of 0.5 gives

$$d = \frac{19{,}326.9}{1480.6} = 13.05 \text{ ft}$$

With $F$ and $d$ known from numerical methods, a free-body diagram is used to develop force and moment balance equations. This free-body diagram is shown in Fig. 15.3. Summing forces in the horizontal and vertical direction and taking moments about point 0 gives

$$\Sigma F_H = 0 = F - T \sin\theta - H \qquad [15.9]$$

$$\Sigma F_V = 0 = V - T \cos\theta \qquad [15.10]$$

$$\Sigma M_0 = 0 = 3V - Fd \qquad [15.11]$$

where $T$ is the tension in the cable. $H$ and $V$ are the unknown reactions on the mast transmitted by the deck. The direction as well as magnitude of $H$ and $V$ are unknown. Equation (15.11) can be solved directly for $V$ because $F$ and $d$ are known.

$$V = \frac{Fd}{3} = \frac{1480.6(13.05)}{3} = 6440.6 \text{ lb}$$

Therefore, from Eq. (15.10),

$$T = \frac{V}{\cos\theta} = \frac{6440.6}{0.995} = 6473 \text{ lb}$$

and from Eq. (15.9),

$$H = F - T\sin\theta = 1480.6 - (6473)(0.0995) = 836.54 \text{ lb}$$

These forces now enable you to proceed with other aspects of the structural design of the boat such as the cables and the deck support system for the mast. This problem illustrates nicely two uses of numerical integration that may be encountered during the engineering design of structures. It is seen that both the trapezoidal rule and Simpson's 1/3 rule are easy to apply and are practical problem-solving tools. Simpson's 1/3 rule is more accurate than the trapezoidal rule for the same step size and thus may often be preferred.

## CASE STUDY 15.4 DETERMINATION OF THE ROOT-MEAN-SQUARE CURRENT BY NUMERICAL INTEGRATION (ELECTRICAL ENGINEERING)

Background: The effective value of a periodically varying electric current is given by the formula for root mean square (see Case Study 12.4):

$$I_{\text{RMS}} = \frac{1}{T}\sqrt{\int_0^T i^2(t)dt} \qquad [15.12]$$

where $T$ is the period, that is, the time for one cycle, and $i(t)$ is the instantaneous current. Calculate the RMS current of the waveform shown in Fig. 15.4 using the trapezoidal rule, Simpson's 1/3 rule, Romberg integration, and Gauss quadrature for $T = 1$ s. Recall that in Case Study 12.4, we solved this problem by analytically integrating a parabola that had been fit to the function to yield an integral estimate of 20.2176887.

Solution: Integral estimates for various applications of the trapezoidal rule and Simpson's 1/3 rule are listed in Table 15.6. One application of Simpson's 1/3 rule yields the same result as was obtained in Case Study 12.4. This is expected because Simpson's 1/3 rule corresponds to the area under a parabola fit to three points. Notice that Simpson's rule is more accurate than the trapezoidal rule.

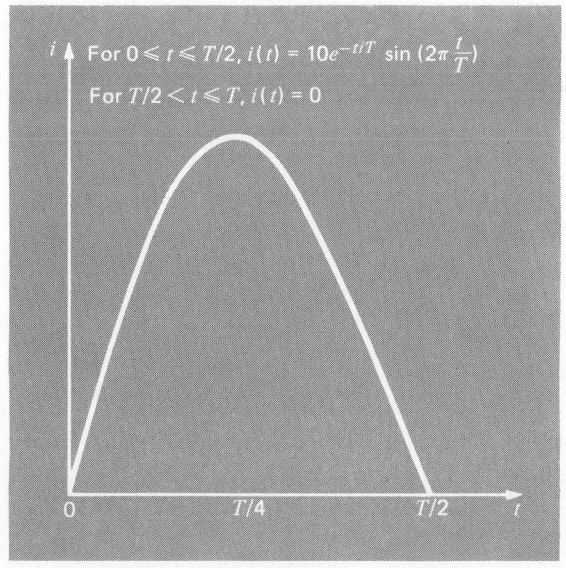


FIGURE 15.4 A periodically varying electric current.

**TABLE 15.6** **Values for the integral calculated using various numerical schemes. The percent relative error $\epsilon_t$ is based on a true value of 15.4126081.**

| Technique | Segments | Integral | $\epsilon_t$, % |
|---|---|---|---|
| Trapezoidal rule | 1 | 0.0 | 100 |
| | 2 | 15.1632665 | 1.62 |
| | 4 | 15.4014291 | 0.0725 |
| | 8 | 15.4119584 | $4.21 \times 10^{-3}$ |
| | 16 | 15.4125682 | $2.59 \times 10^{-4}$ |
| | 32 | 15.4126056 | $1.62 \times 10^{-5}$ |
| | 64 | 15.4126079 | $1.30 \times 10^{-6}$ |
| | 128 | 15.4126081 | 0 |
| Simpson's 1/3 rule | 2 | 20.2176887 | −31.2 |
| | 4 | 15.4808166 | −0.443 |
| | 8 | 15.4154681 | −0.0186 |
| | 16 | 15.4127714 | $-1.06 \times 10^{-3}$ |
| | 32 | 15.4126081 | 0 |

The exact value for the integral is 15.4126081. This result is obtained using a 128-segment trapezoidal rule or a 32-segment Simpson's rule. The same estimate is also determined using Romberg integration (Fig. 15.5).

In addition, Gauss quadrature can be used to make the same estimate. Recall that the determination of the root-mean-square current for Case Study 12.4 involved the evaluation of the integral ($T = 1$)

$$I = \int_0^{1/2} (10e^{-t} \sin 2\pi t)^2 \, dt \qquad [15.13]$$

First, a change in variable is performed by applying Eq. (14.23) and (14.24) to yield

$$t = \frac{1}{4} + \frac{1}{4} t_d$$

| $O(h^2)$ | $O(h^4)$ | $O(h^6)$ | $O(h^8)$ | $O(h^{10})$ | $O(h^{12})$ | $O(h^{14})$ |
|---|---|---|---|---|---|---|
| 0 | 20.2176887 | 15.1650251 | 15.4150177 | 15.4126058 | 15.4126081 | 15.4126081 |
| 15.1632665 | 15.4808166 | 15.4111116 | 15.4126152 | 15.4126081 | 15.4126081 | |
| 15.4014291 | 15.4154682 | 15.4122517 | 15.4126081 | 15.4126081 | | |
| 15.4119584 | 15.4127715 | 15.4126078 | 15.4126081 | | | |
| 15.4125682 | 15.4126180 | 15.4126081 | | | | |
| 15.4126056 | 15.4126087 | | | | | |
| 15.4126079 | | | | | | |

FIGURE 15.5 Result of using Romberg integration to estimate the RMS current.

and

$$dt = \frac{1}{4}\, dt_d$$

These relationships can be substituted into Eq. (15.13) to yield

$$I = \int_{-1}^{1} \left[ 10e^{-[1/4+(1/4)t_d]} \sin 2\pi\left(\frac{1}{4} + \frac{1}{4}t_d\right)\right]^2 \frac{1}{4}\, dt_d$$

For the two-point Gauss-Legendre formula, this function is evaluated at $t_d = 1/\sqrt{3}$ and $-1/\sqrt{3}$, with the results being 7.6840962 and 4.3137280, respectively. These values can be substituted into Eq. (14.17) to yield an integral estimate of 11.9978242, which represents an error of $\epsilon_t = 22$ percent.

The three-point formula is (Table 14.1)

$$\begin{aligned} I &= 0.555555556\ (1.237449345) + 0.888888889\ (15.16326649) \\ &\quad + 0.555555556\ (2.684914679) \\ &= 15.65755021 \qquad \epsilon_t = 1.6\% \end{aligned}$$

The results of using the higher-point formulas are summarized in Table 15.7.

The integral estimate of 15.4126081 can be substituted into Eq. (15.12) to compute an $I_{RMS}$ of 3.9258895 A. This result could then be employed to guide other aspects of the design and operation of the circuit.

**TABLE 15.7 Results of using various-point Gauss quadrature formulas to approximate the integral**

| Points | Estimate | $\epsilon_t$, % |
|---|---|---|
| 2 | 11.9978243 | 22.1 |
| 3 | 15.6575502 | −1.59 |
| 4 | 15.4058023 | $4.42 \times 10^{-2}$ |
| 5 | 15.4126391 | $-2.01 \times 10^{-4}$ |
| 6 | 15.4126109 | $-1.82 \times 10^{-5}$ |

## CASE STUDY 15.5 USE OF NUMERICAL INTEGRATION TO COMPUTE WORK (MECHANICAL ENGINEERING)

Background: Many engineering problems involve the calculation of work. The general formula is

Work = force × distance

When you were introduced to this concept in high school physics, simple applications were presented using forces that remained constant throughout the displacement. For example, if a force of 10 lb was used to pull a block a distance of 15 ft, the work would be calculated as 150 ft · lb.

Although such a simple computation is useful for introducing the concept, realistic problem settings are usually more complex. For example, suppose that the force varies during the course of the calculation. In such cases, the work equation is reexpressed as

$$W = \int_{x_0}^{x_n} F(x)\, dx \tag{15.14}$$

where $W$ is work in foot pounds, $x_0$ and $x_n$ are the initial and final positions, respectively, and $F(x)$ is a force which varies as a function of position. If $F(x)$ is easy to integrate, Eq. (15.14) can be evaluated analytically. However, in a realistic problem setting, the force might not be expressed in such a manner. In fact, when analyzing measured data, the force might only be available

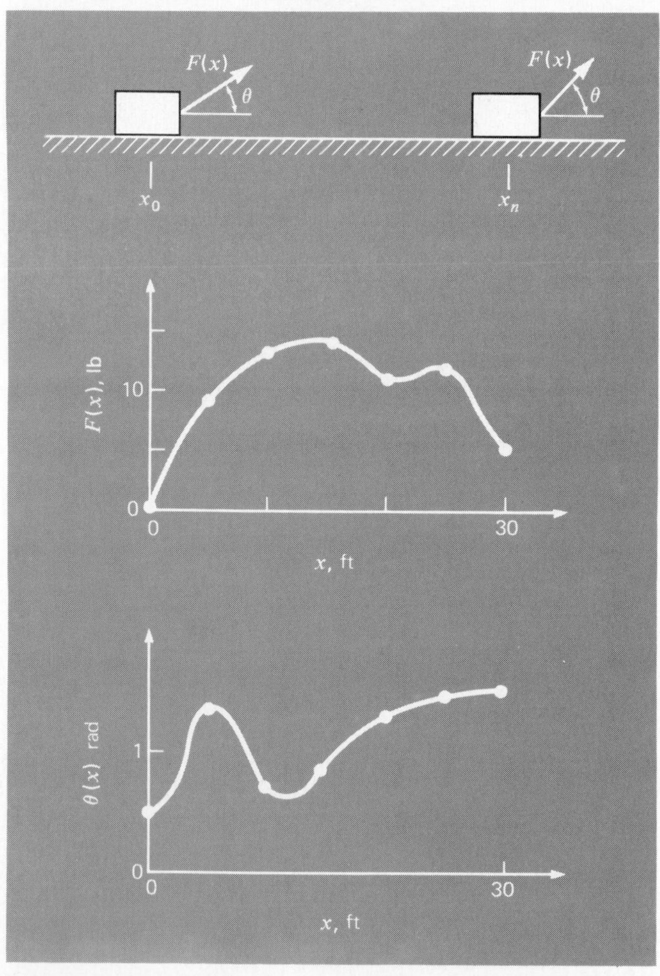


FIGURE 15.6 The case of a variable force acting on a block. For this case, the angle, as well as the magnitude, of the force varies.

in tabular form. For such cases, numerical integration is the only viable option for the evaluation.

Further complexity is introduced if the angle between the force and the direction of movement also varies as a function of position (Fig. 15.6). The work equation can be modified further to account for this effect, as in

$$W = \int_{x_0}^{x_n} F(x) \cos [\theta(x)]\, dx \qquad [15.15]$$

Again, if $F(x)$ and $\theta(x)$ are simple functions, Eq. (15.15) might be solved analytically. However, as in Fig. 15.6, it is more likely that the functional relationship is complicated. For this situation, numerical methods provide the only alternative for determining the integral.

Suppose that you have to perform the computation for the situation depicted in Fig. 15.6. Although the figure shows the continuous values for $F(x)$ and $\theta(x)$, assume that because of experimental constraints, you are provided only with discrete measurements at $x$ = 5-ft intervals (Table 15.8). Use single- and multiple-segment versions of the trapezoidal rule and Simpson's 1/3 and 3/8 rules to compute work for this data.

Solution: The results of the analysis are summarized in Table 15.9. A percent relative error, $\epsilon_t$, was computed in reference to a true value of the integral of

**TABLE 15.8 Data for force $F(x)$ and angle $\theta(x)$ as a function of position $x$**

| $x$, ft | $F(x)$, lb | $\theta$, rad | $F(x) \cos \theta$ |
|---|---|---|---|
| 0 | 0.0 | 0.50 | 0.0000 |
| 5 | 9.0 | 1.40 | 1.5297 |
| 10 | 13.0 | 0.75 | 9.5120 |
| 15 | 14.0 | 0.90 | 8.7025 |
| 20 | 10.5 | 1.30 | 2.8087 |
| 25 | 12.0 | 1.48 | 1.0881 |
| 30 | 5.0 | 1.50 | 0.3537 |

**TABLE 15.9 Estimates of work calculated using the trapezoidal rule and Simpson's rules. The percent relative error ($\epsilon_t$) was computed in reference to a true value of the integral (129.52 ft · lb) that was estimated on the basis of values at 1-ft intervals.**

| Technique | Segments | Work | $\epsilon_t$, % |
|---|---|---|---|
| Trapezoidal | 1 | 5.31 | 95.9 |
| | 2 | 133.19 | −2.84 |
| | 3 | 124.98 | 3.51 |
| | 6 | 119.09 | 8.05 |
| Simpson's 1/3 rule | 2 | 175.82 | −35.75 |
| | 6 | 117.13 | 9.57 |
| Simpson's 3/8 rule | 3 | 139.93 | −8.04 |

129.52 that was estimated on the basis of values taken from Fig. 15.6 at 1-ft intervals.

The results are interesting because the most accurate result occurs for the simple two-segment trapezoidal rule. More refined estimates using more segments, as well as Simpson's rules, yield less accurate results.

The reason for this apparently counterintuitive result is that the coarse spacing of the points is not adequate to capture the variations of the forces and angles. This is particularly evident in Fig. 15.7, where we have plotted the continuous curve for the product of $F(x)$ and $\cos[\theta(x)]$. Notice, how the use of seven points to characterize the continuously varying function misses the two peaks at $x = 2.5$ and 12.5 ft. The omission of these two points effectively limits the accuracy of the numerical integration estimates in Table 15.9. The fact that the two-segment trapezoidal rule yields the most accurate result is due to the chance positioning of the points for this particular problem (Fig. 15.8).

The conclusion to be drawn from Fig. 15.7 is that an adequate number of measurements must be made in order to accurately compute integrals. For the present case, if data were available at $F(2.5)\cos[\theta(2.5)] = 4.3500$ and $F(12.5)\cos[\theta(12.5)] = 11.3600$, we could determine an integral estimate using the algorithm for unequally spaced data described previously in Sec. 13.3. Figure 15.9 illustrates the unequal segmentation for this case. Including the two additional points yields an improved integral estimate of 126.9 ($\epsilon_t = 2.02$ percent). Thus, the inclusion of the additional data would incorporate the peaks that were missed previously and, as a consequence, lead to better results.

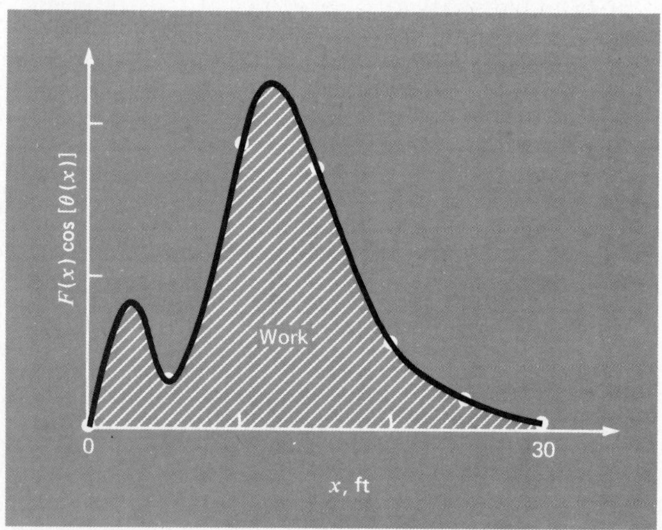


FIGURE 15.7 A continuous plot of $F(x)\cos[\theta(x)]$ versus position, along with the seven discrete points used to develop the numerical integration estimates in Table 15.9. Notice how the use of seven points to characterize this continuously varying function misses two peaks at $x = 2.5$ and 12.5 ft.

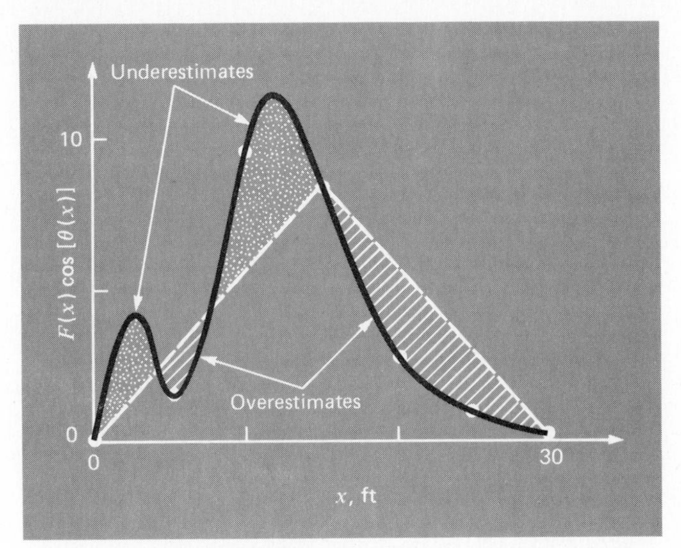


FIGURE 15.8 Graphical depiction of why the two-segment trapezoidal rule yields a good estimate of the integral for this particular case. By chance, the use of two trapezoids happens to lead to an even balance between positive and negative errors.

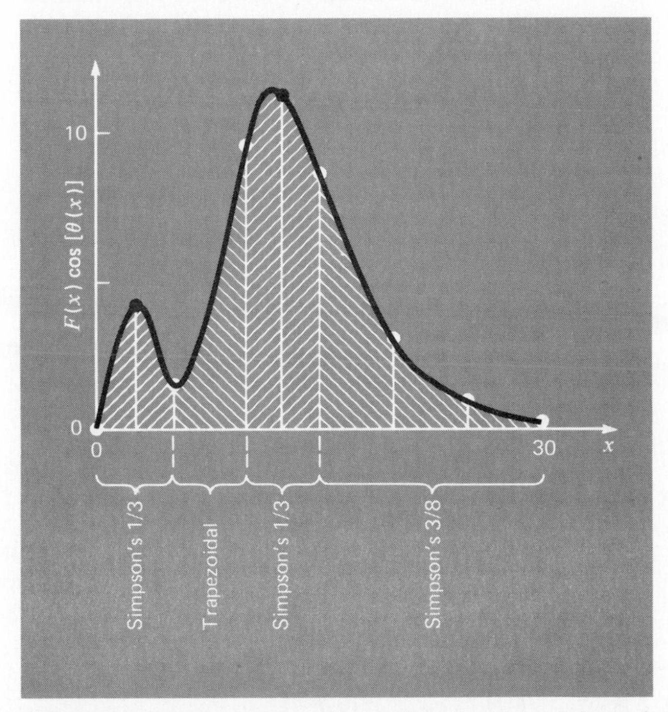


FIGURE 15.9 The unequal segmentation scheme that results from the inclusion of two additional points at $x = 2.5$ and 12.5 in the data in Table 15.8. The numerical integration formulas applied to each set of segments are shown.

# PROBLEMS

## General Engineering:

**15.1** Reproduce the computations in Case Study 15.1 using your own software.

**15.2** Perform the same computations as in Case Study 15.1, but instead of using Eq. (15.1) use the following alternative formulation:

$$\text{Cost per computer (\$)} = 1250 + 1750e^{-5.5\times10^{-5}N}$$

**15.3** You are doing a study of an assembly line at an automobile plant. Over the course of a 24-h period, you visit two points on the line, and at different times during the day you spot-check the number of autos that pass through in a minute. The data is

| Point A | | Point B | |
|---|---|---|---|
| Time | Cars/min | Time | Cars/min |
| Midnight | 3 | Midnight | 3 |
| 2 A.M. | 3 | 1 A.M. | 3 |
| 3 A.M. | 5 | 4 A.M. | 5 |
| 6 A.M. | 4 | 5 A.M. | 2 |
| 9 A.M. | 5 | 7 A.M. | 1 |
| 11 A.M. | 6 | 10 A.M. | 4 |
| 2 P.M. | 2 | 1 P.M. | 3 |
| 5 P.M. | 1 | 3 P.M. | 4 |
| 6 P.M. | 1 | 9 P.M. | 6 |
| 7 P.M. | 3 | 10 P.M. | 1 |
| 8 P.M. | 4 | 11 P.M. | 3 |
| Midnight | 6 | Midnight | 6 |

Use numerical integration and Eq. (V.3) to determine the total number of cars per day that pass through each point.

**15.4** The data listed in Table P15.4 gives hourly measurements of heat flux $q$ at the surface of a solar collector. Estimate the total heat absorbed by a 150,000-cm$^2$ collector panel during a 14-h period. The panel has an absorption efficiency $e_{ab}$ of 45 percent. The total heat absorbed is given by

$$H = e_{ab} \int_0^t q\,A\,dt$$

where $A$ is area and $q$ is heat flux.

**TABLE P15.4 Measurements of solar heat flux**

| Time, h | Heat flux $q$, cal/cm²/h |
|---|---|
| 0 | 0.1 |
| 1 | 1.62 |
| 2 | 5.32 |
| 3 | 6.29 |
| 4 | 7.8 |
| 5 | 8.81 |
| 6 | 8.00 |
| 7 | 8.57 |
| 8 | 8.03 |
| 9 | 7.04 |
| 10 | 6.27 |
| 11 | 5.56 |
| 12 | 3.54 |
| 13 | 1.0 |
| 14 | 0.2 |

## Chemical Engineering

**15.5** Reproduce the computations in Case Study 15.2 using your own software.

**15.6** Perform the same computation as in Case Study 15.2, but compute the amount of heat required to raise the temperature of 2000 g of the material from −200 to 100°C. Use Simpson's rule for your computation, with values of $T$ at 50°C increments.

**15.7** Repeat Prob. 15.6, but use Romberg integration to $\epsilon_s = 0.01$ percent.

**15.8** Repeat Prob. 15.6, but use a two- and a three-point Gauss-Legendre formula. Interpret your results.

**15.9** Use Simpson's rule to compute the total heat for the plate depicted in Case Study 9.2 if the heat capacity is defined by Eq. (15.3).

## Civil Engineering

**15.10** Reproduce the computations in Case Study 15.3 using your own software.

**15.11** Perform the same computation as in Case Study 15.3, but use Romberg integration to evaluate the integral. Employ a stopping criterion of $\epsilon_s = 0.25$ percent.

**15.12** Perform the same computation as in Case Study 15.3, but use Gauss quadrature to evaluate the integral.

**15.13** Perform the same computation as in Case Study 15.3, but change the integral to

$$F = \int_0^{30} \frac{250z}{4+z} e^{-2z/30}\, dz$$

**15.14** Stream cross-sectional areas ($A$) are required for a number of tasks in water resources engineering, including flood forecasting and reservoir design. Unless electronic sounding devices are available to obtain continuous profiles of the channel bottom, the engineer must rely on discrete depth measurements to compute $A$. An example of a typical stream cross section is shown in Fig. P15.14. The data points represent locations where a boat was anchored and depth readings taken. Use two trapezoidal rule applications ($h = 4$ and 2 m), and Simpson's 1/3 rule to estimate the cross-sectional area from this data.

**15.15** During a field survey, you are required to compute the area of the field shown in Fig. P15.15. Use Simpson's rules to determine the area.

**15.16** A transportation engineering study requires the calculation of the total number of cars that pass through an intersection over a 24-h period. An individual visits the intersection at various times during the course of a day and counts the number of cars that pass through the intersection in a minute. Utilize this data, which is summarized in Table P15.16, to estimate the total number of cars that pass through the intersection per day. (Be careful of units.)

**TABLE P15.16 Traffic flow rate for an intersection measured at various times over a 24-h period**

| Time | Rate, cars/min |
|---|---|
| 12:00 Midnight | 10 |
| 2:00 A.M. | 4 |
| 6:00 A.M. | 6 |
| 7:00 A.M. | 40 |
| 8:00 A.M. | 60 |
| 9:00 A.M. | 80 |
| 11:00 A.M. | 25 |
| 1:00 P.M. | 18 |
| 3:00 P.M. | 17 |
| 4:00 P.M. | 28 |
| 5:00 P.M. | 35 |
| 6:00 P.M. | 77 |
| 7:00 P.M. | 40 |
| 8:00 P.M. | 30 |
| 10:00 P.M. | 31 |
| 12:00 Midnight | 15 |

## Electrical Engineering

**15.17** Reproduce the computations in Case Study 15.4 using your own software.

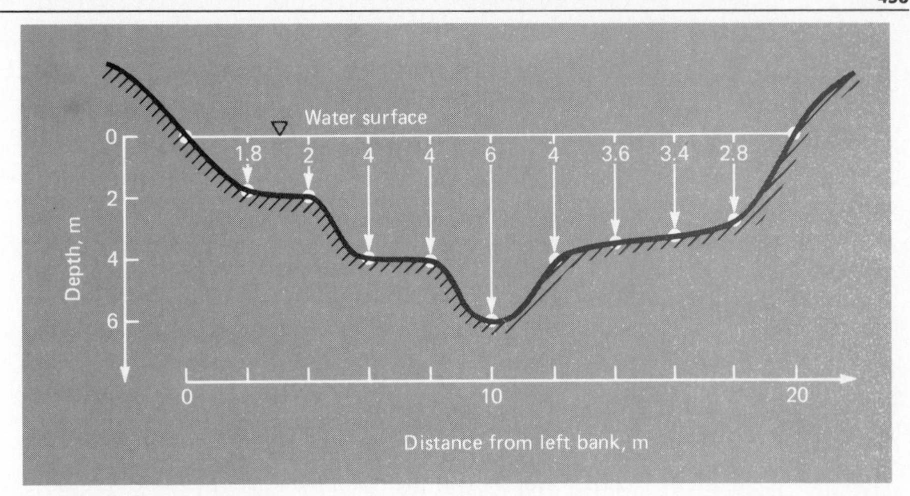


FIGURE P15.14 A stream cross section.

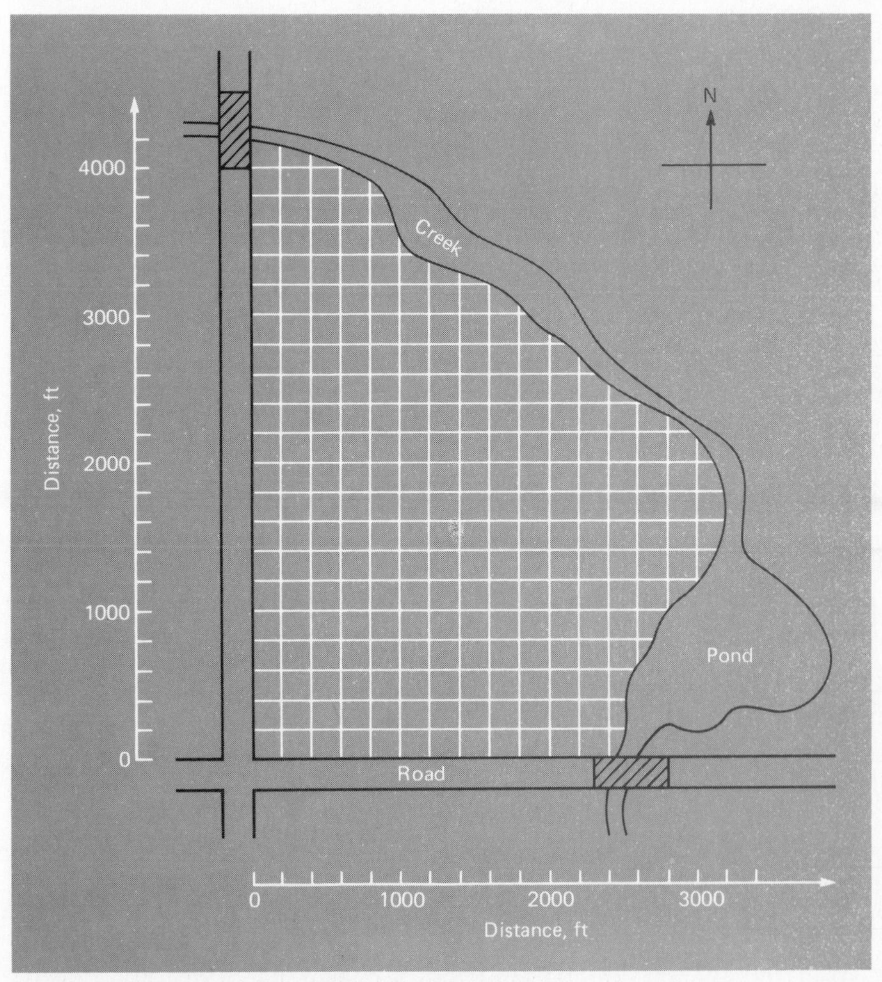


FIGURE P15.15 A field bounded by two roads and a creek.

**15.18** Perform the same computation as in Case Study 15.4, but for the current as specified by

$$i(t) = 5e^{-2t} \sin 2\pi t \qquad \text{for } 0 \le t \le T/2$$

$$i(t) = 0 \qquad \text{for } T/2 \le t \le T$$

where $T = 1$ s. Use a 16-segment Simpson's 1/3 rule to estimate the integral.

**15.19** Repeat Prob. 15.18, but use Gauss quadrature.

**15.20** Repeat Prob. 15.18, but use Romberg integration to $\epsilon_s = 0.1$ percent.

## Mechanical Engineering

**15.21** Reproduce the computations in Case Study 15.5 using your own software.

**15.22** Perform the same computation as in Case Study 15.5, but use the following equation to compute:

$$F(x) = 1.17x - 0.035x^2$$

Employ the values of $\theta$ from Table 15.8.

**15.23** Perform the same computation as in Case Study 15.5, but use the following equation to compute:

$$\theta(x) = 0.5 + 0.1375x - 0.01x^2 + (2.5 \times 10^{-4})x^3$$

Employ the equation from Prob. 15.22 for $F(x)$. Use four-, eight-, and sixteen-segment trapezoidal rules to compute the integral.

**15.24** Repeat Prob. 15.23, but use Simpson's 1/3 rule.

**15.25** Repeat Prob. 15.23, but use Romberg integration to $\epsilon_s = 0.1$ percent.

**15.26** Repeat Prob. 15.23, but use Gauss quadrature.

## Miscellaneous

**15.27** Read all the case studies in Chap. 15. On the basis of your reading and experience make up your own case study for any one of the fields of engineering. This may involve modifying or reexpressing one of our case studies. However, it can also be totally original. As with our examples, it must be drawn from an engineering problem context and must demonstrate the use of numerical methods for integration. Write up your results using our case studies as models.

# EPILOGUE: PART V

## V.4 TRADE-OFFS

Table V.3 provides a summary of the trade-offs involved in numerical integration or quadrature. Most of these methods are based on the simple physical interpretation of an integral as the area under a curve. These techniques are designed to evaluate the integral of two different cases: (1) a continuous mathematical function and (2) discrete data in tabular form.

The Newton-Cotes formulas are the primary methods discussed in Chap. 13. They are applicable to both continuous and discrete functions. Both closed and open versions of these formulas are available. The open forms, which have integration limits that extend beyond the range of the data, are rarely used for the evaluation of definite integrals. However, they have great utility for the solution of ordinary differential equations, as discussed in Chap. 17.

The closed Newton-Cotes formulas are based on replacing a mathematical function or tabulated data by an interpolating polynomial that is easy to integrate. The simplest version is the trapezoidal rule, which is based on taking the area below a straight line joining adjacent values of the function. One way to improve the accuracy of the trapezoidal rule is to divide the integration interval from $a$ to $b$ into a number of segments and apply the method to each segment.

Aside from applying the trapezoidal rule with finer segmentation, another way to obtain a more accurate estimate of the integral is to use higher-order polynomials to connect the points. If a quadratic equation is employed, the result is Simpson's 1/3 rule. If a cubic equation is used, the result is Simpson's 3/8 rule. Because they are much more accurate than the trapezoidal rule, these formulas are usually preferred. Multiple-segment versions are available. For situations with an even number of segments, the multiple application of the 1/3 rule is recommended. For an odd number

**TABLE V.3 Comparison of the characteristics of alternative methods for numerical integration. The comparisons are based on general experience and do not account for the behavior of special functions.**

| Method | Data points required for one application | Data points required for *n* applications | Truncation error | Application | Programming effort | Comments |
|---|---|---|---|---|---|---|
| Trapezoidal rule | 2 | $n + 1$ | $\simeq h^3 f''(\xi)$ | Wide | Easy | |
| Simpson's 1/3 rule | 3 | $2n + 1$ | $\simeq h^5 f^{(4)}(\xi)$ | Wide | Easy | |
| Simpson's rule (1/3 and 3/8) | 3 or 4 | $\geq 3$ | $\simeq h^5 f^{(4)}(\xi)$ | Wide | Easy | |
| Higher-order Newton-Cotes | $\geq 5$ | N/A | $\geq h^7 f^{(6)}(\xi)$ | Rare | Easy | |
| Romberg integration | 3 | | | Requires $f(x)$ be known | Moderate | Inappropriate for tabular data |
| Gauss quadrature | $\geq 2$ | N/A | | Requires $f(x)$ be known | Easy | Inappropriate for tabular data |

of segments, the 3/8 rule can be applied to the last three segments and the 1/3 rule to the remaining segments.

Higher-order Newton-Cotes formulas are also available. However, they are rarely used in practice. Where high accuracy is required, Romberg integration and Gauss quadrature formulas are available. It should be noted that both Romberg integration and Gauss quadrature are of practical value only in cases where the function is available in continuous form. These techniques are ill-suited for tabulated data.

## V.5 IMPORTANT RELATIONSHIPS AND FORMULAS

Table V.4 summarizes important information that was presented in Part V. This table can be consulted to quickly access important relationships and formulas.

## V.6 ADVANCED METHODS AND ADDITIONAL REFERENCES

Although we have reviewed a number of numerical integration techniques, there are other methods that have utility in engineering practice. For example, *adaptive Simpson's integration* is based on dividing the integration interval into a series of subintervals of width $h$. Then Simpson's 1/3 rule is used to evaluate the integral of each subinterval by halving the step size in an iterative fashion, that is, with a step size of $h$, $h/2$, $h/4$, $h/8$, and so forth. The iterations are continued for each subinterval until an approximate error estimate $\epsilon_a$ [Eq. (3.5)] falls below a prespecified stopping criterion $\epsilon_s$. The total integral is then computed as the summation of the integral estimates for the subintervals. This technique is especially valuable for complicated functions that have regions exhibiting both lower- and higher-order variations. Discussions of adaptive integration may be found in Gerald and Wheatley (1984) and Rice (1983).

Another method for obtaining integrals is to fit *cubic splines* to the data. The resulting cubic equations can be integrated easily (Forsythe et al., 1977). Finally, aside from the Gauss-Legendre formulas discussed in Sec. 14.2, there are a variety of other *quadrature formulas*. Carnahan, Luther, and Wilkes (1969) and Ralston and Rabinowitz (1978) summarize many of these approaches.

In summary, the foregoing is intended to provide you with avenues for deeper exploration of the subject. Additionally, all the above references provide descriptions of the basic techniques covered in Part V. We urge you to consult these alternative sources to broaden your understanding of numerical methods for integration.

**TABLE V.4** **Summary of important information presented in Part V.**

| Method | Formulation | Graphic interpretations | Error |
|---|---|---|---|
| Trapezoidal rule | $I \simeq (b-a)\dfrac{f(a)+f(b)}{2}$ | 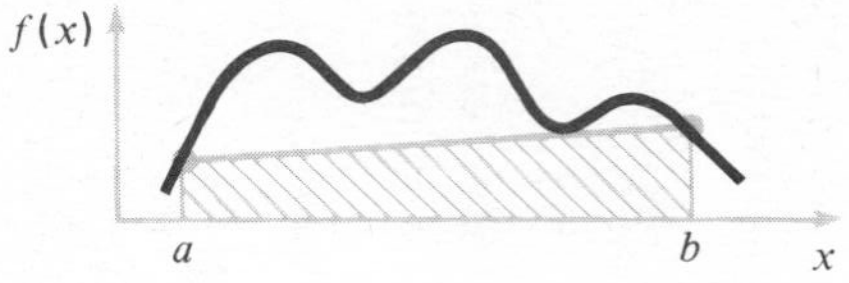 $f(x)$; $a$; $b$; $x$ | $-\dfrac{(b-a)^3}{12}f''(\xi)$ |
| Multiple-segment trapezoidal rule | $I \simeq (b-a)\dfrac{f(x_0)+2\sum_{i=1}^{n-1} f(x_i)+f(x_n)}{2n}$ | 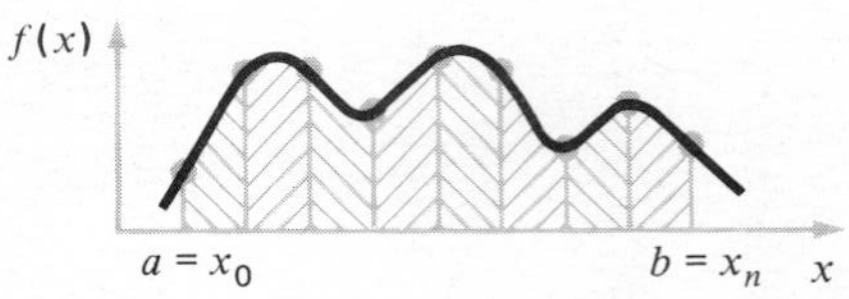 $f(x)$; $a = x_0$; $b = x_n$; $x$ | $-\dfrac{(b-a)^3}{12n^a}\bar{f}''$ |
| Simpson's 1/3 rule | $I \simeq (b-a)\dfrac{f(x_0)+4f(x_1)+f(x_2)}{6}$ | 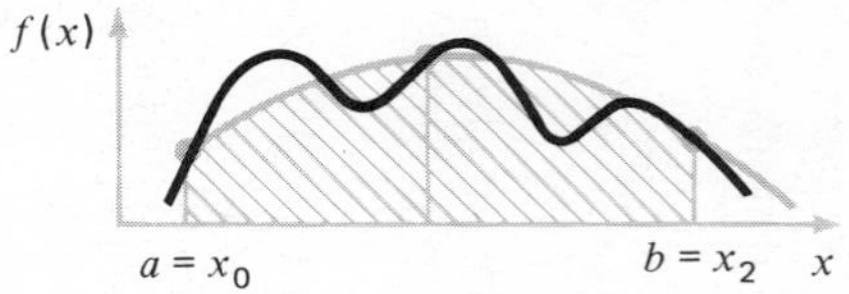 $f(x)$; $a = x_0$; $b = x_2$; $x$ | $-\dfrac{(b-a)^5}{2880}f^{(4)}(\xi)$ |
| Multiple-segment Simpson's 1/3 rule | $I \simeq (b-a)\dfrac{f(x_0)+4\sum_{i=1}^{n-1} f(x_i)+2\sum_{j=2}^{n-2} f(x_j)+f(x_n)}{3n}$ | 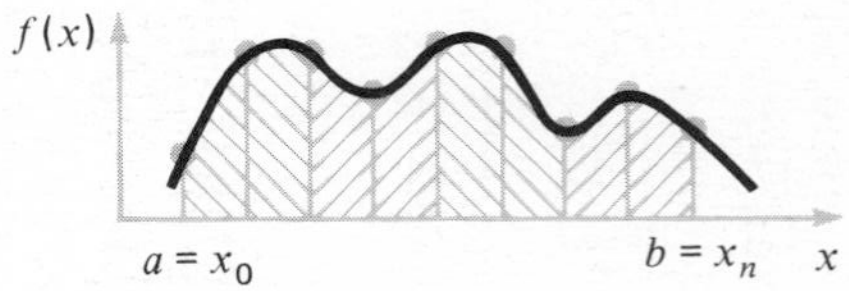 $f(x)$; $a = x_0$; $b = x_n$; $x$ | $-\dfrac{(b-a)^5}{180n^4}\overline{f^{(4)}}$ |
| Simpson's 3/8 rule | $I \simeq (b-a)\dfrac{f(x_0)+3f(x_1)+3f(x_2)+f(x_3)}{8}$ | 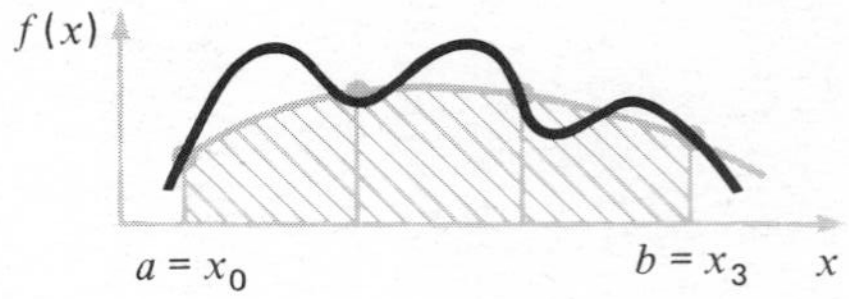 $f(x)$; $a = x_0$; $b = x_3$; $x$ | $-\dfrac{(b-a)^5}{6480}f^{(4)}(\xi)$ |
| Romberg integration | $I_{j,k} = \dfrac{4^{k-1}I_{j+1,k-1}-I_{j,k-1}}{4^{k-1}-1}$ | 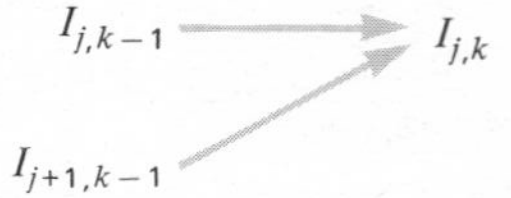 $I_{j,k-1} \rightarrow I_{j,k}$; $I_{j+1,k-1} \nearrow$ | $0(h^{2k})$ |
| Gauss quadrature | $I \simeq c_1 f(x_1) + c_2 f(x_2) + \cdots + c_n f(x_n)$ | 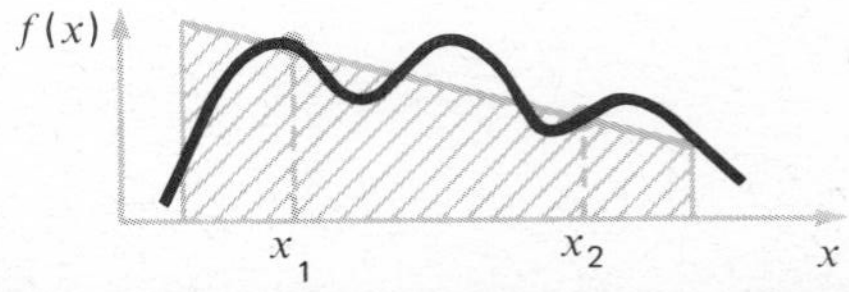 $f(x)$; $x_1$; $x_2$; $x$ | $\simeq f^{(2n+2)}(\xi)$ |

PART SIX

# ORDINARY DIFFERENTIAL EQUATIONS

## VI.1 MOTIVATION

In the first chapter of this book, we derived the following equation based on Newton's second law to compute the velocity $v$ of a falling parachutist as a function of time $t$ [recall Eq. (1.8)]:

$$\frac{dv}{dt} = g - \frac{c}{m}v \qquad \text{[VI.1]}$$

where $g$ is the gravitational constant, $m$ is the mass, and $c$ is a drag coefficient. Such equations, which are composed of an unknown function and its derivatives, are called *differential equations.* Equation (VI.1) is sometimes referred to as a *rate equation* because it expresses the rate of change of a variable as a function of variables and parameters. Such equations play a fundamental role in engineering because many physical phenomena are best formulated mathematically in terms of their rate of change.

In Eq. (VI.1), the quantity being differentiated, $v$, is called the *dependent variable.* The quantity with respect to which $v$ is differentiated, $t$, is called the *independent variable.* When the function involves one independent variable, the equation is called an *ordinary differential equation* (or *ODE*). This is in contrast to a *partial differential equation* (or *PDE*) that involves two or more independent variables.

Differential equations are also classified as to their order. For example, Eq. (VI.1) is called a *first-order equation* because the highest derivative is a first derivative. A *second-order equation* would include a second derivative. For example, the equation describing the position $x$ of a mass-spring system with damping is the second-order equation (recall Case Study 6.5),

$$m\frac{d^2x}{dt^2} + c\frac{dx}{dt} + kx = 0 \qquad \text{[VI.2]}$$

where $c$ is a damping coefficient and $k$ is a spring constant. Similarly, an $n$th-order equation would include an $n$th derivative.

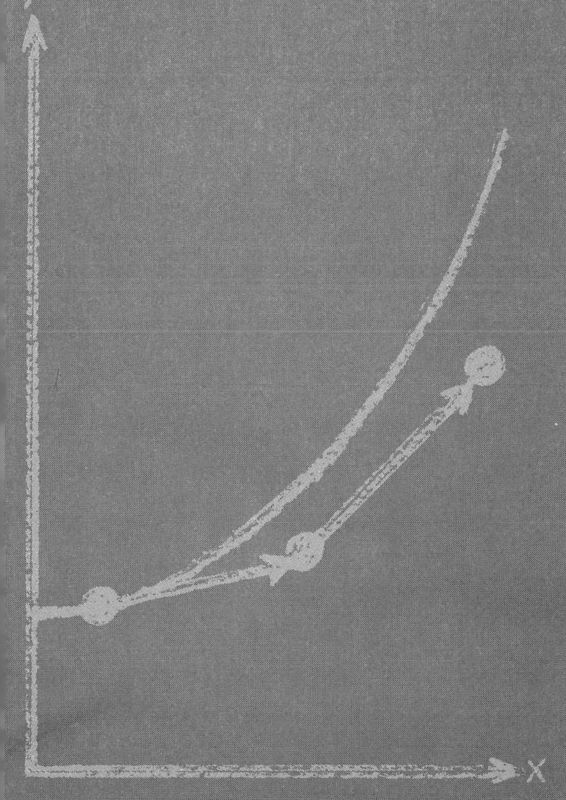

Higher-order equations can be reduced to a system of first-order equations. For Eq. (VI.2), this is done by defining a new variable $y$ where

$$y = \frac{dx}{dt} \tag{VI.3}$$

which itself can be differentiated to yield

$$\frac{dy}{dt} = \frac{d^2x}{dt^2} \tag{VI.4}$$

Equations (VI.3) and (VI.4) can then be substituted into Eq. (VI.2) to give

$$m\frac{dy}{dt} + cy + kx = 0 \tag{VI.5}$$

or

$$\frac{dy}{dt} = -\frac{cy + kx}{m} \tag{VI.6}$$

Thus, Eqs. (VI.3) and (VI.6) are a pair of first-order equations that are equivalent to the original second-order equation. Because other $n$th-order differential equations can be similarly reduced, the present part of our book focuses on the solution of first-order equations. Some of the case studies in Chap. 18 deal with the solution of second-order ODEs by reduction to a pair of first-order equations.

### VI.1.1 Precomputer Methods for Solving ODEs

Before the computer age, ODEs were usually solved with analytical integration techniques. For example, Eq. (V.1) could be multiplied by $dt$ and integrated to yield

$$v = \int \left(g - \frac{c}{m}v\right) dt \tag{VI.7}$$

The right-hand side of this equation is called an *indefinite integral* because the limits of integration are unspecified. This is in contrast to the definite integrals discussed previously in Part V [compare Eq. (VI.7) with Eq. (V.5)].

An analytical solution for Eq. (VI.7) is obtained if the indefinite integral can be evaluated exactly in equation form. For example, recall that for the falling parachutist problem, Eq. (VI.7) was solved analytically by Eq. (1.9): (assuming $v = 0$ at $t = 0$)

$$v = \frac{gm}{c}[1 - e^{-(c/m)t}] \tag{VI.8}$$

The mechanics of deriving such analytical solutions will be discussed in Sec. VI.2. For the time being, the important fact is that, as was the case for definite integration, the analytical evaluation of indefinite integrals usually hinges on foreknowledge of the answer. Unfortunately, exact solutions for many ODEs of practical importance are not available. As is true for most situations discussed in other parts of this book, numerical methods offer the only viable alternative for these cases. Because these numerical methods usually require computers, engineers in the precomputer era were somewhat limited in the scope of their investigations.

One very important method that engineers and applied mathematicians developed to overcome this dilemma was *linearization.* A linear ordinary differential equation is one that fits the general form

$$a_n(x)y^{(n)} + \cdots + a_1(x)y' + a_0(x)y = f(x) \qquad [VI.9]$$

where $y^{(n)}$ is the $n$th derivative of $y$ with respect to $x$ and the $a$'s and $f$'s are specified functions of $x$. This equation is called *linear* because there are no products or nonlinear functions of the dependent variable $y$ and its derivatives. The practical importance of linear ODEs is that they can be solved analytically. In contrast, most nonlinear equations cannot be solved exactly. Thus, in the precomputer era, one tactic to solve nonlinear equations was to linearize them.

A simple example is the application of ODEs to predict the motion of a swinging pendulum (Fig. VI.1). In a manner similar to the derivation of the falling parachutist problem, Newton's second law can be used to develop the following differential equation (see Case Study 18.5 for the complete derivation):

$$\frac{d^2\theta}{dt^2} + \frac{g}{l}\sin\theta = 0 \qquad [VI.10]$$

where $\theta$ is the angle of displacement of the pendulum, $g$ is the gravitational constant and $l$ is the pendulum length. This equation is nonlinear because of the term $\sin\theta$. One way to obtain an analytical solution is to realize that for small displacements of the pendulum from equilibrium (that is, for small values of $\theta$),

$$\sin\theta \simeq \theta \qquad [VI.11]$$

Thus, if it is assumed that we are only interested in cases where $\theta$ is small, Eq. (VI.11) can be substituted into Eq. (VI.10) to give

$$\frac{d^2\theta}{dt^2} + \frac{g}{l}\theta = 0 \qquad [VI.12]$$

We have, therefore, transformed Eq. (VI.10) into a linear form that is easy to solve analytically.

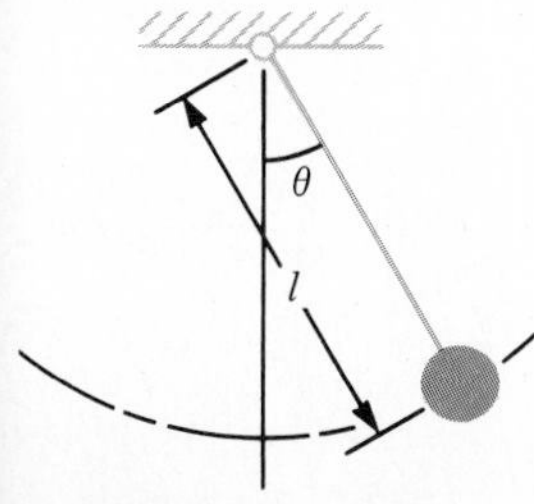

FIGURE VI.1 The swinging pendulum.

Although linearization remains a very valuable tool for engineering problem solving, there are cases where it cannot be invoked. For example, suppose that we were interested in studying the behavior of the pendulum for large displacements from equilibrium. In such instances, numerical methods offer a viable option for obtaining solutions. Today, the widespread availability of computers places this option within reach of all practicing engineers.

### VI.1.2 ODEs and Engineering Practice

The fundamental laws of physics, mechanics, electricity, and thermodynamics are usually based on empirical observations that explain variations in physical properties and states of systems. Rather than describing the *state* of physical systems directly, the laws are usually couched in terms of spatial and temporal *changes.*

Several examples are listed in Table VI.1. These laws define mechanisms of change. When combined with continuity laws for energy, mass, or momentum, differential equations result. Subsequent integration of these differential equations results in mathematical functions that describe the spatial and temporal state of a system in terms of energy, mass, or velocity variations.

The falling parachutist problem introduced in Chap. 1 is an example of the derivation of an ordinary differential equation from a fundamental

**TABLE VI.1 Examples of fundamental laws that are written in terms of the rate of change of variables ($t$ = time and $x$ = position)**

| Law | Mathematical expression | Variables and parameters |
|---|---|---|
| Newton's second law of motion | $\frac{dv}{dt} = \frac{F}{m}$ | Velocity ($v$), force ($F$), and mass ($m$) |
| Fourier's heat law | Heat flux $= -k\frac{\partial T}{\partial x}$ | Thermal conductivity ($k$) and temperature ($T$) |
| Fick's law of diffusion | Mass flux $= -D\frac{\partial c}{\partial x}$ | Diffusion coefficient ($D$) and concentration ($c$) |
| Faraday's law (describes voltage drop across an inductor) | Voltage drop $= L\frac{di}{dt}$ | Inductance ($L$) and current ($i$) |
| Conservation of mass | Accumulation $= V\frac{dc}{dt}$ | Volume ($V$) and concentration ($c$) |

law. Recall that Newton's second law was used to develop an ODE describing the rate of change of velocity of a falling parachutist. By integrating this relationship, we obtained an equation to predict fall velocity as a function of time. This equation could be utilized in a number of different ways, including design purposes.

In fact, such mathematical relationships are the basis of the solution for a great number of engineering problems. However, as described in the previous section, many of the differential equations of practical significance cannot be solved using the analytical methods of calculus. Thus, the methods discussed in the following chapters are extremely important in all fields of engineering.

## VI.2 MATHEMATICAL BACKGROUND

A *solution* of an ordinary differential equation is a specific function of the independent variable and parameters that satisfies the original differential equation. In order to illustrate this concept, let us start with a given function

$$y = -0.5x^4 + 4x^3 - 10x^2 + 8.5x + 1 \qquad \text{[VI.13]}$$

which is a fourth-order polynomial (Fig. VI.2*a*). Now, if we differentiate Eq. (VI.13), we obtain an ODE:

$$\frac{dy}{dx} = -2x^3 + 12x^2 - 20x + 8.5 \qquad \text{[VI.14]}$$

This equation also describes the behavior of the polynomial but in a different manner than does Eq. (VI.13). Rather than explicitly representing the values of $y$ for each value of $x$, Eq. (VI.14) gives the rate of change of $y$ with respect to $x$ (that is, the slope) at every value of $x$. Figure VI.2 shows both the function and the derivative plotted versus $x$. Notice how the zero values of the derivatives correspond to the point at which the original function is flat—that is, has a zero slope. Also, the maximum absolute values of the derivatives are at the ends of the interval where the slopes of the function are greatest.

Although, as just demonstrated, we can determine a differential equation given the original function, the object here is to determine the original function given the differential equation. The original function then represents the solution. For the present case, we can determine this solution analytically by integrating Eq. (VI.14):

$$y = \int [-2x^3 + 12x^2 - 20x + 8.5]\, dx$$

Applying the integration rule (recall Table V.1)

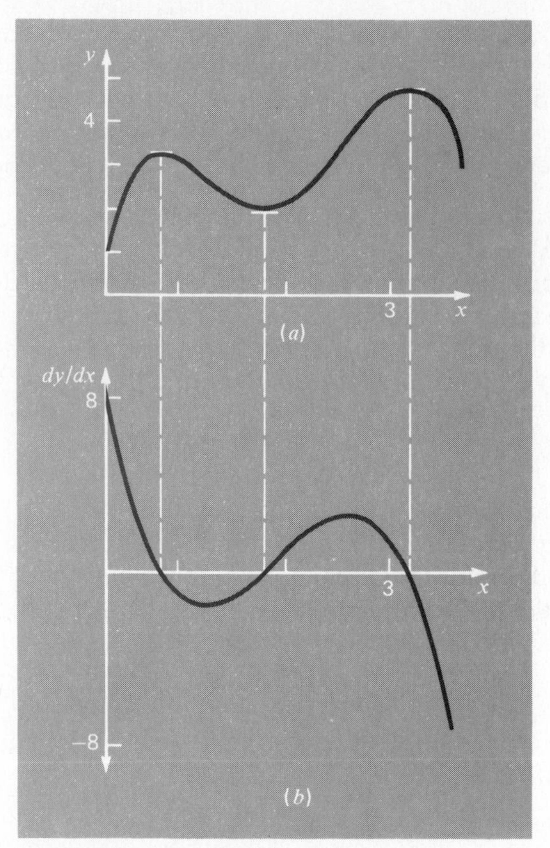


FIGURE VI.2 Plots of (*a*) $y$ versus $x$ and (*b*) $dy/dx$ versus $x$ for the function $y = -0.5x^4 + 4x^3 - 10x^2 + 8.5x + 1$.

$$\int u^n \, du = \frac{u^{n+1}}{n+1} + C \qquad n \neq -1$$

to each term of the equation gives the solution:

$$y = -0.5x^4 + 4x^3 - 10x^2 + 8.5x + C \tag{VI.15}$$

which is identical to the original function with one notable exception. In the course of differentiating and then integrating, we lost the constant value of 1 in the original equation and gained the value $C$. This $C$ is called a *constant of integration.* The fact that such an arbitrary constant appears indicates that the solution is not unique. In fact, it is but one of an infinite number of possible functions (corresponding to an infinite number of possible values of $C$) that satisfy the differential equation. For example, Fig. VI.3 shows six possible functions that satisfy Eq. (VI.14).

Therefore, in order to specify the solution completely, a differential equation is usually accompanied by *auxiliary conditions.* For

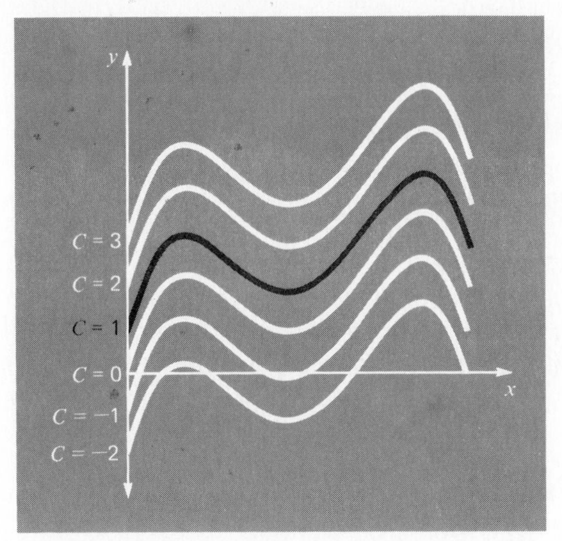


FIGURE VI.3 Six possible solutions for the integral of $-2x^3 + 12x^2 - 20x + 8.5$. Each conforms to a different value of the constant of integration $C$.

first-order ODEs, a type of auxiliary condition called an *initial value* is required to determine the constant and obtain a unique solution. For example, Eq. (VI.14) could be accompanied by the initial condition that at $x = 0$, $y = 1$. These values could be substituted into Eq. (VI.15):

$$1 = -0.5(0)^4 + 4(0)^3 - 10(0)^2 + 8.5(0) + C \qquad [VI.16]$$

to determine $C = 1$. Therefore, the unique solution that satisfies both the differential equation and the specified initial condition is obtained by substituting $C = 1$ into Eq. (VI.15) to yield

$$y = -0.5x^4 + 4x^3 - 10x^2 + 8.5x + 1 \qquad [VI.17]$$

Thus, we have "pinned down" Eq. (VI.15) by forcing it to pass through the initial condition, and in so doing, we have developed a unique solution to the ODE and have come full circle to the original function [Eq. (VI.13)].

Initial conditions usually have very tangible interpretations for differential equations derived from physical problem settings. For example, in the falling parachutist problem the initial condition was reflective of the physical fact that at time zero the vertical velocity was zero. If the parachutist had already been in vertical motion at time zero, the solution would have been modified to account for this initial velocity.

When dealing with an $n$th-order differential equation, $n$ conditions are required to obtain a unique solution. If all conditions are specified at the same value of the independent variable (for example, at $x$ or $t = 0$), then the problem is called an *initial-value problem.* This is in contrast to *boundary-value problems* where specification of conditions occurs at different values of the independent variable. Chapters 16 and 17 will focus on initial-value problems. Boundary-value problems are mentioned at the end of Chap. 16.

# VI.3 ORIENTATION

Before proceeding to numerical methods for solving ordinary differential equations, some orientation might be helpful. The following material is intended to provide you with an overview of the material discussed in Part VI. In addition, we have formulated objectives to focus your studies of the subject area.

## VI.3.1 Scope and Preview

Figure VI.4 provides an overview of Part VI. Two broad categories of numerical methods will be discussed in the present part of this book. One-step methods, which are covered in Chap. 16, permit the calculation of $y_{i+1}$, given the differential equation and $y_i$. Multistep methods, which are covered in Chap. 17, require additional values of $y$ other than at $i$.

With all but a minor exception, the *one-step methods* in *Chap. 16* belong to what are called Runge-Kutta techniques. Although the chapter might have been organized around this theoretical notion, we have opted for a more graphical, intuitive approach to introduce the methods. Thus, we begin the chapter with *Euler's method* which has a very straightforward graphical interpretation. Then, we use visually oriented arguments to develop two improved versions of Euler's method—the *Heun* and the *improved polygon* techniques. After this introduction, we formally develop the concept of *Runge-Kutta* (or RK) approaches and demonstrate how the foregoing techniques are actually first- and second-order RK methods. This is followed by a discussion of the higher-order RK formulations that are frequently used for engineering problem solving. The chapter ends with sections on two applications of one-step methods: *systems of ODEs* and the solution of boundary-value problems using *shooting methods.*

*Chapter 17* is devoted to *multistep methods* that are somewhat more difficult to program for the computer but attain comparable accuracy with less effort than one-step techniques. Again, we initially take a visual, intuitive approach by using a simple method—the *non-self-starting Heun*—to introduce all the essential features of the multistep approaches. Then we launch into a discussion of the *numerical integration formulas* that are at the heart of multistep techniques. This is fol-

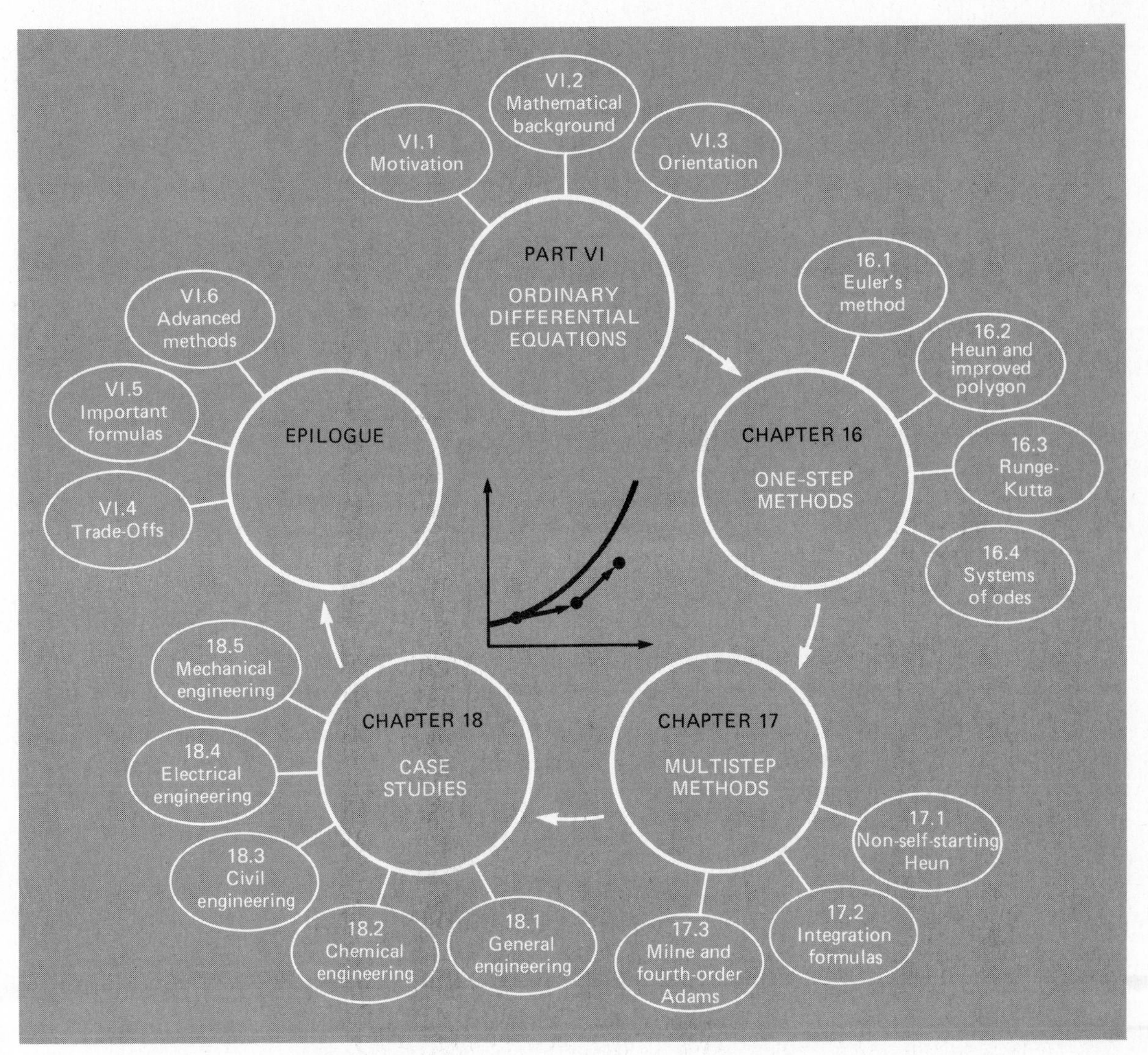


FIGURE VI.4 Schematic representation of the organization of material in Part VI: Ordinary differential equations.

lowed by sections on higher-order versions, including two common approaches—the *Milne* and the *fourth-order Adams methods.*

*Chapter 18* is devoted to case studies from all the fields of engineering. Finally, a short review section is included at the end of Part VI. This *epilogue* summarizes and compares the important formulas and concepts related to ODEs. This comparison includes a discussion of trade-offs that are relevant to their implementation in engineering practice. The epilogue also summarizes important formulas and includes references for advanced topics.

Automatic computation capability is provided in a number of ways. First, the NUMERICOMP software for Euler's method is available on an optional basis for the IBM-PC and Apple II computers. Alternatively, computer code for Euler's method, using both the FORTRAN and BASIC languages, is given directly in the text. This gives you the opportunity to copy this code for implementation on your own personal or on a mainframe computer. Flowcharts, algorithms, or computer programs are provided for most of the other methods described in the text. This information, combined with your own well-written computer code in any language, gives you tools that can be applied to a number of engineering problems.

### VI.3.2 Goals and Objectives

*Study objectives.* After completing Part VI, you should have greatly enhanced your capability to confront and solve ordinary differential equations. General study goals should include mastering the techniques, having the capability to assess the reliability of the answers, and being able to choose the "best" method (or methods) for any particular problem. In addition to these general objectives, the specific study objectives in Table V1.2 should be mastered.

*Computer objectives.* You have been provided with software, simple computer programs, algorithms, and flowcharts to implement the techniques discussed in Part VI. All have utility as learning tools.

The NUMERICOMP personal computer software, which utilizes Euler's method, is user-friendly. The solution can be displayed in either graphical or tabular form. The graphical output will enable you to easily visualize your problem and solution. You can study the efficiency of the method by examining several different step sizes. The software is very easy to apply and can be used to check the results of any computer programs you may develop yourself.

Alternatively, FORTRAN and BASIC programs for Euler's method are supplied directly in the text. In addition, algorithms or flowcharts are provided for most of the other methods discussed in Part VI. This information will allow you to expand your software library to include techniques beyond Euler's method. For example, you may find it useful from a professional viewpoint to have software that employs the fourth-order Runge-Kutta or Adams method, which is usually more efficient and accurate than Euler's method. You may also want to develop software to accommodate systems of ordinary differential equations.

**TABLE VI.2 Specific study objectives for Part VI**

1. Understand the visual representations of Euler's, Heun's, and the improved polygon methods
2. Know the relationship of Euler's method to the Taylor series expansion and the insight it provides regarding the error of the method
3. Understand the difference between local and global truncation errors and how they relate to the choice of a numerical method for a particular problem
4. Know the order and the step-size dependency of the global truncation errors for all the methods described in Part VI; understand how these errors bear on the accuracy of the techniques
5. Understand the basis of predictor-corrector methods. In particular, realize that the efficiency of the corrector is highly dependent on the accuracy of the predictor
6. Know the general form of the Runge-Kutta methods. Understand the derivation of the second-order RK method and how it relates to the Taylor series expansion; realize that there are an infinite number of possible versions for second- and higher-order RK methods
7. Know how to apply any of the RK methods to systems of equations; be able to reduce an $n$th-order ODE to a system of $n$ first-order ODEs.
8. Understand the difference between initial-value and boundary-value problems; be able to implement the shooting method for boundary-value problems
9. Know the difference between multistep and one-step methods; realize that all multistep methods are predictor-correctors but that not all predictor-correctors are multistep methods
10. Understand the role of modifiers in the multistep algorithms
11. Understand the connection between integration formulas and predictor-corrector methods
12. Know the fundamental difference between Newton-Cotes and Adams integration formulas
13. Understand the connection between modifiers and step-size adjustment; recognize the type of problem context where step-size adjustment is important
14. Appreciate the fact that Milne's method is unstable and recognize why this poses difficulties in certain problem contexts

# CHAPTER SIXTEEN

# ONE-STEP METHODS

The present chapter is devoted to solving ordinary differential equations of the form

$$\frac{dy}{dx} = f(x, y)$$

In Chap. 1 we used a numerical method to solve such an equation for the velocity of the falling parachutist. Recall that the equation used to solve this problem was of the general form [Eq. (1.13)]

New value = old value + slope × step size

or, in mathematical terms,

$$y_{i+1} = y_i + \phi h \tag{16.1}$$

According to this equation, the slope estimate $\phi$ is used to extrapolate from an old value $y_i$ to a new value $y_{i+1}$ over a distance $h$ (Fig. 16.1). This formula can be applied step by step to compute out into the future and, hence, trace out the trajectory of the solution.

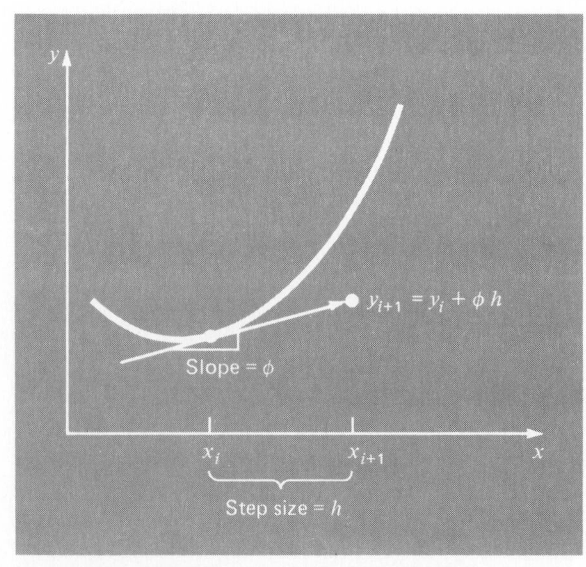


FIGURE 16.1 Graphical depiction of a one-step method.

All one-step methods can be expressed in this general form, with the only difference being the manner in which the slope is estimated. As in the falling parachutist problem, the simplest approach is to use the differential equation to estimate the slope in the form of the first derivative at $x_i$. This approach, which is called *Euler's method,* is discussed in the first part of this chapter. This is followed by other one-step methods that employ alternative slope estimates that result in more accurate predictions.

## 16.1 EULER'S METHOD

The first derivative provides a direct estimate of the slope at $x_i$ (Fig. 16.2):

$$\phi = f(x_i, y_i)$$

where $f(x_i, y_i)$ is the differential equation evaluated at $x_i$ and $y_i$. This estimate can be substituted into Eq. (16.1):

$$\boxed{y_{i+1} = y_i + f(x_i, y_i)h} \tag{16.2}$$

This formula is referred to as *Euler's* (or the *Euler-Cauchy* or the *point-slope) method.* A new value of $y$ is predicted using the slope (equal to the first derivative at the original value of $x$) to extrapolate linearly over the step size $h$ (Fig. 16.2).

EXAMPLE 16.1
Euler's Method

Problem Statement: Use Euler's method to numerically integrate Eq. (VI.14):

$$f(x, y) = -2x^3 + 12x^2 - 20x + 8.5$$

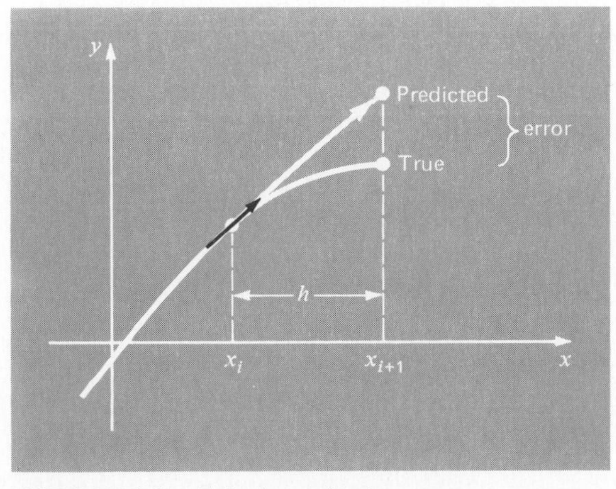


FIGURE 16.2 Euler's method.

from $x = 0$ to $x = 4$ with a step size of 0.5. The initial condition at $x = 0$ is $y = 1$. Recall that the exact solution is given by Eq. (VI.17):

$$y = -0.5x^4 + 4x^3 - 10x^2 + 8.5x + 1$$

Solution: Equation (16.2) can be used to implement Euler's method:

$$y(0.5) = y(0) + f(0, 1)\,0.5$$

where $y(0) = 1$ and the slope estimate at $x = 0$ is

$$f(0, 1) = -2(0)^3 + 12(0)^2 - 20(0) + 8.5 = 8.5$$

Therefore,

$$y(0.5) = 1.0 + 8.5\,(0.5) = 5.25$$

The true solution at $x = 0.5$ is

$$y(0.5) = -0.5(0.5)^4 + 4(0.5)^3 - 10(0.5)^2 + 8.5(0.5) + 1$$
$$= 3.21875$$

Thus, the error is

$$E_t = \text{true} - \text{approximate} = 3.21875 - 5.25 = -2.03125$$

or, expressed as percent relative error, $\epsilon_t = -63.1$ percent. For the second step

$$y(1.0) = y(0.5) + f(0.5, 5.25)\,0.5$$
$$= 5.25 + [-2(0.5)^3 + 12(0.5)^2 - 20(0.5) + 8.5]0.5$$
$$= 5.875$$

**TABLE 16.1 Comparison of true and approximate values of the integral of $y' = -2x^3 + 12x^2 - 20x + 8.5$, with the initial condition that $y = 1$ at $x = 0$. The approximate values were computed using Euler's method with a step size of 0.5. The local error refers to the error incurred over a single step. The global error is the total discrepancy due the past steps as well as the present.**

| | | | $\epsilon_t$ percent relative error | |
|---|---|---|---|---|
| $x$ | $y_{\text{true}}$ | $y_{\text{Euler}}$ | Global | Local |
| 0.0 | 1.00000 | 1.00000 | | |
| 0.5 | 3.21875 | 5.25000 | −63.1 | −63.1 |
| 1.0 | 3.00000 | 5.87500 | −95.8 | −28.0 |
| 1.5 | 2.21875 | 5.12500 | −131.0 | −1.41 |
| 2.0 | 2.00000 | 4.50000 | −125.0 | 20.5 |
| 2.5 | 2.71875 | 4.75000 | −75.7 | 17.3 |
| 3.0 | 4.00000 | 5.87500 | −46.9 | 4.0 |
| 3.5 | 4.71875 | 7.12500 | −51.0 | −11.3 |
| 4.0 | 3.00000 | 7.00000 | −133.0 | −53.0 |

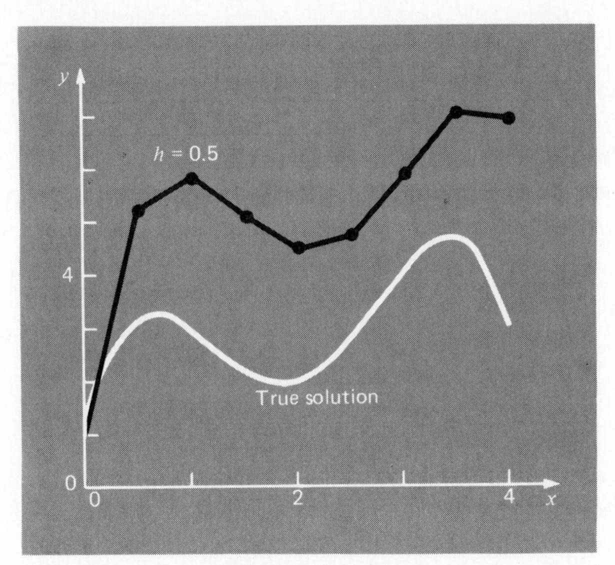


FIGURE 16.3 Comparison of the true solution with a numerical solution using Euler's method for the integral of $y' = -2x^3 + 12x^2 - 20x + 8.5$ from $x = 0$ to $x = 4$ with a step size of 0.5. The initial condition at $x = 0$ is $y = 1$.

The true solution at $x = 1.0$ is 3.0, and therefore the percent relative error is $-95.8$ percent. The computation is repeated, and the results are compiled in Table 16.1 and Fig. 16.3. Note that, although the computation captures the general trend of the true solution, the error is considerable. As discussed in the next section, this error can be reduced by using a smaller step size.

### 16.1.1 Error Analysis for Euler's Method

The numerical solution of ODEs involves two types of error (recall Sec. 3.6):

1. *Truncation or discretization errors* caused by the nature of the techniques employed to approximate values of $y$, and
2. *Round-off errors* caused by the limited numbers of significant digits that can be retained by a computer

The truncation errors are composed of two parts. The first is a *local truncation error* that results from an application of the method in question over a single step. The second is a *propagated truncation error* that results from the approximations produced during the previous steps. The sum of the two is the total or *global truncation error.*

Insight into the magnitude and properties of the truncation error can be gained by deriving Euler's method directly from the Taylor series expansion. In order to do this, realize that the differential equation being integrated will be of the general form

$$y' = f(x, y) \qquad [16.3]$$

where $y' = dy/dx$ and $x$ and $y$ are the independent and the dependent variables, respectively. If the solution—that is, the function describing the behavior of $y$—has continuous derivatives, it can be represented by a Taylor series expansion about a starting value $(x_i, y_i)$, as in [recall Eq. (3.14)]

$$y_{i+1} = y_i + y_i'h + \frac{y_i''}{2}h^2 + \cdots + \frac{y_i^{(n)}}{n!}h^n + R_n \qquad [16.4]$$

where $h = x_{i+1} - x_i$ and $R_n$ is the remainder term defined as

$$R_n = \frac{y^{n+1}(\xi)}{(n+1)!}h^{n+1} \qquad [16.5]$$

where $\xi$ lies somewhere in the interval from $x_i$ to $x_{i+1}$. An alternative form can be developed by substituting Eq. (16.3) into Eqs. (16.4) and (16.5) to yield

$$y_{i+1} = y_i + f(x_i, y_i)\,h + \frac{f'(x_i, y_i)}{2}h^2 + \cdots + \frac{f^{(n-1)}(x_i, y_i)}{n!}h^n + 0(h^{n+1}) \qquad [16.6]$$

where $0(h^{n+1})$ specifies that the local truncation error is proportional to the step size raised to the $(n+1)$th power.

By comparing Eqs. (16.2) and (16.6), it can be seen that Euler's method corresponds to the Taylor series up to and including the term $f(x_i, y_i)h$. Additionally, the comparison indicates that the truncation error is due to the fact that we approximate the true solution using a finite number of terms from the Taylor series. We thus truncate, or leave out, a part of the true solution. For example, the truncation error in Euler's method is attributable to the remaining terms in the Taylor series expansion that were not included in Eq. (16.2). Subtracting Eq. (16.2) from Eq. (16.6) yields

$$E_t = \frac{f'(x_i, y_i)}{2}h^2 + \cdots + 0(h^{n+1}) \qquad [16.7]$$

where $E_t$ is the true local truncation error. For sufficiently small $h$, the errors in the terms in Eq. (16.7) usually decrease as the order increases (recall Example 3.7 and the accompanying discussion), and the result is often represented as

$$E_a = \frac{f'(x_i, y_i)}{2}h^2 \qquad [16.8]$$

or

$$E_a = 0(h^2) \qquad [16.9]$$

where $E_a$ is the approximate local truncation error.

### EXAMPLE 16.2
Taylor Series Estimate for the Error of Euler's Method

Problem Statement: Use Eq. (16.7) to estimate the error of the first step of Example 16.1. Also use it to determine the error due to each higher-order term of the Taylor series expansion.

Solution: Because we are dealing with a polynomial, we can use the Taylor series to obtain exact estimates of the errors in Euler's method. Equation (16.7) can be written as

$$E_t = \frac{f'(x_i, y_i)}{2} h^2 + \frac{f''(x_i, y_i)}{3!} h^3 + \frac{f'''(x_i, y_i)}{4!} h^4 \qquad \text{[E16.2.1]}$$

where $f'(x_i, y_i)$ is the first derivative of the differential equation (that is, the second derivative of the original function). For the present case, this is

$$f'(x_i, y_i) = -6x^2 + 24x - 20 \qquad \text{[E16.2.2]}$$

and $f''(x_i, y_i)$ is the second derivative of the ODE

$$f''(x_i, y_i) = -12x + 24 \qquad \text{[E16.2.3]}$$

and $f'''(x_i, y_i)$ is the third derivative of the ODE

$$f'''(x_i, y_i) = -12 \qquad \text{[E16.2.4]}$$

We can omit additional terms (that is, fourth derivatives and higher) from Eq. (E16.2.1) because for this particular case they equal zero. It should be noted that for other functions (for example, transcendental functions such as sinusoids or exponentials) this would not necessarily be true, and higher-order terms would have nonzero values. However, for the present case, Eqs. (E16.2.1) through (E16.2.4) completely define the truncation error for a single application of Euler's method.

For example, the error due to truncation of the second-order term can be calculated as

$$E_{t,2} = \frac{-6(0.0)^2 + 24(0.0) - 20}{2} (0.5)^2 = -2.5$$

For the third-order term:

$$E_{t,3} = \frac{-12(0.0) + 24}{6} (0.5)^3 = 0.5$$

and the fourth-order term:

$$E_{t,4} = \frac{-12}{24} (0.5)^4 = -0.03125$$

These three results can be added to yield the total truncation error:

$$E_t = E_{t,2} + E_{t,3} + E_{t,4} = -2.5 + 0.5 - 0.03125 = -2.03125$$

which is exactly the error that was incurred in the initial step of Example 16.1. Note how $E_{t,2} > E_{t,3} > E_{t,4}$, which supports the approximation represented by Eq. (16.8).

As illustrated in Example 16.2, the Taylor series provides a means of quantifying the error in Euler's method. However, there are a number of limitations associated with its use for this purpose:

1. The Taylor series only provides an estimate of the local truncation error—that is, the error created during a single step of the method. It does not provide a measure of the propagated and, hence, the global truncation error. In Table 16.1, we have included the local and global truncation errors for Example 16.1. The local error was computed for each time step with Eq. (16.2), but using the true value of $y_i$ (the second column of the table) to compute each $y_{i+1}$ rather than the approximate value (the third column), as is done in the Euler method. As expected, the average local truncation error (25 percent) is less than the average global error (90 percent). The only reason that we can make these exact error calculations is that we know the true value a priori. Such would not be the case in an actual problem. Consequently, as discussed below, you must usually apply techniques such as Euler's method using a number of different step sizes to obtain an indirect estimate of the errors involved.

2. As mentioned above, in actual problems we usually deal with functions that are more complicated than simple polynomials. Consequently, the derivatives that are needed to evaluate the Taylor series expansion would not always be easy to obtain.

Although these limitations preclude exact error analysis for most practical problems, the Taylor series still provides valuable insight into the behavior of Euler's method. According to Eq. (16.8), we see that the local error is proportional to the square of the step size and the first derivative of the differential equation. It can also be demonstrated that the global truncation error is $O(h)$; that is, it is proportional to the step size (Carnahan et al., 1969). These observations lead to some useful conclusions:

1. The error can be reduced by decreasing the step size.

2. The method will provide error-free predictions if the underlying function (that is, the solution of the differential equation) is linear, because for a straight line the second derivative would be zero.

This latter conclusion makes intuitive sense because Euler' method uses straight-line segments to approximate the solution. Hence, Euler's method is referred to as a *first-order method.*

## EXAMPLE 16.3
### Effect of Reduced Step Size on Euler's Method

Problem Statement: Repeat the computation of Example 16.1, but use a step size of 0.25.

Solution: The computation is repeated, and the results are compiled in Fig. 16.4*a*. Halving the step size reduces the absolute value of the average global error to 40 percent and the absolute value of the local error to 6.4 percent. This is compared to global and local errors for Example 16.1 of 90 percent and 24.8 percent. Thus, as expected, the local error is quartered and the global error is halved.

Also, notice how the local error changes sign for intermediate values along the range. This is due primarily to the fact that the first derivative of the

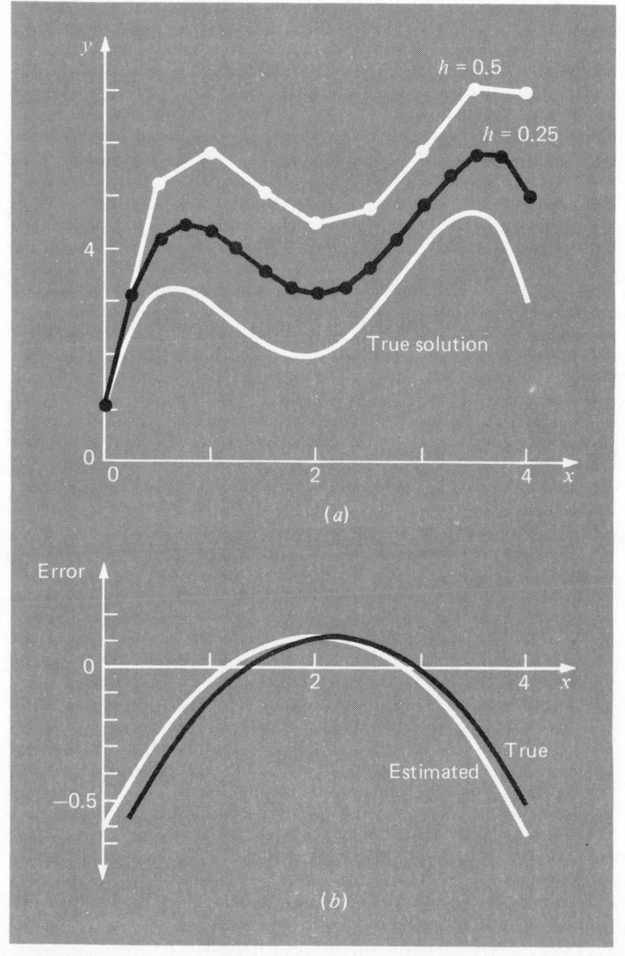


FIGURE 16.4 (*a*) Comparison of two numerical solutions with Euler's method using step sizes of 0.5 and 0.25. (*b*) Comparison of true and estimated local truncation error.

differential equation is a parabola that changes sign [recall Eq. (E16.2.2) and see Fig. 16.4*b*)]. Because the local error is proportional to this function, the net effect of the oscillation in sign is to keep the global error from continuously growing as the calculation proceeds. Thus, from $x = 0$ to $x = 1.25$, the local errors are all negative, and consequently, the global error increases over this interval. In the intermediate section of the range, positive local errors begin to reduce the global error. Near the end, the process is reversed and the global error again inflates. If the local error continuously changes sign over the computation interval, the net effect is usually to minimize the global error. However, where the local errors are of the same sign, the numerical solution may diverge farther and farther from the true solution as the computation proceeds. Such results are said to be *unstable*.

The effect of further step-size reductions on the global truncation error of Euler's method is illustrated in Fig. 16.5. This plot shows the absolute percent relative error at $x = 5$ as a function of step size for the problem we have been examining in Examples 16.1 through 16.3. Notice that even when $h$ is reduced to 0.001, the error still exceeds 0.1 percent. Because this step size corresponds to 5000 steps to proceed from $x = 0$ to $x = 5$, the plot suggests that a first-order technique such as Euler's method demands great computational effort to obtain acceptable error levels. The following sections

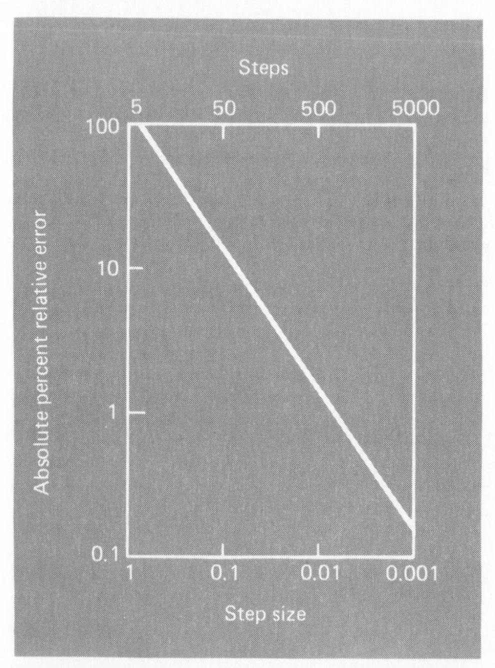


FIGURE 16.5 Effect of step size on the global truncation error of Euler's method for the integral of $y' = -2x^3 + 12x^2 - 20x + 8.5$. The plot shows the absolute percent relative error at $x = 5$ as a function of step size.

present higher-order techniques that attain much better accuracy for the same computational effort. However, it should be noted that despite its inefficiency, the simplicity of Euler's method makes it an extremely attractive option for many engineering problems. Because it is very easy to program, the technique is particularly useful for quick initial computations prior to full-scale analysis. In the next section, a computer program for Euler's method is developed.

### 16.1.2 Computer Program for Euler's Method

Algorithms for one-step techniques such as Euler's method are extremely simple to program on a personal computer. As specified previously at the beginning of this chapter, all one-step methods have the general form:

$$\text{New value} = \text{old value} + \text{slope} \times \text{step size} \qquad [16.10]$$

The only way in which the methods differ is in the calculation of the slope.

Although the program in Fig. 16.6 is specifically designed to implement Euler's method, it is cast in the general form of Eq. (16.10). All that is required to apply this code to the other one-step methods is to modify the computation of the slope in the subroutine (line 1000).

The program in Fig. 16.6 is not user-friendly; it is designed strictly to come up with the answer. In Prob. 16.12 you will have the task of making

FORTRAN

```
      COMMON X,Y
      F(X,Y)=4*EXP(.8*X)-.5*Y
      READ(5,1)XO,X1
    1 FORMAT(2F10.0)
      READ(5,2)YO
    2 FORMAT(F10.0)
      READ(5,2)H
      READ(5,2)PI
      NP=(X1-XO)/PI
      NC=PI/H
      X=XO
      Y=YO
      WRITE(6,3)X,Y
    3 FORMAT(' ',2F20.3)
      DO 270 I=1,NP
      DO 250 J=1,NC
      CALL EUL(SL)
      Y=Y+SL*H
      X=X+H
  250 CONTINUE
      WRITE(6,3)X,Y
  270 CONTINUE
      STOP
      END

      SUBROUTINE EUL(SL)
      COMMON X,Y
      F(X,Y)=4*EXP(.8*X)-.5*Y
      SL=F(X,Y)
      RETURN
      END
```

BASIC

```
100  DEF  FN F(Y) = 4 * EXP (.8 *     (Function specifying
     X) - .5 * Y                       differential equation)
110  INPUT XO,X1                       XO, X1 = initial and final values
                                          of independent variable
120  INPUT YO                          YO = initial value of
                                          dependent variable
130  INPUT H                           H = step size
140  INPUT PI                          PI = print interval
150 NP = (X1 - XO) / PI                NP = number of print steps
160 NC = PI / H                        NC = number of computation steps
170 X = XO
180 Y = YO
190  PRINT X,Y
200  FOR I = 1 TO NP
210  FOR J = 1 TO NC
220  GOSUB 1000
230 Y = Y + SL * H
240 X = X + H
250  NEXT J
260  PRINT X,Y
270  NEXT I
280  END

1000 SL = FN F(Y)                      (Subroutine to compute slope)
1010  RETURN
```

FIGURE 16.6 Annotated computer programs in FORTRAN and BASIC for Euler's method.

this skeletal computer code easier to use and understand. An example of a user-friendly program for Euler's method is included in the supplementary NUMERICOMP software associated with this text. The following example demonstrates the use of this software for solving ODEs. It also provides a reference for assessing and testing your own software.

## EXAMPLE 16.4
### Solving ODEs with the Computer

Problem Statement: A user-friendly computer program to implement Euler's method is contained in the NUMERICOMP software associated with the text. We can use this software to solve another problem associated with the falling parachutist. You recall from Part I that our mathematical model for the velocity was based on Newton's second law in the form

$$\frac{dv}{dt} = g - \frac{c}{m} v$$

This differential equation was solved both analytically (Example 1.1) and numerically using Euler's method (Example 1.2). The objective of the present example is to repeat these numerical computations employing a more complicated model for the velocity based on a more complete mathematical description of the drag force caused by wind resistance. This model is given by

$$\frac{dv}{dt} = g - \frac{c}{m}\left[v + a\left(\frac{v}{v_{max}}\right)^b\right] \qquad \text{[E16.4.1]}$$

where $a$, $b$, and $v_{max}$ are empirical constants. Note that this model is more capable of accurately fitting empirical measurements of drag forces versus velocity than is the simple linear model of Example 1.1. However, this increased flexibility is gained at the expense of evaluating three coefficients rather than one. Furthermore, the resulting mathematical model is more difficult to solve analytically. In this case, Euler's method provides a convenient alternative to obtain an approximate numerical solution.

Solution: We will use NUMERICOMP to tackle Eq. (E16.4.1). Figure 16.7*a* shows the solution of the model with an integration step size of 0.1 s. The plot in Fig. 16.7*b* also shows an overlay of the solution of the linear model for comparison purposes. Note that the computer can plot only one solution at a time.

The results of the two simulations indicate how increasing the complexity of the formulation of the drag force affects the velocity of the parachutist. In this case, the terminal velocity is lowered because of resistance caused by the higher-order terms in Eq. (E16.4.1).

Alternative models could be tested in a similar fashion. The combination of the NUMERICOMP software and your personal computer makes this an

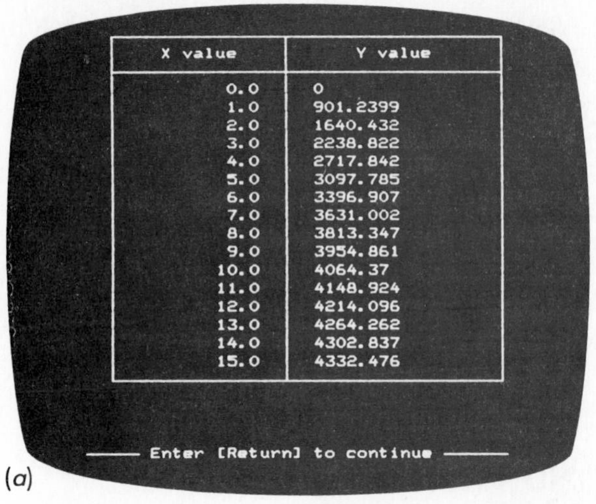

| X value | Y value |
|---|---|
| 0.0 | 0 |
| 1.0 | 901.2399 |
| 2.0 | 1640.432 |
| 3.0 | 2238.822 |
| 4.0 | 2717.842 |
| 5.0 | 3097.785 |
| 6.0 | 3396.907 |
| 7.0 | 3631.002 |
| 8.0 | 3813.347 |
| 9.0 | 3954.861 |
| 10.0 | 4064.37 |
| 11.0 | 4148.924 |
| 12.0 | 4214.096 |
| 13.0 | 4264.262 |
| 14.0 | 4302.837 |
| 15.0 | 4332.476 |

—— Enter [Return] to continue ——

(a)

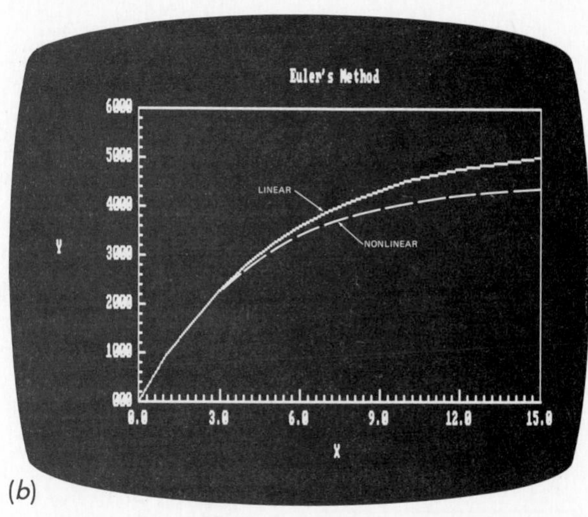


(b)

FIGURE 16.7 (a) Tabular results of the computation, and (b) graphical results for the solution of the nonlinear ODE [Eq. (E16.4.1)]. Notice that (b) also shows the solution for the linear model for comparative purposes. In fact, the software is not designed to superimpose plots in this manner.

easy and efficient task. This convenience should allow you to devote more of your time to considering creative alternatives and holistic aspects of the problem rather than to tedious manual computations.

### 16.1.3 Higher-Order Taylor Series Methods

One method for reducing the error of Euler's method would be to include higher-order terms of the Taylor series expansion in the solution. For example, including the second-order term from Eq. (16.6) yields

$$y_{i+1} = y_i + f(x_i, y_i)h + \frac{f'(x_i, y_i)}{2} h^2 \qquad [16.11]$$

with a local truncation error of

$$E_a = \frac{f''(x_i, y_i)}{6} h^3$$

Although the incorporation of higher-order terms is simple enough to implement for polynomials, their inclusion is not so trivial when the ODE is complicated. In particular, ODEs that are a function of both the dependent and independent variable require chain-rule differentiation. For example, the first derivative of $f(x, y)$ is

$$f'(x, y) = \frac{\partial f(x, y)}{\partial x} + \frac{\partial f(x, y)}{\partial y}\frac{dy}{dx}$$

The second derivative is

$$f''(x, y) = \frac{\partial[\partial f/\partial x + (\partial f/\partial y)(dy/dx)]}{\partial x} + \frac{\partial[\partial f/\partial x + (\partial f/\partial y)(dy/dx)]}{\partial y}\frac{dy}{dx}$$

Higher-order derivatives become increasingly more complicated.

Consequently, as described in the following sections, alternative one-step methods have been developed. These schemes are comparable in performance to the higher-order Taylor series approaches but require only the calculation of first derivatives.

## 16.2 MODIFICATIONS AND IMPROVEMENTS OF EULER'S METHOD

A fundamental source of error in Euler's method is that the derivative at the beginning of the interval is assumed to apply across the entire interval. Two simple modifications are available to help circumvent this shortcoming. As will be demonstrated in Sec. 16.3, both modifications actually belong to a larger class of solution techniques called Runge-Kutta methods. However, because they have a very straightforward graphical interpretation, we will present them prior to their formal derivation as Runge-Kutta methods.

### 16.2.1 Heun's Method

One method to improve the estimate of the slope involves the determination of two derivatives for the interval—one at the initial point and another at the end point. The two derivatives are then averaged to obtain an improved estimate of the slope for the entire interval. This approach, called *Heun's method,* is depicted graphically in Fig. 16.8.

Recall that in Euler's method, the slope at the beginning of an interval

$$y_i' = f(x_i, y_i) \qquad [16.12]$$

is used to extrapolate linearly to $y_{i+1}$:

$$y_{i+1}^0 = y_i + f(x_i, y_i)\, h \qquad [16.13]$$

For the standard Euler method we would stop at this point. However, in Heun's method the $y_{i+1}^0$ calculated in Eq. (16.13) is not the final answer but an intermediate prediction. This is why we have distinguished it with a superscripted 0. Equation (16.13) is called a *predictor equation.* It provides an estimate of $y_{i+1}$ that allows the calculation of an estimated slope at the end of the interval:

$$y_{i+1}' = f(x_{i+1}, y_{i+1}^0) \qquad [16.14]$$

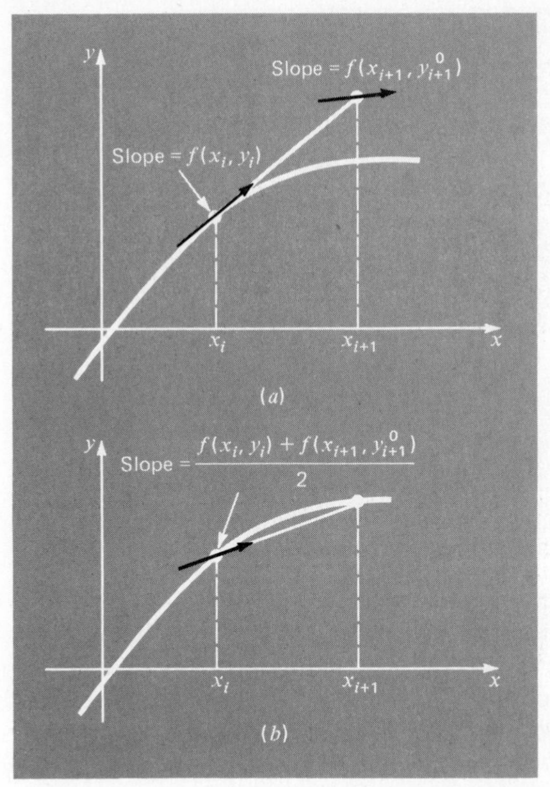


FIGURE 16.8 Graphical depiction of Heun's method. (*a*) Predictor and (*b*) corrector.

Thus, the two slopes [Eqs. (16.12) and (16.14)] can be combined to obtain an average slope for the interval:

$$\bar{y}' = \frac{y_i' + y_{i+1}'}{2} = \frac{f(x_i, y_i) + f(x_{i+1}, y_{i+1}^0)}{2}$$

This average slope is then used to extrapolate linearly from $y_i$ to $y_{i+1}$ using Euler's method:

$$y_{i+1} = y_i + \frac{f(x_i, y_i) + f(x_{i+1}, y_{i+1}^0)}{2} h$$

which is called a *corrector equation.*

The Heun method is a *predictor-corrector approach.* All the multistep methods to be discussed subsequently in Chap. 17 are of this type. The Heun method is the only one-step, predictor-corrector method described in this book. As derived above, it can be expressed concisely as

| | | |
|---|---|---|
| Predictor (Fig 16.8*a*): | $y_{i+1}^0 = y_i + f(x_i, y_i)\, h$ | [16.15] |
| Corrector (Fig. 16.8*b*): | $y_{i+1} = y_i + \frac{f(x_i, y_i) + f(x_{i+1}, y_{i+1}^0)}{2} h$ | [16.16] |

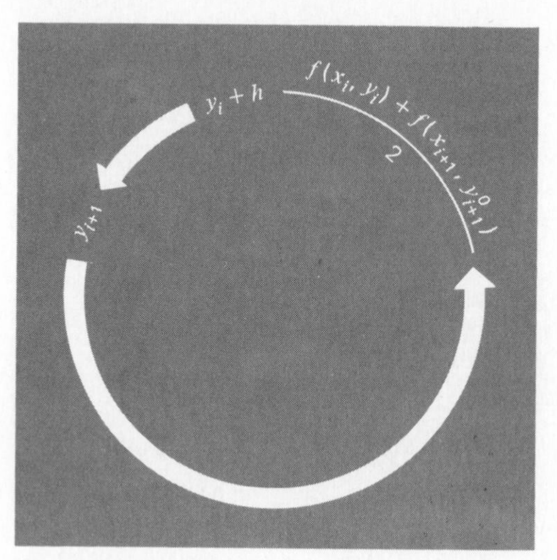


FIGURE 16.9 Graphical representation of iterating the corrector of Heun's method to obtain an improved estimate.

Note that because Eq. (16.16) has $y_{i+1}$ on both sides of the equal sign, it can be applied to "correct" in an iterative fashion. That is, an old estimate can be used repeatedly to provide an improved estimate of $y_{i+1}$. The process is depicted in Fig. 16.9. It should be understood that this iterative process will not necessarily converge on the true answer but will converge on an estimate with a finite truncation error, as demonstrated in the following example.

As with similar iterative methods discussed in previous sections of the book, a termination criterion for convergence of the corrector is provided by [recall Eq. (3.5)]

$$|\epsilon_a| = \left| \frac{y^j_{i+1} - y^{j-1}_{i+1}}{y^j_{i+1}} \right| 100\% \qquad [16.17]$$

where $y^{j-1}_{i+1}$ and $y^j_{i+1}$ are the result from the prior and the present iteration of the corrector, respectively.

## EXAMPLE 16.5
Heun's Method

Problem Statement: Use Heun's method to integrate $y' = 4e^{0.8x} - 0.5y$ from $x = 0$ to $x = 4$ with a step size of 1. The initial condition at $x = 0$ is $y = 2$.

Solution: Before solving the problem numerically, calculus can be used to determine the following analytical solution:

$$y = \frac{4}{1.3}(e^{0.8x} - e^{-0.5x}) + 2e^{-0.5x} \qquad [E16.5.1]$$

This formula can be used to generate the true-solution values in Table 16.2.

The numerical solution is obtained by using the predictor [Eq. (16.15)] to obtain an estimate of $y$ at 0.5:

$$y_1^0 = 2 + [4e^0 - 0.5(2)]\,1 = 5$$

Note that this is the result that would be obtained by the standard Euler method. Using the true value in Table 16.2, it corresponds to a percent relative error of 19.3 percent.

The slope at $(x_0, y_0)$ is

$$y_0' = 4e^0 - 0.5(2) = 3$$

This result is quite different from the actual average slope for the interval from 0 to 1.0, which is equal to 4.1946, as calculated from the original differential equation using Eq. (V.3). Therefore, in order to improve the estimate of the slope, we use the value $y_1^0$ to predict the slope at the end of the interval:

$$y_1' = f(x_1, y_1^0) = 4e^{0.8(1)} - 0.5(5) = 6.40216371$$

which can be combined with the initial slope to yield

$$y' = \frac{3 + 6.40216371}{2} = 4.70108186$$

which is closer to the true average slope of 4.1946. This result can then be substituted into the corrector [Eq. (16.16)] to give the prediction at $x = 1$:

$$y_1 = 2 + (4.70108186)1 = 6.70108186$$

which represents a percent relative error of −8.18 percent. Thus, the Heun method reduces the absolute value of the error by a factor of 2.4 as compared with Euler's method.

Now this estimate can be used to refine or correct the prediction of $y_1$ by substituting the new result back into the right-hand side of Eq. (16.16):

**TABLE 16.2 Comparison of true and approximate values of the integral of $y' = 4e^{0.8x} - 0.5y$ with the initial condition that $y = 2$ at $x = 0$. The approximate values were computed using the Heun method with a step size of 1. Two cases, corresponding to different numbers of corrector iterations, are shown, along with the absolute percent relative error.**

| | | Iterations of Heun's method | | | |
|---|---|---|---|---|---|
| | | 1 | | 15 | |
| $x$ | $y_{true}$ | $y_{heun}$ | $\lvert\epsilon_t\rvert$ % | $y_{heun}$ | $\lvert\epsilon_t\rvert$ % |
| 0 | 2.00000000 | 2.00000000 | 0.00 | 2.0000000 | 0.00 |
| 1 | 6.19463138 | 6.70108186 | 8.18 | 6.36086549 | 2.68 |
| 2 | 14.8439219 | 16.3197819 | 9.94 | 15.3022367 | 3.09 |
| 3 | 33.6771718 | 37.1992489 | 10.46 | 34.7432761 | 3.17 |
| 4 | 75.3389626 | 83.3377674 | 10.62 | 77.7350962 | 3.18 |

$$y_1 = 2 + \frac{[3 + 4e^{0.8(1)} - 0.5(6.70108186)]}{2} 1 = 6.27581139$$

which represents an absolute percent relative error of 1.31 percent. This result, in turn, can be substituted back into Eq. (16.16) to further correct $y_1$:

$$y_1 = 2 + \frac{[3 + 4e^{0.8(1)} - 0.5(6.27581139)]}{2} 1 = 6.38212901$$

which represents an $|\epsilon_t|$ of 3.03 percent. Notice how the errors sometimes grow as the iterations proceed. For example, for three iterations the error grows to 3.03 percent. Such increases can occur, especially for large step sizes, and they prevent us from drawing the general conclusion that an additional iteration will always improve the result. However, for a sufficiently small step size, the iterations should eventually converge on a single value. For our case, 6.36086549, which represents a relative error of 2.68 percent, is attained after 15 iterations. Table 16.2 shows results for the remainder of the computation using the method with 1 and 15 iterations per step.

In the previous example, the derivative is a function of both the dependent variable $y$ and the independent variable $x$. For cases such as polynomials, where the ODE is solely a function of the independent variable, the predictor step [Eq. (16.15)] is not required and the corrector is only applied once for each iteration. For such cases, the technique is expressed concisely as

$$y_{i+1} = y_i + \frac{f(x_i) + f(x_{i+1})}{2} h \qquad [16.18]$$

Notice the similarity between the right-hand side of Eq. (16.18) and the trapezoidal rule [Eq. (13.3)]. The connection between the two methods can be formally demonstrated by starting with the ordinary differential equation

$$\frac{dy}{dx} = f(x)$$

This equation can be solved for $y$ by integration:

$$\int_{y_i}^{y_{i+1}} dy = \int_{x_i}^{x_{i+1}} f(x)\, dx \qquad [16.19]$$

which yields

$$y_{i+1} - y_i = \int_{x_i}^{x_{i+1}} f(x)\, dx \qquad [16.20]$$

or

$$y_{i+1} = y_i + \int_{x_i}^{x_{i+1}} f(x)\, dx \qquad [16.21]$$

Now, recall from Sec. 13.1 that the trapezoidal rule [Eq. (13.3)] is defined as

$$\int_a^b f(x)dx \simeq (b - a)\frac{f(b) + f(a)}{2}$$

or, for the present case,

$$\int_{x_i}^{x_{i+1}} f(x)dx \simeq \frac{f(x_i) + f(x_{i+1})}{2} h \qquad [16.22]$$

where $h = x_{i+1} - x_i$. Substituting Eq. (16.22) into Eq. (16.21) yields

$$y_{i+1} = y_i + \frac{f(x_i) + f(x_{i+1})}{2} h \qquad [16.23]$$

which is equivalent to the corrector [Eq. (16.16)].

Because Eq. (16.23) is a direct expression of the trapezoidal rule, the local truncation error is given by [recall Eq. (13.6)]

$$E_t = -\frac{f''(\xi)}{12} h^3 \qquad [16.24]$$

where $\xi$ is between $x_i$ and $x_{i+1}$. Thus, the method is second order because the second derivative of the ODE is zero when the true solution is a quadratic. In addition, the local and global errors are $0(h^3)$ and $0(h^2)$, respectively. Therefore, decreasing the step size decreases the error at a faster rate than for Euler's method. Figure 16.10, which shows the result of using Heun's method to solve the polynomial from Example 16.1, demonstrates this behavior.

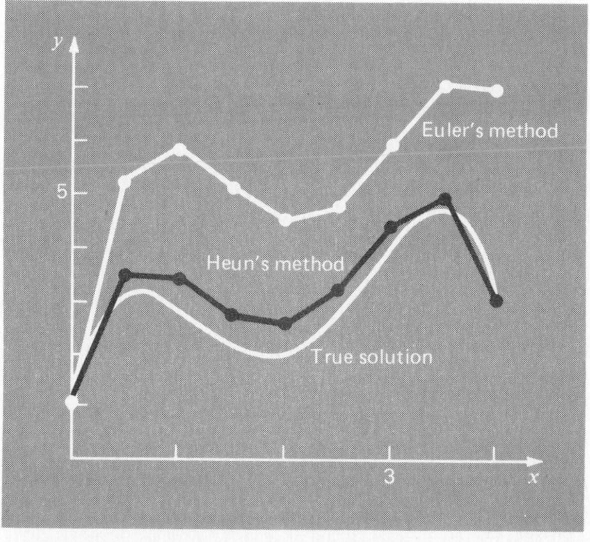


FIGURE 16.10 Comparison of the true solution with a numerical solution using Euler's and Heun's method for integral of $y' = -2x^3 + 12x^2 - 20x + 8.5$.

### 16.2.2 The Improved Polygon Method (Modified Euler)

Figure 16.11 illustrates another simple modification of Euler's method. Called the *improved polygon* (or the *modified Euler*), this technique uses Euler's method to predict a value of $y$ at the midpoint of the interval (Fig. 16.11*a*):

$$y_{i+1/2} = y_i + f(x_i, y_i)\frac{h}{2} \tag{16.25}$$

Then this predicted value is used to estimate a slope at the midpoint:

$$y'_{i+1/2} = f(x_{i+1/2}, y_{i+1/2}) \tag{16.26}$$

which is assumed to represent a valid approximation of the average slope for the entire interval. This slope is then used to extrapolate linearly from $x_i$ to $x_{i+1}$ using Euler's method (Fig. 16.11*b*):

$$y_{i+1} = y_i + f(x_{i+1/2}, y_{i+1/2})h \tag{16.27}$$

Notice that because $y_{i+1}$ is not on both sides, the corrector [Eq. (16.27)] cannot be applied iteratively to improve the solution.

The improved polygon method is superior to Euler's method because it utilizes a slope estimate at the midpoint of the prediction interval. Recall from

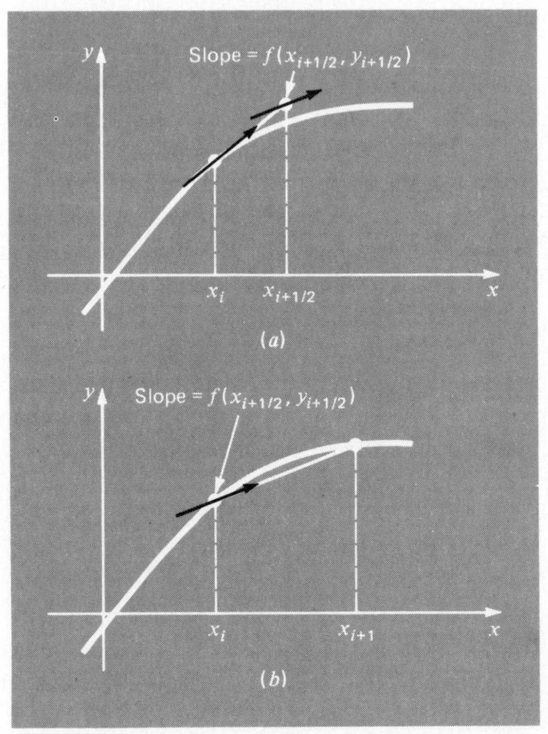


FIGURE 16.11 Graphical depiction of the improved polygon method. (*a*) Equation (16.25) and (*b*) Eq. (16.27).

our discussion of numerical differentiation in Sec. 3.5.4 that centered finite divided differences are better approximations of derivatives than either forward or backward versions. In the same sense, a centered approximation such as Eq. (16.26) has a local truncation error of $0(h^2)$ in comparison with the forward approximation of Euler's method that has an error of $0(h)$. Consequently, the local and global errors of the improved polygon method are $0(h^3)$ and $0(h^2)$, respectively.

### 16.2.3 Computer Algorithms for Improved and Modified Euler Methods

Both the Heun method with a single corrector and the improved polygon method can be easily programmed using the general structure depicted in Fig. 16.6. It is a relatively straightforward task to modify the subroutine of the general program in order to compute the slope in conformance with these methods.

However, when the iterative version of the Heun method is to be implemented, the modifications are a bit more involved. We have developed a subroutine for this purpose in Fig. 16.12. This subroutine can be combined with Fig. 16.6 to develop software for the iterative Heun method.

### 16.2.4 Summary

By tinkering with Euler's method, we have derived two new second-order techniques. Even though these versions require more computational effort to determine the slope, the accompanying reduction in error will allow us to conclude in a subsequent section (Sec. 16.3.4) that the improved accuracy is usually worth the effort. Although there are certain cases where easily programmable techniques such as Euler's method can be applied to advan-

FORTRAN

```
      SUBROUTINE HEUN(X,Y)
      COMMON H,IM,ES
      F(X,Y)=4*EXP(.8*X)-.5*Y
      S1=F(X,Y)
      X=X+H
      Y1=Y+S1*H
      DO 1100 IT=1,IM
      S2=F(X,Y1)
      SL=(S1+S2)/2
      Y2=Y+SL*H
      EA=ABS((Y2-Y1)/Y2)*100
      IF(EA.LE.ES)GO TO 1120
      Y1=Y2
 1100 CONTINUE
      WRITE(6,4)EA
    4 FORMAT(' ','MAX ITER EXCEEDED - EA = ',
     CF10.5)
 1120 X=X-H
      RETURN
      END
```

BASIC

```
1000 S1 =  FN F(Y)                    S1 = slope at beginning of interval
1010 X = X + H                        Y1 = prediction at end of interval
1020 Y1 = Y + S1 * H                  IM = maximum iterations of corrector
1030  FOR IT = 1 TO IM                S2 = slope at end of interval
1040 S2 =  FN F(Y1)                   SL = average slope
1050 SL = (S1 + S2) / 2               (Corrector)
1060 Y2 = Y + SL * H                  EA = estimated error %
1070 EA =  ABS ((Y2 - Y1) / Y2) *
     100
1080  IF EA <  = ES THEN 1120         (Error check where
1090 Y1 = Y2                          ES = acceptable error)
1100  NEXT IT
1110  PRINT "MAX ITER EXCEEDED-EA
     =";EA
1120 X = X - H
1130  RETURN
```

FIGURE 16.12 FORTRAN and BASIC versions of a subroutine to implement the iterative Heun method.

tage, the Heun and improved polygon methods are generally superior and should be implemented if consistent with the problem objectives.

As noted at the beginning of this section, the Heun (without iterations), the improved polygon method, and, in fact, the Euler technique itself are versions of a broader class of one-step approaches called Runge-Kutta methods. We now turn to a formal derivation of these techniques.

## 16.3 RUNGE-KUTTA METHODS

Runge-Kutta methods achieve the accuracy of a Taylor series approach without requiring the calculation of higher derivatives. Many variations exist but all can be cast in the generalized form of Eq. (16.1):

$$y_{i+1} = y_i + \phi(x_i, y_i, h)\, h \tag{16.28}$$

where $\phi(x_i, y_i, h)$ is called an *increment function* which can be interpreted as an average slope over the interval. The increment function can be written in general form as

$$\phi = a_1 k_1 + a_2 k_2 + \cdots + a_n k_n \tag{16.29}$$

where the $a$'s are constants and the $k$'s are

$$k_1 = f(x_i, y_i) \tag{16.29a}$$

$$k_2 = f(x_i + p_1 h, y_i + q_{11} k_1 h) \tag{16.29b}$$

$$k_3 = f(x_i + p_2 h, y_i + q_{21} k_1 h + q_{22} k_2 h) \tag{16.29c}$$

$$\vdots$$

$$k_n = f(x_i + p_{n-1} h, y_i + q_{n-1,1}\, k_1 h + q_{n-1,2}\, k_2 h + \cdots + q_{n-1,n-1}\, k_{n-1} h) \tag{16.29d}$$

Notice that the $k$'s are recurrence relationships. That is, $k_1$ appears in the equation for $k_2$, which appears in the equation for $k_3$, and so forth. This recurrence makes RK methods efficient for computer calculations.

Various types of Runge-Kutta methods can be devised by employing different numbers of terms in the increment function as specified by $n$. Note that the first-order RK method with $n = 1$ is, in fact, Euler's method. Once $n$ is chosen, values for the $a$'s, $p$'s and $q$'s are evaluated by setting Eq. (16.28) equal to terms in a Taylor series expansion (Box 16.1). Thus, at least for the lower order versions, the number of terms $n$ usually represents the order of the approach. For example, in the next section, second-order RK methods use an increment function with two terms ($n = 2$). These second-order methods will be exact if the solution to the differential equation is quadratic. In addition, because terms with $h^3$ and higher are dropped during the deri-

**BOX 16.1 Derivation of the Second-Order Runge-Kutta Methods**

The second-order version of Eq. (16.28) is

$$y_{i+1} = y_i + (a_1 k_1 + a_2 k_2)h \qquad \text{[B16.1.1]}$$

where

$$k_1 = f(x_i, y_i) \qquad \text{[B16.1.2]}$$

and

$$k_2 = f(x_i + p_1 h, y_i + q_{11} k_1 h) \qquad \text{[B16.1.3]}$$

In order to use Eq. (B16.1.1) we have to determine values for the constants $a_1$, $a_2$, $p_1$, and $q_{11}$. To do this, we recall that the second-order Taylor series for $y_{i+1}$ in terms of $y_i$ and $f(x_i, y_i)$ is written as [Eq. (16.11)]

$$y_{i+1} = y_i + f(x_i, y_i)h + f'(x_i, y_i)\frac{h^2}{2} \qquad \text{[B16.1.4]}$$

where $f'(x_i, y_i)$ must be determined by chain-rule differentiation (Sec. 16.1.3):

$$f'(x_i, y_i) = \frac{\partial f}{\partial x} + \frac{\partial f}{\partial y}\frac{dy}{dx} \qquad \text{[B16.1.5]}$$

Substituting Eq. (B16.1.5) into (B16.1.4) gives

$$y_{i+1} = y_i + f(x_i, y_i)h + \left(\frac{\partial f}{\partial x} + \frac{\partial f}{\partial y}\frac{dy}{dx}\right)\frac{h^2}{2} \qquad \text{[B16.1.6]}$$

The basic strategy underlying Runge-Kutta methods is to use algebraic manipulations to solve for values of $a_1$, $a_2$, $p_1$, and $q_{11}$ that make Eqs. (B16.1.1) and (B16.1.6) equivalent.

To do this, we first use a Taylor series to expand Eq. (B16.1.3). The Taylor series for a two-variable function is defined as

$$g(x+r, y+s) = g(x, y) + r\frac{\partial g}{\partial x} + s\frac{\partial g}{\partial y} + \cdots$$

Applying this method to expand Eq. (B16.1.3) gives

$$f(x_i + p_1 h, y_i + q_{11} k_1 h) = f(x_i, y_i) + p_1 h\frac{\partial f}{\partial x} + q_{11} k_1 h\frac{\partial f}{\partial y} + 0(h^2)$$

This result can be substituted along with Eq. (B16.1.2) into Eq. (B16.1.1) to yield

$$y_{i+1} = y_i + a_1 h f(x_i, y_i) + a_2 h f(x_i, y_i) + a_2 p_1 h^2\frac{\partial f}{\partial x} + a_2 q_{11} h^2 f(x_i, y_i)\frac{\partial f}{\partial y} + 0(h^3)$$

or, by collecting terms,

$$y_{i+1} = y_i + [a_1 f(x_i, y_i) + a_2 f(x_i, y_i)]h + \left[a_2 p_1\frac{\partial f}{\partial x} + a_2 q_{11} f(x_i, y_i)\frac{\partial f}{\partial y}\right]h^2 + 0(h^3) \qquad \text{[B16.1.7]}$$

Now, comparing like terms in Eqs. (B16.1.6) and (B16.1.7), we determine that in order for the two equations to be equivalent, the following must hold:

$$a_1 + a_2 = 1$$

$$a_2 p_1 = \tfrac{1}{2}$$

$$a_2 q_{11} = \tfrac{1}{2}$$

These three simultaneous equations contain the four unknown constants. Because there is one more unknown than the number of equations, there is no unique set of constants that satisfy the equations. However, by assuming a value for one of the constants, we can determine the other three. Consequently, there is a family of second-order methods rather than a single version.

vation, the local truncation error is $0(h^3)$ and the global error is $0(h^2)$. In subsequent sections, the third- and fourth-order RK methods ($n = 3$ and 4) are developed. For these cases, the global truncation errors are $0(h^3)$ and $0(h^4)$, respectively.

### 16.3.1 Second-Order Runge-Kutta Methods

The second-order version of Eq. (16.28) is

$$y_{i+1} = y_i + (a_1 k_1 + a_2 k_2)\, h \tag{16.30}$$

where

$$k_1 = f(x_i, y_i) \tag{16.30a}$$

$$k_2 = f(x_i + p_1 h, y_i + q_{11} k_1 h) \tag{16.30b}$$

As described in Box 16.1, values for $a_1$, $a_2$, $p_1$, and $q_{11}$ are evaluated by setting Eq. (16.30) equal to a Taylor series expansion to the second-order term. By doing this, we derive three equations to evaluate the four unknown constants. The three equations are

$$a_1 + a_2 = 1 \tag{16.31}$$

$$a_2 p_1 = \tfrac{1}{2} \tag{16.32}$$

$$a_2 q_{11} = \tfrac{1}{2} \tag{16.33}$$

Because we have three equations with four unknowns, we must assume a value of one of the unknowns in order to determine the other three. Suppose that we specify a value for $a_2$. Then Eqs. (16.31) through (16.33) can be solved simultaneously for

$$a_1 = 1 - a_2 \tag{16.34}$$

$$p_1 = q_{11} = \frac{1}{2a_2} \tag{16.35}$$

Because we can choose an infinite number of values for $a_2$, there are an infinite number of second-order RK methods. Every version would yield exactly the same results if the solution to the ODE were quadratic, linear, or a constant. However, they yield different results when (as is typically the case) the solution is more complicated. We present three of the most commonly used and preferred versions:

**Heun Method with a Single Corrector ($a_2 = 1/2$).** If $a_2$ is assumed to be $1/2$, Eqs. (16.34) and (16.35) can be solved for $a_1 = 1/2$ and $p_1 = q_{11} = 1$. These parameters, when substituted into Eq. (16.30), yield

$$\boxed{y_{i+1} = y_i + (\tfrac{1}{2} k_1 + \tfrac{1}{2} k_2)h} \tag{16.36}$$

where

$$k_1 = f(x_i, y_i) \tag{16.36a}$$

$$k_2 = f(x_i + h, y_i + hk_1) \tag{16.36b}$$

Note that $k_1$ is the slope at the beginning of the interval and $k_2$ is the slope at the end of the interval. Consequently, this second-order Runge-Kutta method is actually Heun's technique with a single iteration of the corrector.

**The Improved Polygon Method ($a_2 = 1$).** If $a_2$ is assumed to be 1, then $a_1 = 0$, $p_1 = q_{11} = 1/2$, and Eq. (16.30) becomes

$$y_{i+1} = y_i + k_2 h \tag{16.37}$$

where

$$k_1 = f(x_i, y_i) \tag{16.37a}$$

$$k_2 = f(x_i + \tfrac{1}{2}h, y_i + \tfrac{1}{2}hk_1) \tag{16.37b}$$

This is the improved polygon method.

**Ralston's Method ($a_2 = 2/3$).** Ralston (1962) and Ralston and Rabinowitz (1978) determined that choosing $a_2 = 2/3$ provides a minimum bound on the truncation error for the second-order RK algorithms. For this version, $a_1 = 1/3$ and $p_1 = q_{11} = 3/4$:

$$y_{i+1} = y_i + (\tfrac{1}{3}k_1 + \tfrac{2}{3}k_2)h \tag{16.38}$$

where

$$k_1 = f(x_i, y_i) \tag{16.38a}$$

$$k_2 = f(x_i + \tfrac{3}{4}h, y_i + \tfrac{3}{4}hk_1) \tag{16.38b}$$

### EXAMPLE 16.6
Comparison of Various Second-Order RK Schemes

Problem Statement: Use the improved polygon [Eq. (16.37)] and Ralston's method [Eq. (16.38)] to numerically integrate Eq. (VI.14):

$$f(x, y) = -2x^3 + 12x^2 - 20x + 8.5$$

from $x = 0$ to $x = 4$ using a step size of 0.5. The initial condition at $x = 0$ is $y = 1$. Compare the results with the values obtained using another second-order RK algorithm: the Heun method with one corrector iteration (Fig. 16.10 and Table 16.3).

Solution: The first step in the improved polygon method is to use Eq. (16.37*a*) to compute

$$k_1 = -2(0)^3 + 12(0)^2 - 20(0) + 8.5 = 8.5$$

However, because the ODE is a function of $x$ only, this result has no bearing on the second step—the use of Eq. (16.37*b*) to compute

$$k_2 = -2(0.25)^3 + 12(0.25)^2 - 20(0.25) + 8.5 = 4.21875$$

Notice that this estimate of the slope is much closer to the average value for

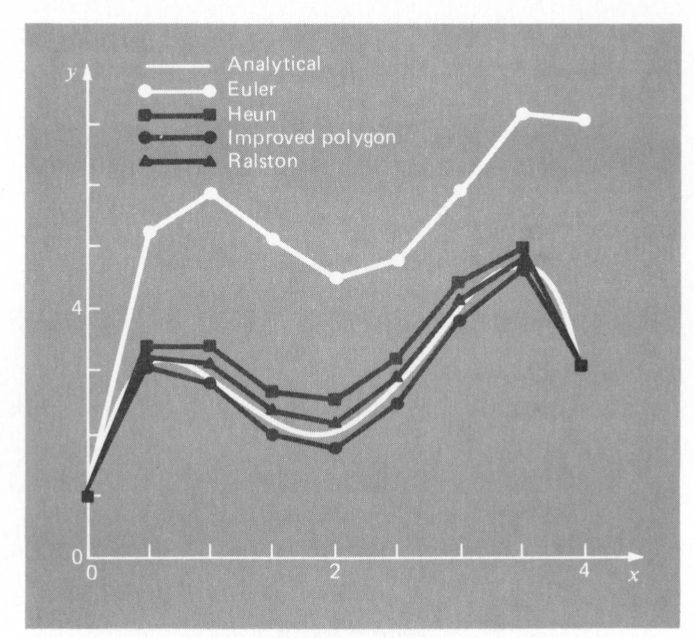


FIGURE 16.13 Comparison of the true solution with numerical solutions using three second-order RK methods and Euler's method.

the interval (4.4375) than the slope at the beginning of the interval (8.5) that would have been used for Euler's approach. The slope at the midpoint can then be substituted into Eq. (16.37) to predict

$$y(0.5) = 1 + 4.21875(0.5) = 3.109375 \qquad \epsilon_t = 3.4\%$$

The computation is repeated, and the results are summarized in Fig. 16.13 and Table 16.3.

**TABLE 16.3 Comparison of true and approximate values of the integral of $y' = -2x^3 + 12x^2 - 20x + 8.5$, with the initial condition that $y = 1$ at $x = 0$. The approximate values were computed using three versions of second-order RK methods with a step size of 0.5.**

| | | Single-corrector Heun | | Improved polygon | | Second-order Ralston RK | |
|---|---|---|---|---|---|---|---|
| $x$ | $y_{true}$ | $y$ | $\|\epsilon_t\|$ % | $y$ | $\|\epsilon_t\|$ % | $y$ | $\|\epsilon_t\|$ % |
| 0.0 | 1.00000 | 1.00000 | 0 | 1.00000 | 0 | 1.00000 | 0 |
| 0.5 | 3.21875 | 3.43750 | 6.8 | 3.109375 | 3.4 | 3.27734375 | 1.8 |
| 1.0 | 3.00000 | 3.37500 | 12.5 | 2.81250 | 6.3 | 3.1015625 | 3.4 |
| 1.5 | 2.21875 | 2.68750 | 21.1 | 1.984375 | 10.6 | 2.34765625 | 5.8 |
| 2.0 | 2.00000 | 2.50000 | 25.0 | 1.75 | 12.5 | 2.140625 | 7.0 |
| 2.5 | 2.71875 | 3.18750 | 17.2 | 2.484375 | 8.6 | 2.85546875 | 5.0 |
| 3.0 | 4.00000 | 4.37500 | 9.4 | 3.81250 | 4.7 | 4.1171875 | 2.9 |
| 3.5 | 4.71875 | 4.93750 | 4.6 | 4.609375 | 2.3 | 4.80078125 | 1.7 |
| 4.0 | 3.00000 | 3.00000 | 0 | 3 | 0 | 3.03125 | 1.0 |

For Ralston's method, $k_1$ for the first interval also equals 8.5 and [Eq. (16.38*b*)]

$$k_2 = -2(0.375)^3 + 12(0.375)^2 - 20(0.375) + 8.5 = 2.58203125$$

The average slope is computed by

$$\phi = \tfrac{1}{3}(8.5) + \tfrac{2}{3}(2.58203125) = 4.5546875$$

which can be used to predict

$$y(0.5) = 1 + 4.5546875(0.5) = 3.27734375 \qquad \epsilon_t = -1.82\%$$

The computation is repeated, and the results are summarized in Fig. 16.13 and Table 16.3. Notice how all the second-order RK methods are superior to Euler's method.

### 16.3.2 Third-Order Runge-Kutta Methods

For $n = 3$, a derivation similar to the one for the second-order method can be performed. The result of this derivation is six equations with eight unknowns. Therefore, values for two of the unknowns must be specified a priori in order to determine the remaining parameters. One common version that results is

$$\boxed{y_{i+1} = y_i + [\tfrac{1}{6}(k_1 + 4k_2 + k_3)]h} \qquad [16.39]$$

where

$$k_1 = f(x_i, y_i) \qquad [16.39a]$$

$$k_2 = f(x_i + \tfrac{1}{2}h, y_i + \tfrac{1}{2}hk_1) \qquad [16.39b]$$

$$k_3 = f(x_i + h, y_i - hk_1 + 2hk_2) \qquad [16.39c]$$

Note that if the derivative is a function of $x$ only, this third-order method reduces to Simpson's 1/3 rule. Ralston (1962) and Ralston and Rabinowitz (1978) have developed an alternative version that provides a minimum bound on the truncation error. In any case, the third-order RK methods have local and global errors of $O(h^4)$ and $O(h^3)$, respectively, and yield exact results when the solution is a cubic. As shown in the following example, when dealing with polynomials, Eq. (16.39) will also be exact when the differential equation is cubic and the solution is quartic. This is because Simpson's 1/3 rule provides exact integral estimates for cubics (recall Box 13.3).

#### EXAMPLE 16.7
Third-Order RK Method

Problem Statement: Use Eq. (16.39) to integrate

(*a*) An ODE that is solely a function of $x$ [Eq. (VI.14)]:

$$\frac{dy}{dx} = -2x^3 + 12x^2 - 20x + 8.5$$

with $y(0) = 1$ and step size of 0.5.

(*b*) An ODE that is a function of both $x$ and $y$:

$$\frac{dy}{dx} = 4e^{0.8x} - 0.5y$$

with $y(0) = 2$ from $x = 0$ to 1 with a step size of 1.

Solution:
(*a*) Equation (16.39*a*) through (16.39*c*) can be used to compute

$$k_1 = -2(0)^3 + 12(0)^2 - 20(0) + 8.5 = 8.5$$

$$k_2 = -2(0.25)^3 + 12(0.25)^2 - 20(0.25) + 8.5 = 4.21875$$

$$k_3 = -2(0.5)^3 + 12(0.5)^2 - 20(0.5) + 8.5 = 1.25$$

which can be substituted into Eq. (16.39) to yield

$$y(0.5) = 1 + \{\tfrac{1}{6}[8.5 + 4(4.21875) + 1.25]\}0.5 = 3.21875$$

which is exact. Thus, because the true solution is a fourth-order polynomial [Eq. (VI.13)], Simpson's 1/3 rule provides an exact result.

(*b*) Equation (16.39*a*) through (16.39*c*) can be used to compute

$$k_1 = 4e^{0.8(0)} - 0.5(2) = 3$$

$$k_2 = 4e^{0.8(0.5)} - 0.5[2 + 0.5(1)3] = 4.21729879$$

$$k_3 = 4e^{0.8(1.0)} - 0.5[2 - 1(3) + 2(1)4.21729879] = 5.184864924$$

which can be substituted into Eq. (16.39) to yield

$$y(1.0) = 2 + \{\tfrac{1}{6}[3 + 4(4.21729879) + 5.184864924]\}\ 1$$

$$= 6.175676681$$

which represents an $\epsilon_t = 0.31$ percent (true value $= 6.19463138$), which is far superior to the results obtained previously with the second-order RK methods (that is, Heun without iterations) in Example 16.5.

### 16.3.3 Fourth-Order Runge-Kutta Methods

The most popular RK methods are fourth order. As with the second-order approaches, there are an infinite number of versions. The following is sometimes called the *classical fourth-order RK method:*

$$\boxed{y_{i+1} = y_i + [\tfrac{1}{6}(k_1 + 2k_2 + 2k_3 + k_4)]h} \qquad [16.40]$$

where

$$k_1 = f(x_i, y_i) \qquad [16.40a]$$

$$k_2 = f(x_i + \tfrac{1}{2}h, y_i + \tfrac{1}{2}hk_1) \qquad [16.40b]$$

$$k_3 = f(x_i + \tfrac{1}{2}h, y_i + \tfrac{1}{2}hk_2) \qquad [16.40c]$$

$$k_4 = f(x_i + h, y_i + hk_3) \qquad [16.40d]$$

Notice that for ODEs that are a function of $x$ alone, the classical fourth-order RK method is also equivalent to Simpson's 1/3 rule.

### EXAMPLE 16.8
Classical Fourth-Order RK Method

Problem Statement: Use the classical fourth-order RK method [Eq. (16.40)] to integrate

$$f(x, y) = -2x^3 + 12x^2 - 20x + 8.5$$

using a step size of 0.5 and an initial condition of $y = 1$ at $x = 0$.

Solution: Equation (16.40$a$) through (16.40$d$) can be used to compute

$$k_1 = -2(0)^3 + 12(0)^2 - 20(0) + 8.5 = 8.5$$

$$k_2 = -2(0.25)^3 + 12(0.25)^2 - 20(0.25) + 8.5 = 4.21875$$

$$k_3 = 4.21875$$

$$k_4 = -2(0.5)^3 + 12(0.5)^2 - 20(0.5) + 8.5 = 1.25$$

which can be substituted into Eq. (16.40) to yield

$$y(0.5) = 1 + \{\tfrac{1}{6}[8.5 + 2(4.21875) + 2(4.21875) + 1.25]\}0.5$$
$$= 3.21875$$

which is exact. Thus, because the true solution is a quartic [Eq. (VI.13)], the fourth-order method gives an exact result.

## 16.3.4 Higher-Order Runge-Kutta Methods

Where more accurate results are required, *Butcher's* (1964) *fifth-order RK method* is recommended:

$$\boxed{y_{i+1} = y_i + h[\tfrac{1}{90}(7k_1 + 32k_3 + 12k_4 + 32k_5 + 7k_6)]} \qquad [16.41]$$

where

$$k_1 = f(x_i, y_i) \qquad [16.41a]$$

$$k_2 = f(x_i + \tfrac{1}{4}h, y_i + \tfrac{1}{4}hk_1) \qquad [16.41b]$$

$$k_3 = f(x_i + \tfrac{1}{4}h,\ y_i + \tfrac{1}{8}hk_1 + \tfrac{1}{8}hk_2) \qquad [16.41c]$$

$$k_4 = f(x_i + \tfrac{1}{2}h,\ y_i - \tfrac{1}{2}hk_2 + hk_3) \qquad [16.41d]$$

$$k_5 = f(x_i + \tfrac{3}{4}h,\ y_i + \tfrac{3}{16}hk_1 + \tfrac{9}{16}hk_4) \qquad [16.41e]$$

$$k_6 = f(x_i + h,\ y_i - \tfrac{3}{7}hk_1 + \tfrac{2}{7}hk_2 + \tfrac{12}{7}hk_3 - \tfrac{12}{7}hk_4 + \tfrac{8}{7}hk_5) \qquad [16.41f]$$

Note the similarity between Butcher's method and the fifth-order Newton-Cotes formula in Table 13.3. Higher-order RK formulas such as Butcher's method are available but, in general, beyond fourth-order methods the gain in accuracy is offset by the added computational effort and complexity.

## EXAMPLE 16.9
### Comparison of Runge-Kutta Methods

Problem Statement: Use first- through fifth-order RK methods to solve

$$\frac{dy}{dx} = 4e^{0.8x} - 0.5y$$

with $y(0) = 2$ from $x = 0$ to $x = 4$ with various step sizes. Compare the accuracy of the various methods for the result at $x = 4$ based on the exact answer of $y(4) = 75.33896261$.

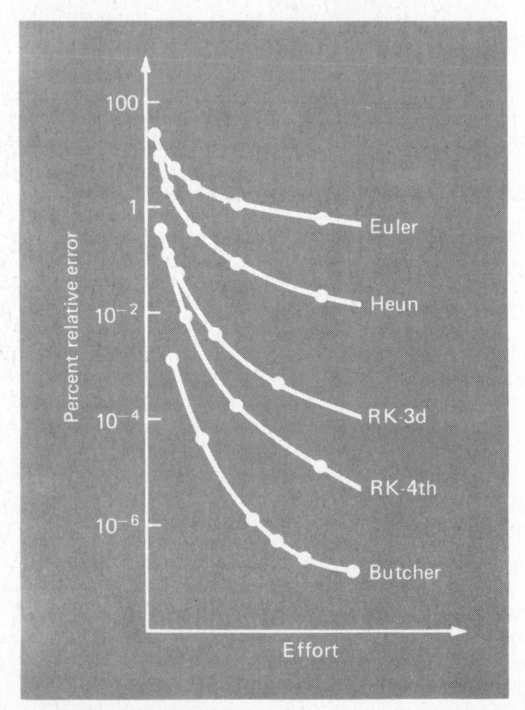


FIGURE 16.14 Comparison of percent relative error versus computational effort for first- through fifth-order RK methods.

Solution: The computation is performed using Euler's, the uncorrected Heun, the third-order RK [Eq. (16.39)], the classical fourth-order RK, and Butcher's fifth-order RK methods. The results are presented in Fig. 16.14, where we have plotted the absolute value of the percent relative error versus the computational effort. This latter quantity is equivalent to the number of function evaluations required to attain the result, as in

$$\text{Effort} = n_f \frac{b - a}{h} \qquad [16.42]$$

where $n_f$ is the number of function evaluations involved in the particular RK computation. For orders $\leq 4$, $n_f$ is equal to the order of the method. However, note that Butcher's fifth-order technique requires six function evaluations [Eq. (16.41*a*) through (16.41*f*)]. The quantity $(b - a)/h$ is the total integration interval divided by the step size—that is, it is the number of applications of the RK technique required to obtain the result. Thus, because the function evaluations are usually the primary time-consuming steps, Eq. (16.42) provides a rough measure of the run time required to attain the answer.

Inspection of Fig. 16.14 leads to a number of conclusions: first, that the higher-order methods attain better accuracy for the same computational effort and, second, that the gain in accuracy for the additional effort tends to diminish after a point. (Notice that the curves drop rapidly at first and then tend to level off.)

Example 16.9 and Fig. 16.14 might lead one to conclude that higher-order RK techniques are always the methods of preference. However, other factors such as programming costs and the accuracy requirements of the problem also must be considered when choosing a solution technique. Such trade-offs will be explored in detail in the case studies in Chap. 18 and in the epilogue for Part VI.

### 16.3.5 Local Truncation Error of Runge-Kutta Methods

Because an $n$th-order Runge-Kutta method is determined by setting terms of Eq. (16.28) equal to a Taylor series expansion through the term containing $h^n$, the local truncation error can be expressed as

$$E_a = 0(h^{n+1}) \qquad [16.43]$$

where the exact value of $E_a$ depends upon $f(x, y)$ and its higher derivatives. It is not generally feasible to estimate $E_a$ on the basis of Eq. (16.43) because the computations are too complicated. At best, if $h$ is small and, hence, if the equation is dominated by the first term of the Taylor series, the coefficients of the RK method [that is, the $a$'s, $p$'s, and $q$'s of Eq. (16.29)] can be chosen to minimize the upper bound on $E_a$. Beyond that, an error analysis of RK methods becomes more involved.

For example, the *Runge-Kutta–Fehlberg method* is based on computing two RK estimates of different order and subtracting the results to obtain an error estimate. The technique consists of a fourth-order formula:

$$y_{i+1} = y_i + \left(\frac{25}{216}k_1 + \frac{1408}{2565}k_3 + \frac{2197}{4104}k_4 - \frac{1}{5}k_5\right)h \qquad [16.44]$$

along with a fifth-order formula:

$$y_{i+1} = y_i + \left(\frac{16}{135}k_1 + \frac{6656}{12{,}825}k_3 + \frac{28{,}561}{56{,}430}k_4 - \frac{9}{50}k_5 + \frac{2}{55}k_6\right)h \qquad [16.45]$$

where

$$k_1 = f(x_i, y_i)$$

$$k_2 = f\left(x_i + \frac{1}{4}h,\ y_i + \frac{1}{4}hk_1\right)$$

$$k_3 = f\left(x_i + \frac{3}{8}h,\ y_i + \frac{3}{32}hk_1 + \frac{9}{32}hk_2\right)$$

$$k_4 = f\left(x_i + \frac{12}{13}h,\ y_i + \frac{1932}{2197}hk_1 - \frac{7200}{2197}hk_2 + \frac{7296}{2197}hk_3\right)$$

$$k_5 = f\left(x_i + h,\ y_i + \frac{439}{216}hk_1 - 8hk_2 + \frac{3680}{513}hk_3 - \frac{845}{410}hk_4\right)$$

$$k_6 = f\left(x_i + \frac{1}{2}h,\ y_i - \frac{8}{27}hk_1 + 2hk_2 - \frac{3544}{2565}hk_3 + \frac{1859}{4104}hk_4 - \frac{11}{40}hk_5\right)$$

The error estimate is obtained by subtracting Eq. (16.44) from Eq. (16.45) to yield

$$E_a = \left(\frac{1}{360}k_1 - \frac{128}{4275}k_3 - \frac{2197}{75{,}240}k_4 + \frac{1}{50}k_5 + \frac{2}{55}k_6\right)h \qquad [16.46]$$

Thus, the ODE can be solved with Eq. (16.44) and the error estimated from Eq. (16.46). However, the error estimate is attained at the expense of extra complexity and computational effort. Note, that after each step, Eq. (16.46) can be added to Eq. (16.44) to make the result fifth order.

Although the RK-Fehlberg method is somewhat more unwieldy than the classical fourth-order RK method, there are situations where the error esti-

FORTRAN

```
SUBROUTINE RALST(SL)
COMMON X,Y,H
F(X,Y)=4*EXP(.8*X)-.5*Y
K1=F(X,Y)
X=X+3*H/4
K2=F(X,Y+3*H/4*K1)
SL=K1/3+2*K2/3
X=X-3*H/4
RETURN
END
```

BASIC

```
1000 K1 =  FN F(Y)
1010 X = X + 3 * H / 4
1020 K2 =  FN F(Y + 3 * H / 4 * K
     1)
1030 SL = K1 / 3 + 2 * K2 / 3 ——— SL = increment function
                                    or average slope
1040 X = X - 3 * H / 4
1050  RETURN
```

FIGURE 16.15 FORTRAN and BASIC subroutines to determine the slope of Ralston's second-order RK method.

mate makes it the preferred method. Error estimation is of particular importance when dealing with functions that require small step sizes for some regions and large step sizes for others. For such functions, an error estimate could provide a basis for changing the step size during the computation. Otherwise, the step size must be chosen conservatively—that is, it must be smaller than needed to achieve the desired accuracy—in order to accommodate the region requiring the smallest step size. This limitation will be considered in further detail when we discuss multistep methods in Chap. 17 for which error estimates are more easily obtainable.

### 16.3.6 Computer Algorithms for Runge-Kutta Methods

As with all the methods covered in this chapter, the RK techniques fit nicely into the general algorithm of Fig. 16.6. Figure 16.15 presents FORTRAN and BASIC subroutines to determine the slope of Ralston's second-order RK method [Eq. (16.38)]. Subroutines to compute slopes for all the other versions can be easily programmed in a similar fashion.

For the RK-Fehlberg method, the variable step size can be incorporated in a number of different ways. One approach (Maron, 1982) is to specify a lower and an upper limit for the error. The object is to employ a step size that results in an error estimate within the acceptable range. If the estimated error is greater than the upper limit, the step size is halved until the error falls within the acceptable range. If the estimated error is less than the lower limit, the step size is doubled until the error is raised to within the acceptable range.

## 16.4 SYSTEMS OF EQUATIONS

Many practical problems in engineering and science require the solution of a system of simultaneous differential equations rather than a single equation. Such systems may be represented generally as

$$\frac{dy_1}{dx} = f_1(x, y_1, y_2, \ldots, y_n)$$

$$\frac{dy_2}{dx} = f_2(x, y_1, y_2, \ldots, y_n) \qquad [16.47]$$

$$\vdots$$

$$\frac{dy_n}{dx} = f_n(x, y_1, y_2, \ldots, y_n)$$

The solution of such a system requires that $n$ initial conditions be known at the starting value of $x$.

All the methods discussed in this chapter for single equations can be extended to the system shown above. Engineering applications can involve several hundred simultaneous equations. In each case, the procedure for solving a system of equations simply involves applying the one-step technique for every equation at each step before proceeding to the next step. This is best illustrated by the following example.

### EXAMPLE 16.10
Solving Systems of ODEs Using Euler's method

Problem Statement: Solve the following set of differential equations using Euler's method, assuming that at $x = 0$, $y_1 = 4$, and $y_2 = 6$. Integrate to $x = 2$ with a step size of 0.5.

$$\frac{dy_1}{dx} = -0.5y_1$$

$$\frac{dy_2}{dx} = 4 - 0.3y_2 - 0.1y_1$$

Solution: Euler's method is implemented as in Eq. (16.2):

$$y_1(0.5) = 4 + [-0.5(4)]0.5 = 3$$

$$y_2(0.5) = 6 + [4 - 0.3(6) - 0.1(4)]0.5 = 6.9$$

Note that $y_1(0) = 4$ is used in the second equation rather than the $y_1(0.5) = 3$ computed with the first equation. Proceeding in a like manner gives

| $x$ | $y_1$ | $y_2$ |
|---|---|---|
| 0 | 4 | 6 |
| 0.5 | 3 | 6.9 |
| 1.0 | 2.25 | 7.715 |
| 1.5 | 1.6875 | 8.44525 |
| 2.0 | 1.265625 | 9.0940875 |

### 16.4.1 Computer Algorithm for Solving Systems of ODEs

The computer code for solving a single ODE with Euler's method (Fig. 16.6) can be easily extended to systems of equations. The modifications include:

1. Inputting the number of equations, $n$
2. Inputting the initial values for each of the $n$ dependent variables
3. Modifying the subroutine so that it computes slopes for each of the dependent variables
4. Including additional functions to compute derivative values for each of the ODEs
5. Including additional equations (of the type in line 230 of the BASIC version) to compute a new value for each dependent variable

Note that any of the one-step methods in the present chapter could be used for such an algorithm. The only difference would be the formulation of the subroutine to compute the slopes. The classic fourth-order RK method is a good choice for this purpose as it provides excellent accuracy yet is relatively easy to program. An important feature of a computer program to solve systems of ODEs with an RK method is the sequencing of the calculation of the $k$'s as demonstrated in the following example.

EXAMPLE 16.11
Solving Systems of ODEs Using the Fourth-Order RK Method

Problem Statement: Use the fourth-order RK method to solve the ODEs from Example 16.10.

Solution: First, we must solve for all the $k_1$'s:

$$k_{1,1} = f(0, 4, 6) = -0.5(4) = -2$$

$$k_{1,2} = f(0, 4, 6) = 4 - 0.3(6) - 0.1(4) = 1.8$$

where $k_{i,j}$ is the $i$th value of $k$ for the $j$th dependent variable. Next, we must calculate the values of $y_1$ and $y_2$ that are needed to determine the $k_2$'s:

$$y_1 + \tfrac{1}{2}hk_{1,1} = 4 + \tfrac{1}{2}(0.5)(-2) = 3.5$$

$$y_2 + \tfrac{1}{2}hk_{1,2} = 6 + \tfrac{1}{2}(0.5)(1.8) = 6.45$$

which can be used to compute

$$k_{2,1} = f(0.25, 3.5, 6.45) = -1.75$$

$$k_{2,2} = f(0.25, 3.5, 6.45) = 1.715$$

The process can be continued to calculate the remaining $k$'s:

$$k_{3,1} = f(0.25, 3.5625, 6.42875) = -1.78125$$

$$k_{3,2} = f(0.25, 3.5625, 6.42875) = 1.715125$$

$$k_{4,1} = f(0.5, 3.109375, 6.8575625) = -1.5546875$$

$$k_{4,2} = f(0.5, 3.109375, 6.8575625) = 1.63179375$$

The values of $k$ can then be used to compute [Eq. (16.40)]:

$$y_1(0.5) = 4 + \tfrac{1}{6}[-2 + 2(-1.75 - 1.78125) - 1.5546875]0.5$$
$$= 3.11523438$$

$$y_2(0.5) = 6 + \tfrac{1}{6}[1.8 + 2(1.715 + 1.715125) + 1.63179375]0.5$$
$$= 6.85767032$$

Proceeding in a like manner for the remaining steps yields

| $x$ | $y_1$ | $y_2$ |
|---|---|---|
| 0 | 4 | 6 |
| 0.5 | 3.1152344 | 6.8576703 |
| 1.0 | 2.4261713 | 7.6321057 |
| 1.5 | 1.8895231 | 8.3268860 |
| 2.0 | 1.4715768 | 8.9468651 |

### 16.4.2 Boundary-Value Problems: Shooting Methods

Solution of boundary-value problems using the shooting method is one example of a problem context where systems of ODEs must be solved. Recall from our discussion at the beginning of Part VI that an ordinary differential equation is accompanied by auxiliary conditions. These conditions are used to evaluate the constants of integration that result during the solution of an ODE. For an $n$th-order equation, $n$ constants must be evaluated, and therefore, $n$ conditions are required. If all the conditions are specified at the same value of the independent variable, then we are dealing with an *initial-value problem.* Most of Part VI deals with this type of problem.

In contrast, there is another kind of ODE for which the conditions are not known at a single point but rather at different values of the dependent variable. Because these conditions are often specified at the extreme points or boundaries, these are commonly referred to as *boundary-value problems.* A variety of significant engineering problems fall within this class. In the present chapter, we discuss one general approach for solving these problems: the shooting method.

The *shooting method* is based on converting the boundary-value problem into an equivalent initial-value problem. A trial-and-error approach is then implemented to solve the initial-value version. The approach can be illustrated by an example.

### EXAMPLE 16.12
The Shooting Method

Problem Statement: Use the shooting method to solve:

$$\frac{d^2y}{dx^2} + 0.2y = 2$$

with the boundary conditions $y(0) = 0$ and $y(10) = 0$.

Solution: Using the same approach as was employed to transform Eq. (VI.2) into Eqs. (VI.3) and (VI.6), the second-order equation can be expressed as two first-order ODEs:

$$\frac{dy}{dx} = z \qquad \text{[E16.12.1]}$$

and

$$\frac{dz}{dx} = 2 - 0.2y \qquad \text{[E16.12.2]}$$

In order to solve these equations, we require an initial value for $z$. For the shooting method, we guess a value—say, $z(0) = 1$. The solution is then obtained by integrating Eqs. (E16.12.1) and (E16.12.2) simultaneously. For example, using a fourth-order RK method for systems of ODEs, we obtain a value at the end of the interval of $y(10) = 10.208$ (Fig. 16.16*a*), which differs from the true value of $y(10) = 0$. Therefore, we make another guess, $z(0) = 2$, and perform the computation again. This time, the result $y(10) = 8.035$ is somewhat closer to the true value of $y(10) = 0$ but is still in error (Fig. 16.16*b*).

Now, because the original ODE is linear, the values

$$z(0) = 1 \qquad y(10) = 10.208$$

and

$$z(0) = 2 \qquad y(10) = 8.035$$

are linearly related. As such, they can be used to compute the value of $z(0)$ that conforms to $y(10) = 0$. A linear interpolation formula [recall Eq. (11.2)] can be employed for this purpose:

$$z(0) = 1 + \frac{2 - 1}{8.035 - 10.208}(0 - 10.208) = 5.7$$

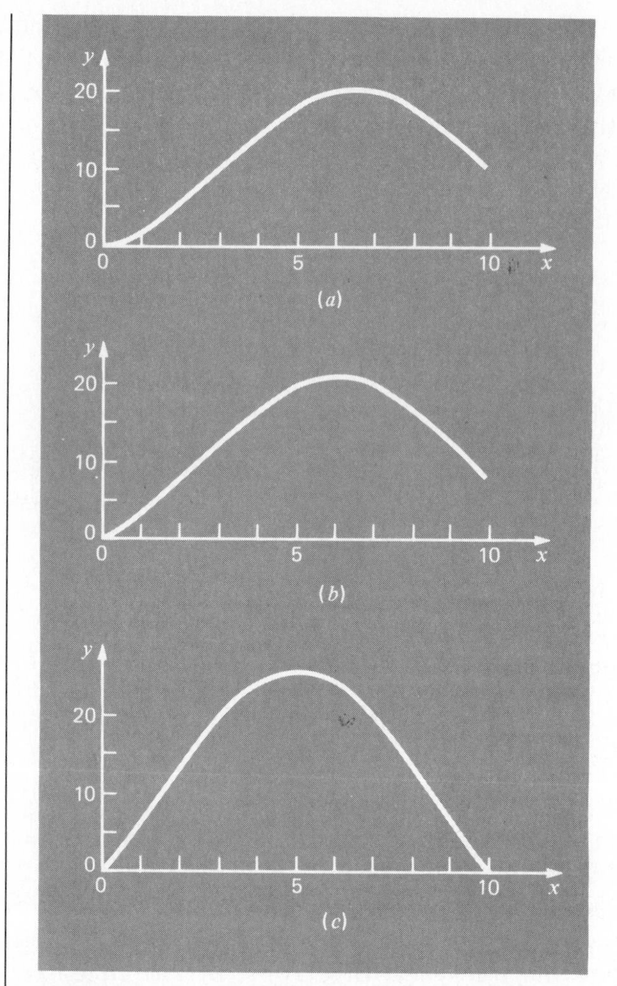


FIGURE 16.16 The shooting method: (*a*) The first "shot"; (*b*) the second "shot"; and (*c*) the final exact "hit."

This value can then be used to determine the correct solution as depicted in Fig. 16.16*c*.

For nonlinear boundary-value problems, linear interpolation or extrapolation through two solution points will not necessarily result in an accurate estimate of the required boundary condition to attain an exact solution. An alternative approach is to perform three simulations and use a quadratic interpolating polynomial to estimate the boundary condition. However, it is unlikely that such an approach would yield the exact answer, and additional iterations would usually be necessary to obtain the solution.

Because this is a fairly inefficient process, alternative methods are available for such cases. The most common are *finite difference methods.* These methods are appropriate for both linear and nonlinear boundary-value problems. In these approaches, finite divided differences are substituted for the derivatives in the original equation. In this way, the differential equation is transformed into a set of simultaneous algebraic equations that can be solved using the methods from Part III. This is the approach that was used in Case Study 9.2 to solve for the distribution of temperature on a heated plate. Problems 9.8, 16.9, and 18.10 relate to solving boundary-value problems.

# PROBLEMS

## Hand Calculations

**16.1** Solve the following initial-value problem analytically over the interval from $x = 0$ to $x = 2$:

$$\frac{dy}{dx} = yx^2 - y$$

where $y(0) = 1$. Plot the solution.

**16.2** Use Euler's method with $h = 0.5$ and $0.25$ to solve Prob. 16.1. Plot the results on the same graph to visually compare the accuracy for the two step sizes.

**16.3** Use Heun's method with $h = 0.5$ and $0.25$ to solve Prob. 16.1. Iterate the corrector to $\epsilon_s = 1$ percent. Plot the results on the same graph to visually compare the accuracy for the two step sizes with the analytical solution. Interpret your results.

**16.4** Use the improved polygon method with $h = 0.5$ and $0.25$ to solve Prob. 16.1.

**16.5** Use the Ralston's second-order RK method with $h = 0.5$ to solve Prob. 16.1.

**16.6** Use the classical fourth-order RK method with $h = 0.5$ to solve Prob. 16.1.

**16.7** Use the fourth-order RK-Fehlberg method with $h = 0.5$ to solve Prob. 16.1. Compute the estimated error for each step.

**16.8** Repeat Probs. 16.1 through 16.7 but for the following initial-value problem over the interval from $x = 0$ to $x = 1$:

$$\frac{dy}{dx} = x\sqrt{y} \qquad y(0) = 1$$

**16.9** Use the shooting method to solve

$$8\frac{d^2y}{dx^2} + 16\frac{dy}{dx} - 4y = 20$$

with the boundary conditions, $y(0) = 5$ and $y(20) = 2$.

**16.10** Use Euler's method with a step size of 1.0 to solve the following system of equations from $x = 0$ to $x = 10$:

$$\frac{dy_1}{dx} = y_1 - 0.1y_1 y_2$$

$$\frac{dy_2}{dx} = -0.5y_2 + 0.02y_1 y_2$$

where $y_1 = 25$ and $y_2 = 7$ at $x = 0$.

**16.11** Use the fourth-order RK method to solve Prob. 16.10 using $h = 1.0$ from $x = 0$ to 1.

## Computer-Related Problems

**16.12** Reprogram Fig. 16.6 so that it is user-friendly. Among other things,
(*a*) Place documentation statements throughout the program to identify what each section is intended to accomplish.
(*b*) Label the input and output.

**16.13** Test the program you developed in Prob. 16.12 by duplicating the computations from Examples 16.1, 16.3, and 16.4.

**16.14** Use the program you developed in Prob. 16.12 to repeat Probs. 16.1 and 16.2.

**16.15** Repeat Probs. 16.13 and 16.14, except use the NUMERICOMP software available with the text.

**16.16** Develop a user-friendly program for the Heun method with an iterative corrector. Base your program on Figs. 16.6 and 16.12. Test the program by duplicating the results in Table 16.3.

**16.17** Develop a user-friendly computer program for Ralston's second-order RK method based on Figs. 16.6 and 16.15. Test the program by duplicating Example 16.6.

**16.18** Develop a user-friendly computer program for the classical fourth-order RK method. Test the program by duplicating Example 16.8 and Problem 16.6.

**16.19** Develop a user-friendly computer program for systems of equations using Euler's method. Base your program on the discussion in Sec. 16.4.1. Use this program to duplicate the computation in Example 16.12.

**16.20** Repeat Prob. 16.19, but use the fourth-order RK method.

# CHAPTER SEVENTEEN

# MULTISTEP METHODS

The *one-step methods* described in the previous chapter utilize information at a single point $x_i$ to predict a value of the dependent variable $y_{i+1}$ at a future point $x_{i+1}$ (Fig. 17.1*a*). Alternative approaches, called *multistep methods* (Fig. 17.1*b*), are based on the insight that once the computation has begun, valuable information from previous points is at our command. The curvature of the lines connecting these previous values provides information regarding the trajectory of the solution. The multistep methods explored in the present chapter exploit this information in order to solve ODEs and evaluate their error. Before describing the higher-order versions, we will present a simple second-order method that serves to demonstrate the general characteristics of multistep approaches.

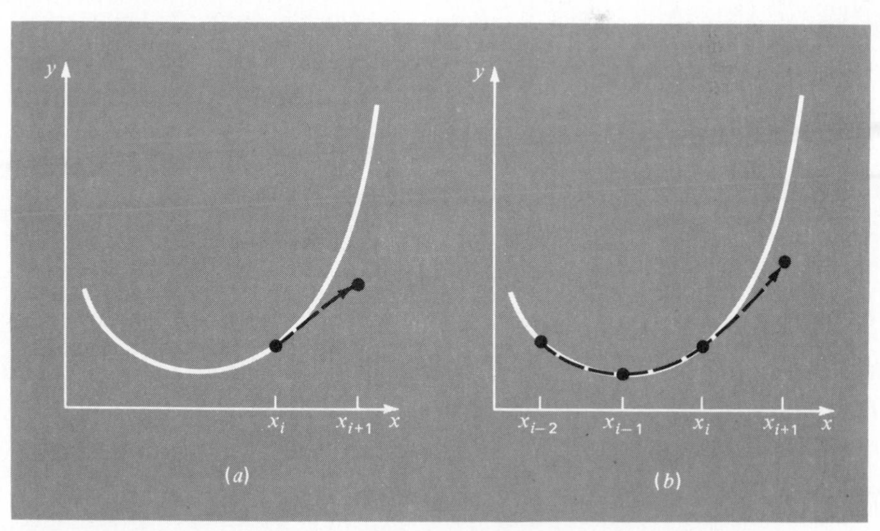


FIGURE 17.1 Graphical depiction of the fundamental difference between (*a*) one-step and (*b*) multistep methods for solving ODEs.

## 17.1 A SIMPLE MULTISTEP APPROACH: THE NON-SELF-STARTING HEUN METHOD

Recall that the Heun approach uses *Euler's method* as a *predictor:*

$$y_{i+1}^{0} = y_i + f(x_i, y_i)h \qquad [17.1]$$

and the *trapezoidal rule* as a *corrector:*

$$y_{i+1} = y_i + \frac{f(x_i, y_i) + f(x_{i+1}, y_{i+1}^{0})}{2} h \qquad [17.2]$$

Thus, the predictor and the corrector have local truncation errors of $O(h^2)$ and $O(h^3)$, respectively. This suggests that the predictor is the weak link in the method because it has the greater error. This weakness is significant because the efficiency of the iterative corrector step depends on the accuracy of the initial prediction. Consequently, one way to improve Heun's method is to develop a predictor that has a local error of $O(h^3)$. This can be accomplished by using Euler's method and the slope at $y_i$, but making the prediction from a previous point $y_{i-1}$, as in

$$y_{i+1}^{0} = y_{i-1} + f(x_i, y_i)2h \qquad [17.3]$$

Equation (17.3) is not self-starting because it involves a previous value of the dependent variable $y_{i-1}$. Such a value would not be available in a typical initial-value problem. Because of this fact, Eqs. (17.3) and (17.2) are called the *non-self-starting Heun method.*

Notice that, as depicted in Fig. 17.2, the derivative estimate in Eq. (17.3) is now located at the midpoint rather than at the beginning of the interval over which the prediction is made. As demonstrated subsequently, this centering improves the error of the predictor to $O(h^3)$. However, before proceeding to a formal derivation of the non-self-starting Heun, we will summarize the method and express it using a slightly modified nomenclature:

Predictor: $$y_{i+1}^{0} = y_{i-1}^{m} + f(x_i, y_i^{m})\, 2h \qquad [17.4]$$

Corrector: $$y_{i+1}^{j} = y_i^{m} + \frac{f(x_i, y_i^{m}) + f(x_{i+1}, y_{i+1}^{j-1})}{2} h$$

$$(\text{for } j = 1, 2, \ldots, m) \qquad [17.5]$$

where the superscripts have been added to denote that the corrector is applied iteratively from $j = 1$ to $m$ in order to obtain refined solutions. Note that $y_i^m$ and $y_{i-1}^m$, are the final results of the iterations of the corrector at the previous time steps. The iterations are terminated at any time step on the basis of the stopping criterion

$$|\epsilon_a| = \left| \frac{y_{i+1}^{j} - y_{i+1}^{j-1}}{y_{i+1}^{j}} \right| 100\% \qquad [17.6]$$

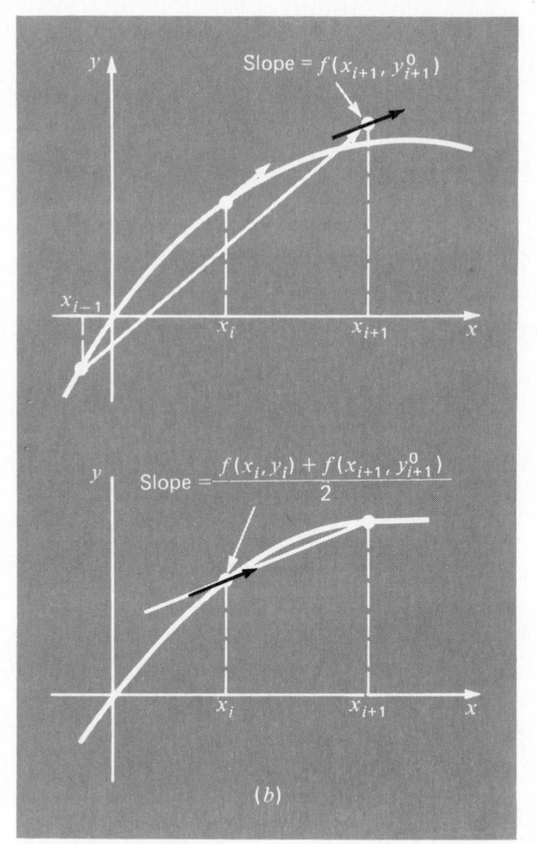


FIGURE 17.2 A graphical depiction of the non-self-starting Heun method. (*a*) The midpoint method that is used as a predictor; (*b*) the trapezoidal rule that is employed as a corrector.

When $\epsilon_a$ is less than a prespecified error tolerance, $\epsilon_s$, the iterations are terminated. At this point, $j = m$. The use of Eqs. (17.4) through (17.6) to solve an ODE is demonstrated in the following example.

## EXAMPLE 17.1
### Non-Self-Starting Heun Method

Problem Statement: Use the non-self-starting Heun method to perform the same computations as was done previously in Example 16.5 using Heun's method. That is, integrate $y' = 4e^{0.8x} - 0.5y$ from $x = 0$ to $x = 4$ using a step size of 1.0. As with Example 16.5, the initial condition at $x = 0$ is $y = 2$. However, because we are now dealing with a multistep method, we require the additional information that $y$ is equal to $-0.392995325$ at $x = -1$.

Solution: The predictor [Eq. (17.4)] is used to extrapolate linearly from $x = -1$ to $x = 1$:

$$y_1^0 = -0.392995325 + [4e^{0.8(0)} - 0.5(2)]2 = 5.607004675$$

The corrector [Eq. (17.5)] is then used to compute the value:

$$y_1^1 = 2 + \frac{[4e^{0.8(0)} - 0.5(2) + 4e^{0.8(1)} - 0.5(5.607004675)]}{2}1$$

$$= 6.549330688$$

which represents a percent relative error of −5.73 percent (true value = 6.194631377). This error is somewhat smaller than the value of −8.18 percent incurred in the self-starting Heun.

Now Eq. (17.5) can be applied iteratively to improve the solution:

$$y_1^2 = 2 + \frac{3 + 4e^{0.8(1)} - 0.5(6.549330688)}{2}1 = 6.313749185$$

which represents an $\epsilon_t$ of −1.92 percent. An approximate estimate of the error can also be determined using Eq. (17.6):

$$|\epsilon_a| = \left|\frac{6.313749185 - 6.549330688}{6.313749185}\right| 100\% = 3.7\%$$

Equation (17.5) can be applied iteratively until $\epsilon_a$ falls below a prespecified value of $\epsilon_s$. As was the case with the Heun method (recall Example 16.5), the iterations converge on a value of 6.36086549 ($\epsilon_t$ = −2.68 percent). However, because the initial predictor value is more accurate, the multistep method converges at a somewhat faster rate.

For the second step, the predictor is

$$y_2^0 = 2 + [4e^{0.8(1)} - 0.5(6.36086549)]\,2$$

$$= 13.4434612 \qquad \epsilon_t = 9.43\%$$

which is superior to the prediction of 12.08259646 ($\epsilon_t$ = 18 percent) that was computed with the original Heun method. The first corrector yields 15.95539553 ($\epsilon_t$ = −6.8 percent) and subsequent iterations converge on the same result as was obtained with the self-starting Heun method: 15.3022367 ($\epsilon_t$ = −3.1 percent). As with the previous step, the rate of convergence of the corrector is somewhat improved because of the better initial prediction.

### 17.1.1 Derivation and Error Analysis of Predictor-Corrector Formulas

We have just employed graphical concepts to derive the non-self-starting Heun. We will now show how the same equations can be derived mathematically. This derivation is particularly interesting because it ties together ideas from curve fitting, numerical integration, and ODEs. The derivation is also useful because it provides a simple stepping-stone for developing higher-order multistep methods and estimating their errors.

The derivation is based on solving the general ODE

$$\frac{dy}{dx} = f(x, y)$$

This equation can be solved by multiplying both sides by $dx$ and integrating between limits at $i$ and $i + 1$:

$$\int_{y_i}^{y_{i+1}} dy = \int_{x_i}^{x_{i+1}} f(x, y)\, dx$$

The left side can be integrated and evaluated using the fundamental theorem [recall Eq. (16.21)]:

$$y_{i+1} = y_i + \int_{x_i}^{x_{i+1}} f(x, y)\, dx \tag{17.7}$$

Equation (17.7) represents a solution to the ODE if the integral can be evaluated. That is, it provides a means to compute a new value of the dependent variable $y_{i+1}$ on the basis of a prior value $y_i$ and the differential equation.

Numerical integration formulas such as those developed in Chap. 13 provide one way to make this evaluation. For example, the trapezoidal rule [Eq. (13.3)] can be used to evaluate the integral, as in

$$\int_{x_i}^{x_{i+1}} f(x, y)\, dx \simeq \frac{f(x_i, y_i) + f(x_{i+1}, y_{i+1})}{2} h \tag{17.8}$$

where $h = x_{i+1} - x_i$ is the step size. Substituting Eq. (17.8) into Eq. (17.7) yields

$$y_{i+1} = y_i + \frac{f(x_i, y_i) + f(x_{i+1}, y_{i+1})}{2} h$$

which is the corrector equation for the Heun method. Because this equation is based on the trapezoidal rule, the truncation error can be taken directly from Table 13.2:

$$E_c = -\tfrac{1}{12} h^3 y'''(\xi_c) = -\tfrac{1}{12} h^3 f''(\xi_c) \tag{17.9}$$

where the subscript $c$ designates that this is the error of the corrector.

A similar approach can be used to derive the predictor. For this case, the integration limits are from $i - 1$ to $i + 1$:

$$\int_{y_{i-1}}^{y_{i+1}} dy = \int_{x_{i-1}}^{x_{i+1}} f(x, y)\, dx$$

which can be integrated and rearranged to yield

$$y_{i+1} = y_{i-1} + \int_{x_{i-1}}^{x_{i+1}} f(x, y)\, dx \tag{17.10}$$

Now, rather than using a closed formula from Table 13.2, the first Newton-Cotes open integration formula (see Table 13.4) can be used to evaluate the integral, as in

$$\int_{x_{i-1}}^{x_{i+1}} f(x, y)\, dx = 2h\, f(x_i, y_i) \tag{17.11}$$

which is called the *midpoint method.* Substituting Eq. (17.11) into Eq. (17.10) yields

$$y_{i+1} = y_{i-1} + f(x_i, y_i)2h$$

which is the predictor for the non-self-starting Heun. As with the corrector, the local truncation error can be taken directly from Table 13.4

$$E_p = \tfrac{1}{3} h^3\, y'''(\xi_p) = \tfrac{1}{3} h^3\, f''(\xi_p) \tag{17.12}$$

where the subscript $p$ designates that this is the error of the predictor.

Thus, the predictor and the corrector for the non-self-starting Heun method have truncation errors of the same order. Aside from upgrading the accuracy of the predictor, this fact has additional benefits related to error analysis, as elaborated in the next section.

### 17.1.2 Error Estimates

If the predictor and the corrector of a multistep method are of the same order, the local truncation error may be estimated during the course of a computation. This is a tremendous advantage because it establishes a criterion for adjustment of the step size.

The local truncation error for the predictor is estimated by Eq. (17.12). This error estimate can be combined with the estimate of $y_{i+1}$ from the predictor step to yield [recall our basic definition of Eq. (3.1)]

$$\text{True value} = y_{i+1}^0 + \tfrac{1}{3} h^3\, y'''(\xi_p) \tag{17.13}$$

Using a similar approach, the error estimate for the corrector [Eq. (17.9)] can be combined with the true value and the corrector result $y_{i+1}$ to give

$$\text{True value} = y_{i+1}^m - \tfrac{1}{12} h^3\, y'''(\xi_c) \tag{17.14}$$

Equation (17.13) can be subtracted from Eq. (17.14) to yield

$$0 = y_{i+1}^m - y_{i+1}^0 - \tfrac{5}{12} h^3\, y'''(\xi) \tag{17.15}$$

where $\xi$ is between $x_{i-1}$ and $x_{i+1}$. Now dividing Eq. (17.15) by 5 and rearranging the result gives

$$\frac{y_{i+1}^0 - y_{i+1}^m}{5} = -\tfrac{1}{12} h^3\, y'''(\xi) \tag{17.16}$$

Notice that the right-hand sides of Eqs. (17.9) and (17.16) are identical, with the exception of the argument of the third derivative. If the third derivative does not vary appreciably over the interval in question, we can assume that the right-hand sides are equal, and, therefore, the left-hand sides should also be equivalent, as in

$$E_c \simeq -\frac{y_{i+1}^m - y_{i+1}^0}{5} \qquad [17.17]$$

Thus, we have arrived at a relationship that can be used to estimate the per-step truncation error on the basis of two quantities—the predictor ($y_{i+1}^0$) and the corrector ($y_{i+1}^m$), which are routine by-products of the computation.

### EXAMPLE 17.2

Estimate of Per-Step Truncation Error for the Non-Self-Starting Heun Method

Problem Statement: Use Eq. (17.17) to estimate the per-step truncation error of Example 17.1. Note that the true values at $x = 1$ and 2 are 6.19463138 and 14.8439219, respectively.

Solution: At $x_{i+1} = 1$, the predictor gives 5.607004675 and the corrector yields 6.36086549. These values can be substituted into Eq. (17.17) to give

$$E_c = -\frac{6.36086549 - 5.607004675}{5} = -0.150772163$$

which compares well with the exact error,

$$E_t = 6.19463138 - 6.36086549 = -0.166234110$$

At $x_{i+1} = 2$, the predictor gives 13.4434619 and the corrector yields 15.3022367, which can be used to compute

$$E_c = -\frac{15.3022367 - 13.4434619}{5} = -0.371754960$$

which also compares favorably with the exact error, $E_t = 14.8439219 - 15.3022367 = -0.4583148$.

The ease with which the error can be estimated using Eq. (17.17) represents a decided advantage of the multistep over the one-step methods. Among other things, it provides a rational basis for step-size adjustment during the course of a computation. For example, if Eq. (17.17) indicates that the error is greater than an acceptable level, the step size could be decreased. In a subsequent section (Sec. 17.4), we will delineate how such step-size adjustments might be incorporated into a computer algorithm.

### 17.1.3 Modifiers

Before developing computer algorithms, we must note two other ways in which the non-self-starting Heun method can be made more accurate and efficient. First, you should realize that besides providing a criterion for step-size adjustment, Eq. (17.17) represents a numerical estimate of the discrepancy between the final corrected value at each step $y_{i+1}$ and the true value. Thus, it can be added directly to $y_{i+1}$ to refine the estimate further:

$$y_{i+1}^m \leftarrow y_{i+1}^m - \frac{y_{i+1}^m - y_{i+1}^0}{5} \qquad [17.18]$$

Equation (17.18) is called a *corrector modifier.* (The symbol $\leftarrow$ is read "is replaced by.") The left-hand side is the modified value of $y_{i+1}^m$.

A second improvement, one that relates more to program efficiency, is a *predictor modifier*, which is designed to adjust the predictor result so that it is closer to the final convergent value of the corrector. This is advantageous because, as noted previously at the beginning of this section, the number of iterations of the corrector is highly dependent on the accuracy of the initial prediction. Consequently, if the prediction is modified properly, we might reduce the number of iterations required to converge on the ultimate value of the corrector.

Such a modifier can be derived simply by assuming that the third derivative is relatively constant from step to step. Therefore, using the result of the previous step at $i$, Eq. (17.16) can be solved for

$$h^3 y'''(\xi) = -\tfrac{12}{5}\,(y_i^0 - y_i^m) \qquad [17.19]$$

which, assuming that $y'''(\xi) \simeq y'''(\xi_p)$, can be substituted into Eq. (17.12) to give

$$E_p = \tfrac{4}{5}\,(y_i^m - y_i^0) \qquad [17.20]$$

which can then be used to modify the predictor result:

$$y_{i+1}^0 \leftarrow y_{i+1}^0 + \tfrac{4}{5}\,(y_i^m - y_i^0) \qquad [17.21]$$

## EXAMPLE 17.3
Effect of Modifiers on Predictor-Corrector Results

Problem Statement: Recompute Example 17.1 using the modifiers as specified in Fig. 17.3.

Solution: As in Example 17.1, the initial predictor result is 5.607004675. Because the predictor modifier [Eq. (17.21)] requires values from a previous iteration, it cannot be employed to improve this initial result. However, Eq. (17.18) can be used to modify the corrected value of 6.36086549 ($\epsilon_t = -2.684$ percent), as in

$$y_1^m = 6.36086549 - \frac{6.36086549 - 5.607004675}{5} = 6.210093327$$

which represents an $\epsilon_t = -0.25$ percent. Thus, the error is reduced over an order of magnitude.

*Predictor:*

$$y_{i+1}^0 = y_{i-1}^m + f(x_i, y_i^m)2h$$

(Save result as $y_{i+1,u}^0 = y_{i+1}^0$ where the subscript $u$ designates that the variable is unmodified.)

*Predictor modifier:*

$$y_{i+1}^0 \leftarrow y_{i+1,u}^0 + \tfrac{4}{5}(y_{i,u}^m - y_{i,u}^0)$$

*Corrector:*

$$y_{i+1}^j = y_i^m + \frac{f(x_i, y_i^m) + f(x_{i+1}, y_{i+1}^{j-1})}{2}h$$

(for $j = 1$ to maximum iterations $m$)

*Error check:*

$$|\epsilon_a| = \left|\frac{y_{i+1}^j - y_{i+1}^{j-1}}{y_{i+1}^j}\right| 100\%$$

(If $|\epsilon_a| >$ error criterion, set $j = j+1$ and repeat corrector; if $\epsilon_a \le$ error criterion, save result as $y_{i+1,u}^m = y_{i+1}^m$.)

*Corrector error estimate:*

$$E_c = -\tfrac{1}{5}(y_{i+1,u}^m - y_{i+1,u}^0)$$

(If computation is to continue, set $i = i+1$ and return to predictor.)

FIGURE 17.3 The sequence of formulas used to implement the non-self-starting Heun method. Note that the corrector error estimates can be used to modify the corrector. However, because this can affect the stability of the corrector, the modifier is not included in this algorithm. The corrector error estimate is included because of its utility for step-size adjustment.

For the next iteration, the predictor [Eq. (17.4)] is used to compute

$$y_2^0 = 2 + [4e^{0.8(1)} - 0.5(6.210093327)]2$$
$$= 13.59423410 \qquad \epsilon_t = 8.42\%$$

which is about half the error of the predictor for the second iteration of Example 17.1, which was $\epsilon_t = 18.6$ percent. This improvement is due to the fact that we are using a superior estimate of $y$ (6.210093327, as opposed to 6.36086549) in the predictor. In other words, the propagated and global errors are reduced by the inclusion of the corrector modifier.

Now because we have information from the prior iteration, Eq. (17.21) can be employed to modify the predictor, as in

$$y_2^0 = 13.59423410 + \tfrac{4}{5}(6.36086549 - 5.607004675)$$
$$= 14.19732275 \qquad \epsilon_t = -4.36\%$$

which, again, halves the error.

This modification has no effect on the final outcome of the subsequent corrector step. Regardless of whether the unmodified or modified predictors are used, the corrector will ultimately converge on the same answer. However, because the rate or efficiency of convergence depends on the accuracy of the initial prediction, the modification can reduce the number of iterations required for convergence.

Implementing the corrector yields a result of 15.21177723 ($\epsilon_t = -2.48$ percent) which represents an improvement over Example 17.1 because of the reduction of global error. Finally, this result can be modified using Eq. (17.18):

$$y_2^m = 15.21177723 - \frac{15.21177723 - 13.59423410}{5}$$

$$= 14.88826860 \qquad \epsilon_t = -0.30\%$$

Again, the error has been reduced an order of magnitude.

As in the previous example, the addition of the modifiers increases both the efficiency and accuracy of multistep methods. In particular, the corrector modifier effectively increases the order of the technique. Thus, the non-self-starting Heun with modifiers is third-order rather than second-order as is the case for the unmodified version. However, it should be noted that there are cases where the corrector modifier will affect the stability of the corrector. As a consequence, the modifier is not included in the algorithm for the non-self-starting Heun delineated in Fig. 17.3. Nevertheless, the corrector modifier can still have utility for step-size control as discussed in Sec. 17.1.5.

### 17.1.4 Computer Program for Multistep Methods

A program for the constant step-size version of the non-self-starting Heun is contained in Fig. 17.4. Note that the program includes the predictor modifier delineated in Fig. 17.3.

Because this algorithm employs a constant step size, a value for $h$ must be chosen at the beginning of the computation. In general, experience indicates that an optimal step size should be small enough to ensure convergence within two iterations of the corrector (Hull and Creemer, 1963). In addition, it must be small enough to yield a sufficiently small truncation error. As with other methods for ODEs, the only practical way to assess the magnitude of the global error is to compare the results for the same problem but with a halved step size.

Notice that a fourth-order RK method is used to generate the additional point needed at the start of the computation. We chose a fourth-order RK

## FORTRAN

```
      DIMENSION X(100),Y(100)
      COMMON X1,Y1,H
      F(X,Y)=4*EXP(.8*X)-.5*Y
      READ(5,1)X(1),XF
      X1=X(1)
    1 FORMAT(2F10.0)
      READ(5,1)H
      READ(5,2)MX
    2 FORMAT(I5)
      READ(5,1)ES
      READ(5,1)Y(1)
      Y1=Y(1)
      WRITE(6,3)X(1),Y(1)
    3 FORMAT(' ',2F10.3)
      CALL RK(Y2)
      X(2)=X(1)+H
      Y(2)=Y2
      NC=(XF-X(1))/H
      DO 450 I=2,NC
      K=I+1
      L=I-1
      WRITE(6,3)X(I),Y(I)
      X(K)=X(I)+H
      XX=X(I)
      S1=F(XX,Y(I))
      Y(K)=Y(L)+2*H*S1
      PU=Y(K)
      IF(I.EQ.2)GO TO 330
      Y(K)=Y(K)-4./5.*(PU-CU)
  330 DO 400 J=1,MX
      XX=X(K)
      S2=F(XX,Y(K))
      YP=Y(K)
      Y(K)=Y(I)+H*(S1+S2)/2
      EA=ABS((Y(K)-YP)/Y(K))*100
      IF(EA.LE.ES)GO TO 410
  400 CONTINUE
  410 CU=Y(K)
      PI=PU
      CI=CU
  450 CONTINUE
      WRITE(6,3)X(I),Y(I)
      STOP
      END
```

## BASIC

```
100  DIM X(100),Y(100)
110  DEF  FN F(YY) = 4 * EXP (.8
     * XX) - .5 * YY
120  INPUT X(1),XF
130  INPUT H
140  INPUT MX
150  INPUT ES
160  INPUT Y(1)
170  PRINT X(1),Y(1)
180  GOSUB 1000
190 X(2) = X(1) + H
200 Y(2) = Y2
210 NC = (XF - X(1)) / H
220  FOR I = 2 TO NC
230 K = I + 1
240 L = I - 1
250  PRINT X(I),Y(I)
260 X(K) = X(I) + H
270 XX = X(I)
280 S1 =  FN F(Y(I))
290 Y(K) = Y(L) + 2 * H * S1
300 PU = Y(K)
310  IF I = 2 THEN 330
320 Y(K) = Y(K) - 4 / 5 * (PI - C
    I)
330  FOR J = 1 TO MX
340 XX = X(K)
350 S2 =  FN F(Y(K))
360 YP = Y(K)
370 Y(K) = Y(I) + H * (S1 + S2) /
    2
380 EA =  ABS ((Y(K) - YP) / Y(K)
    ) * 100
390  IF EA < = ES THEN 410
400  NEXT J
410 CU = Y(K)
430 PI = PU
440 CI = CU
450  NEXT I
460  PRINT X(I),Y(I)
470  END
```

(Function specifying differential equation)
X(1), XF = initial and final values of independent variable
H = step size
MX = maximum iterations of corrector
ES = acceptable error (%) of corrector
Y(1) = initial value of dependent variable
(Subroutine to compute second value of dependent variable using a fourth-order RK method
NC = number of steps from X(1) to XF
(Predictor)
(Predictor modifier)
(Corrector)

FIGURE 17.4 Annotated programs in FORTRAN and BASIC for the non-self-starting Heun method.

method for this purpose because, although it is slightly more difficult to program than lower-order methods, its much greater accuracy justifies its inclusion.

### 17.1.5 The Issue of Step-Size Control

With the exception of the RK-Fehlberg discussed in Sec. 16.3.5, we have employed a constant step size to numerically integrate ordinary differential equations. Although such an approach has high utility for many engineering problems, there are certain situations where it is highly inefficient. For example, suppose that we are integrating an ODE with a solution of the type depicted in Fig. 17.5. For most of the range, the solution changes gradually. Such behavior suggests that a fairly large step size could be employed to obtain adequate results. However, for a localized region from $x = 1.75$ to

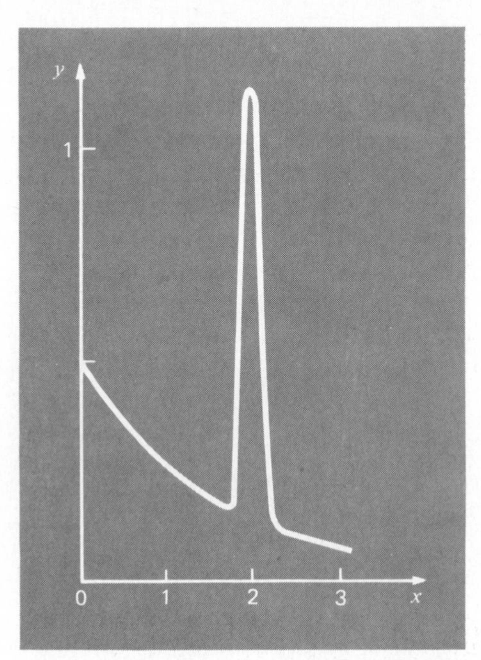


FIGURE 17.5 An example of a solution of an ODE that exhibits impulsive behavior. Automatic step-size adjustment has great advantages for such cases.

$x = 2.25$, the solution undergoes an abrupt change in the form of an impulse, or spike, function. ODEs with solutions consisting of rapidly and slowly varying components are called *stiff equations.*

The practical consequence of dealing with such equations is that a very small step size would be required to accurately capture the impulsive behavior. If a constant step-size algorithm were employed, the smaller step size required for the region of abrupt change would have to be applied to the entire range of the computation. As a consequence, a much smaller step size than necessary—and, therefore, many more calculations—would be applied to the regions of gradual change. For such cases, an algorithm that automatically adjusts the step size would avoid such overkill and hence be of great advantage.

As stated previously, the multistep methods described in this chapter provide a basis for such an algorithm. Thus, it might seem odd that the computer program described in the foregoing section employed a constant step size. The reason that we have separated this issue from the general algorithm is that step-size adjustment is not a trivial programming task. In fact, the cost (whether in terms of programming time or software development cost) could be a deciding factor in whether you choose to incorporate this option. With this as background, we will describe the mechanics of step-size control. This discussion should make it clear why including this capability is not a trivial exercise.

The choice of a step size is predicated on a number of opposing factors. In general, the step size should be small enough so that the corrector converges and that it does so within as few iterations as possible. Additionally, it should be small enough so that the results are accurate enough for problem requirements. At the same time, the step size should be as large as possible in order to minimize run-time cost and round-off error.

Two criteria are typically used to decide whether a change in step size is warranted. First, if Eq. (17.17) is greater than some prespecified error criterion, the step size is decreased. Second, the step size is chosen so that the convergence criterion of the corrector is satisfied in two iterations. This criterion is intended to account for the trade-off between the rate of convergence and the total number of steps in the calculation. For smaller values of $h$, convergence will be more rapid but more steps are required. For larger $h$, convergence is slower but fewer steps result. Experience (Hull and Creemer,

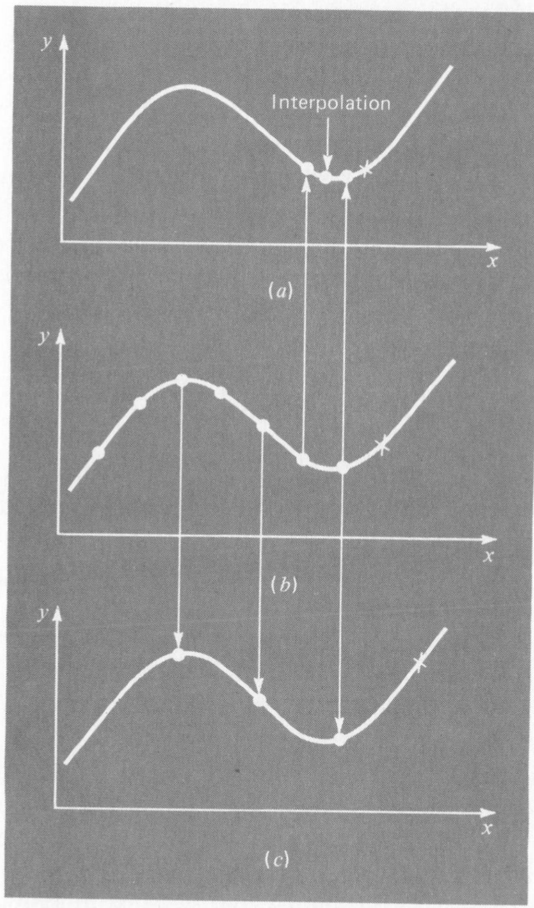


FIGURE 17.6 A plot indicating how a halving-doubling strategy allows the use of (*b*) previously calculated values for a third-order multistep method. (*a*) Halving and (*c*) doubling.

1963) suggests that the total steps will be minimized if $h$ is chosen so that the corrector converges within two iterations. Therefore, if over two iterations are required, the step size is decreased, and if less than two iterations are required, the step size is increased.

Although the above strategy specifies when step-size modifications are in order, it does not indicate *how* they should be changed. This is a critical question because multistep methods by definition require several points in order to compute a new point. Once the step size is changed, a new set of these points must be determined. One approach is to restart the computation and use the one-step method to generate a new set of starting points.

A more efficient strategy that makes use of presently existing information is to increase and decrease by doubling and halving the step size. As depicted in Fig. 17.6*b*, if a sufficient number of previous values have been generated, increasing the step size by doubling is a relatively straightforward task (Fig. 17.6*c*). All that is necessary is to keep track of subscripts so that old values of $x$ and $y$ become the appropriate new values. Halving the step size is somewhat more complicated because some of the new values will be unavailable (Fig. 17.6*a*). However, interpolating polynomials of the type developed in Chap. 11 can be used to determine these intermediate values.

In any event, the decision to incorporate step-size control represents a trade-off between initial investment in program complexity versus the long-term return because of increased efficiency. Obviously, the magnitude and importance of the problem itself will have a strong bearing on this trade-off.

## 17.2 INTEGRATION FORMULAS

The non-self-starting Heun method is characteristic of most multistep methods. It employs an open integration formula (the midpoint method) to make an initial estimate. This predictor step requires a previous data point. Then, a closed integration formula (the trapezoidal rule) is applied iteratively to improve the solution.

It should be obvious that a strategy for improving multistep methods would be to use higher-order integration formulas as predictors and correctors. For example, the higher-order Newton-Cotes formulas developed in Chap. 13 could be used for this purpose.

Before describing these methods, we will review the most common integration formulas upon which they are based. As mentioned above, the first of these are the Newton-Cotes formulas. However, there is a second class called the Adams formulas which we will also review and which are often preferred. As depicted in Fig. 17.7, the fundamental difference between the Newton-Cotes and the Adams formulas relates to the manner in which the integral is applied to obtain the solution. As depicted in Fig. 17.7*a*, the Newton-Cotes formulas estimate the integral over an interval spanning several points. This integral is then used to project from the beginning of the

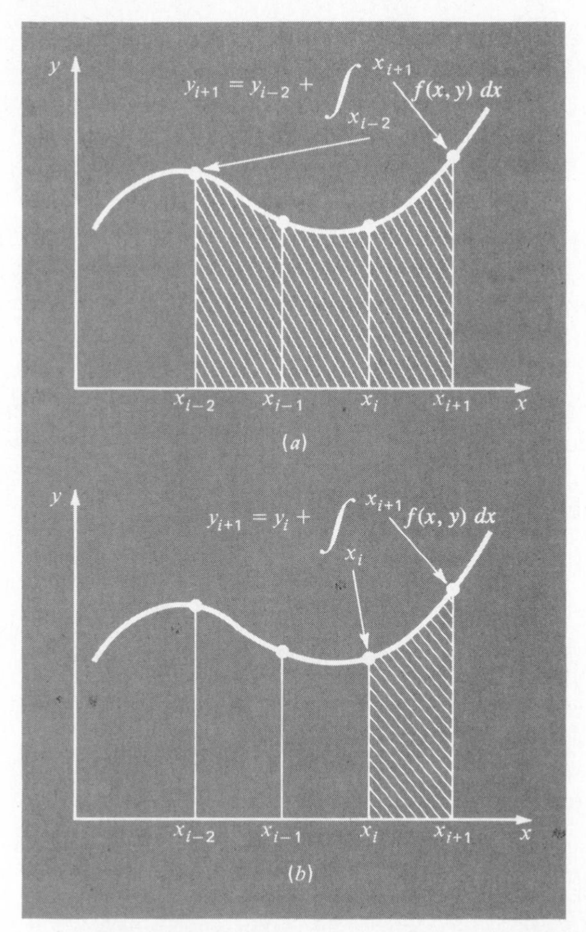


FIGURE 17.7 Illustration of the fundamental difference between the Newton-Cotes and the Adams integration formulas. (*a*) The Newton-Cotes formulas use a series of points to obtain an integral estimate over a number of segments. The estimate is then used to project across the entire range. (*b*) The Adams formulas use a series of points to obtain an integral estimate for a single segment. The estimate is then used to project across this segment.

interval to the end. In contrast, the Adams formulas (Fig. 17.7*b*) use a set of points from an interval to estimate the integral solely for the last segment in the interval. This integral is then used to project across this last segment.

### 17.2.1 Newton-Cotes Formulas

Some of the most common formulas for solving ordinary differential equations are based on fitting an $n$th-degree interpolating polynomial to $n + 1$ known values of $y$ and then using this equation to compute the integral. As discussed previously in Chap. 13, the Newton-Cotes integration formulas are based on such an approach. These formulas are of two types: open and closed forms.

**Open Formulas.** For $n$ equally spaced data points, the open formulas can be expressed in the form of a solution of an ODE, as was done previously for Eq. (17.10). The general equation for this purpose is

$$y_{i+1} = y_{i-n} + \int_{x_{i-n}}^{x_{i+1}} f_n(x)dx \qquad [17.22]$$

where $f_n(x)$ is an $n$th-order interpolating polynomial. The evaluation of the integral yields the $n$th-order Newton-Cotes open integration formula (Table 13.4). For example, if $n = 1$,

$$y_{i+1} = y_{i-1} + 2hf_i \qquad [17.23]$$

where $f_i$ is an abbreviation for $f(x_i, y_i)$—that is, the differential equation evaluated at $x_i$ and $y_i$. Equation (17.23) is referred to the *midpoint method* and was used previously as the predictor in the non-self-starting Heun method. For $n = 2$,

$$y_{i+1} = y_{i-2} + \frac{3h}{2}(f_i + f_{i-1})$$

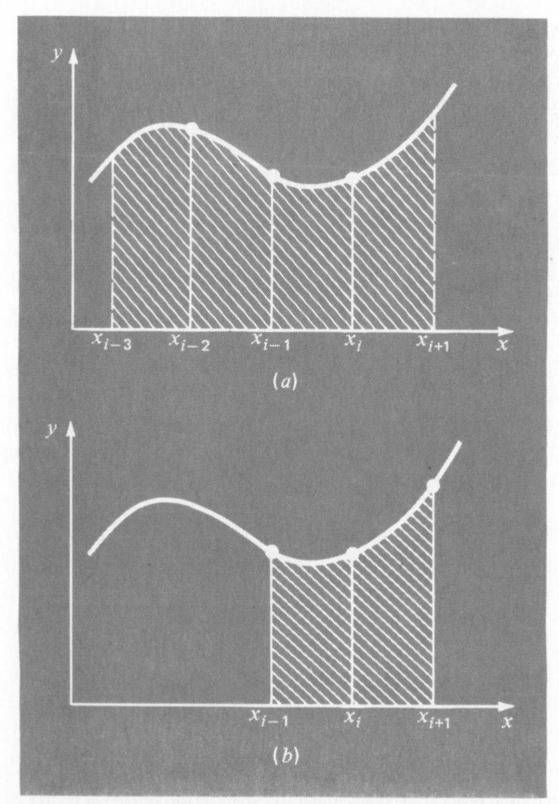


FIGURE 17.8 Graphical depiction of open and closed Newton-Cotes integration formulas. (*a*) The third open formula [Eq. (17.24)] and (*b*) Simpson's 1/3 rule [Eq. (17.26)].

and for $n = 3$,

$$y_{i+1} = y_{i-3} + \frac{4h}{3}(2f_i - f_{i-1} + 2f_{i-2}) \qquad [17.24]$$

Equation (17.24) is depicted graphically in Fig. 17.8*a*.

**Closed Formulas.** The closed form can be expressed generally as

$$y_{i+1} = y_{i-n+1} + \int_{x_{i-n+1}}^{x_{i+1}} f_n(x)dx \qquad [17.25]$$

where the integral is approximated by an $n$th-order Newton-Cotes closed integration formula (Table 13.2). For example, for $n = 1$,

$$y_{i+1} = y_i + \frac{h}{2}(f_i + f_{i+1})$$

which is equivalent to the trapezoidal rule. For $n = 2$,

$$y_{i+1} = y_{i-1} + \frac{h}{3}(f_{i-1} + 4f_i + f_{i+1}) \qquad [17.26]$$

which is equivalent to Simpson's 1/3 rule. Equation (17.26) is depicted in Fig. 17.8*b*.

### 17.2.2 Adams Formulas

The other type of integration formulas that can be used to solve ODEs are the Adams formulas. Many popular computer algorithms for multistep solution of ODEs are based on these methods.

**Open Formulas (Adams-Bashforth).** The Adams formulas can be derived in a variety of ways. One technique is to write a forward Taylor series expansion around $x_i$:

$$y_{i+1} = y_i + f_i h + \frac{f_i'}{2}h^2 + \frac{f_i''}{3!}h^3 + \cdots$$

which can also be written as

$$y_{i+1} = y_i + h\left(f_i + \frac{h}{2}f_i' + \frac{h^2}{6}f_i'' + \cdots\right) \qquad [17.27]$$

Recall from Sec. 3.5.4 that a backward difference can be used to approximate the derivative:

$$f_i' = \frac{f_i - f_{i-1}}{h} + \frac{f_i''}{2}h + 0(h^2)$$

which can be substituted into Eq. (17.27) to yield

$$y_{i+1} = y_i + h\left\{f_i + \frac{h}{2}\left[\frac{f_i - f_{i-1}}{h} + \frac{f_i''}{2}h + 0(h^2)\right] + \frac{h^2}{6}f_i'' + \cdots\right\}$$

or, grouping terms,

$$y_{i+1} = y_i + h\left(\tfrac{3}{2}f_i - \tfrac{1}{2}f_{i-1}\right) + \tfrac{5}{12}h^3 f_i'' + 0(h^4) \qquad [17.28]$$

This formula is called the *second-order open Adams formula.* Open Adams formulas are also referred to as *Adams-Bashforth* formulas. Consequently, Eq. (17.28) is sometimes called the second Adams-Bashforth formula.

Higher-order Adams-Bashforth formulas can be developed by substituting higher-difference approximations into Eq. (17.27). The $n$th-order open Adams formula can be represented generally as

$$y_{i+1} = y_i + h\sum_{k=0}^{n-1}\beta_k f_{i-k} + 0(h^{n+1}) \qquad [17.29]$$

The coefficients $\beta_k$ are compiled in Table 17.1. The fourth-order version is depicted in Fig. 17.9*a*. Notice that the first-order version is Euler's method.

Closed Formulas (Adams-Moulton). A backward Taylor series around $x_{i+1}$ can be written as

$$y_i = y_{i+1} - f_{i+1}h + \frac{f_{i+1}'}{2}h^2 - \frac{f_{i+1}''}{3!}h^3 + \cdots$$

Solving for $y_{i+1}$ yields

$$y_{i+1} = y_i + h\left(f_{i+1} - \frac{h}{2}f_{i+1}' + \frac{h^2}{6}f_{i+1}'' - \cdots\right) \qquad [17.30]$$

A difference can be used to approximate the derivative:

**TABLE 17.1 Coefficients and truncation error for Adams-Bashforth predictors.**

| Order | $\beta_0$ | $\beta_1$ | $\beta_2$ | $\beta_3$ | $\beta_4$ | $\beta_5$ | Local truncation error |
|---|---|---|---|---|---|---|---|
| 1 | $1$ | | | | | | $\frac{1}{2}h^2 f'(\xi)$ |
| 2 | $\frac{3}{2}$ | $-\frac{1}{2}$ | | | | | $\frac{5}{12}h^3 f''(\xi)$ |
| 3 | $\frac{23}{12}$ | $-\frac{16}{12}$ | $\frac{5}{12}$ | | | | $\frac{9}{24}h^4 f^{(3)}(\xi)$ |
| 4 | $\frac{55}{24}$ | $-\frac{59}{24}$ | $\frac{37}{24}$ | $-\frac{9}{24}$ | | | $\frac{251}{720}h^5 f^{(4)}(\xi)$ |
| 5 | $\frac{1901}{720}$ | $-\frac{2774}{720}$ | $\frac{2616}{720}$ | $-\frac{1274}{720}$ | $\frac{251}{720}$ | | $\frac{475}{1440}h^6 f^{(5)}(\xi)$ |
| 6 | $\frac{4277}{720}$ | $-\frac{7923}{720}$ | $\frac{9982}{720}$ | $-\frac{7298}{720}$ | $\frac{2877}{720}$ | $-\frac{475}{720}$ | $\frac{19{,}087}{60{,}480}h^7 f^{(6)}(\xi)$ |

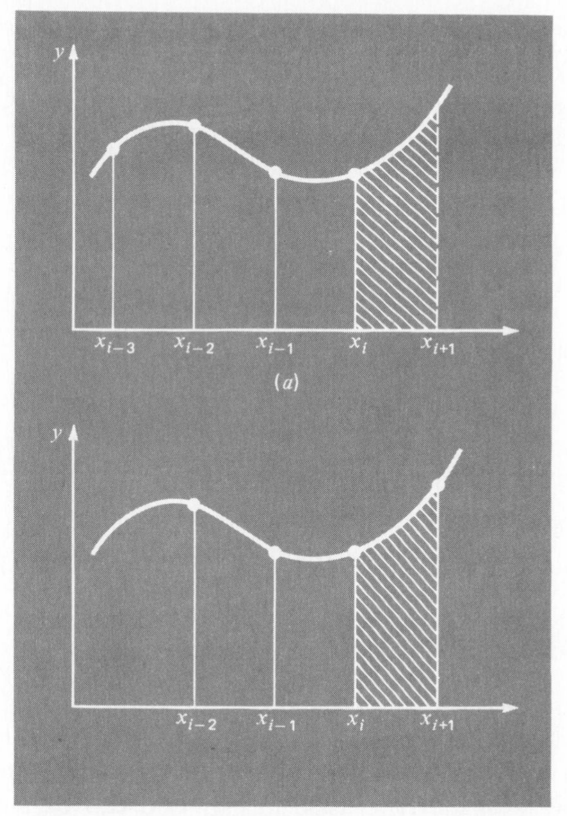


FIGURE 17.9 Graphical depiction of open and closed Adams integration formulas. (*a*) The fourth Adams-Bashforth open formula and (*b*) the fourth Adams-Moulton closed formula.

$$f'_{i+1} = \frac{f_{i+1} - f_i}{h} + \frac{f''_{i+1}}{2}h + 0(h^2)$$

which can be substituted into Eq. (17.30) to yield

$$y_{i+1} = y_i + h\left[f_{i+1} - \frac{h}{2}\left(\frac{f_{i+1} - f_i}{h} + \frac{f''_{i+1}}{2}h + 0(h^2)\right) + \frac{h^2}{6}f''_{i+1} - \cdots\right]$$

or, grouping terms,

$$y_{i+1} = y_i + h(\tfrac{1}{2}f_{i+1} + \tfrac{1}{2}f_i) - \tfrac{1}{12}h^3 f''_{i+1} - 0(h^4)$$

This formula is called the *second-order closed Adams formula* or the *second Adams-Moulton formula.* Also, notice that it is the trapezoidal rule.

The $n$th-order closed Adams formula can be written generally as

$$y_{i+1} = y_i + h\sum_{k=0}^{n-1} \beta_k f_{i+1-k} + 0(h^{n+1})$$

The coefficients $\beta_k$ are listed in Table 17.2. The fourth-order method is depicted in Fig. 17.9*b*.

**TABLE 17.2 Coefficients and truncation error for Adams-Moulton correctors.**

| Order | $\beta_0$ | $\beta_1$ | $\beta_2$ | $\beta_3$ | $\beta_4$ | $\beta_5$ | Local truncation error |
|---|---|---|---|---|---|---|---|
| 2 | $\frac{1}{2}$ | $\frac{1}{2}$ | | | | | $-\frac{1}{12}h^3f''(\xi)$ |
| 3 | $\frac{5}{12}$ | $\frac{8}{12}$ | $-\frac{1}{12}$ | | | | $-\frac{1}{24}h^4f^{(3)}(\xi)$ |
| 4 | $\frac{9}{24}$ | $\frac{19}{24}$ | $-\frac{5}{24}$ | $\frac{1}{24}$ | | | $-\frac{19}{720}h^5f^{(4)}(\xi)$ |
| 5 | $\frac{251}{720}$ | $\frac{646}{720}$ | $-\frac{264}{720}$ | $\frac{106}{720}$ | $-\frac{19}{720}$ | | $-\frac{27}{1440}h^6f^{(5)}(\xi)$ |
| 6 | $\frac{475}{1440}$ | $\frac{1427}{1440}$ | $-\frac{798}{1440}$ | $\frac{482}{1440}$ | $-\frac{173}{1440}$ | $\frac{27}{1440}$ | $-\frac{863}{60{,}480}h^7f^{(6)}(\xi)$ |

## 17.3 HIGHER-ORDER MULTISTEP METHODS

Now that we have formally developed the Newton-Cotes and Adams integration formulas, we can use them to derive higher-order multistep methods. As was the case with the non-self-starting Heun method, the integration formulas are applied in tandem as predictor-corrector methods. In addition, if the open and closed formulas have local truncation errors of the same order, modifiers of the type listed in Fig. 17.3 can be incorporated to improve accuracy and allow step-size control. Box 17.1 provides general equations for these modifiers. In the following section, we present two of the most common higher-order multistep approaches: Milne's method and the fourth-order Adams method.

### BOX 17.1 Derivation of General Relationships for Modifiers

The relationship between the true value, the approximation, and the error of a predictor can be represented generally as

$$\text{True value} = y_{i+1}^0 + \frac{\eta_p}{\delta_p} h^{n+1} y^{(n+1)}(\xi_p) \qquad \text{[B17.1.1]}$$

where $\eta_p$ and $\delta_p$ are the numerator and denominator of the constant of the truncation error for either an open Newton-Cotes (Table 13.4) or an Adams-Bashforth (Table 17.1) predictor and $n$ is the order.

A similar relationship can be developed for the corrector

$$\text{True value} = y_{i+1}^m - \frac{\eta_c}{\delta_c} h^{n+1} y^{(n+1)}(\xi_c) \qquad \text{[B17.1.2]}$$

where $\eta_c$ and $\delta_c$ are the numerator and denominator of the constant of the truncation error for either a closed Newton-Cotes (Table 13.2) or an Adams-Moulton (Table 17.2) corrector. As was done in the derivation of Eq. (17.15), Eq. (B17.1.1) can be subtracted from Eq. (B17.1.2) to yield

$$0 = y_{i+1}^m - y_{i+1}^0 - \frac{\eta_c + \eta_p \delta_c/\delta_p}{\delta_c} h^{n+1} y^{(n+1)}(\xi) \qquad \text{[B17.1.3]}$$

Now dividing the equation by $\eta_c + \eta_p \delta_c/\delta_p$, multiplying the last term by $\delta_p/\delta_p$, and rearranging provides an estimate of the local truncation error of the corrector:

$$E_c \simeq \frac{y_{i+1}^0 - y_{i+1}^m}{\eta_c + \eta_p \delta_c/\delta_p}$$

$$= -\frac{\delta_p}{\eta_c \delta_p + \eta_p \delta_c}(y_{i+1}^m - y_{i+1}^0) \qquad [B17.1.4]$$

For the predictor modifier, Eq. (B17.1.3) can be solved at the previous step for

$$h^n y^{(n+1)}(\xi) = -\frac{\delta_c \delta_p}{\eta_c \delta_p + \eta_p \delta_c}(y_i^0 - y_i^m)$$

which can be substituted into the error term of Eq. (B17.1.1) to yield

$$E_p = \frac{\eta_p \delta_c}{\eta_c \delta_p + \eta_p \delta_c}(y_i^m - y_i^0) \qquad [B17.1.5]$$

Equations (B17.1.4) and (B17.1.5) are general versions of modifiers that can be used to improve multistep algorithms. For example, Milne's method has $\eta_p = 14$, $\delta_p = 45$, $\eta_c = 1$, and $\delta_c = 90$. Substituting these values into Eqs. (B17.1.4) and (B17.1.5) yields Eqs. (17.33) and (17.34). Similar modifiers can be developed for other pairs of open and closed formulas that have local truncation errors of the same order.

### 17.3.1 MILNE'S METHOD

*Milne's method* is the most common multistep method based on Newton-Cotes integration formulas. It uses the three-point Newton-Cotes open formula as a predictor:

$$y_{i+1}^0 = y_{i-3}^m + \tfrac{4h}{3}(2f_i^m - f_{i-1}^m + 2f_{i-2}^m) \qquad [17.31]$$

and the three-point Newton-Cotes closed formula (Simpson's 1/3 rule) as a corrector:

$$y_{i+1}^j = y_{i-1}^m + \tfrac{h}{3}(f_{i-1}^m + 4f_i^m + f_{i+1}^{j-1}) \qquad [17.32]$$

The predictor and corrector modifiers for Milne's method can be developed from the formulas in Box 17.1 and the error coefficients in Tables 13.2 and 13.4:

$$E_p \simeq \tfrac{28}{29}(y_i^m - y_i^0) \qquad [17.33]$$

and

$$E_c \simeq -\tfrac{1}{29}(y_{i+1}^m - y_{i+1}^0) \qquad [17.34]$$

#### EXAMPLE 17.4
Milne's Method

Problem Statement: Use Milne's method to integrate $y' = 4e^{0.8x} - 0.5y$ from $x = 0$ to $x = 4$ using a step size of 1. The initial condition at $x = 0$ is $y = 2$. Because we are dealing with a multistep method, previous points are required. In an actual application, a one-step method such as a fourth-order

RK would be used to compute the required points. For the present example, we will use the analytical solution [recall Eq. (E16.5.1) from Example 16.5] to compute exact values at $x_{i-3} = -3$, $x_{i-2} = -2$, and $x_{i-1} = -1$ of $y_{i-3} = -4.547302219$, $y_{i-2} = -2.306160375$, and $y_{i-1} = -0.392995325$, respectively.

Solution: The predictor [Eq. (17.31)] is used to calculate a value at $x = 1$:

$$y_1^0 = -4.547302219 + \tfrac{4}{3}[2(3) - 1.993813519 + 2(1.960666259)]$$
$$= 6.02272313 \qquad \epsilon_t = 2.8\%$$

The corrector [Eq. (17.32)] is then employed to compute

$$y_1^1 = -0.392995325 + \tfrac{1}{3}[1.993813519 + 4(3) + 5.890802157]$$
$$= 6.235209902 \qquad \epsilon_t = -0.66\%$$

This result can be substituted back into Eq. (17.32) to iteratively correct the estimate. This process converges on a final corrected value of 6.20485465 ($\epsilon_t = -0.17$ percent).

This value is more accurate than the comparable estimate of 6.36086549 ($\epsilon_t = -2.68$ percent) obtained previously with the non-self-starting Heun method (Examples 17.1 through 17.3). The results for the remaining steps are $y(2) = 14.8603072$ ($\epsilon_t = -0.11$ percent), $y(3) = 33.7242601$ ($\epsilon_t = -0.14$ percent), and $y(4) = 75.4329487$ ($\epsilon_t = -0.12$ percent).

As in the previous example, Milne's method usually yields results of high accuracy. However, there are certain cases where it performs poorly. Before elaborating on these cases, we will describe another higher-order multistep approach—the fourth-order Adams method.

### 17.3.2 Fourth-Order Adams Method

A popular multistep method based on the Adams integration formulas uses the fourth-order Adams-Bashforth formula (Table 17.1) as the predictor:

$$y_{i+1}^0 = y_i^m + h\left(\tfrac{55}{24}f_i^m - \tfrac{59}{24}f_{i-1}^m + \tfrac{37}{24}f_{i-2}^m - \tfrac{9}{24}f_{i-3}^m\right) \qquad [17.35]$$

and the fourth-order Adams-Moulton formula (Table 17.2) as the corrector:

$$y_{i+1}^j = y_i^m + h\left(\tfrac{9}{24}f_{i+1}^{j-1} + \tfrac{19}{24}f_i^m - \tfrac{5}{24}f_{i-1}^m + \tfrac{1}{24}f_{i-2}^m\right) \qquad [17.36]$$

The predictor and the corrector modifiers for the fourth-order Adams method can be developed from the formulas in Box 17.1 and the error coefficients in Tables 17.1 and 17.2 as

$$E_p \simeq \tfrac{251}{270}(y_i^m - y_i^0) \qquad [17.37]$$

and

$$E_c \simeq -\tfrac{19}{270}(y_{i+1}^m - y_{i+1}^0) \qquad [17.38]$$

EXAMPLE 17.5
Fourth-Order Adams Method

Problem Statement: Use the fourth-order Adams method to solve the same problem as in Example 17.4.

Solution: The predictor [Eq. (17.35)] is used to compute a value at $x = 1$.

$$y_1^0 = 2 + 1(\tfrac{55}{24}3 - \tfrac{59}{24}1.993813519 + \tfrac{37}{24}1.960666259$$
$$- \tfrac{9}{24}2.649382908)$$
$$= 6.002716992 \qquad \epsilon_t = 3.1\%$$

which is comparable to but somewhat less accurate than the result using the Milne method. The corrector [Eq. (17.38)] is then employed to calculate

$$y_1^1 = 2 + 1(\tfrac{9}{24}5.900805218 + \tfrac{19}{24}3 - \tfrac{5}{24}1.993813519$$
$$+ \tfrac{1}{24}1.960666259)$$
$$= 6.254118568 \qquad \epsilon_t = -0.96\%$$

which again is comparable to but less accurate than the result using Milne's method. This result can be substituted back into Eq. (17.38) to iteratively correct the estimate. The process converges on a final corrected value of 6.214423582 ($\epsilon_t = 0.32$ percent) which is an accurate result but again somewhat inferior to that obtained with the Milne method.

### 17.3.3 Stability of Multistep Methods

The superior accuracy of the Milne method exhibited in Examples 17.4 and 17.5 would be anticipated on the basis of the error terms for the predictors [Eqs. (17.33) and (17.37)] and the correctors [Eqs. (17.34) and (17.38)]. The coefficients for the Milne method, $\frac{14}{45}$ and $\frac{1}{90}$, are smaller than for the fourth-order Adams, $\frac{251}{720}$ and $\frac{19}{720}$. Additionally, the Milne method employs fewer function evaluations to attain these higher accuracies. At face value, these results might lead to the conclusion that the Milne method is superior and, therefore, preferable to the fourth-order Adams. Although this conclusion holds for many cases, there are instances where the Milne method performs unacceptably. Such behavior is exhibited in the following example.

## EXAMPLE 17.6
### Stability of Milne's and Fourth-Order Adams Methods

Problem Statement: Employ Milne's and the fourth-order Adams methods to solve

$$\frac{dy}{dx} = -y$$

with the initial condition that $y = 1$ at $x = 0$. Solve this equation from $x = 0$ to $x = 10$ using a step size of $h = 0.5$. Note that the analytical solution is $y = e^{-x}$.

Solution: The results, as summarized in Fig. 17.10, indicate problems with Milne's method. Shortly after the onset of the computation, the errors begin to grow and oscillate in sign. By $t = 10$, the relative error has inflated to 2831 percent and the predicted value itself has started to oscillate in sign.

In contrast, the results for the Adams method would be much more acceptable. Although the error also grows, it would do so at a slow rate. Additionally, the discrepancies would not exhibit the wild swings in sign exhibited by the Milne method.

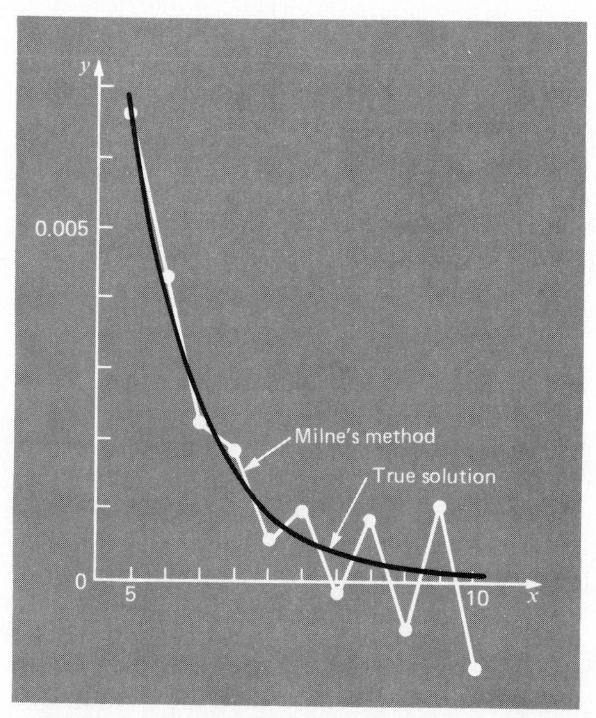


FIGURE 17.10 Graphical depiction of the instability of Milne's method.

The unacceptable behavior manifested in the previous example by the Milne method is referred to as *instability*. Although it does not always occur, its possibility leads to the conclusion that Milne's approach should be avoided. Thus, the fourth-order Adams method is normally preferred.

The instability of Milne's method is due to the corrector. Consequently, attempts have been made to rectify the shortcoming by developing stable correctors. One commonly used alternative that employs this approach is *Hamming's method,* which uses the Milne predictor and a stable corrector:

$$y_{i+1}^{j} = \frac{9y_i^m - y_{i-2}^m + 3h(f_{i+1}^{j-1} + 2f_i^m - f_{i-1}^m)}{8}$$

which has a local truncation error:

$$E_c = \tfrac{1}{40}h^5 f^{(4)}(\xi_c)$$

Hamming's method also includes modifiers of the form

$$E_p \simeq \tfrac{9}{121}(y_i^m - y_i^0)$$

and

$$E_c \simeq -\tfrac{112}{121}(y_{i+1}^m - y_{i+1}^0)$$

The reader can obtain additional information on this and other multistep methods elsewhere (Hamming, 1973; Lapidus and Seinfield, 1971).

# PROBLEMS

## Hand Calculations

**17.1** Solve the following initial-value problem over the interval from $x = 2$ to $x = 3$:

$$\frac{dy}{dx} = -0.5y$$

Use the non-self-starting Heun method with a step size of 0.5 and initial conditions of $y(1.5) = 4.72367$ and $y(2.0) = 3.67879$. Iterate the corrector to $\epsilon_s = 1$ percent. [*Note:* The exact results obtained analytically are $y(2.5) = 2.86505$ and $y(3.0) = 2.23130$.] Compute the true percent relative errors $\epsilon_t$ for your results.

**17.2** Repeat Prob. 17.1, but use Milne's method. [*Note:* $y(0.5) = 7.78801$ and $y(1.0) = 6.06531$.] Iterate the corrector to $\epsilon_s = 0.01$ percent.

**17.3** Repeat Prob. 17.2, but use the fourth-order Adams method ($\epsilon_s = 0.01$ percent).

**17.4** Solve the following initial-value problem from $x = 4$ to $x = 5$:

$$\frac{dy}{dx} = -\frac{y}{x}$$

Use a step size of 0.5 and initial values of $y(2.5) = 1.2$, $y(3) = 1$, $y(3.5) = 0.857142857$, and $y(4) = 0.75$. Obtain your solutions using the following techniques: (*a*) the non-self-starting Heun Method ($\epsilon_s = 1$ percent) (*b*) Milne's method ($\epsilon_s = 0.01$ percent), and (*c*) the fourth-order Adams method ($\epsilon_s = 0.01$ percent). [*Note:* The exact answers obtained analytically are $y(4.5) = 0.66666667$ and $y(5) = 0.6$.] Compute the true percent relative errors $\epsilon_t$ for your results.

**17.5** Solve the following initial-value problem from $y = 0$ to $y = 0.5$:

$$\frac{dy}{dx} = yx^2 - y$$

Use the non-self-starting Heun method with a step size of 0.25. If $y(-0.25) = 1.277355170$, employ a fourth-order RK method with a step size of 1 to predict the starting value at $y(0)$.

**17.6** Solve the following initial-value problem from $x = 1.5$ to $x = 2.5$:

$$\frac{dy}{dx} = \frac{-y}{1 + x}$$

Use the fourth-order Adams method. Employ a step size of 0.5 and the fourth-order RK method to predict the start-up values if $y(0) = 2$.

**17.7** Repeat Prob. 17.6, but use Milne's method.

**17.8** Determine the predictor, corrector, and modifiers for a second-order Adams method. Use it to solve Prob. 17.1.

**17.9** Determine the predictor, corrector, and modifiers for a third-order Adams method. Use it to solve Prob. 17.4.

## Computer-Related Problems

**17.10** Develop a user-friendly program for the non-self-starting Heun method with modifiers based on Sec. 17.1.3. Employ a fourth-order RK method to compute starter values. Test the program by duplicating Example 17.3.

**17.11** Use the program developed in Prob. 17.10 to solve Prob. 17.5.

**17.12** Develop a user-friendly program for the fourth-order Adams method with modifiers. Use a fourth-order RK method to compute starter values. Test the program by duplicating Example 17.5.

**17.13** Employ the program developed in Prob. 17.12 to solve Prob. 17.6.

# CHAPTER EIGHTEEN

# CASE STUDIES: ORDINARY DIFFERENTIAL EQUATIONS

The purpose of this chapter is to solve some ordinary differential equations using the numerical methods presented in Chaps. 16 and 17. The equations originate from practical engineering applications. Many of these applications result in nonlinear differential equations that cannot be solved using analytic techniques. Therefore, numerical methods are usually required. Thus, the techniques for the numerical solution of ordinary differential equations are fundamental capabilities that characterize good engineering practice. The problems in this chapter illustrate some of the trade-offs associated with various methods developed in Chaps. 16 and 17.

In *Case Study 18.1,* a differential equation is used to predict trends of computer sales. Among other things, this example illustrates how a parameter of a mathematical model is calibrated to data. The fourth-order RK method is used for this application.

*Case Study 18.2* originates from a chemical-engineering problem context. It provides a demonstration of how a proper step size is chosen and how differential equations can be used to improve a chemical production process. A second-order Runge-Kutta is used for this example.

Case Studies 18.3 and 18.4, which are taken from civil and electrical engineering, respectively, both deal with the solution of systems of equations. In *Case Study 18.3,* Euler's method is used because the problem does not require highly accurate results. *Case Study 18.4,* on the other hand, demands high accuracy, and as a consequence, a fourth-order RK scheme is used.

Finally, *Case Study 18.5* employs a variety of different approaches to investigate the behavior of a swinging pendulum. This problem also utilizes two simultaneous equations. An important aspect of this example is that it illustrates how numerical methods allow nonlinear effects to be incorporated easily into engineering analysis.

## CASE STUDY 18.1 MATHEMATICAL MODEL FOR COMPUTER SALES PROJECTIONS (GENERAL ENGINEERING)

Background: Operations and profitability at a computer company are very dependent on management's knowledge of the number of computers available on the market at any time. The extrapolation techniques discussed in Case Study 12.1 have proved unreliable and inaccurate. You have, therefore, been asked to derive a mathematical model that is capable of simulating and predicting the number of unsold computers available on the market as a function of time $t$. An ordinary differential equation can be developed for this purpose.

The marketing department of the company has determined from long experience and empirical observation that the expected sales rate of the computers can be described by

$$\begin{array}{c}\text{Sales rate}\\ \text{(computers sold per day)}\end{array} \propto \frac{\text{number of computers on the market}}{\text{cost of individual computer}} \tag{18.1}$$

That is, the more computers that are exposed to the public, the faster they sell; and the higher their cost, the slower they sell. Furthermore, the cost of an individual computer is related to the number of individual computers on the market, as in [recall Eq. (15.1)]

$$\text{Cost per computer (\$)} = 3000 - 1750\frac{N}{10{,}000 + N} \tag{18.2}$$

where $N$ is the number of computers.

The time rate of change of the number of computers remaining on the market is equal to the negative of the sales rate:

$$\frac{dN}{dt} = -\text{sales rate} \tag{18.3}$$

where the sales rate is derived by combining Eqs. (18.1) and (18.2):

$$\text{Sales rate} = k\frac{N}{3000 - 1750N/(10{,}000 + N)} \tag{18.4}$$

where $k$ is a proportionality constant having units of dollars per time. Substituting Eq. (18.4) into Eq. (18.3) yields

$$\frac{dN}{dt} = -k\frac{N}{3000 - 1750N/(10{,}000 + N)} \tag{18.5}$$

Planning considerations require that an estimate be obtained of how long 50,000 new computers will remain on the market as a function of time.

You have at your disposal the data from Table 12.1. Use this information to estimate the parameter $k$. Then use a fourth-order Runge-Kutta method to solve Eq. (18.5) from $t = 0$ to $t = 90$.

Solution: The first step in this analysis will be to determine a value for $k$. To do this, we can solve Eq. (18.5) for

$$k = -\frac{dN}{dt}\frac{3 \times 10^7 + 1250N}{N(10{,}000 + N)}$$

On the basis of this equation, we could evaluate $k$ if we had an estimate of $dN/dt$. This can be done on the basis of the data in Table 18.1, using finite divided differences to estimate $dN/dt$, as in (recall Sec. 3.5.4),

$$\left.\frac{dN}{dt}\right|_i \simeq \frac{N_{i+1} - N_{i-1}}{2\Delta t}$$

The results are contained in Table 18.1 and can be used to determine a mean value of $k$ = \$49.3 per day.

**TABLE 18.1** **Estimates of $k$ obtained from computer sales data. The mean of $k$ values is 49.3.**

| $t$ days | $N$ | $dN/dt$ | $k$ |
|---|---|---|---|
| 0 | 50,000 | | |
| 10 | 35,000 | −950 | 44.5 |
| 20 | 31,000 | −750 | 40.6 |
| 30 | 20,000 | −600 | 55.0 |
| 40 | 19,000 | −397.5 | 38.8 |
| 50 | 12,050 | −400 | 67.8 |
| 60 | 11,000 | | |

Now this value can be substituted into Eq. (18.5) to yield

$$\frac{dN}{dt} = -49.3\,\frac{N}{3000 - 1750\,[N/(10{,}000 + N)]}$$

which can be integrated using a fourth-order RK method with an initial condition of $N$ = 50,000 and a time step of 1 day. Note that we also performed the simulation using a time step of 0.5 day and obtained almost identical results, indicating that the accuracy using a step size of 1.0 is acceptable.

The results are depicted in Fig. 18.1 along with the data. Just as in regression, we can compute the sum of the squares of the residuals to quantify the goodness of fit. The result is $2.85 \times 10^7$. Although the fit appears to be satisfactory, we perform the computation again using values of

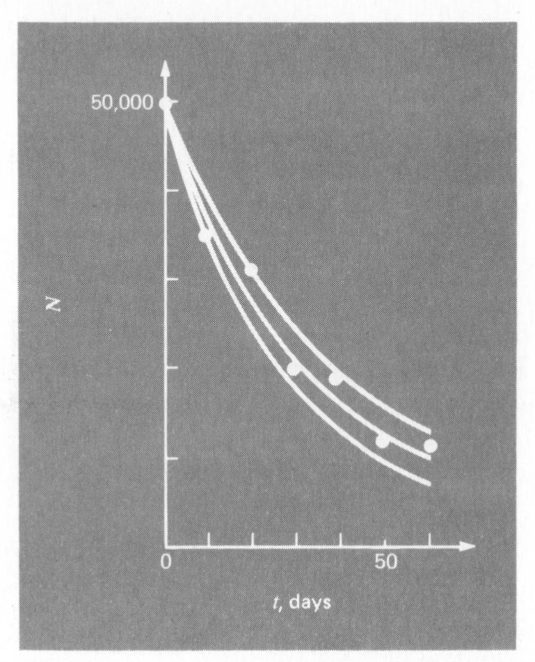


FIGURE 18.1 Plot of the number of computers $N$ on the market versus time $t$ in days. Three simulations using an ordinary differential equation model [Eq. (18.5)] are depicted for the case where $N = 50{,}000$ at $t = 0$. The three simulations correspond to different values of a model parameter $k$.

$k$ that are $\pm$ 20 percent of the original value of \$49.3 per day. Using these $k$'s of 59.2 and 39.4 results in residual sum of the squares of $1.05 \times 10^8$ and $5.35 \times 10^7$, respectively. These simulations are also depicted in Fig. 18.1.

Next we plot the sum of the squares of the residuals versus the $k$ values (Fig. 18.2) and fit a parabola through the points using an interpolating polynomial. We then determine the $k$ corresponding to the minimum sum of the squares by differentiating the second-order equation, setting it equal to zero, and solving for $k$. The resulting value of $k$ = \$46.8 per day can then be substituted into Eq. (18.5) to give

$$\frac{dN}{dt} = -46.8 \frac{N}{3000 - 1750[N/(10{,}000 + N)]}$$

This model yields a sum of the squares of the residuals of $2.24 \times 10^7$. It can then be used for predictive purposes. The predictions are shown along with the original data in Fig. 18.3. The results for $t$ = 55, 65, and 90 are 11,720, 9383, and 5596, respectively. This information, which is superior to that obtained by curve fitting in Chap. 12, can then be utilized by management to guide decisions regarding the marketing of these computers.

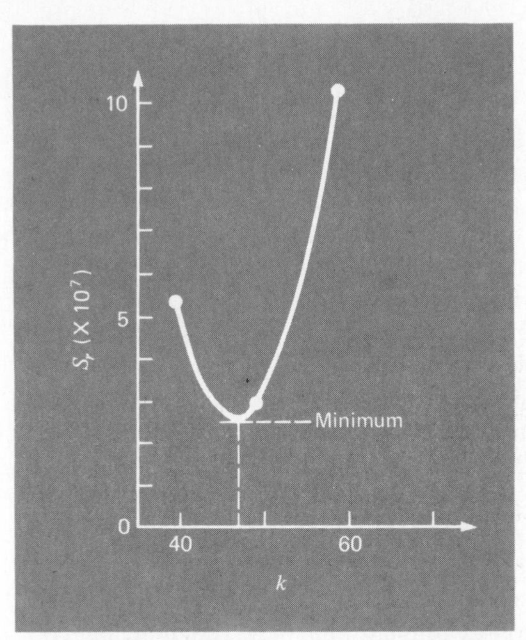


FIGURE 18.2 Plot of the sum of the squares of the residuals ($S_r$) versus values of the model parameter $k$. The curve is a parabola that was fit to the three points. The point of zero slope of this curve represents an estimate of the $k$ value ($46.8/day) that corresponds to a minimum value of $S_r$.

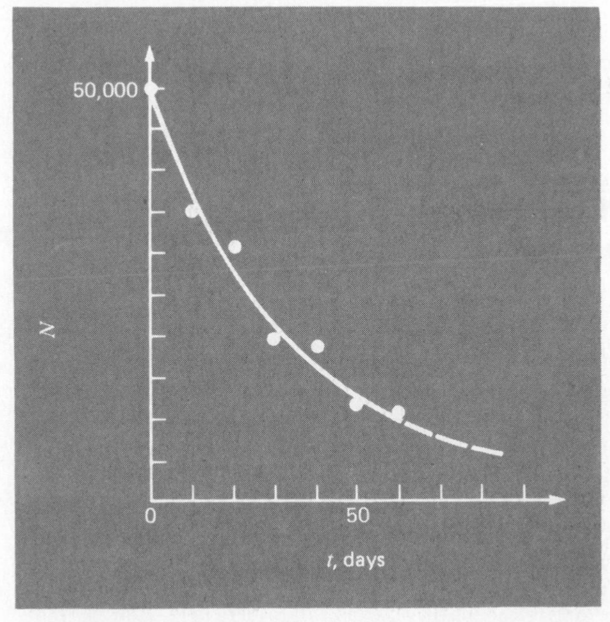


FIGURE 18.3 Model predictions using Eq. (18.5) with a $k$ of $46.8/day.

## CASE STUDY 18.2 REACTOR DESIGN FOR PHARMACEUTICAL PRODUCTION (CHEMICAL ENGINEERING)

Background: Chemical engineers design reaction vessels to grow populations of microbial organisms (recall Case Study 12.2). The by-products of growth can be useful pharmaceutical products. Figure 18.4 is a schematic representation of a reaction vessel that operates on a continuous-flow basis. The input flow contains few microbes and by-product, but is high in nutrient content. The inflow resides in the reactor for a certain time while biochemical transformations occur and then overflows from the tank. The overflow contains high numbers of newly grown microbes and high concentrations of growth by-products. Nutrients are lower than in the input because of microbial utilization. The reactor contents are vigorously mixed so that the composition of the overflow and the tank are identical.

If the flow rate and nutrient content are constant, the growth of microbes is balanced by the loss of organisms from the tank, and a stable population density is eventually attained. The time interval while the organisms are adjusting and increasing their density is called the *start-up period.* The length of the start-up period is important because it is wasted time that costs the company money.

You are asked to develop a mathematical model for the microbes in the reactor in order to predict the start-up period. The company's biochemical research laboratory has determined that the microbes grow according to the logistic growth model (recall Case Study 6.3):

$$\text{Growth rate} = K\,(p_{\max} - p)\,p$$

where $p_{\max} = 2 \times 10^6$ cells per liter is the maximum microbe density and $K = 2 \times 10^{-7}$ liter per cell per day is the growth-rate coefficient. You are asked to estimate the start-up period for a case where $p(t = 0) = 100{,}000$ cells per liter, the tank inflow rate $Q = 100$ L/day, and the tank volume

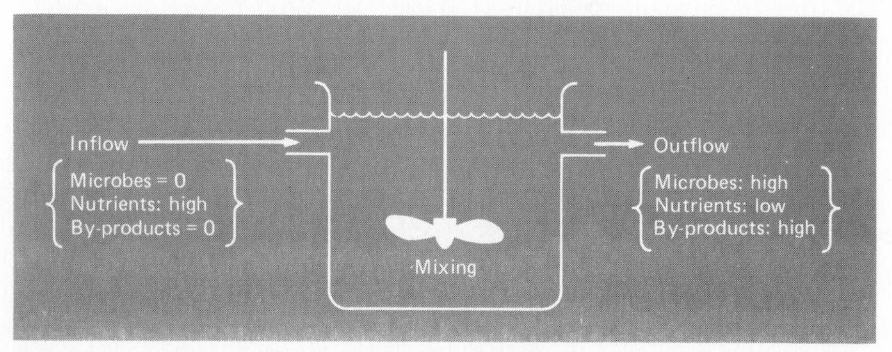


FIGURE 18.4 Schematic representation of a completely mixed, continuous-flow reactor used to grow populations of microbial organisms.

$V = 700$ L. The start-up period is defined as the time required for the population to grow to $6 \times 10^5$ cells per liter. At this point pharmaceutical production can begin.

After you are satisfied that you have developed a reliable computation, you are asked to use your model to help the plant's operators decide on the optimal number of cells to be used at $t = 0$. The more organisms that exist at $t = 0$, the shorter the start-up period. This is important because it costs the company $1000 for every day that the tank is out of production. Thus, there is an advantage to shortening the start-up period by using more organisms at $t = 0$.

On the other hand, new organisms are quite expensive to purchase. At present, the firm obtains stocks from a biological laboratory at a cost of $3000 per 100 million cells. Thus, the cost of 100,000 cells per liter used in the present analysis would be

$$\text{Cost} = 100{,}000 \text{ cells/L}(700 \text{ L}) \frac{\$3000}{100 \times 10^6 \text{ cells}} = \$2100$$

The cost of 200,000 cells per liter would be twice as much. Consequently, there is a trade-off between shortening the start-up time and the cost of new organisms. Your job is to use your model to provide guidance to the plant operators regarding the proper number of organisms at $t = 0$.

Solution: First, you must develop the capability to simulate the number of organisms as a function of time. Mass balance considerations suggest that

$$\frac{dp}{dt} = [K\,(p_{\max} - p)\,p] - \left(\frac{Q}{V}\right)p \qquad [18.6]$$

| Microbial accumulation in tank | = | growth of microbial biomass | − | loss of microbial mass through outlet |
|---|---|---|---|---|

Substituting the parameters into Eq. (18.6) yields

$$\frac{dp}{dt} = 2 \times 10^{-7}(2 \times 10^6 - p)\,p - \frac{100}{700}\,p$$

or, collecting terms,

$$\frac{dp}{dt} = 0.25714\,p - 2 \times 10^{-7}\,p^2$$

This equation can be solved analytically, but we will use numerical methods to obtain our solutions. First, we use Euler's method with a step size of 1 day to compute the results shown in Fig. 18.5. We use Euler's method for this purpose because it is extremely simple to program and provides a quick estimate of the general behavior of the solution. As can be seen, the organisms take about 10 days for the start-up period; by $t = 20$ days they have reached a nearly stable population. This stable period is called a *steady state*.

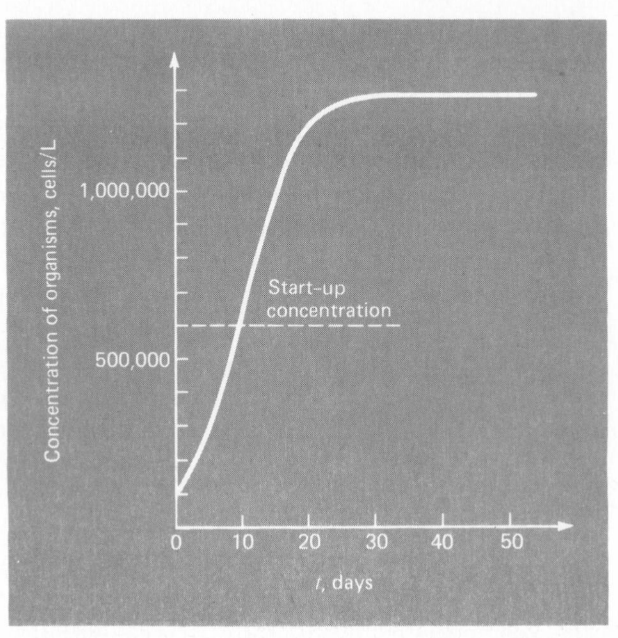


FIGURE 18.5 A simulation for microbial growth in a chemical production process. Euler's method is used for this simulation in order to make a quick appraisal of the behavior of the solution. Note that within 10 days, the start-up period is completed. Within 20 days, the reactor has almost reached a steady state.

On the basis of the foregoing result, we decide to perform our simulations for a 20-day period. We also decide that because of the ease with which it can be programmed and its increased accuracy, we will use Ralston's second-order RK method for the remainder of our computations. Table 18.2 lists the results for step sizes of 2, 1, and 0.5 days. Although the exact,

**TABLE 18.2 Microbial growth as simulated using an ODE with Ralston's second-order RK method. Results for different step sizes as well as the true solution are shown.**

| | Ralston's second-order RK method | | | |
|---|---|---|---|---|
| **$t$, days** | **$h = 2$** | **$h = 1$** | **$h = 0.5$** | **True solution** |
| 0 | 100,000 | 100,000 | 100,000 | 100,000 |
| 2 | 157,389 | 158,482 | 158,810 | 158,931 |
| 4 | 241,459 | 244,265 | 245,097 | 245,403 |
| 6 | 356,983 | 361,805 | 363,218 | 363,736 |
| 8 | 502,124 | 508,550 | 510,415 | 511,095 |
| 10 | 664,649 | 671,699 | 673,738 | 674,479 |
| 12 | 824,332 | 831,161 | 833,149 | 833,867 |
| 14 | 961,864 | 968,558 | 970,419 | 971,080 |
| 16 | 1,068,231 | 1,074,745 | 1,076,459 | 1,077,050 |
| 18 | 1,144,048 | 1,150,200 | 1,151,723 | 1,152,233 |
| 20 | 1,195,245 | 1,200,719 | 1,207,002 | 1,202,420 |

analytical results would not be available in most real-world applications, we have included them in Table 18.2 for comparative purposes. Notice that all the numerical results are quite good, with even the $t = 2$-h time step exhibiting errors of less than 5 percent. In the absence of knowledge of the true result, the accuracy of the computation can be appraised by comparing the results for various step sizes. For example, differences between the results for $h = 1$ and 0.5 are occurring in the third significant figure. Consequently, additional accuracy is unwarranted because more precision would not be discernible on a graph. Therefore, we decide that $h = 0.5$ is adequate for our purposes.

Using this step size and the Ralston model we perform two additional simulations with initial conditions of 200,000 and 400,000 cells per liter. These results, along with the case of 100,000 cells per liter, are plotted in Fig. 18.6. As expected, the use of more seed organisms shortens the start-up time with the results summarized in Table 18.3. Note that using more seed organisms reduces the delay cost from \$9200 to \$2500. However, the purchase cost of the organisms increases from \$2100 to \$8400. The total cost, which is plotted in Fig. 18.7, suggests a minimum somewhere in the vicinity of 250,000 cells per liter. The minimum point can be approximated by fitting a parabola to the three points. This function can be differentiated, set to zero, and solved for a value of 264,000 cells per liter. This level corresponds to a total cost of \$10,000 and represents the cheapest overall cost considering both start-up and seed organism costs.

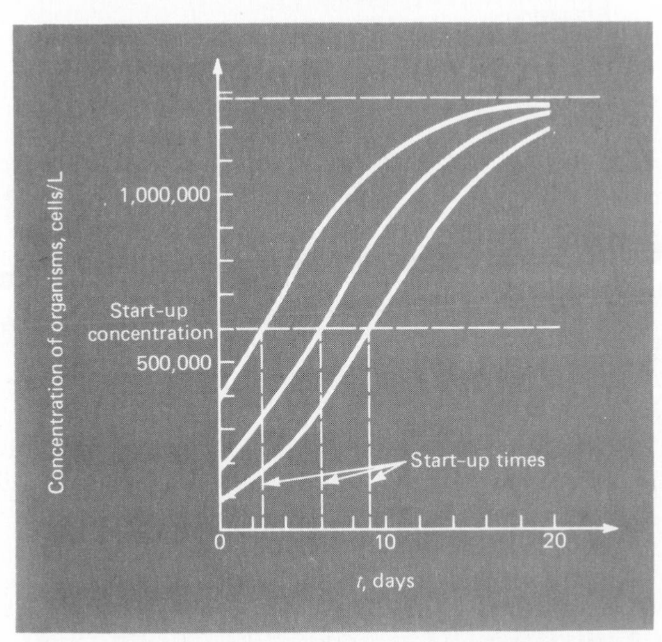


FIGURE 18.6 Simulations of microbial growth using three different initial conditions. These cases demonstrate that, as the number of seed organisms is increased, the start-up period becomes shorter.

**TABLE 18.3 Cost trade-offs for various initial levels of organisms used in a chemical production process**

| Initial concentration of organisms, cells/L | Purchase cost of organisms, $ | Start-up time, h | Cost of delay, $ | Total cost, $ |
|---|---|---|---|---|
| 100,000 | 2100 | 9.2 | 9200 | 11,300 |
| 200,000 | 4200 | 6.0 | 6000 | 10,200 |
| 400,000 | 8400 | 2.5 | 2500 | 10,900 |

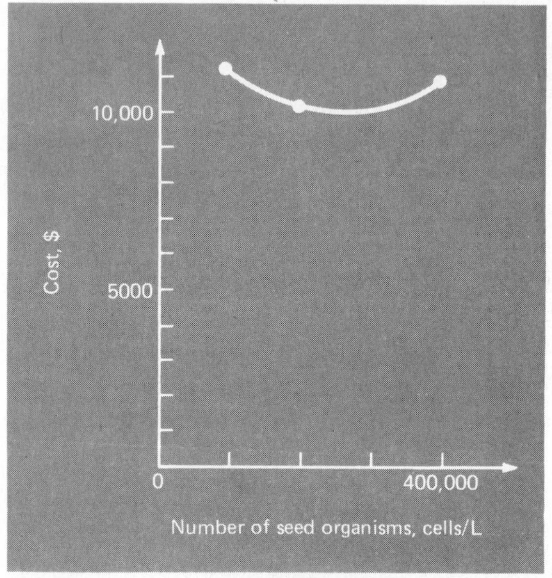


FIGURE 18.7 Plot of cost versus number of seed organisms (that is, number of organisms at $t = 0$). The fact that the curve is flat suggests that although a minimum exists at 264,000 cells per liter, this result is relatively insensitive to the number of seed organisms.

## CASE STUDY 18.3 DEFLECTION OF A SAILBOAT MAST (CIVIL ENGINEERING)

Background: A sailboat similar to that of Case Studies 12.3 and 15.3 is shown in Fig. 18.8, with a uniform force $f$ distributed along the mast. In this case, the cables supporting the mast have been removed, but the mast is mounted solidly to the deck for support.

The wind force causes the mast to deflect as depicted in Fig. 18.9. The deflection is similar to that of a cantilever beam. The following differential equation, based on the laws of mechanics, can be used to characterize this deflection:

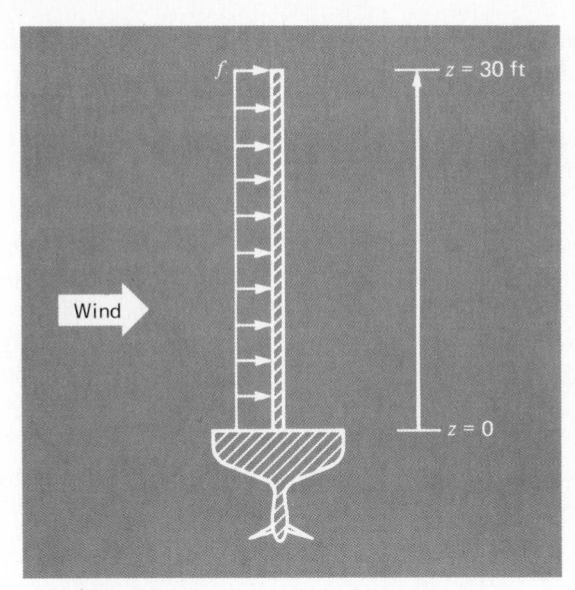


FIGURE 18.8 Sailboat mast subject to a uniform force *f*.

$$\frac{d^2y}{dz^2} = \frac{f}{2EI}(L - z)^2 \qquad [18.7]$$

where $E$ is the modulus of elasticity, $L$ is the height of the mast, and $I$ is the moment of inertia. At $z = 0$, $y = 0$ and $dy/dz = 0$. Calculate the deflection of the top of the mast where $z = L$ using both analytical and numerical methods. Assume that the hull does not rotate.

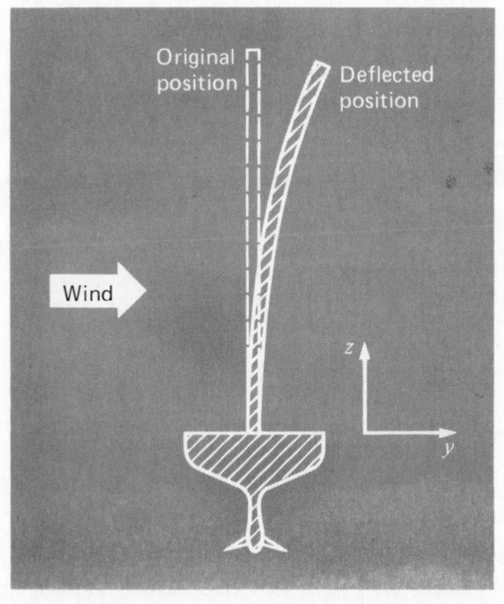


FIGURE 18.9 Deflection of a mast subjected to a uniform force.

Solution: Equation (18.7) can be solved analytically for the deflection at $z = L$:

$$y(z = L) = \frac{fL^4}{8EI} \tag{18.8}$$

This problem involves a simple differential equation that has a solution with smooth characteristics. Furthermore, the integration interval is relatively short, and the mast deflection is small. Also, the values of $f$ and $E$ are based on experimental data that is variable and difficult to measure accurately. Therefore, it seems satisfactory to use a low-order, simple method to solve the differential equation. Only one starting value will be needed, and we can probably use a small step size without accumulation of excessive round-off error.

Equation (18.7) can be written as a system of two first-order equations by transforming variables. Let

$$\frac{dy}{dz} = u \tag{18.9}$$

and, therefore, Eq. (18.7) can be expressed as

$$\frac{du}{dz} = \frac{f}{2EI}(L - z)^2 \tag{18.10}$$

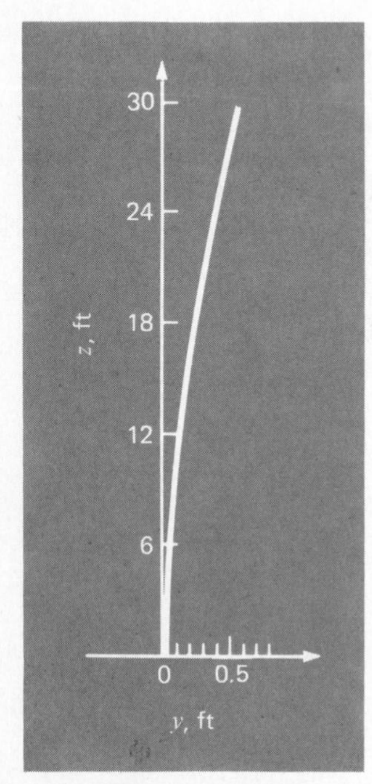


FIGURE 18.10 Plot of deflection of a sailboat mast as computed with Euler's method.

This pair of differential equations can be solved simultaneously using Euler's method.

First, however, we can obtain the analytical solution for comparison. Given a uniform load of $f = 50$ lb/ft, $L = 30$ ft, $E = 1.5 \times 10^8$ lb/ft$^2$ and $I = 0.06$ ft$^4$, Eq. (18.8) can be solved for

$$y(30) = \frac{50(30)^4}{8(1.5 \times 10^8)0.06} = 0.5625 \text{ ft}$$

Next we can solve Eqs. (18.9) and (18.10) using Euler's method. The results for a number of step sizes are

| **y(30)** | **Euler's step size** |
|---|---|
| 0.5744 | 1.0 |
| 0.5637 | 0.1 |
| 0.5631 | 0.05 |

Therefore, the answers obtained appear satisfactory. The deflection of the mast is depicted in Fig. 18.10.

This computation can be employed for design purposes. It is especially valuable for cases where the wind stress is not constant but varies in a complicated manner as a function of the height above the deck. Problem 18.13 provides an example of such a situation.

## CASE STUDY 18.4 SIMULATING TRANSIENT CURRENT FOR AN ELECTRIC CIRCUIT (ELECTRICAL ENGINEERING)

Background: Electric circuits where the current is time-variable rather than constant are common. A transient current is established in the right-hand loop of the circuit shown in Fig. 18.11 when the switch is suddenly closed.

Equations that describe the transient behavior of the circuit in Fig. 18.11 are based on Kirchhoff's law, which states that the algebraic sum of the voltage drops around a closed loop is zero (recall Case Study 6.4). Thus,

$$L\frac{di}{dt} + Ri + \frac{q}{C} - E(t) = 0 \qquad [18.11]$$

where $L(di/dt)$ is the voltage drop across the inductor, $L$ is inductance (in henrys), $R$ is resistance (in ohms), $q$ is the charge on the capacitor (in coulombs), $C$ is capacitance (in farads), $E(t)$ is the time-variable voltage source (in volts), and

$$i = \frac{dq}{dt} \qquad [18.12]$$

Equations (18.11) and (18.12) are a pair of first-order linear differential equations that can be solved analytically. For example, if $E(t) = E_0 \sin wt$ and $R = 0$,

$$q(t) = \frac{-E_0}{L(p^2 - w^2)}\frac{w}{p}\sin pt + \frac{E_0}{L(p^2 - w^2)}\sin wt \qquad [18.13]$$

where $p = 1/\sqrt{LC}$. The values of $q$ and $dq/dt$ are zero for $t = 0$. Use a numerical approach to solve Eqs. (18.11) and (18.12) and compare the results with Eq. (18.13).

Solution: This problem involves a rather long integration interval and demands the use of a highly accurate scheme to solve the differential equation if good results are expected. Let's assume that $L = 1$ H, $E_0 = 1$ V, $C = 0.25$ C, and $W^2 = 3.5$ s$^2$. This gives $p = 2$, and Eq. (18.13) becomes

$$q(t) = -1.8708 \sin 2t + 2 \sin(1.8708\, t)$$

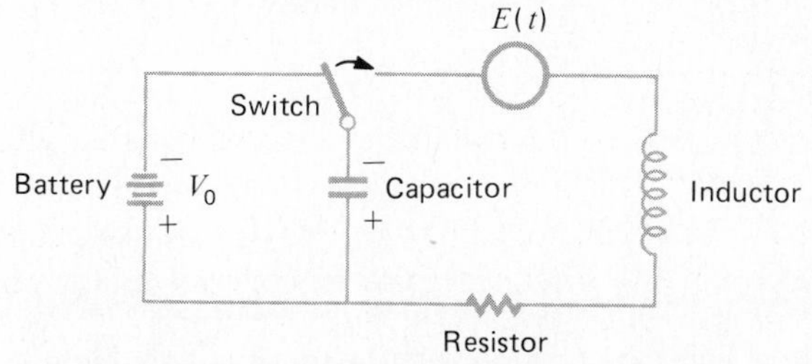


FIGURE 18.11 An electric circuit where the current varies with time.

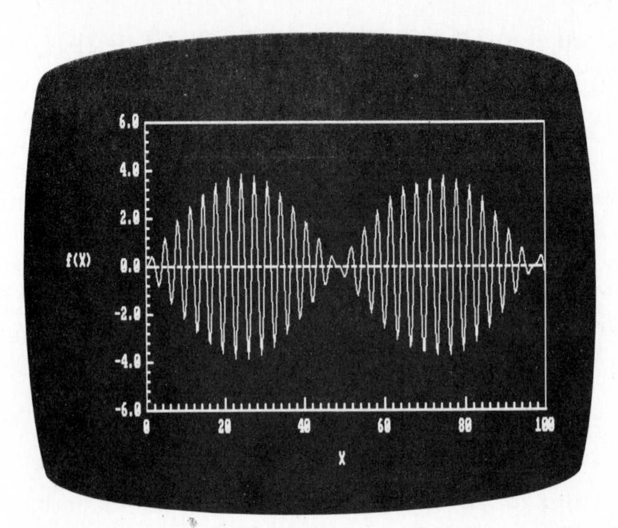


FIGURE 18.12 Computer screen showing the plot of the function [Eq. (18.13)].

for the analytical solution. This function is plotted in Fig. 18.12. The rapidly changing nature of the function places a severe requirement on any numerical procedure to find $q(t)$. Furthermore, because the function exhibits a slowly varying periodic nature as well as a rapidly varying component, long integration ranges are necessary to portray the solution. Thus, we expect that a high-order method is preferred for this problem.

However, we can try both Euler and fourth-order Runge-Kutta methods and compare the results. Using a step size of 0.1 s gives a value for $q$ at $t = 10$ s of $-6.638$ with Euler's method and a value of $-1.9897$ with the

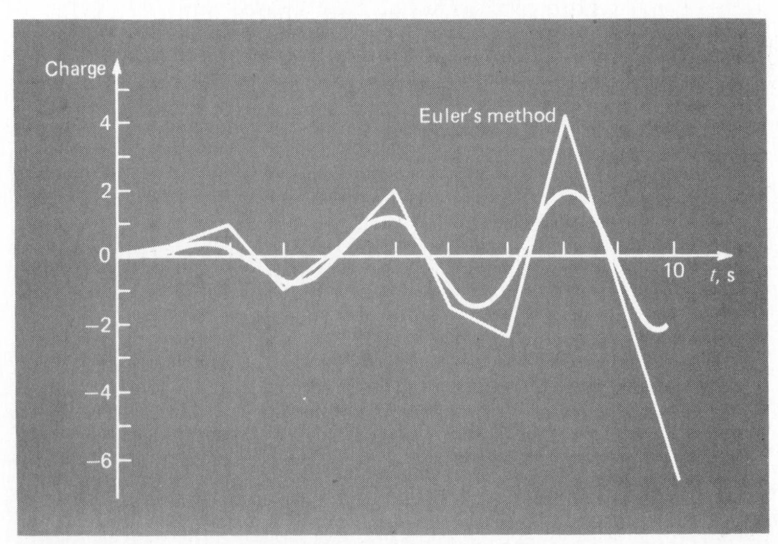


FIGURE 18.13 Results of Euler integration versus exact solution. Note that only every tenth output point is plotted.

fourth-order Runge-Kutta method. These results compare to an exact solution of −1.996 C.

Figure 18.13 shows the results of Euler integration every 1.0 s compared to the exact solution. Note that only every tenth output point is plotted. It is seen that the global error increases as $t$ increases. This divergent behavior intensifies as $t$ approaches infinity.

## CASE STUDY 18.5 THE SWINGING PENDULUM (MECHANICAL ENGINEERING)

Background: Mechanical engineers (as well as all others) are frequently faced with problems concerning the periodic motion of free bodies (recall Case Study 6.5). The engineering approach to such problems ultimately requires that the position and velocity of the body be known as a function of time. These functions of time invariably are the solution of ordinary differential equations. The differential equations are usually based on Newton's laws of motion.

As an example, consider the simple pendulum shown previously in Fig. VI.1. The particle of weight $W$ is suspended on a weightless rod of length $l$. The only forces acting on the particle are its weight and the tension $R$ in the rod. The position of the particle at any time is completely specified in terms of the angle $\theta$ and $l$.

The free-body diagram in Fig. 18.14 shows the forces on the particle and the acceleration. It is convenient to apply Newton's laws of motion in the $x$ direction tangent to the path of the particle:

$$\sum F = -W \sin \theta = \frac{W}{g} a$$

where $g$ is the gravitational constant (32.2 ft/s$^2$) and $a$ is the acceleration in the $x$ direction. The angular acceleration of the particle ($\alpha$) becomes

$$\alpha = \frac{a}{l}$$

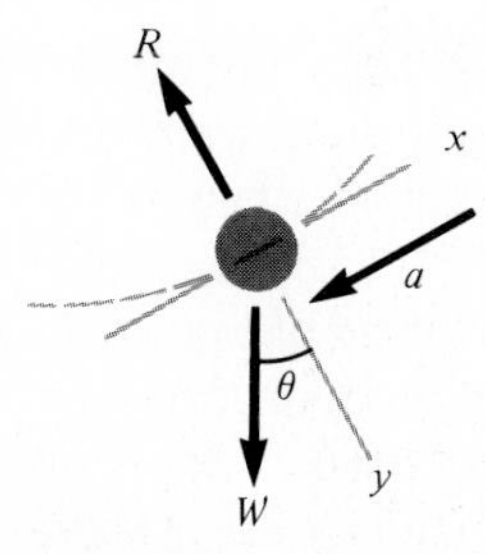

FIGURE 18.14 A free-body diagram of the swinging pendulum showing the forces on the particle and the acceleration.

Therefore, in polar coordinates ($\alpha = d^2\theta/dt^2$),

$$-W \sin \theta = \frac{Wl}{g} \alpha = \frac{Wl}{g} \frac{d^2\theta}{dt^2}$$

or,

$$\frac{d^2\theta}{dt^2} + \frac{g}{l} \sin \theta = 0 \qquad [18.14]$$

This apparently simple equation is a second-order nonlinear differential equation. In general, such equations are difficult or impossible to solve analytically. You have two choices regarding further progress. First, the differential equation might be reduced to a form that can be solved analytically (recall Sec. VI.1.1). Or a numerical approximation technique can be used to solve the differential equation directly. We will examine both of these alternatives in this example.

Solution: Proceeding with the first approach, we note that the series expansion for $\sin \theta$ is given by

$$\sin \theta = \theta - \frac{\theta^3}{3!} + \frac{\theta^5}{5!} - \frac{\theta^7}{7!} + \cdots \qquad [18.15]$$

For small angular displacements, $\sin \theta$ is approximately equal to $\theta$ when expressed in radians. Therefore, for small displacements, Eq. (18.14) becomes

$$\frac{d^2\theta}{dt^2} + \frac{g}{l}\theta = 0 \qquad [18.16]$$

which is a second-order linear differential equation. This approximation is very important because Eq. (18.16) is easy to solve analytically. The solution, based on the theory of differential equations, is given by

$$\theta(t) = \theta_0 \cos \sqrt{\frac{g}{l}}\, t \qquad [18.17]$$

where $\theta_0$ is the displacement at $t = 0$ and where it is assumed that the velocity ($v = d\theta/dt$) of the particle is zero at $t = 0$. The time required for the particle to complete one cycle of oscillation is called the period and is given by

$$T = 2\pi \sqrt{\frac{l}{g}}$$

Figure 18.15 shows a plot of the displacement ($\theta$) and velocity ($d\theta/dt$) as a function of time, as calculated from Eq. (18.17) with $\theta_0 = \pi/4$ and $l = 2$ ft. The period, as calculated from Eq. (18.17), is 1.5659 s.

The above calculations essentially are a complete solution of the motion of the particle. However, you must also consider the accuracy of the results because of the assumptions inherent in Eq. (18.16). In order to evaluate the accuracy, it is necessary to obtain a numerical solution for Eq. (18.14), which is a more complete physical representation of the motion. Any of the methods discussed in Chaps. 16 and 17 could be used for this purpose—for example, the Euler and fourth-order Runge-Kutta methods. Equation (18.16) must be transformed into two first-order equations to be compatible with the above methods. This is accomplished as follows. The velocity $v$ is defined by

$$\frac{d\theta}{dt} = v \qquad [18.18]$$

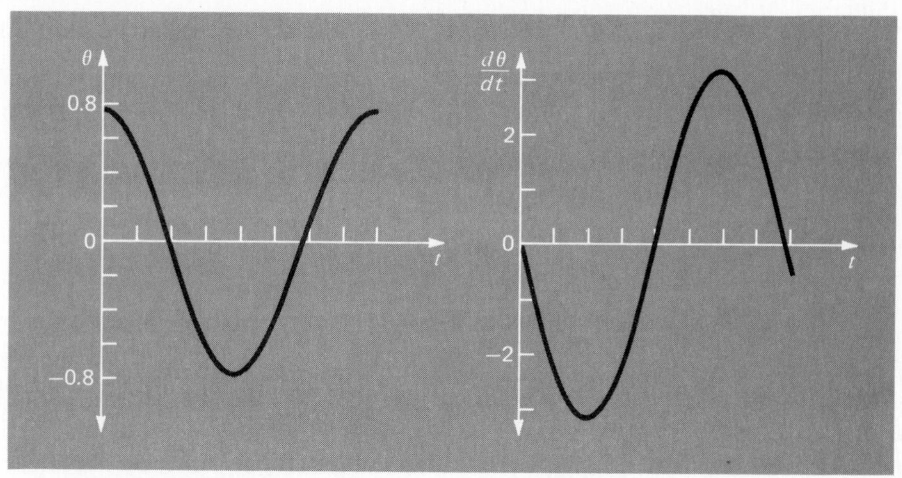


FIGURE 18.15 Plot of displacement ($\theta$) and velocity ($d\theta/dt$) as a function of time ($t$), as calculated from Eq. (18.17). $\theta_0$ is $\pi/4$ and length is 2 ft.

and, therefore, Eq. (18.14) can be expressed as

$$\frac{dv}{dt} = -\frac{g}{l}\sin\theta \tag{18.19}$$

Equations (18.18) and (18.19) are a coupled system of two ordinary differential equations. The numerical solutions by the Euler method and the fourth-order Runge-Kutta method give the results shown in Table 18.4. Table 18.4 compares the analytic solution for the linear equation of motion [Eq. (18.17)] in column (*a*) with the numerical solutions in columns (*b*), (*c*), and (*d*).

**TABLE 18.4 Comparison of a linear analytical solution of the swinging pendulum problem with three nonlinear numerical solutions**

| | | Nonlinear numerical solutions | | |
|---|---|---|---|---|
| Time, s | Linear analytical solution (*a*) | Euler ($h$ = 0.05) (*b*) | 4th-order RK ($h$ = 0.05) (*c*) | 4th-order RK ($h$ = 0.01) (*d*) |
| 0.0 | 0.785398 | 0.785398 | 0.785398 | 0.785398 |
| 0.2 | 0.545784 | 0.615453 | 0.566582 | 0.566579 |
| 0.4 | −0.026852 | 0.050228 | 0.021895 | 0.021882 |
| 0.6 | −0.0583104 | −0.639652 | −0.535802 | −0.535820 |
| 0.8 | −0.783562 | −1.050679 | −0.784236 | −0.784242 |
| 1.0 | −0.505912 | −0.940622 | −0.595598 | −0.595583 |
| 1.2 | 0.080431 | −0.299819 | −0.065611 | −0.065575 |
| 1.4 | 0.617698 | 0.621700 | 0.503352 | 0.503392 |
| 1.6 | 0.778062 | 1.316795 | 0.780762 | 0.780777 |

**TABLE 18.5. Comparison of the period of an oscillating body calculated from linear and nonlinear models.**

| | Period, s | |
|---|---|---|
| **Initial displacement, $\theta_0$** | **Linear model ($T = 2\pi\sqrt{l/g}$)** | **Nonlinear model [numerical solution of Eq. (18.14)]** |
| $\pi/16$ | 1.5659 | 1.57 |
| $\pi/4$ | 1.5659 | 1.63 |
| $\pi/2$ | 1.5659 | 1.85 |

The Euler and fourth-order RK yield different results and both disagree with the analytic solution, although the fourth-order Runge-Kutta method for the nonlinear case is closer to the analytic solution than is the Euler. To properly evaluate the difference between the linear and nonlinear models it is important to determine the accuracy of the numerical results. This is accomplished in three ways. First, the Euler numerical solution is easily recognized as inadequate because it overshoots the initial condition at $t = 0.8$ s. This clearly violates conservation of energy. Second, columns ($c$) and ($d$) in Table 18.4 show the solution of the fourth-order Runge-Kutta method for step sizes of 0.05 and 0.01. Because these vary in the fourth decimal place, it is reasonable to assume that the solution with a step size of 0.01 is also accurate with this degree of certainty. Third, for the 0.01-s step-size case, $\theta$ obtains a local maximum value of 0.785385 at $t = 1.63$ s (not shown in Table 18.4). This indicates that the particle returns to its original position with four-place accuracy with a period of 1.63 s. These considerations allow you to safely assume that the difference between columns ($a$) and ($d$) in Table 18.4 truly represent the difference between the linear and nonlinear model.

Another way to characterize the difference between the linear and the nonlinear model is on the basis of period. Table 18.5 shows the period of oscillation as calculated by the linear model and nonlinear model for three different initial displacements. It is seen that the calculated periods agree closely when $\theta$ is small because $\theta$ is a good approximation for sin $\theta$ in Eq. (18.15). This approximation deteriorates when $\theta$ becomes large.

These analyses are typical of cases you will routinely encounter as an engineer. The utility of the numerical techniques becomes particularly significant in nonlinear problems, and in many cases real-world problems are nonlinear.

# PROBLEMS

## General Engineering

**18.1** Reproduce computations performed in Case Study 18.1 using your own software.

**18.2** Perform the same computation as in Case Study 18.1, but with a $k = \$60/\text{day}$.

**18.3** Perform the same computation as in Case Study 18.1, but with a new equation for computer cost [replacing Eq. (18.2)]

$$\text{Cost of individual computer (\$)} = 1500\ (1 + e^{-4.4\times10^{-5}N})$$

**18.4** Repeat the falling parachutist problem (Example 1.2), but with the upward force due to drag as a second-order rate:

$$F_u = -cv^2$$

where $c = 2.4$ g/cm. Plot your results and compare with those of Example 1.1.

## Chemical Engineering

**18.5** Reproduce the computation performed in Case Study 18.2 using your own software.

**18.6** Perform the same computation as in Case Study 18.2, but for the case where $p(t = 0) = 50{,}000$ cells per liter.

**18.7** Perform the same computation as in Case Study 18.2, but for $p(t = 0) = 100{,}000$ cells per liter and $k = 3 \times 10^{-7}$ liter per cell per day.

**18.8** In Case Study 12.2, we developed Eq. (12.5) to model the growth of yeast employed in the commercial production of beer. If the yeast decay at a rate of $0.8p$ and if the rate of change of $f$ is described by

$$\frac{df}{dt} = -\frac{dp}{dt}$$

solve for $f$ and $p$ as a function of time if $f(0) = 100$ and $p(0) = 1$. Integrate the pair of ODEs until $p$ and $f$ reach stable levels. Plot your results.

**18.9** A mass balance for a chemical in a completely mixed reactor can be written as

$$V\frac{dc}{dt} = F - Qc - kVc^2$$

$$\text{Accumulation} = \underset{\text{rate}}{\text{feed}} - \text{outflow} - \text{reaction}$$

where $V$ is volume (10 $m^3$), $c$ is concentration, $F$ is feed rate (200 g/min), $Q$ is flow rate (1 $m^3$/min), and $k$ is reaction rate (0.1 $m^3$/g/min). If $c(0) = 0$, solve the ODE until concentration reaches a stable level. Plot your results.

**18.10** Repeat Prob. 9.8, but use the shooting method.

## Civil Engineering

**18.11** Reproduce the computation performed in Case Study 18.3, using your own software.

**18.12** Perform the same computation as in Case Study 18.3, but use a uniform load of 80 lb/ft and an $E = 2 \times 10^8$ lb/ft$^2$. Check your result by comparing it with the analytical solution.

**18.13** Perform the same computation as in Case Study 18.3, but rather than using a constant wind force $f$ employ a force that varies with height according to (recall Case Study 15.3)

$$f(z) = 200 \frac{z}{5 + z} e^{-2z/30}$$

Plot $y(z)$ versus $z$ and compare the results with those for Case Study 18.3.

**18.14** Duplicate Fig. 6.4 by numerically integrating the ODE in Case Study 6.3. Check your results by comparing them with the analytical solution [Eq. (6.9)].

**18.15** The logistic growth model from Case Study 6.3 can be applied to human as well as microbial populations. Suppose you are planning a water supply system for an island. If $p_{max} = 100{,}000$ people and $K = 10^{-6}$/people · year and if the initial population is 10,000 people, how long will it take the population to reach 90,000 people?

## Electrical Engineering

**18.16** Reproduce the computations performed in Case Study 18.4 using your own software.

**18.17** Perform the same computation as in Case Study 18.4, but with $R = 20\ \Omega$.

**18.18** Solve the ODE in Case Study 6.4 using numerical techniques if $q = 0.1$ and $i = -3.281515$ at $t = 0$.

**18.19** For a simple RL circuit, Kirchhoff's voltage law requires that (if Ohm's law holds)

$$L \frac{di}{dt} + Ri = 0$$

where $i$ is current, $L$ is inductance, and $R$ is resistance. Solve for $i$, if $L = R = 1$ and $i(0) = 10^{-3}$ amp. Solve this problem analytically and with a numerical method.

**18.20** In contrast to Prob. 18.19, real resistors may not always obey Ohm's law. For example, the voltage drop may be nonlinear and the circuit dynamics described by a relationship such as

$$L \frac{di}{dt} + \left[\frac{-i}{I} + \left(\frac{i}{I}\right)^3\right] R = 0$$

where all other parameters are as defined in Prob. 18.19 and $I$ is a known reference current equal to 1. Solve for $i$ as a function of time under the same conditions as specified in Prob. 18.19.

## Mechanical Engineering

**18.21** Reproduce the computations performed in Case Study 18.5 using your own software.

**18.22** Perform the same computation as in Case Study 18.5, but for a 3-ft-long pendulum.

**18.23** Use a numerical method to duplicate the computation displayed in Fig. 6.10.

**18.24** The rate of cooling a body can be expressed as

$$\frac{dT}{dt} = -k(T - T_a)$$

where $T$ is the temperature of the body (in degrees Celsius), $T_a$ is the temperature of the surrounding medium (also in degrees Celsius) and $k$ is proportionality constant (per minute). Thus, this equation specifies that the rate of cooling is proportional to the difference in temperature between the body and the surrounding medium. If a metal ball heated to 90°C is dropped into water that is held at a constant value of $T_a = 20°C$, use a numerical method to compute how long it takes the ball to cool to 30°C if $k = 0.1 \text{ min}^{-1}$.

## Miscellaneous

**18.25** Read all the case studies in Chap. 18. On the basis of your reading and experience, make up your own case study for any one of the fields of engineering. This may involve modifying or reexpressing one of our case studies. However, it can also be totally original. As with our examples, it must be drawn from an engineering problem context and must demonstrate the use of numerical methods for solving ODEs. Write up your results using our case studies as models.

# EPILOGUE: PART VI

## VI.4 TRADE-OFFS

Table VI.3 contains trade-offs associated with numerical methods for the solution of ordinary differential equations. The factors in this table must be evaluated by the engineer when selecting a method for each particular applied problem.

Simple self-starting techniques such as Euler's method can be used if the problem requirements involve a short range of integration. In this case, adequate accuracy may be obtained using small step sizes to avoid large truncation errors, and the round-off errors may be acceptable. Euler's method may also be appropriate for cases where the mathematical model has an inherently high level of uncertainty or has coefficients or forcing functions with significant errors as might arise during a measurement process. In this case, the accuracy of the model itself simply does not justify the effort involved to employ a more complicated numerical method. Finally, the simpler techniques may be best when the problem or simulation need only be performed a few times. In these applications, it is probably best to use a simple method that is easy to program and understand, despite the fact that the method may be computationally inefficient and relatively time consuming to run on the computer.

If the range of integration of the problem is long enough to involve a large number of steps (say, $>1000$), then it may be necessary and appropriate to use a more accurate technique than Euler's method. The fourth-order Runge-Kutta method and the fourth-order Adams method are popular and reliable for many engineering problems. In these cases, it may also be advisable to estimate the truncation error for each step as a guide to selecting the best step size. This can be accomplished with the fourth-order Adams or RK-Fehlberg approaches. If the truncation errors are extremely small, it may be wise to increase the step size to save computer time. On the other hand, if the truncation error is large, the step size should be

**TABLE VI.3**
**Comparison of alternative methods for the numerical solution of ordinary differential equations. The comparisons are based on general experience and do not account for the behavior of special functions.**

| Method | Starting values | Iterations required | Global error | Ease of changing step size | Programming effort | Comments |
|---|---|---|---|---|---|---|
| One step | | | | | | |
| Euler's | 1 | No | $0(h)$ | Easy | Easy | Good for quick estimates |
| Heun's | 1 | Yes | $0(h^2)$ | Easy | Moderate | — |
| Improved polygon | 1 | No | $0(h^2)$ | Easy | Moderate | — |
| Second-order Ralston | 1 | No | $0(h^2)$ | Easy | Moderate | The second-order RK method that minimizes truncation error |
| Fourth-order RK | 1 | No | $0(h^4)$ | Easy | Moderate | Widely used |
| RK-Fehlberg | 1 | No | $0(h^5)$* | Easy | Moderate to difficult† | Error estimate allows step-size adjustment |
| Multistep | | | | | | |
| Non-self-starting Heun | 2 | Yes | $0(h^3)$* | Difficult | Moderate to difficult† | Simple multistep method |
| Milne's | 4 | Yes | $0(h^5)$* | Difficult | Moderate to difficult† | Sometimes unstable |
| Fourth-order Adams | 4 | Yes | $0(h^5)$* | Difficult | Moderate to difficult† | |

*Provided error estimate is used to modify the solution.
†With variable step size.

decreased to avoid accumulation of error. Milne's method should be avoided if significant stability problems are expected. The Runge-Kutta method is simple to program and convenient to use but may be less efficient than the multistep methods. However, the Runge-Kutta method is usually employed in any event to obtain starting values for the multistep methods.

If extremely accurate answers are required or if the function has large higher-order derivatives, Butcher's fifth-order Runge-Kutta method is sometimes desirable.

A large number of engineering problems may fall into an intermediate range of integration interval and accuracy requirement. In these cases, the second-order RK and the non-self-starting Heun methods are simple to use and are relatively efficient and accurate.

## VI.5 IMPORTANT RELATIONSHIPS AND FORMULAS

Table VI.4 summarizes important information that was presented in Part VI. This table can be consulted to quickly access important relationships and formulas.

## VI.6 ADVANCED METHODS AND ADDITIONAL REFERENCES

Although we have reviewed a number of techniques for solving ordinary differential equations, there is additional information that is important in engineering practice. The question of *stability* was introduced in Sec. 17.3.3. This topic has general relevance to all methods for solving ODEs. Further discussion of the topic can be pursued in Carnahan, Luther, and Wilkes (1969), Gear (1971), and Hildebrand (1974).

The issue of stability has special significance to a topic that we addressed briefly in Sec. 17.1.5 and Case Study 18.4—the solution of *stiff equations.* These are equations with slowly and rapidly varying components. Although using a variable step size or higher-order methods can sometimes be helpful, special techniques are usually required for the adequate solution of stiff equations. You can consult Enright et al. (1975), Gear (1971), and Shampine and Gear (1979) for additional information regarding these techniques.

In Sec. 16.4.2, we introduced the shooting method for solving *boundary-value problems.* We also alluded to the fact that

**TABLE VI.4 Summary of important information presented in Part V**

| Method | Formulation | Graphic interpretation | Errors or modifiers |
|---|---|---|---|
| Euler (first-order RK) | $y_{i+1} = y_i + hk_1$<br>$k_1 = f(x_i, y_i)$ | 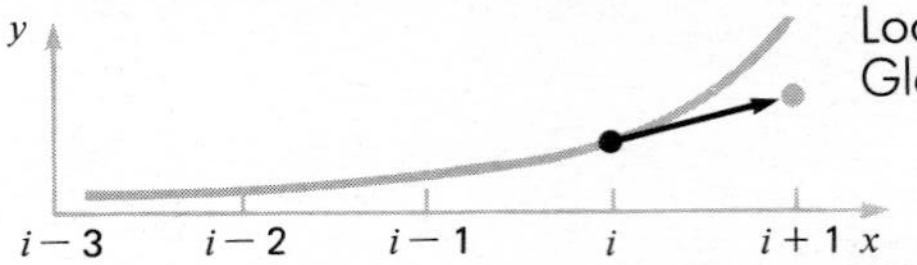 | Local error $\simeq 0(h^2)$<br>Global error $\simeq 0(h)$ |
| Ralston's second-order RK | $y_{i+1} = y_i + h\,(\frac{1}{3}k_1 + \frac{2}{3}k_2)$<br>$k_1 = f(x_i, y_i)$<br>$k_2 = f(x_i + \frac{3}{4}h,\ y_i + \frac{3}{4}hk_1)$ | 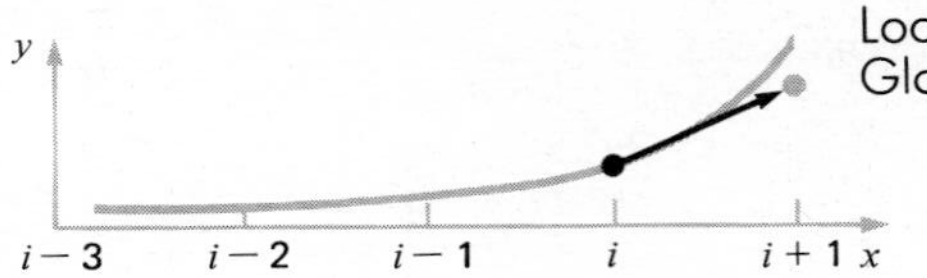 | Local error $\simeq 0(h^3)$<br>Global error $\simeq 0(h^2)$ |
| Classic fourth-order RK | $y_{i+1} = y_i + h(\frac{1}{6}k_1 + \frac{1}{3}k_2 + \frac{1}{3}k_3 + \frac{1}{6}k_4)$<br>$k_1 = f(x_i, y_i)$<br>$k_2 = f(x_i + \frac{1}{2}h,\ y_i + \frac{1}{2}hk_1)$<br>$k_3 = f(x_i + \frac{1}{2}h,\ y_i + \frac{1}{2}hk_2)$<br>$k_4 = f(x_i + h_1, y_i + hk_3)$ | 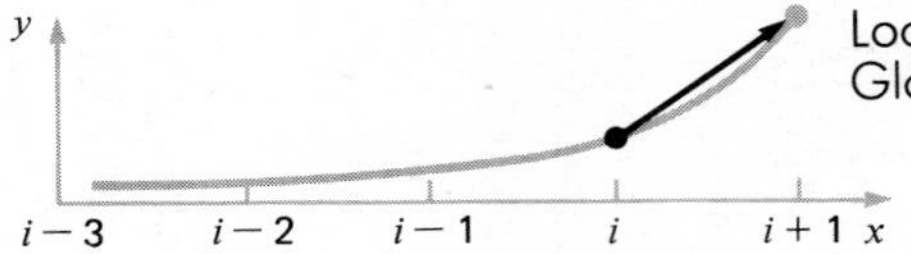 | Local error $\simeq 0(h^5)$<br>Global error $\simeq 0(h^4)$ |
| Non-self-starting Heun | Predictor: (midpoint method)<br>$y_{i+1}^{0} = y_{i-1}^{m} + 2hf(x_i, y_i^{m})$ | 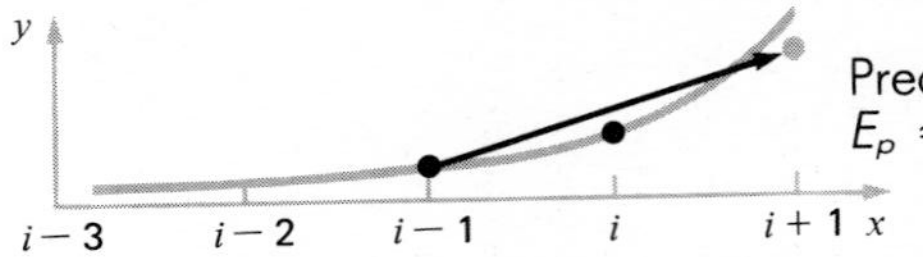 | Predictor modifier:<br>$E_p \simeq \frac{4}{5}(y_{i,u}^{m} - y_{i,u}^{0})$ |
| | Corrector: (trapezoidal rule)<br>$y_{i+1}^{j} = y_i^{m} + h\dfrac{f(x_i, y_i^{m}) + f(x_{i+1}, y_{i+1}^{j-1})}{2}$ | 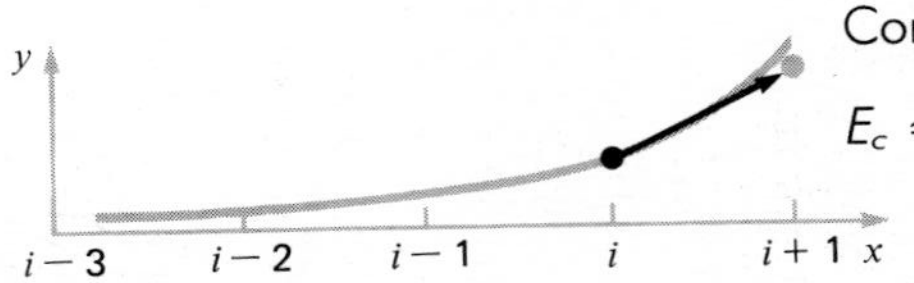 | Corrector modifier:<br>$E_c \simeq -\dfrac{y_{i+1,u}^{m} - y_{i+1,u}^{0}}{5}$ |
| 4th-order Adams | Predictor: (fourth Adams-Bashforth)<br>$y_{i+1}^{0} = y_i^{m} + h(\frac{55}{24}f_i^{m} - \frac{59}{24}f_{i-1}^{m} + \frac{37}{24}f_{i-2}^{m} - \frac{9}{24}f_{i-3}^{m})$ | 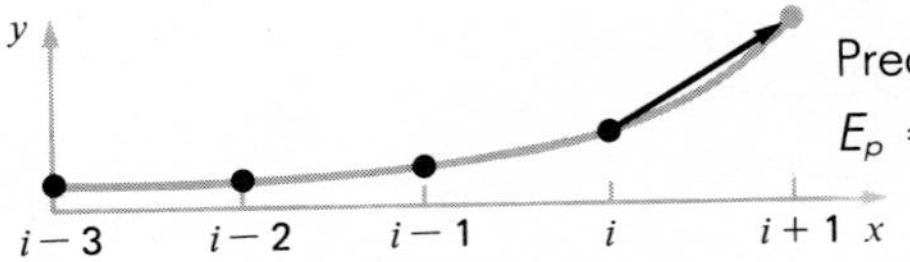 | Predictor modifier:<br>$E_p \simeq \frac{251}{270}(y_{i,u}^{m} - y_{i,u}^{0})$ |
| | Corrector: (fourth Adams-Moulton)<br>$y_{i+1}^{j} = y_i^{m} + h(\frac{9}{24}f_{i+1}^{j-1} + \frac{19}{24}f_i^{m} - \frac{5}{24}f_{i-1}^{m} + \frac{1}{24}f_{i-2}^{m})$ | 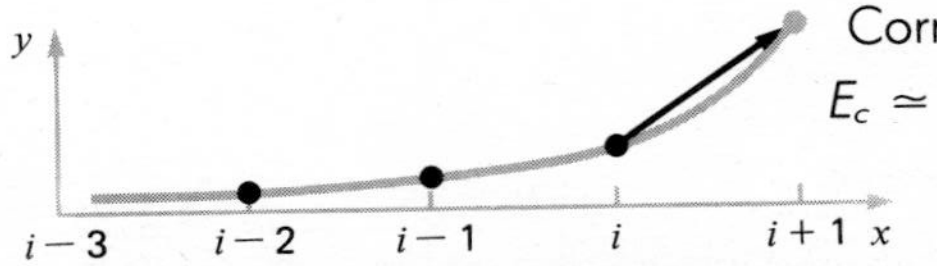 | Corrector modifier:<br>$E_c \simeq -\frac{19}{270}(y_{i+1,u}^{m} - y_{i+1,u}^{0})$ |

finite-difference methods of the sort used in Case Study 9.2 can be employed for such problems. Isaacson and Keller (1966), Keller (1968), Na (1979), and Scott and Watts (1976) can be consulted for additional information on boundary-value problems.

Finally, numerical techniques are available for the solution of partial differential equations. Carnahan, Luther, and Wilkes (1969), Gerald and Wheatley (1984), and Rice (1983) provide good introductions to the topic. You can also consult Ames (1977), Gladwell and Wait (1979), Vichnevetsky (1981, 1982), and Zienkiewicz (1971) for more in-depth treatments.

In summary, the foregoing is intended to provide you with avenues for deeper exploration of the subject. Additionally, all the above references provide descriptions of the basic techniques covered in Part VI. We urge you to consult these alternative sources to broaden your understanding of numerical methods for the solution of differential equations.

# BIBLIOGRAPHY


Ames, W. F., *Numerical Methods for Partial Differential Equations,* Academic Press, New York, 1977.

Ang, A. H-S., and W. H. Tang, *Probability Concepts in Engineering Planning and Design, Vol. 1: Basic Principles,* Wiley, New York, 1975.

Bent, R. J., and G. C. Sethares, *An Introduction to Computer Programming,* 2d ed., Brooks/Cole, Monterey, Calif., 1982.

Brigham, E. O., *The Fast Fourier Transform,* Prentice-Hall, Englewood Cliffs, N.J., 1974.

Butcher, J. C., "On Runge-Kutta Processes of Higher Order," *J. Australian Math. Soc.,* **4**:179 (1964).

Carnahan, B., H. A. Luther, and J. O. Wilkes, *Applied Numerical Methods,* Wiley, New York, 1969.

Cheney, W., and D. Kincaid, *Numerical Mathematics and Computing,* Brooks/Cole, Monterey, Calif., 1980.

Davis, P. J., and P. Rabinowitz, *Methods of Numerical Integration,* Academic Press, New York, 1975.

Draper, N. R., and H. Smith, *Applied Regression Analysis,* 2d ed., Wiley, New York, 1981.

Enright, W. H., T. E. Hull, and B. Lindberg, "Comparing Numerical Methods for Stiff Systems of ODE's," *BIT,* **15**:10 (1975).

Forsythe, G. E., M. A. Malcolm, and C. B. Moler, *Computer Methods for Mathematical Computation,* Prentice-Hall, Englewood Cliffs, N.J., 1977.

Gear, C. W., *Numerical Initial-Value Problems in Ordinary Differential Equations,* Prentice-Hall, Englewood Cliffs, N.J., 1971.

Gerald, C. F., and P. O. Wheatley, *Applied Numerical Analysis,* 3d ed., Addison-Wesley, Reading, Mass., 1984.

Gladwell, J., and R. Wait, *A Survey of Numerical Methods for Partial Differential Equations,* Oxford University Press, New York, 1979.

Guest, P. G., *Numerical Methods of Curve Fitting,* Cambridge University Press, New York, 1961.

Hamming, R. W., *Numerical Methods for Scientists and Engineers,* 2d ed., McGraw-Hill, New York, 1973.

Henrici, P. H., *Elements of Numerical Analysis,* Wiley, New York, 1964.

Hildebrand, F. B., *Introduction to Numerical Analysis,* 2d ed., McGraw-Hill, New York, 1974.

Hornbeck, R. W., *Numerical Methods,* Quantum, New York, 1975.

Householder, A. S., *The Theory of Matrices in Numerical Analysis,* Blaisdell, New York, 1964.

Hull, T. E., and A. L. Creemer, "The Efficiency of Predictor-Corrector Procedures," *J. Assoc. Comput. Mach.,* **10**:291 (1963).

Isaacson, E., and H. B. Keller, *Analysis of Numerical Methods,* Wiley, New York, 1966.

James, M. L., G. M. Smith, and J. C. Wolford, *Applied Numerical Methods for Digital Computations with FORTRAN and CSMP,* Harper & Row, New York, 1977.

Keller, H. B., *Numerical Methods for Two-Point Boundary-Value Problems,* Wiley, New York, 1968.

Lapidus, L., and J. H. Seinfield, *Numerical Solution of Ordinary Differential Equations,* Academic Press, New York, 1971.

Lapin, L. L., *Probability and Statistics for Modern Engineering,* Brooks/Cole, Monterey, Calif., 1983.

Lyness, J. M., "Notes on the Adaptive Simpson Quadrature Routine," *J. Assoc. Comput. Mach.,* **16**:483 (1969).

McCracken, D. D., *A Guide to FORTRAN IV Programming,* Wiley, New York, 1965.

Malcolm, M. A., and R. B. Simpson, "Local Versus Global Strategies for Adaptive Quadrature," *ACM Trans. Math. Software,* **1**:129 (1975).

Maron, M. J., *Numerical Analysis, A Practical Approach,* Macmillan, New York, 1982.

Merchant, M. J., *The ABC's of Computer Programming,* Wadsworth, Belmont, Calif., 1979.

———and J. R. Sturgul, *Applied FORTRAN Programming with Standard FORTRAN, WATFOR, WATFIV and Structural WATFIV,* Wadsworth, Belmont, Calif., 1977.

Muller, D. E., "A Method for Solving Algebraic Equations Using a Digital Computer," *Math. Tables Aids Comput.,* **10**:205 (1956).

Na, T. Y., *Computational Methods in Engineering Boundary Value Problems,* Academic Press, New York, 1979.

Noyce, R. N., "Microelectronics," *Scientific American,* **237**:62 (1977).

Ortega, J., and W. Rheinboldt, *Iterative Solution of Nonlinear Equations in Several Variables,* Academic Press, New York, 1970

Ralston, A., "Runge-Kutta Methods with Minimum Error Bounds," *Match. Comp.,* **16**:431 (1962).

———and P. Rabinowitz, *A First Course in Numerical Analysis,* 2d ed., McGraw-Hill, New York, 1978.

Rice, J. R., *Numerical Methods, Software and Analysis,* McGraw-Hill, New York, 1983.

Ruckdeschel, F. R., *BASIC Scientific Subroutines, Vol. 2,* Byte/McGraw-Hill, Peterborough, N.H., 1981.

Scarborough, J. B., *Numerical Mathematical Analysis,* 6th ed., Johns Hopkins Press, Baltimore, Md., 1966.

Scott, M. R., and H. A. Watts, "A Systematized Collection of Codes for Solving Two-Point Boundary-Value Problems," in *Numerical Methods for Differential Equations,* L. Lapidus and W. E. Schiesser, eds., Academic Press, New York, 1976.

Shampine, L. F., and R. C. Allen, Jr., *Numerical Computing: An Introduction,* Saunders, Philadelphia, 1973.

———and C. W. Gear, "A User's View of Solving Stiff Ordinary Differential Equations," *SIAM Review,* **21**:1 (1979).

Stark, P. A., *Introduction to Numerical Methods,* Macmillan, New York, 1970.

Swokowski, E. W., *Calculus with Analytical Geometry,* 2d ed., Prindle, Weber and Schmidt, Boston, 1979.

Thomas, G. B., Jr., and R. L. Finney, *Calculus and Analytical Geometry,* 5th ed., Addison-Wesley, Reading, Mass., 1979.

Vichnevetsky, R., *Computer Methods for Partial Differential Equations, Vol. 1: Elliptical Equations and the Finite Element Method,* Prentice-Hall, Englewood Cliffs, N.J., 1981.

———,*Computer Methods for Partial Differential Equations, Vol. 2: Initial Value Problems,* Prentice-Hall, Englewood Cliffs, N.J., 1982.

Wilkonson, J. H., *The Algebraic Eigenvalue Problem,* Oxford University Press, Fair Lawn, N.J., 1965.

Zienkiewicz, O. C., *The Finite Element Method in Engineering Science,* McGraw-Hill, London, 1971.

# INDEX